T0413710

NEUROMETHODS

Series Editor
Wolfgang Walz
University of Saskatchewan
Saskatoon, SK, Canada

For further volumes:
http://www.springer.com/series/7657

Animal Models of Movement Disorders

Volume I

Edited by

Emma L. Lane

The Brain Repair Group, Welsh School of Pharmacy, Cardiff University, Cardiff, Wales, UK

Stephen B. Dunnett

The Brain Repair Group, School of Biosciences, Cardiff University, Cardiff, Wales, UK

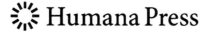 Humana Press

Editors
Emma L. Lane
The Brain Repair Group
Welsh School of Pharmacy
Cardiff University
Cardiff, Wales, UK
LaneEL@cardiff.ac.uk

Stephen B. Dunnett
The Brain Repair Group
School of Biosciences
Cardiff University
Cardiff, Wales, UK
DunnettSB@cardiff.ac.uk

ISSN 0893-2336   e-ISSN 1940-6045
ISBN 978-1-61779-297-7   e-ISBN 978-1-61779-298-4
DOI 10.1007/978-1-61779-298-4
Springer New York Dordrecht Heidelberg London

Library of Congress Control Number: 2011935379

© Springer Science+Business Media, LLC 2011
All rights reserved. This work may not be translated or copied in whole or in part without the written permission of the publisher (Humana Press, c/o Springer Science+Business Media, LLC, 233 Spring Street, New York, NY 10013, USA), except for brief excerpts in connection with reviews or scholarly analysis. Use in connection with any form of information storage and retrieval, electronic adaptation, computer software, or by similar or dissimilar methodology now known or hereafter developed is forbidden.
The use in this publication of trade names, trademarks, service marks, and similar terms, even if they are not identified as such, is not to be taken as an expression of opinion as to whether or not they are subject to proprietary rights.

Printed on acid-free paper

Humana Press is part of Springer Science+Business Media (www.springer.com)

Preface to the Series

Under the guidance of its founders Alan Boulton and Glen Baker, the Neuromethods series by Humana Press has been very successful since the first volume appeared in 1985. In about 17 years, 37 volumes have been published. In 2006, Springer Science + Business Media made a renewed commitment to this series. The new programme will focus on methods that are either unique to the nervous system and excitable cells or which need special consideration to be applied to the neurosciences. The programme will strike a balance between recent and exciting developments like those concerning new animal models of disease, imaging, in vivo methods, and more established techniques. These include immunocytochemistry and electro-physiological technologies. New trainees in neurosciences still need a sound footing in these older methods in order to apply a critical approach to their results. The careful application of methods is probably the most important step in the process of scientific inquiry. In the past, new methodologies led the way in developing new disciplines in the biological and medical sciences. For example, Physiology emerged out of Anatomy in the nineteenth century by harnessing new methods based on the newly discovered phenomenon of electricity. Nowadays, the relationships between disciplines and methods are more complex. Methods are now widely shared between disciplines and research areas. New developments in electronic publishing also make it possible for scientists to download chapters or protocols selectively within a very short time of encountering them. This new approach has been taken into account in the design of individual volumes and chapters in this series.

Wolfgang Walz

Preface

Movement is controlled by the interaction of many component parts of the central nervous system, from myelinated motor neurons projecting from the spinal cord to the higher neural processes in cerebellum and basal ganglia. This produces a highly controllable, functional system. However, this finely integrated network can be disrupted by injury and a range of disease processes that lead to significant motor dysfunction. Damage to single elements of this circuitry, which result from both idiopathic and genetic conditions, can cause profound alterations in motor function.

In order to understand these disorders and thereby facilitate recovery and repair, it is necessary to translate in vitro findings and hypotheses into animal representations of both functional and dysfunctional systems. These animal models range in species from lower orders, such as *drosophila* and *Caenorhabditis elegans*, through vertebrate species including fish, to mammals, such as rodents and nonhuman primates. Each model has its own profile of face, construct, and predictive validities, all of which must be considered when selecting the most appropriate for the experiment in hand. Similarly, the assessment methods used will depend on the species and the outcome variables that need to be assessed and must be similarly scrutinized for validity to answer the postulated hypothesis.

This volume introduces the variety of tools used in the assessment of motor function, highlighting their advantages and limitations and noting important technical considerations. We first take a look through the clinician's perspective on animal models of disease, before exploring both simple (e.g., *drosophila*) and more complex (rodent and nonhuman primate) model systems and reviewing the use of genetic manipulations, behavioural assessments, and the increasing use of imaging techniques. We then take a journey, descending through the central nervous system, describing animal models of disorders that target different levels of motor control. One interesting development found through the process of formulating this volume was the overlap in rodent behavioural techniques that are used across a range of motor disorders. Importantly, despite their wide use, each laboratory has its own approach to each behavioural technique. Many of the standard tests appear simple on first inspection, but a critical eye is required, and seemingly insignificant manipulations can produce critical differences in the outcomes and interpretation of the data produced.

The dopaminergic influence of motor control is well known, typified in the motor disorder, Parkinson's disease. The interest in this disorder, aided by significant developments such as the accidental discovery of a toxin, MPTP, that produces pathology similar to the disease process and identification of specific genes involved in the familial form of the disease have led to an extraordinary array of animal models. Furthermore, we not only need to model the disease, but also the consequences of pharmacological treatment, the development of another dopaminergic motor phenotype, and l-dopa-induced dyskinesia. From here, we expand to cover animal models of other basal ganglia disorders such as Huntington's disease and multiple systems atrophy, then through neo- and allo-cortical systems to describe models of ischemia and eye movement control. Descending through the cerebellum, there is a description of the role of this complex nucleus in the control of fine motor function, disorders that affect this "little brain," and how they are represented in vivo. The last section will consider the role of spinal cord systems, including the modelling of spinal cord injury, demyelinating disorders, and amyotrophic lateral sclerosis.

We would like to take this opportunity to thank the teams that contributed chapters and in particular to acknowledge the more junior members who are often those actually at the coalface of these experiments. We also regret the loss of some chapters (thankfully, very few), a consequence of the ever-increasing demands on the time of researchers. We hope that this text will be a valuable reference for those studying motor disorders by covering methodologies in detail and providing the information necessary to consider both the appropriate models and assessment tools that can most informatively answer the key experimental issues in our field.

Wales, UK   *Emma L. Lane*
                                                                                Stephen B. Dunnett

Contents

Volume I

Preface to the Series . . . . *v*
Preface . . . . *vii*
Contributors . . . . *xi*

PART I   GENERIC METHODS OF ASSESSMENT

 1 Why Cannot a Rodent Be More Like a Man? A Clinical Perspective   3
    Anne E. Rosser

 2 Zebrafish as a Vertebrate Model Organism for Studying Movement
    Disorders   11
    Maria Sundvik and Pertti Panula

 3 Methodological Strategies to Evaluate Functional Effectors
    Related to Parkinson's Disease Through Application of
    Caenorhabditis elegans Models   31
    Kim A. Caldwell and Guy A. Caldwell

 4 Effects of Alpha-Synuclein Expression on Behavioral Activity in *Drosophila*:
    A Simple Model of Parkinson's Disease   55
    Robert G. Pendleton, Xiaoyun C. Yang, Natalie Jerome,
    Ornela Dervisha, and Ralph Hillman

 5 Neurological Evaluation of Movement Disorders in Mice   65
    Simon P. Brooks

 6 Rodent Skilled Reaching for Modeling Pathological Conditions
    of the Human Motor System   87
    Jenni M. Karl and Ian Q. Whishaw

 7 High-Throughput Mouse Phenotyping.   109
    Sabine M. Hölter and Lisa Glasl

 8 MRI of Neurological Damage in Rats and Mice   135
    Mathias Hoehn

 9 Quantification of Brain Function and Neurotransmission System
    In Vivo by Positron Emission Tomography: A Review of Technical
    Aspects and Practical Considerations in Preclinical Research   151
    Nadja Van Camp, Yann Bramoullé, and Philippe Hantraye

10 Optical Approaches to Studying the Basal Ganglia.   191
    Joshua L. Plotkin, Jaime N. Guzman, Nicholas Schwarz,

Geraldine Kress, David L. Wokosin, and D. James Surmeier

11   Electrophysiological Analysis of Movement Disorders in Mice................ 221
     Shilpa P. Rao, Véronique M. André, Carlos Cepeda,
     and Michael S. Levine

PART II   DOPAMINE SYSTEMS

12   Genetic Models of Parkinson's Disease 243
     Ralf Kühn, Daniela Vogt-Weisenhorn, and Wolfgang Wurst

13   6-OHDA Lesion Models of Parkinson's Disease in the Rat 267
     Eduardo M. Torres and Stephen B. Dunnett

14   6-OHDA Toxin Model in Mouse 281
     Gaynor A. Smith and Andreas Heuer

15   Rotation in the 6-OHDA-Lesioned Rat 299
     Stephen B. Dunnett and Eduardo M. Torres

16   Of Rats and Patients: Some Thoughts About Why Rats Turn in Circles
     and Parkinson's Disease Patients Cannot Move Normally 317
     Gordon W. Arbuthnott

17   Comparing Behavioral Assessment of Sensorimotor Function in Rat
     and Mouse Models of Parkinson's Disease and Stroke...................... 325
     Sheila M. Fleming and Timothy Schallert

18   Rodent Models of L-DOPA-Induced Dyskinesia 337
     Hanna S. Lindgren and Emma L. Lane

19   Using the MPTP Mouse Model to Understand Neuroplasticity:
     A New Therapeutic Target for Parkinson's Disease? 353
     Giselle M. Petzinger, Beth E. Fisher, Garnik Akopian, Ruth Wood,
     John P. Walsh, and Michael W. Jakowec

20   The MPTP-Treated Primate, with Specific Reference to the Use
     of the Common Marmoset (*Callithrix jacchus*)........................... 371
     Michael J. Jackson and Peter Jenner

21   Behavioral Assessment in the African Green Monkey
     After MPTP Administration 401
     D. Eugene Redmond Jr.

Index... 437

Volume II

Preface to the Series.   v
Preface   vii
Contributors.   xi

PART I   BASAL GANGLIA

 1   Behavioral Assessment of Genetic Mouse Models of Huntington's Disease   3
     Miriam A. Hickey and Marie-Françoise Chesselet

 2   Excitotoxic Lesions of the Rodent Striatum   21
     Máté D. Döbrössy, Fabian Büchele, and Guido Nikkhah

 3   Combination Lesion Models of MSA   37
     Daniela Kuzdas and Gregor K. Wenning

 4   The Role of the Dorsal Striatum in Instrumental Conditioning.   55
     Mark A. Rossi and Henry H. Yin

 5   3-Nitropropionic Acid and Other Metabolic Toxin Lesions of the Striatum.   71
     Cesar V. Borlongan and Paul R. Sanberg

 6   Functional Assessment of Subcortical Ischemia   91
     Tracy D. Farr and Rebecca C. Trueman

PART II   NEO- AND ALLO-CORTICAL SYSTEMS

 7   Functional Organization of Rat and Mouse Motor Cortex   117
     G. Campbell Teskey and Bryan Kolb

 8   Forebrain Circuits Controlling Whisker Movements   139
     Kevin D. Alloway and Jared B. Smith

 9   An Approach to Understanding the Neural Circuitry of Saccade Control
     in the Cerebral Cortex Using Antidromic Identification in the
     Awake Behaving Macaque Monkey Model   161
     Kevin Johnston and Stefan Everling

10   Photothrombotic Infarction of Caudate Nucleus and Parietal Cortex   183
     Toshihiko Kuroiwa and Richard F. Keep

11   Models of Rodent Cortical Traumatic Brain Injury   193
     Frances Corrigan, Jenna M. Ziebell, and Robert Vink

12   The Use of Commissurotomy in Studies of Interhemispheric Communication   211
     Ian Steele-Russell

PART III   CEREBELLAR AND BRAIN STEM SYSTEMS

13   Genetic Models of Cerebellar Dysfunction 241
     Robert Lalonde and Catherine Strazielle

14   Cerebellar Control of Fine Motor Function 263
     Rachel M. Sherrard

15   Cerebellum and Classical Conditioning............................... 281
     Richard F. Thompson

16   Assessments of Visual Function..................................... 287
     Ma'ayan Semo, Carlos Gias, Anthony Vugler, and Peter John Coffey

17   The Role of the Pedunculopontine Tegmental Nucleus in Motor Disorders 321
     Nadine K. Gut and Philip Winn

PART IV   SPINAL CORD SYSTEMS

18   Contusion Models of Spinal Cord Injury in Rats......................... 345
     Kelly A. Dunham and Candace L. Floyd

19   Demyelination Models in the Spinal Cord.............................. 363
     *Paul A. Felts, Damineh Morsali, Mona Sadeghian, Marija Sajic,
     and Kenneth J. Smith*

20   Preparation of Spinal Cord Injured Tissue for Light
     and Electron Microscopy Including Preparation for Immunostaining 381
     Margaret L. Bates, Raisa Puzis, and Mary Bartlett Bunge

21   Assessing Spinal Cord Injury....................................... 401
     Gillian D. Muir and Erin J. Prosser-Loose

22   Precise Finger Movements in Monkeys 419
     Roger Lemon

Index 435

Contributors

GARNIK AKOPIAN • *Andrus Gerontology Center, University of Southern California, Los Angeles, CA, USA*

VÉRONIQUE M. ANDRÉ • *IDDRC, Semel Institute for Neuroscience and Human Behaviour, David Geffen School of Medicine, UCLA, Los Angeles, CA, USA*

GORDON W. ARBUTHNOTT • *Brain Mechanisms for Behaviour Unit, Okinawa Institute of Science & Technology Promotion Corporation, Okinawa, Japan*

YANN BRAMOULLÉ • *Molecular Imaging Research Center and URA CEA-CNRS 2210, Institute of Biomedical Imaging, CEA, France*

SIMON P. BROOKS • *The Brain Repair Group, School of Biosciences, Cardiff University, Cardiff, Wales, UK*

GUY A. CALDWELL • *Department of Biological Sciences, University of Alabama, Tuscaloosa, AL, USA*

KIM A. CALDWELL • *Department of Biological Sciences, University of Alabama, Tuscaloosa, AL, USA*

NADJA VAN CAMP • *Molecular Imaging Research Center and URA CEA-CNRS 2210, Institute of Biomedical Imaging, CEA, France*

CARLOS CEPEDA • *IDDRC, Semel Institute for Neuroscience and Human Behaviour, David Geffen School of Medicine, UCLA, Los Angeles, CA, USA*

ORNELA DERVISHA • *Department of Biology, Temple University, Philadelphia, PA, USA*

STEPHEN B. DUNNETT • *The Brain Repair Group, School of Biosciences, Cardiff University, Cardiff, Wales, UK*

BETH E. FISHER • *Division of Biokinesiology & Physical Therapy, University of Southern California, Los Angeles, CA, USA*

SHEILA M. FLEMING • *Departments of Psychology and Neurology, University of Cincinnati, Cincinnati, IL, USA*

LISA GLASL • *Helmholtz Zentrum München, Institute of Developmental Genetics, Technical University Munich, Neuherberg, Germany*

JAIME N. GUZMAN • *Robert H. Lurie Medical Research Center, Northwestern University, Chicago, IL, USA*

PHILIPPE HANTRAYE • *Molecular Imaging Research Center and URA CEA-CNRS 2210, Institute of Biomedical Imaging, CEA, France*

ANDREAS HEUER • *The Brain Repair Group, School of Biosciences, Cardiff University, Cardiff, Wales, UK*

RALPH HILLMAN • *Department of Biology, Temple University, Philadelphia, PA, USA*

MATHIAS HOEHN • *In-vivo-NMR Laboratory, Max Planck Institute for Neurological Research, Köln, Germany*

SABINE M. HÖLTER • *Helmholtz Zentrum München, Institute of Developmental Genetics, Technical University Munich, Neuherberg, Germany*

MICHAEL J. JACKSON • *Neurodegenerative Diseases Research Center, Institute of Pharmaceutical Sciences, King's College, London, UK*

MICHAEL W. JAKOWEC • *Departments of Neurology, University of Southern California, Los Angeles, CA, USA; Division of Biokinesiology & Physical Therapy, University of Southern California, Los Angeles, CA, USA; Department of Cell and Neurobiology, University of Southern California, Los Angeles, CA, USA*

PETER JENNER • *Neurodegenerative Diseases Research Centre, Institute of Pharmaceutical Sciences, School of Biomedical Sciences, King's College, London, UK*

NATALIE JEROME • *Department of Biology, Temple University, Philadelphia, PA, USA*

JENNI M. KARL • *Department of Neuroscience, Canadian Centre for Behavioural Neuroscience, University of Lethbridge, Lethbridge, AB, Canada*

GERALDINE KRESS • *Robert H. Lurie Medical Research Center, Northwestern University, Chicago, IL, USA*

RALF KÜHN • *Helmholtz Zentrum München, Institute of Developmental Genetics, Technical University Munich, Neuherberg, Germany*

EMMA L. LANE • *The Brain Repair Group, Welsh School of Pharmacy, Cardiff University, Cardiff, Wales, UK*

MICHAEL S. LEVINE • *IDDRC, Semel Institute for Neuroscience and Human Behaviour, David Geffen School of Medicine, UCLA, Los Angeles, CA, USA*

HANNA S. LINDGREN • *The Brain Repair Group, School of Biosciences, Cardiff University, Cardiff, Wales, UK*

PERTTI PANULA • *Neuroscience Center and Institute of Biomedicine/Anatomy, University of Helsinki, Helsinki, Finland*

ROBERT G. PENDLETON • *Department of Biology, Temple University, Philadelphia, PA, USA*

GISELLE M. PETZINGER • *Department of Neurology, University of Southern California, Los Angeles, CA, USA; Division of Biokinesiology & Physical Therapy, University of Southern California, Los Angeles, CA, USA*

JOSHUA L. PLOTKIN • *Robert H. Lurie Medical Research Center, Northwestern University, Chicago, IL, USA*

SHILPA P. RAO • *IDDRC, Semel Institute for Neuroscience and Human Behaviour, David Geffen School of Medicine, UCLA, Los Angeles, CA, USA*

D. EUGENE REDMOND JR. • *Yale University School of Medicine, New Haven, CT, USA*

ANNE E. ROSSER • *The Brain Repair Group, School of Biosciences, Cardiff University, Cardiff, Wales, UK*

TIMOTHY SCHALLERT • *Department of Psychology, University of Texas at Austin, Austin, TX, USA*

NICHOLAS SCHWARZ • *Robert H. Lurie Medical Research Center, Northwestern University, Chicago, IL, USA*

GAYNOR A. SMITH • *The Brain Repair Group, School of Biosciences, Cardiff University, Cardiff, Wales, UK*

MARIA SUNDVIK • *Neuroscience Center and Institute of Biomedicine/Anatomy, University of Helsinki, Helsinki, Finland*

D. JAMES SURMEIER • *Robert H. Lurie Medical Research Center, Northwestern University, Chicago, IL, USA*

EDUARDO M. TORRES • *The Brain Repair Group, School of Biosciences, Cardiff University, Cardiff, Wales, UK*

DANIELA VOGT-WEISENHORN • *Helmholtz Zentrum München, Institute of Developmental Genetics, Technical University Munich, Neuherberg, Germany*

JOHN P. WALSH • *Andrus Gerontology Center, University of Southern California, Los Angeles, CA, USA*

IAN Q. WHISHAW • *Department of Neuroscience, Canadian Centre for Behavioural Neuroscience, University of Lethbridge, Lethbridge, AB, Canada*

DAVID L. WOKOSIN • *Robert H. Lurie Medical Research Center, Northwestern University, Chicago, IL, USA*

RUTH WOOD • *Department of Cell and Neurobiology, University of Southern California, Los Angeles, CA, USA*

WOLFGANG WURST • *Helmholtz Zentrum München, Institute of Developmental Genetics, Technical University Munich, Neuherberg, Germany*

XIAOYUN C. YANG • *Department of Biology, Temple University, Philadelphia, PA, USA*

Part I

Generic Methods of Assessment

Why Cannot a Rodent Be More Like a Man?
A Clinical Perspective

Anne E. Rosser

Abstract

Neurodegeneration is largely limited to humans, with spontaneous neurodegenerative conditions being extremely rare in animals. However, whole animal models are crucial for a proper understanding of the neurodegenerative process as well as essential for preclinical assessment of novel therapies. Thus, it has been necessary to generate animal models of neurodegeneration using a combination of techniques, including injectable toxins and genetic manipulation. Given the constraints inherent in these approaches, how successful are animal models of neurodegeneration and how can such models be refined in the future?

Key words: Experimental medicine, Translation, Clinical studies, Animal studies, Validity

1. Introduction

The clinician's perspective on the value, suitability and desirability of animal models does not differ markedly from that of the neuro-biologist. From a clinical perspective, the need is to use animal models to understand more about the disease mechanisms, to test therapeutic options and to investigate disease clinical features in animals in a way that cannot be undertaken in humans for both ethical and practical reasons. For simplicity, the focus here is largely on rodent models of neurodegeneration, although it is acknowledged that a wide range of species, including worms, flies, toads, pigs, sheep and primates have all played a role in furthering our understanding of the brain more generally and neurodegenerative processes more specifically.

What one would really like of animal models is for them to represent all the pathological and clinical features of specific diseases seen in man. Ideally, this would utilize naturally occurring

Emma L. Lane and Stephen B. Dunnett (eds.), *Animal Models of Movement Disorders: Volume I*, Neuromethods, vol. 61, DOI 10.1007/978-1-61779-298-4_1, © Springer Science+Business Media, LLC 2011

models (but there are very few of these indeed) that have "face" validity such that the abnormal movements and behaviours we see in man are accurately reflected in the animal. Moreover, the model should have "construct" validity such that it reflects the underlying pathological changes that are faithful to the disease process in man. The model should allow assessment of the full range of symptoms seen in man, including psychiatric, as well as cognitive and motor dysfunction. It should be accurate enough for testing the efficacy of new and emerging therapies. However, most of us are realistic enough to understand that this is unachievable and that it is inevitable that animal models are incomplete representations of the human condition. So the question is how good are the models to date, and how useful is an incomplete model?

First and foremost, by definition, the genetic backgrounds of rodents and humans are not identical and, as a complete understanding of all the genetic elements of any condition (even for a dominant condition with complete penetrance) is a long way off, it is not realistic at this stage to think in terms of manipulating the genetics to compensate for this. Furthermore, the brain of rodent models, and even of most primate models, may follow a similar structural plan to that of man, but is not identical. There are also well-documented differences in terms of the precise cellular content and neuronal connections of homologous structures – an example being the change in functional significance of the red nucleus in lower, compared to higher mammals (1) (see also Lemon, volume II of this series). There are many other differences, such as the rodent brain being set up with olfaction and whisker touch as the predominant special senses, as opposed to vision in man. Another important difference that almost certainly impacts on the validity of rodent–human extrapolations is the difference in lifespan of over 70 years. This may play a major role in the lack of clinical disease seen in many rodent disease models despite apparently appropriate disease processes at the cellular level, and this is not surprising when one takes into account that age is a major risk factor for many neurodegenerative conditions in adult humans (2). Given that most neurodegenerative diseases take decades to develop in man, it may simply be the case that the rodent lifespan is too short to manifest the disease phenotype.

The reality, of course, is that we must make do with models that are partial but nevertheless can provide valuable insights providing that we use them intelligently. A major route to using a model successfully is that it is well-characterized so that its strengths and limitations are well-understood. This can be a labour-intensive and expensive process, but is essential for understanding the appropriate use of a model for a specific need and also for using that model most efficiently. For example, *Drosophila* models exist for a number of neurodegenerative conditions, and although some of them may have behavioural phenotypes (3), the major value is their rapid life cycle, so allowing research that requires examination of multiple generations and their suitability for sophisticated genetic manipulation. In contrast, genetic manipulation of rodents is more time-consuming, expensive and complex, but rodents lend themselves much more readily to more sophisticated movement or cognitive analysis. Another example of the requirement for proper characterization is for testing of therapeutic agents, where the reliability, timing and nature of functional deficits need to be carefully defined in detail so that effects of the agent can be accurately assessed (4).

2. Do Models Have Face Validity?

In general, face validity in rodent neurodegenerative models is variable. Some animals with what appears to be the appropriate pathological lesion may demonstrate no discernible functional phenotype, whereas others may have a variety of functional deficits that have features suggesting that they are close correlates of the human condition. Given that rodents are nocturnal animals that walk on all fours, use their whiskers and olfactory system to sense their environment and for social interaction, it is perhaps not surprising that many neurodegenerative models have limited face validity, as it could be argued that rodents are set up rather differently to meet different challenges to those faced by man. This may explain to some extent why some of the deficits seen in rodent neurodegenerative models may be in the same domain as those in human, but are not precisely recognizable as the human counterpart. For example, in many of the available models of Huntington's disease (HD), the animals display both motor and cognitive deficits that appear to approximate to those seen in man, but are not identical. One of the more striking features of the human condition is the chorea: purposeless involuntary movements that commonly increase in frequency and severity throughout the disease, but may wane in advanced disease as rigidity and dystonia come to dominate. Rodent models of HD do not show chorea-like movements of the type seen in the patients for reasons that are not yet clear. There are numerous possibilities. For example, toxin models, such as striatal injection of quinolinic acid or systemic injection of 3NP that replicate the selective loss of medium spiny neurons seen in the human condition, may be too acute and some symptoms may only emerge from gradual and progressive cell loss. Furthermore most models that show progressive cell loss are created using transgenic or knock-in technology and in order to see a phenotype within the lifetime of the rodent very large CAG repeats are used that may mean that the models are more representative of the juvenile disease (in which repeat numbers are high) than adult onset HD in which repeat numbers are typically between 40 and 50 (see Hickey and

Chesselet, volume II of this series). The juvenile form of the human disease usually presents with a rigid/dystonic variant of the condition with chorea being uncommon. Other possibilities exist, for example the anatomical distribution of pathology induced in rodents may be sufficiently different to that in man to explain the difference in motor symptopmatology. Equally, one could speculate that the basal ganglia could have a different, although overlapping, functional profile in rodents and man, and the function in man that leads to striatal damage producing chorea (perhaps the capacity of the basal ganglia to "focus" movements and suppress extraneous movements) may be more prominent in man and of less importance in rodents.

However, it should also be emphasized that sometimes face validity may be remarkably accurate. One example of this is the paw reaching deficit seen in HD rodent models (5). In both humans and rodents, qualitative analysis of the precise sequences of arm/forelimb and hand/paw movement during a reaching task demonstrated remarkable similarities between the deficits in man and rodent.

3. Do Models Have Content Validity?

In particular, do the functional deficits seen in animal models map to the pathological changes and is the pathology meaningful in terms of the human condition? This can be extremely hard to assess and can weaken even a model with good face validity. An example is the α-synuclein model of Parkinson's disease (PD). α-Synuclein lesions are pathological hallmarks of Parkinson's disease, α-synuclein gene mutations have been discovered in some types of familial PD and certain common haplotypes of α-synuclein are associated with sporadic PD (6). It has been shown that over-expression of human wild-type α-synuclein as well as the mutated form can lead to Lewy body-like structures in PD models (7). These mice do show a phenotype in that there are locomotor and cognitive deficits, but it is not clear that these deficits correlate with the α-synuclein deposits either in terms of numbers of deposits or their location (8). In particular, there are a few models in which the α-synuclein deposits are found largely in the substantia nigra, but unfortunately this is not associated with dopamine loss therein (8). Thus, it is not clear what is causing the functional deficits in the mouse and so the relationship between the deficits and the pathological changes (are the deposits part of the pathogenic process or simply a co-morbid phenomenon?) is hard to use experimentally at the current time.

Likewise, the amyloid precursor protein (APP) mutation-transgenic models of Alzheimer's disease generally show behavioural impairment before deposition of amyloid, and even once an amyloid cerebropathy has developed neuronal loss tends to be small (9, 10). This makes it difficult to know how much the cognitive deficits seen in these mice are equivalent to the ones seen in humans. Thus, in both of these examples, findings in the animal models may be of interest in terms of understanding brain function generally, but are of questionable relevance in terms of understanding a particular human disease.

4. What Is the Value of Partial Models?

So it seems that all available animal models are at best partial, in which case how does this affect their value in neurodegenerative research?

First, a partial model may be valuable provided it is relevant to the hypotheses being addressed. In this respect, the R6/2 model of HD, one of the first transgenics available for this condition, has been invaluable in furthering our understanding of the human disease, in particular understanding of the cellular and molecular machinery (11). It also demonstrates important elements of HD pathology, such as the intranuclear inclusions (12). However, it has to be recognized that the distribution of inclusions is not typical of adult onset HD and neural loss is surprisingly low it is a very rapidly progressive model with early death; and it commonly shows some features, such as frank diabetes, that are not common in HD. Thus, whether the R6/2 mouse is a model of the human disease or rather a model of severe CAG repeat toxicity is debatable. Some of the following generations of transgenics and knock-in mutations have less dramatic phenotypes, and some may be regarded as more representative of the presymptomatic/early phase with respect to their functional profile (13).

Thus, specific animal models may be appropriate to address specific questions. Injection of quinolinic acid (an excitotoxic agent) into the striatum selectively destroys medium spiny neurons (the major neuronal phenotype lost in Huntington's disease) leaving the interneurons largely intact. This imitates the pathology seen in the striatum in HD, albeit acutely, rather than in a progressive fashion. This model has been important in highlighting the potential role of glutamate in the pathogenesis of HD and is also extremely useful in producing a relatively stable lesion for the assessment of neural transplantation strategies. It does not, of course, replicate other cell and molecular features of HD and so is of little use to explore the cellular features of mutant HD.

The partial nature of the model may also be a tool for furthering understanding of the elements important in disease. Using one of the examples given above, the fact that APP does not alone produce a satisfactory model of Alzheimer's may help to lead to an understanding of other elements that are also important by a process of combing this model with other disease-modifying candidates.

5. Can Partial Models Be Used for Assessment of Potential Therapeutic Agents?

Proof of principle that animal models can demonstrate efficacy that is translatable to man can be found in the cell transplantation field. Here, models, such as the 6-OHDA lesion model, have been used to demonstrate that transplants of fetal ventral mesencephalon (the source of developing nigral dopamine cells) can result in amelioration of functional deficits produced by the toxic loss of dopamine (14). This has been translated clinically, at least in preliminary studies, in patients with Parkinson's disease (15). As a second example, immunization in the APP mouse model of Alzheimer's disease cleared amyloid and this was also translated into human studies, confirmed by post-mortem, although as outlined above it is not clear that this was associated with clinical benefit.

However, a host of therapeutic agents have been notable to date in having a poor record of being translated into positive clinical findings, a result that has been debated most intensively in the context of a widespread failure to develop effective therapeutics for stroke (16). There are multiple possible reasons for this. Some may be related to deficiencies in the animal model, such as the link between the functional readouts and the underlying pathology not being secure as outlined above or the outcome measures not measuring equivalent changes, and some may be related to deficits in the translation process, such as dose equivalents in man being incorrect, and design flaws in the clinical trial, such as insensitive outcome measures and trials being too short (usually for economic reasons) to see the changes demonstrated in the rodent.

The validity of a model for preclinical testing of therapeutic agents can be tested by attempting to produce tools that test the same things in rodents and in man. The animal learning literature has already developed tests for rodents that appear to depend on many of the same underlying functional processes implicated in humans, in other words "parallel" forms of tasks (17).

6. What Is the Future?

Given the inherent differences between rodent and man, it seems unlikely that complete animal models of neurodegeneration will ever exist, but is it possible to produce more satisfactory models that better replicate the elements of a disease that we may wish to study? This seems a wholly achievable aim. In particular, as more knowledge is acquired about neurodegeneration in man and as the relevant genetic features of a condition are uncovered, this information can be applied back to animal models to improve their accuracy and relevance. Thus, it is likely that the production of improved models of neurodegeneration will be an evolving and iterative process.

References

1. Massion J (1988) Red nucleus: past and future. Behav Brain Res 28: 1–8

2. de Oliveira RM, Pais TF, Outeiro TF (2010) Sirtuins: common targets in aging and in neurodegeneration. Curr Drug Targets 11: 1270–80

3. Whitworth AJ (2011) Drosophila models of Parkinson's disease. Adv Genet 73: 1–50

4. Brooks SP, Janghra N, Workman VL, Bayram-Weston Z, Jones L, Dunnett SB (2011) Longitudinal analysis of the behavioural phenotype in R6/1 (C57BL/6J) Huntington's disease transgenic mice. Brain Res Bull, in press

5. Klein A, Sacrey LA, Dunnett SB, Whishaw IQ, Nikkhah G (2011) Proximal movements compensate for distal forelimb movement impairments in a reach-to-eat task in Huntington's disease: new insights into motor impairments in a real-world skill. Neurobiol Dis, in press

6. Venda LL, Cragg SJ, Buchman VL, Wade-Martins R (2010) α-Synuclein and dopamine at the crossroads of Parkinson's disease. Trends Neurosci 33: 559–68

7. Magen I, Chesselet MF (2010) Genetic mouse models of Parkinson's disease: The state of the art. Prog Brain Res 184: 53–87

8. Dawson TM, Ko HS, Dawson VL (2010) Genetic animal models of Parkinson's disease. Neuron. 66: 646–61

9. Ashe KH (2001) Learning and memory in transgenic mice modeling Alzheimer's disease. Learn Mem 8: 301–8

10. Chen G, Chen KS, Knox J, Inglis J, Bernard A, Martin SJ, Justice A, McConlogue L, Games D, Freedman SB, Morris RG (2000) A learning deficit related to age and beta-amyloid plaques in a mouse model of Alzheimer's disease. Nature. 408: 975–9

11. Li JY, Popovic N, Brundin P (2005) The use of the R6 transgenic mouse models of Huntington's disease in attempts to develop novel therapeutic strategies. NeuroRx 2: 447–64

12. Sathasivam K, Hobbs C, Mangiarini L, Mahal A, Turmaine M, Doherty P, Davies SW, Bates GP (1999) Transgenic models of Huntington's disease. Philos Trans R Soc Lond B Biol Sci 354: 963–9

13. Harvey BK, Richie CT, Hoffer BJ, Airavaara M (2011) Transgenic animal models of neurodegeneration based on human genetic studies. J Neural Transm 118: 27–45

14. Fricker-Gates RA, Lundberg C, Dunnett SB (2001) Neural transplantation: restoring complex circuitry in the striatum. Restor Neurol Neurosci 19:119–38

15. Brundin P, Barker RA, Parmar M (2010) Neural grafting in Parkinson's disease Problems and possibilities. Prog Brain Res 184: 265–94

16. Macleod MR, Fisher M, O'Collins V, Sena ES, Dirnagl U, Bath PMW, Buchan A, van der Worp H, Traystman R, Minematsu K, Donnan GA, Howells DW (2009) Good laboratory practice: preventing introduction of bias at the bench. Stroke 40: e50–52

17. Chudasama Y, Robbins TW (2006) Functions of frontostriatal systems in cognition: comparative neuropsychopharmacological studies in rats, monkeys and humans. Biol Psychiat 73: 19–38

Chapter 2

Zebrafish as a Vertebrate Model Organism for Studying Movement Disorders

Maria Sundvik and Pertti Panula

Abstract

Zebrafish, *Danio rerio*, a subtropical vertebrate has, during the last decades, emerged as an important model organism in neurobiological and biomedical research. The zebrafish neurotransmitter systems, including major small molecular substances and neuropeptides and their receptors, are very similar with those of mammals. The small size of the fish at the time when active movement starts allows three-dimensional visualization of essentially all neurons, which contain specific markers. This allows detailed quantification of neurons of all important nuclei. This information can be correlated to fast kinematic analysis of, e.g., sensorimotor responses, and tracking of swimming episodes over longer periods of time. A large number of mutant fish with motor disturbances have been produced in genetic screens with alkylating agents and with retroviruses. Development of gene knock-out and transgenic fish has also become possible with TILLING or zinc finger nuclease methods. The easiest methods to modify gene expression include morpholino-modified oligonucleotides, which enable translation inhibition at best for several days. This method requires stringent controls and knowledge of non-specific off-target effects, which are common. Several genes that cause autosomal hereditary Parkinson's disease have been identified and inactivated in zebrafish. The effects range from very mild to severe, some of which are likely non-specific. Well-controlled studies have given valuable information of basic functions of genes important in Parkinson's disease and other human neurological diseases. The emergence of advanced gene modification methods, most obviously the gene knock-out and transgenic methods, is about to render zebrafish a very fast, quantifiable, and economic model to reveal basic functions of genes important for human neurological diseases.

This chapter deals with the neurotransmitter and CNS systems underlying locomotion, methods used to assess different behaviors in developing and adult zebrafish, and published research on genes relevant for movement disorders.

Key words: Zebrafish, *Danio rerio*, Vertebrate, Locomotion, Movement, Neurotransmitter systems

1. Introduction

Zebrafish reach sexual maturation at 3 months of age and have a short generation cycle. This species spawns regularly with the onset of light each morning. The zebrafish embryos undergo ex utero

Emma L. Lane and Stephen B. Dunnett (eds.), *Animal Models of Movement Disorders: Volume I*, Neuromethods, vol. 61, DOI 10.1007/978-1-61779-298-4_2, © Springer Science+Business Media, LLC 2011

fertilization, are transparent for the first few days, and develop rapidly (1). Although spontaneous tail coilings are observed as early as at 17 h post-fertilization (hpf) (2), the larval zebrafish utilize the contents of the yolk sacks for their development and do not move actively. At 5 days post-fertilization (dpf), zebrafish larvae show active beat-and-glide swimming pattern as at that age they have consumed most of the nutrients of the yolk sack and have to start hunting in order to find food. This age is an appropriate time to start to observe the motor functions of zebrafish as the larvae move actively, and several methods are available for both observing fast movements and analyzing locomotor behavior.

2. Zebrafish CNS Systems and Methods for Assessing Motor Behaviors

Neuronal circuits underlying behavior, which varies from simple spinal reflexes to behaviors controlled by the higher centers such as the brainstem, cerebellum, and telencephalon, are well known in mammals due to long-standing research interests in the field. The upper motor neuron functions in mammals are modulated by the cortico-striato-pallido-thalamic system and cerebellum. The first system is regulated by afferent excitatory glutamatergic corticostriatal inputs from cortex to the caudate–putamen (striatum), and the dopaminergic nigrostriatal input from neurons in substantia nigra pars compacta is a functionally important modulatory input in mammals. The caudate–putamen in turn sends efferent projections via either the direct pathway to the pallidum internum or via the indirect pathway through pallidum externum and subthalamic nucleus to thalamus. The thalamus functions as a relay station that filters somatosensory information and motor signals. The thalamus also has reciprocal connections to the cerebral cortex. The most common movement disorder, Parkinson's disease, is the result of a significantly reduced number of dopaminergic neurons in the substantia nigra, which in turn reduces the dopaminergic input to the caudate–putamen. These systems are not yet well characterized in zebrafish.

In zebrafish, the neuronal circuits underlying behavior have been studied quite intensely in the spinal cord and brainstem, whereas the zebrafish forebrain is an area which has only recently started to gain interest. Thus, the upper motor neurons connecting the telencephalon with the brainstem have not yet been identified in zebrafish, and the main focus of zebrafish behavior has been on the role of motoneurons and interneurons in the brain stem and spinal cord.

The forebrain, telencephalon, develops in Actinopterygii, ray-finned fish, via eversion, whereas the forebrain in other vertebrates develops via evagination (3). The difference in development is

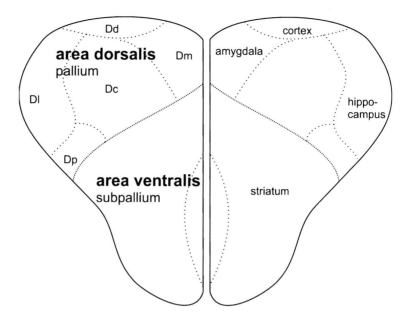

Fig. 1. A schematic drawing of the telencephalon of ray-finned fish illustrating the subdivisions of the telencephalon (*left*) in a coronal section. On the *right* is a presentation of how the fish anatomical structures are thought to correspond to mammalian structures. *Dd* dorsal region of area dorsalis telencephali, *Dm* medial region of area dorsalis telencephali, *Dl* lateral region of area dorsalis telencephali, *Dc* central region of area dorsalis telenchphali.

postulated to lie behind the non-lamilar telencephalon in ray-finned fish (4). As in ray-finned fish, zebrafish telencephalon can also be grossly divided into the area dorsalis, pallium, and area ventralis telencephali, subpallium (5, 6). An area that corresponds to the mammalian basal ganglia has been, based on gene expression data, localized to area ventralis telencephali (7) in zebrafish, whereas the dorsal part of the area dorsalis telencephali (Dd, Fig. 1) in goldfish has been suggested to correspond to mammalian cerebral cortex (8). Based on lesion studies of the medial pallium the corresponding area to mammalian amygdala was suggested to be located in the medial region of area dorsalis telencephali (Dm, Fig. 1) in goldfish (9). Similar lesion studies of the lateral pallium suggest that the phylogenetically conserved area that corresponds to mammalian hippocampus in goldfish is in the lateral region of area dorsalis telencephali (Dl, Fig. 1) (10).

The brain structure in zebrafish that is laminar is the optic tectum, which corresponds to the superior colliculus in the midbrain, mesencephalon, of mammals. In both mammals (11) and zebrafish (12), this structure integrates visual information with motor output, and in zebrafish this area drives prey capture behavior (13).

The brainstem nuclei are in a critical position to regulate lower motor neurons and mediate modulatory inputs from telencephalic and cerebellar systems in mammals. The neurons and neuronal networks in the brainstem and spinal cord have been the focus of movement studies in zebrafish; and in the following chapters, we will review the work further.

The role of the systems that modulate motor responses, such as basal ganglia and cerebellum in zebrafish are largely unknown. Based on a tracing study and gene expression data, basal ganglia, including the striatum, are suggested to be located in the ventral telencephalon (7, 14), whereas toxicity studies, neuronal connectivity, and tracing studies suggest substantia nigra to be located in the posterior tuberculum of zebrafish (15–17). In mammals, the thalamic complex is the largest structure of the diencephalon and consists of the prethalamus, zona limitans intrathalamica, and thalamus. Comparative analyses of mouse, chick, and zebrafish revealed that the development of the thalamus is evolutionarily conserved and that sonic hedgehog plays a crucial role in the formation and differentiation of thalamic nuclei (reviewed in (18)). A double labeling study in zebrafish showed that the main dopaminergic ascending innervation from the posterior tuberculum to the ventral telencephalon was from the periventricular nucleus of the posterior tuberculum (14). The cerebellum of teleosts differs from the mammalian cerebellum, as in teleosts eurydendroid cells interact with the Purkinje cells, whereas in mammals the axon of the Purkinje cells interact with the deep cerebellar nuclei (19). To study the role of the cerebellum in the modulation of higher brain functions and behavior, the anatomy of the cerebellum in developing and adult zebrafish as well as mutations that affect the development of the cerebellum was recently compiled (20).

Genetic approaches combined with the transparency and the small size of the zebrafish larvae are features that render it a good model organism for studying the neuronal circuits underlying movement. Targeting and stimulating and/or ablating single neurons in an intact vertebrate organism and assessing the role of specific neurons in different types of behaviors is possible in zebrafish larvae (21). Calcium imaging can be used to assess which neurons mediate specific behaviors. An increase in calcium levels, which can be detected by several Ca^{2+} indicators, reflects the activity of a neuron in response to specific stimuli (22, 23). Optogenetics is another approach that allows activation of neurons within millisecond range providing a tool to study the role of targeted neurons in specific behaviors (24, 25) (see also Chap. 6 this volume). Quantification of the behavioral responses can be recorded by live imaging of the zebrafish and correlated to neuron activation measured with optical or electrophysiological methods. High-throughput drug screening combined with a behavioral readout (26) is another appealing approach for studying behavior and pharmacological correlates in this vertebrate model organism.

2.1. Neurotransmitter Systems in Zebrafish Brain

The neurotransmitter systems relevant for motor control and movement disorders are very similar in zebrafish and mammals. Thus, the major biogenic amine systems are well conserved and the CNS domains, which harbor cell body groups, are similar, with a few exceptions (27). Contrary to invertebrates such as *D. melanogaster* and *C. elegans*, where receptor mechanisms are very different from mammals despite the similar nature of many neurotransmitters, the zebrafish neurotransmitter receptors are in general very alike those of mammals. Genome duplication has resulted in expansions of some receptor families, and the total number of G-protein-coupled receptors in zebrafish is larger than that of mammals (for a detailed description, see (27)). For example, the three α_2 receptor subtypes (α_{2a-c}) are similar to those of mammals, but there is a fourth duplicated receptor that is not found in mammals (α_{2da} and α_{db}) (28, 29). The potentially duplicated α_{2a-c} receptors have been lost during zebrafish evolution. Genes corresponding to the three histamine receptors (hrh$_{1-3}$), known to be important in mammalian brain functions (30), are found in zebrafish and expressed in the brain (31), and their ligands have significant effects on zebrafish locomotor activity.

The rate-limiting enzyme tyrosine hydroxylase (TH) is duplicated in zebrafish (32), and this phenomenon has not been noted in most papers that deal with the zebrafish dopaminergic system. Thus, most studies which deal with neurotoxicological or genetic manipulation of the dopamine system, only TH1 has been examined, since also available antibodies detect this form efficiently (33). There is now evidence that TH2 expressing neurons, in addition to those which express TH1, are affected by translation inhibition of PINK1, a mitochondrial kinase important in autosomal hereditary Parkinson's disease (34). The expression patterns of TH1 and TH2 are complementary. TH2 is expressed in four major bilateral clusters in the preoptic region and hypothalamus (33), whereas a total of 17 cell groups in the whole brain express TH1 (17). Recently, we have developed an antibody against TH2, which shows that the TH2-expressing cells produce TH protein and the cells display a distinct morphology (Semenova et al., in preparation).

Whereas mammals have two distinct monoamine oxidase genes, the zebrafish genome encodes for one enzyme which appears functionally as a hybrid of the two mammalian forms (35), with eight of the 12 residues of the active site identical with human MAO A and MAO B, two with human MAO A and two appear unique for zebrafish (35). MAO mRNA is expressed early (at 24 hpf) in developing zebrafish, and enzyme activity is measurable at 42 hpf (36). Inhibition of zebrafish MAO with deprenyl strongly elevates brain 5-HT content and decreases swimming activity sharply in a dose-dependent manner (36). This effect is due to release and uptake of serotonin in non-serotonergic neurons in basal hypothalamus (36). It is important to take this into account in, e.g., toxicology experiments with 1-methyl-4-phenyl-1,2,3,6-tetrahydropyridine (MPTP), which require conversion to 1-methyl-4-phenylpyridinium (MPP+) by MAO to induce a decline in motility in both larval (17) and adult (37) zebrafish.

The cholinergic neurons in zebrafish brain can be found in the telencephalon, preoptic region, dorsal thalamus, pretectal nuclei, hypothalamus, optic tectum, and tegmentum (38–40).

2.2. Brainstem and Spinal Cord Systems in Zebrafish

The brainstem of zebrafish larvae contains approximately 150 reticulospinal neurons that project to spinal cord to regulate execution of an array of different behaviors and movement patterns. Of these reticulospinal neurons, the Mauthner neuron, discovered in 1859 by Ludwig Mauthner, is the most famous and is thought to be the sole driver of the escape response, which also is called the startle response. The startle response of freely moving larval zebrafish has been characterized and elegantly demonstrated to include a short latency and a long latency startle response (21). Both types of startle response are short lasting and completed in less than 1 s after stimulation. Laser ablation of the Mauthner neuron abolished the short latency startle response, whereas the long latency startle responses were unaltered, indicating that only the short latency startle response is mediated by the Mauthner neuron. Other neurons of the brainstem such as MID2cm (Mauthner neuron homologue, middle dorsal 2 contralateral medial longitudinal fasciculus interneuron in rhombomere 5) and MID3cm (Mauthner neuron homologue, middle dorsal 3 contralateral medial longitudinal fasciculus interneuron in rhombomere 6) cooperate with the Mauthner neuron to provide the escape response with a direction (22). Stimulation of the head versus the tail elicits activity in a specific but different manner in Mauthner neurons and MID2cm and/or MID3cm neurons. In the case of a head stimulus both Mauthner neuron and MID2cm or MID3 neuron respond, whereas when the tail is stimulated only the Mauthner neurons are activated. Orger et al. (23) studied the role of specific reticulospinal neurons in turning behaviors and slow swimming, and found forward-preferring neurons among the neurons of the medial longitudinal fasciculus, nucMLF. In the same study, neurons that execute right or left movements were also identified. A schematic drawing of the anatomical position of the described neurons is presented in Fig. 2, and also their connections via commisural interneurons to lower motor neurons in the spinal cord are illustrated.

2.3. Movement Analysis of Zebrafish Larvae

Zebrafish larvae show spontaneous movement as early as 17 hpf (2). At 3 dpf, the fish show bursts of activity, but the continuous beat-and-glide swim pattern is observed at the earliest at 4 dpf. As the fish grows older the time spent swimming increases. By 5 dpf larvae have almost consumed the nutrients in their yolk sack and need to start hunting for food in order to survive. At this stage

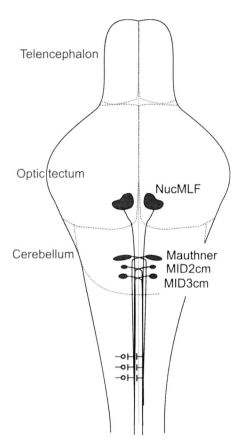

Fig. 2. A schematic picture of some reticulospinal neurons in zebrafish brainstem. NucMLF, Mauthner neuron, MiD2cm, and MiD3cm are illustrated as examples of reticulospinal upper motor neurons, because the behaviors driven by these neurons have been identified (21–23). Their location is correlated to telencephalon, optic tectum, and cerebellum. In the spinal cord commissural interneurons and connections to lower motor neurons are illustrated.

different behaviors can be observed, such as prey capture (41) and sleep (42, 43).

In order to link neurons to circuits and behaviors, behaviors can be dissected into smaller units or movement patterns. Kinematic analysis, which includes dissection of specific behaviors into different movement patterns, provides a detailed description of the repertoire of movements that the zebrafish combines to produce specific behaviors. For dissecting the kinematics, it is essential to observe different behaviors or movement patterns of zebrafish using high frequency imaging (starting at 500–1,000 frames/s). Behaviors that need to be assessed with high frequency imaging are very rapid, lasting only for milliseconds, e.g., startle response and prey capture. For other movement patterns and behaviors such as place preference, sleep, or motility assays, which can be followed for long periods (from seconds to days), live imaging with standard video tracking frequency (starting at five samples/s or less) is adequate.

Several studies have addressed the movement pattern repertoire of zebrafish larvae and classified the movement patterns in different manner (41, 44, 45). Zebrafish exhibits an escape response consisting of a C-start turn (a Mauthner cell-driven activation of motor neurons on the contralateral side of the spinal cord enabling the fish to turn and swim away from the stimuli) in response to acoustic/vibrational stimuli (21) and O-bends (a Mauthner cell-independent movement that is slower than the C-start turn) in response to dark flashes (46). For navigational purposes, adjustment moves, turns, and scoots (45), and slow-swim-like patterns with the pectoral fins can be detected (41). These movement patterns can be detected with high frequency imaging, which requires specific cameras that can acquire images at a high speed (500–1,000 frames/s) and lenses with resolution to image larvae. Strong light, either the infrared or the visual light wavelength range, is also needed as the contrast between the larvae and background needs to be high enough to distinguish the larvae. As the behaviors that are detected are very rapid, it is an advantage if the stimuli and recording of the behavior can be synchronized. Several individuals can be recorded in the same group, and when the images have been captured, the different movement patterns can be identified from the images.

Above, we already reviewed how the escape response is divided into two different startle responses. Another robust response of the larval zebrafish is the prey capture behavior, which consists of combining the movement patterns described above. *Paramecia* are often used as live food for rearing zebrafish, and the prey capture behavior of zebrafish larvae has been studied by high frequency imaging when exposing paramecia to the fish (41). As a consequence, the larvae orient themselves in response to the position of the paramecia by small discrete movements, including routine-like turns to align the body axis of the larvae with the paramecia and slow swim-like patterns with the fins to slowly move closer to the prey before engulfing it (41). Prey capture has been shown to be mediated by two specific neurons as bilateral ablation of two neurons, MeLc and MeLr, in the medial longitudinal fasciculus (nuc-MLF) completely abolished the prey capture without affecting other behaviors (47). Recently, filtering of small visual cues in the optic tectum was demonstrated to be essential to enable prey capture in zebrafish larvae (13).

2.4. Locomotor Analysis of Larval and Adult Zebrafish

Motility assays are the most common behavioral assays used for both larval and adult zebrafish. We and other groups have established methods for detecting the movement and motor functions of both adult (37, 48, 49) and larval zebrafish (17, 26, 31, 36).

We have used a computerized video tracking system successfully for almost a decade. A number of modifications have been done to the basic setup in order to observe different aspects of zebrafish behavior. Automated video tracking that converts the live image to graphical representations allows quantification of the movement of fish. In this manner, the behavior of larval and adult zebrafish can be monitored from seconds to days. If the tracking lasts for several days, it is of utmost importance that the conditions, noise, light, humidity, and temperature of the observation room is held at constant and appropriate levels. The basic setup is robust and relatively simple, utilizing both wavelengths from the visual range and/or infrared light for detection of the freely moving pigmented zebrafish. Furthermore, this system offers high throughput, because as many as 96 larvae can be observed simultaneously in a single system.

We have focused on the locomotor functions of zebrafish, such as total distance moved, velocity and frequency of movement. We have also observed the circular movement of zebrafish, such as turn angle, angular velocity, and meander (31, 36, 37, 48). Furthermore, we have assessed behaviors such as dark flash response in larval zebrafish (Sundvik et al., in preparation) and place preference in both adult (37, 48) and larval zebrafish (unpublished observation).

The locomotor behavioral method with its several modifications, has limitations that have to be taken into account, but can easily be mastered. In the following, we deal with the major variables. To start with, it is important to take into account how many individuals are necessary to be included in the experiment, and if both the control and treatment group should be monitored simultaneously or if they can be analyzed subsequently. If a large number of individuals per group is essential, this reduces the number of parameters that can be assessed, as the well size is reduced. If place preference is a priority parameter, it is important to have arenas spacious enough for the fish to be able to show a preference. Usually, for larval zebrafish, we used standard well plates of different size, ranging from 6- to 96-well plates, which result in arena diameter of 36–7 mm, respectively. Most often, we have used the 48-well plates, with arena diameter of 13 mm, as the different parameters of movement and different behaviors can still be distinguished. This format also offers high throughput as the behavior of 48 larvae can be assessed simultaneously. The standard well plates have one major drawback, as they are made of transparent instead of white plastic, the inner surface of the wells need to be modified in order to inhibit formation of mirror images of the larvae on the walls of the wells. Therefore, we have also made arenas of white plastic (this offers better contrast between the larvae and background and does not produce mirror images of the larvae) and have a larger diameter (40 mm) where the place preference behavior of zebrafish larvae is easily monitored. When assessing movement and place preference in adult zebrafish, the same basic principles as for larvae apply. It is important to have spacious enough arenas, the animals should be easily distinguishable from the background, and the method should offer options for tracking the behaviors of several individuals simultaneously. As the adult zebrafish tend to jump, it is important that the walls of the arenas are high enough (at least 8 cm over water level) so that the fish cannot move from one arena to the other. The last aspect is that the tracking systems that utilize only one camera create a two-dimensional representation of the movement pattern of the fish. Therefore, the depth of the water column in which the fish is swimming should be as small as possible when horizontal movement is tracked. Recently, some groups have reported that the anxiolytic effect of drugs can be assessed by observing whether the fish spend time in the upper compartment of the water column or at the bottom of the tank (50). Taken together, our approach for assessing behavior offers a robust and crude measurement of motor functions in developing and adult zebrafish.

Apart from studying the locomotor activity, it is feasible both in larval and adult zebrafish to assess the response the fish show to changes in the environment. This can either be studied by moving white and black stripes unidirectionally under the arena where freely moving fish are held, to analyze optomotor response, or by fixing the zebrafish with, e.g., agar and monitor eye movements (optokinetic response). These two assays have successfully been used to screen for visual system defect mutants (51, 52).

2.5. Zebrafish Mutants with Movement Disorders

Fifteen years ago, the large-scale screen in Tübingen identified 166 zebrafish mutants with abnormal movement (44). The identified mutants that were identified by an abnormal touch response fell into 14 phenotypical groups affecting 48 genes in total, of which 18 genes affected somatic muscle development and the remaining 30 genes were suggested to affect neuronal development. The latter group comprised mutants with reduced motility, circling behavior, and motor circuit deficits. A circling behavior mutant, *twitch twice*, was identified in this screen as a mutant that only showed right or left turns, rather than alternating turning behavior to both sides of the trunk as wild-type zebrafish to accomplish straightforward locomotion. The *twitch twice* mutant carries a mutation in the roundabout receptor family gene, robo3 (53). This class of genes is important for axonal guidance, and in the case of the *twitch twice* mutant, the axons of the Mauthner neuron did not cross the midline as in normal larvae (53). As a consequence, the Mauthner neuron axons are either on left or right side of the trunk and thereby the larva exhibits a unidirectional behavior. A few years later in another mutagenesis screen, the *touchtone* mutant was identified. This mutant showed reduced motility and pigmentation defects, and the mutation was shown to be located in the transient receptor potential melastatin 7 (trpm7) gene (54). TRPM7 is an ion channel permeable for Mg^{2+} and Ca^{2+} and widely expressed in murine cells. Interestingly, loss of trpm7 has been correlated to parkinsonian dementia (55). In zebrafish trpm7 mutants the pigmentation deficit was induced by necrosis of melanophores. The cell death could be inhibited by inhibition of melanin synthesis. As melanin and catecholamines share the same metabolic pathway, this finding gives new insight about the mechanisms underlying cell death of pigmented neurons and can offer a new approach for protecting pigmented neurons from cell death.

3. Zebrafish Models of Parkinson's Disease

3.1. Neurotoxins to Model PD and Other Neural Diseases

A Parkinson's disease-like condition can be induced by MPTP in humans and some experimental animals including monkeys and mice. It has been used in several species (56–58), and it gives more consistent results than other current toxin models (59). In astrocytes or neurons, MPTP is converted to the toxic metabolite MPP+ by monoamine oxidase, and this metabolite enters dopaminergic neurons by dopamine transporter to cause death of cells in substantia nigra in mammals by inhibition of complex 1 of the mitochondrial respiratory chain. In adult zebrafish, MPTP causes a delayed, partly transient decline in dopamine and noradrenaline levels, and a locomotor defect characterized by slower movement and abnormal swim tracks (37). 6-Hydroxydopamine, which acts as a false transmitter in dopaminergic neurons, causes grossly similar alterations in dopamine levels and locomotor behavior (37). These motor effects of can be analyzed in detail with a video tracking system, which takes 5–25 frames/s (37, 60), but the fast fine movements and startle responses of the fish are beyond the sensitivity of this method. The histaminergic system is another diffuse projection system also in zebrafish like in other vertebrates (61, 62), which has cell bodies in the posterior basal hypothalamus of the fish (63), and widespread projections throughout the brain, most extensively to the dorsal telencephalon and optic tectum (15, 39). An irreversible inhibitor of the synthesizing enzyme histidine decarboxylase, alpha-fluoromethylhistidine, induces a significant decline in brain histamine in adult zebrafish and an alteration in swim tracks in a spherical aquarium, without affecting swimming speed (48). To detect this behavioral alteration, the swimming arena needs to be large enough to be divided into zones.

MPTP has been reported to cause cell death in larval zebrafish of the pretectal dopaminergic cell population at 5 dpf (64), the putative posterior tuberal nucleus at 2 dpf (65), and the diencephalic dopaminergic neurons at 5 dpf (66). These studies do not define the diencephalic dopaminergic cell populations anatomically, which renders it challenging to correlate the findings to motor behavioral abnormalities and compare the results of different studies. In one study, specific identified cell populations were affected, and exact decreases in distinct diencephalic clusters were reported following MPTP or MPP+ administration (17). Lam et al. (65) describe cell loss at 2–3 dpf in their study. At this early stage, the MAO activity is not yet detectable in all brain areas (36), so that the mechanism of toxicity may not require conversion to MPP+ and is thus different from that of later larval stages. This is supported by a study by Thirumalai and Cline (67), which found that dopamine inhibits initiation of fictive swimming at 3 dpf, and the effect was antagonized by MPTP and D2 receptor antagonist. Several dopaminergic cell populations are not yet fully developed at 2–3 dpf, and the actions of dopamine are different at different early stages of zebrafish. One of the slowly developing cell groups is the cluster containing the putative population corresponding to mammalian mesencephalic substantia nigra neurons (15, 68), referred to in a detailed anatomical study as population 5,6,11 (17). Now that the *th2* expressing cells are known to reside in this area as well (33, 69), the cells targeted by MPTP could also belong to this group. There are currently no published reports on the effect of MPTP on *th2* cells, but the diencephalic *th2* neurons seem to be sensitive to MPTP (Chen et al., in preparation).

Decreases in dopamine, noradrenaline, and serotonin levels in the larval zebrafish after MPTP exposure (17) suggest that both catecholaminergic and serotonergic systems are affected, which agrees with findings in rodents after MPTP exposure (70, 71) and human PD patients (70, 72). MPP+ administration in zebrafish larvae also affected one cell population in the posterior tuberculum (17). In line with the immunohistochemical results, MPP+ did not affect serotonin levels significantly, while it decreased dopamine and noradrenaline levels. Actual cell death in the zebrafish brain following MPTP has not been observed consistently in different studies. Instead, the loss of TH-ir was found to recover rapidly and completely in some studies (17, 37). The reported declines in dopaminergic neurons are based on the smaller size of *th1* mRNA expressing cell clusters without counting the cells (64), reduction of in situ hybridization signal for dopamine transporter mRNA at 48 and 72 hpf (65), reduction of fluorescence intensity in transgenic fish expressing EGFP under the vesicular monoamine transporter 2 promoter (73), or counting *th1* mRNA expressing cells in the diencephalon of fish larvae at 5 dpf. In the last study, the total number of cells in control fish diencephalon at 5 dpf was under 40, whereas it in another study, carried out with a TH antibody and three-dimensional imaging, was much larger at 7 dpf (17). This suggests that the significant differences may in part depend on methodological issues. Long-term studies are needed to establish if developmental exposure leads to permanent changes in cell numbers.

3.2. Genetic PD Models   The etiology and pathogenesis of Parkinson's disease are still not well understood despite extensive studies on the mechanisms underlying neuronal degeneration of dopaminergic and other, e.g., serotonergic neurons in the disease. Several of the genes involved in the rare genetic forms of PD have been cloned in zebrafish, and their translation has been inactivated using morpholino oligonucleotides (MOs). Four genes, Park2 (Parkin), Park6 (Pink1), Park7 (DJ-1), and Park8 (LRRK2) have been studied in detail by inactivating their translation (Table 1).

Table 1
Zebrafish studies using morpholino oligonucleotides for translation inhibition of Parkinson's disease related genes

Gene	Morpholino type	Cells affected	Phenotype	Reference
Parkin (Park2)	spMO	Diencephalic DA cells (th1)	Reduced complex I act Increase sensitivity to MPP+ Normal swimming beh Electron dens mat. in t-tubules	(82)
PINK1 (Park6)	trMO	Diencephalic DA cells (th1)	Small eyes, curved tails Enlarged brain ventricles Elevated GSK3β activity Early larval death	(77)
PINK1 (Park6)	spMO	Diencephalic DA cells (th1 and th2, DAT)	Increased sensitivity to MPTP Decreased swimming following low MPTP	(34)
PINK1 (Park6)	trMO	Diencephalic DA cells	Increased p53 and Δtp113 Decreased swimming Dispersed th1 neurons in diencephalon Delayed swim bladder devl Craniofacial malformation	(34)
PINK1 (park6)	trMO	Diencephalic DA cells (th1)	Several morphological defects Dispersed DA neurons Decreased tactile responses	(78)
DJ1 (Park7)	spMO	None (th1)	Increased SOD1 mRNA Increased p53 mRNA Increased Bax mRNA	(74)
LRRK2 (Park8)	trMO	Diencephalic DA cells	Lethality Heart edema, reduced brain size	(75)
LRRK2 (Park8)	spMO	Diencephalic DA neurons (th1 and DAT)	Widespread increased apoptosis Decreased locomotion	(75)

spMO splicing site targeting MO, *trMO* translation initiation site targeting MO, *DA cells* dopamine-producing cells

Inactivation of DJ-1 expression with a MO resulted in increased p53 expression, and after exposure to hydrogen peroxide and the proteasome inhibitor MG132 to loss of diencephalic dopamine neurons (74). Severe developmental defects were observed when LRRK2 translation was inhibited in zebrafish (75), whereas deletion of the WP40 domain resulted in loss of some dopaminergic neurons and a locomotor phenotype, which could be rescued with LRRK2 mRNA or L-DOPA. Whereas the LRRK2 (Park8) mutations are responsible for most of the almost 10% of the Mendelian genetic forms of PD in humans, inactivation of functional LRRK2 in mice reveals no abnormalities in the dopaminergic system, and these KO mice are not more susceptible to MPTP than control mice (76). In zebrafish, translation inhibiting MOs have been reported to cause early lethality and gross morphological abnormalities, and inhibition of splicing of the 45th exon caused developmental abnormalities in the brain, including decreased numbers of TH1 and DAT expressing neurons in the diencephalon, and decreased swimming speed which responded to L-DOPA (75).

Studies on PINK1 have exemplified some complexities in using the MO approach. A recent study reported a modest decline in neuron numbers of zebrafish larvae in only a few dopaminergic cell groups expressing either TH1 or TH2 (34), and increased sensitivity to MPTP that resulted in locomotor abnormalities following a splice site targeting MO which did not induce p53 or Δtp113. Application of two other MOs, which inhibited translation of PINK1, induced p53 along with a characteristic off-target phenotype and some changes in dopaminergic neurons, which may be due to specific inhibition of PINK1 translation. Even in this case, some phenotypic features were rescued by PINK1 mRNA, suggesting that either translation inhibiting MO is more effective in inhibiting formation of PINK1 protein, or that PINK1 is generally neuroprotective against toxic substances like oligonucleotides. Another study has reported a robust phenotype with grossly abnormal body structure and modest decline in dopaminergic neurons (77). In this study, and another one carried out with essentially similar MOs (78), induction of p53 was not assessed, and the phenotype resembled the off-target phenotype described by others. Rescue by overexpression of human PINK1 mRNA may be due to either a specific normalization of PINK1 protein levels, or general neuroprotective properties of PINK1. One study has reported a survival rate below 20% at 10 dpf, gross morphological defects, and no significant decline in the number of ventral diencephalic dopamine neurons (78). In one subgroup the number of neurons was reduced, and the ventral diencephalic neurons were disorganized in comparison to control larvae. These effects were reported to be normalized by PINK1 mRNA (78). The PINK1 morphants also suffered from tactile insensitivity, which was normalized by dopamine D_1 receptor agonist SKF-38393. Swimming speed was reduced in PINK1 morphants, who also had problems balancing. This is feasible, because the larvae injected with translation inhibition MOs seem to lack a normal swim bladder (78). In another fish species, medaka (*Oryzias latipes*), permanent PINK1 mutant fish with inactivated PINK1 kinase domain were grossly normal and fertile, exhibited decreased life span, normal numbers of dopaminergic neurons, and elevated rather than decreased dopamine levels (79). The mild effects observed with splice site targeting MOs and in the medaka mutant are more in agreement with the human PINK1 mutant phenotype (80) than those observed with translation blocking MO in zebrafish. In PINK1 mouse KO model, the phenotype is very subtle (81), with no obvious abnormalities in DA content and cell number, but with an altered DA release.

Knockdown of the parkin gene with a splice site targeting MO results in a decline in dopaminergic th1 expressing, but not presumably serotonergic aromatic amino acid decarboxylase expressing neurons in larval zebrafish (82). In this study, larval swimming behavior was not affected by MO treatment, but the knockdown rendered the fish susceptible to MPP+ toxicity characterized by a further decline in diencephalic DA neurons and impaired motility. As all the DA cell populations have different projections and morphology they can be analyzed separately. It is difficult to compare the results from this study and other MO studies (75, 77, 78) to, e.g., toxin studies (17) and MO studies (34), where the cell groups are analyzed separately.

In summary, the effects observed with translation blocking MOs for PD genes in zebrafish are often associated with robust phenotypes and embryonic lethality. Controls for known off-target effects (e.g., activation of p53 pathway, characteristic phenotype with small brain, enlarged brain ventricles, small eyes, craniofacial defects, and swim bladder development delay) are often not carried out or analyzed in detail, and results with mRNA injections have been variable. On the other hand, it is feasible that p53 is part of the signaling pathway specifically regulated by the gene of interest. In that case, its induction can be expected and inhibition of p53 translation would lead to a false negative result. Although injection of p53 MO along with the MO for the gene of interest has been recommended (83), in our hands, this MO has lead to severe phenotypes with other MOs, even at low doses (unpublished observation). Obviously, designing experiments which can be controlled with state-of-the-art methods is difficult, as the MOs can be expected to have significant effects during the first few days, and the motility assays – and associated neuron counts and related morphological studies – are done at 5–7 dpf, when the fish are moving freely. At this time, there is little evidence that mRNA injected at 1-cell-stage is still sufficiently active in all relevant circuits in the CNS.

4. Future Prospects

In the near future several zebrafish mutants for many disease genes will be available. Unpublished data from some of these mutants indicate that the phenotypes are mild rather than robust, and revision of some MO studies will be unavoidable. However, the current tools are already very good when properly controlled. The TILLING mutants, produced by the advanced selection of mutants from a live library of mutagenized fish (84), need to be backcrossed to make sure that single mutations lie behind the phenotypes, and the MO approach is very suitable for this. Hundreds of mutants produced with the zinc finger nuclease method will also be available, although the method currently is limited to a few large laboratories and consortia, as will transgenic methods for inducible mutants. Locomotor analysis of about 100 larvae can already be carried out in 10 min using rather simple video tracking system, and new algorithms for analysis of group behavior are being developed. The transparency of zebrafish larvae, with a possibility to visualize neurons labeled with neuron type-specific fluorescent markers and activity indicators enable very detailed studies on functioning neural circuits activated during specific behaviors. New methods to produce stable transgenic fish, e.g., the Tol2 transposon system (85, 86) can produce very specific expression patterns in the nervous system. The Gal4-UAS system has proven very useful to drive expression of more than one gene from the same construct for several purposes (87). The small size of the larval brain renders it possible to count neurons and map projections accurately. These methods, when applied together with the powerful genetic methods, will make zebrafish an increasingly attractive organism for both basic neurobiology and as a disease model.

References

1. Westerfield M (2007) THE ZEBRAFISH BOOK, 5th Edition; A guide for the laboratory use of zebrafish (*Danio rerio*), vol. 5th. University of Oregon Press, Eugene.

2. Brustein E, Saint-Amant L, Buss RR, Chong M, McDearmid JR, Drapeau P (2003) Steps during the development of the zebrafish locomotor network. J Physiol Paris 97:77–86.

3. Nieuwenhuys R, ten Donkelar HJ, Nicholson C (1998) The Central Nervous System of Vertebrates. Springer, Berlin.

4. Ito H, Yamamoto N (2009) Non-laminar cerebral cortex in teleost fishes? Biol Lett 5: 117–121.

5. Nieuwenhuys R (1963) The Comparative Anatomy of the Actinopterygian Forebrain. J Hirnforsch 13:171–192.

6. Nieuwenhuys R (2010) The development and general morphology of the telencephalon of actinopterygian fishes: synopsis, documentation and commentary. Brain Struct Funct.

7. Mueller T, Wullimann MF, Guo S (2008) Early teleostean basal ganglia development visualized by zebrafish Dlx2a, Lhx6, Lhx7, Tbr2 (eomesa), and GAD67 gene expression. J Comp Neurol 507:1245–1257.

8. Vargas JP, Lopez JC, Portavella M (2009) What are the functions of fish brain pallium? Brain Res Bull 79:436–440.

9. Portavella M, Torres B, Salas C, Papini MR (2004) Lesions of the medial pallium, but not of the lateral pallium, disrupt spaced-trial avoidance learning in goldfish (*Carassius auratus*). Neurosci Lett 362:75–78.

10. Portavella M, Torres B, Salas C (2004) Avoidance response in goldfish: emotional and temporal involvement of medial and lateral telencephalic pallium. J Neurosci 24:2335–2342.

11. Isa T (2002) Intrinsic processing in the mammalian superior colliculus. Curr Opin Neurobiol 12:668–677.

12. Sato T, Hamaoka T, Aizawa H, Hosoya T, Okamoto H (2007) Genetic single-cell mosaic analysis implicates ephrinB2 reverse signaling in projections from the posterior tectum to the hindbrain in zebrafish. J Neurosci 27: 5271–5279.

13. Del Bene F, Wyart C, Robles E, Tran A, Looger L, Scott EK, Isacoff EY, Baier H (2010) Filtering of visual information in the tectum by an identified neural circuit. Science 330:669–673.

14. Rink E, Wullimann MF (2001) The teleostean (zebrafish) dopaminergic system ascending to the subpallium (striatum) is located in the basal diencephalon (posterior tuberculum). Brain Res 889:316–330.

15. Kaslin J, Panula P (2001) Comparative anatomy of the histaminergic and other aminergic systems in zebrafish (*Danio rerio*). J Comp Neurol 440:342–377.

16. Rink E, Wullimann MF (2004) Connections of the ventral telencephalon (subpallium) in the zebrafish (*Danio rerio*). Brain Res 1011: 206–220.

17. Sallinen V, Torkko V, Sundvik M, Reenila I, Khrustalyov D, Kaslin J, Panula P (2009) MPTP and MPP+ target specific aminergic cell populations in larval zebrafish. J Neurochem 108:719–731.

18. Scholpp S, Lumsden A (2010) Building a bridal chamber: development of the thalamus. Trends Neurosci 33:373–380.

19. Ikenaga T, Yoshida M, Uematsu K (2006) Cerebellar efferent neurons in teleost fish. Cerebellum 5:268–274.

20. Bae YK, Kani S, Shimizu T, Tanabe K, Nojima H, Kimura Y, Higashijima S, Hibi M (2009) Anatomy of zebrafish cerebellum and screen for mutations affecting its development. Dev Biol 330:406–426.

21. Burgess HA, Granato M (2007) Sensorimotor gating in larval zebrafish. J Neurosci 27: 4984–4994.

22. O'Malley DM, Kao YH, Fetcho JR (1996) Imaging the functional organization of zebrafish hindbrain segments during escape behaviors. Neuron 17:1145–1155.

23. Orger MB, Kampff AR, Severi KE, Bollmann JH, Engert F (2008) Control of visually guided behavior by distinct populations of spinal projection neurons. Nat Neurosci 11:327–333.

24. Szobota S, Gorostiza P, Del Bene F, Wyart C, Fortin DL, Kolstad KD, Tulyathan O, Volgraf M, Numano R, Aaron HL, Scott EK, Kramer RH, Flannery J, Baier H, Trauner D, Isacoff EY (2007) Remote control of neuronal activity with a light-gated glutamate receptor. Neuron 54:535–545.

25. Douglass AD, Kraves S, Deisseroth K, Schier AF, Engert F (2008) Escape behavior elicited by single, channelrhodopsin-2-evoked spikes in zebrafish somatosensory neurons. Curr Biol 18:1133–1137.

26. Rihel J, Prober DA, Arvanites A, Lam K, Zimmerman S, Jang S, Haggarty SJ, Kokel D, Rubin LL, Peterson RT, Schier AF (2010) Zebrafish behavioral profiling links drugs to biological targets and rest/wake regulation. Science 327:348–351.

27. Panula P, Chen YC, Priyadarshini M, Kudo S, Semenova S, Sundvik M, Sallinen V (2010) The comparative neuroanatomy and neurochemistry of zebrafish CNS systems of relevance to human neuropsychiatric diseases. Neurobiol Dis 40:46–57.

28. Ruuskanen JO, Xhaard H, Marjamaki A, Salaneck E, Salminen T, Yan YL, Postlethwait JH, Johnson MS, Larhammar D, Scheinin M (2004) Identification of duplicated fourth alpha2-adrenergic receptor subtype by cloning and mapping of five receptor genes in zebrafish. Mol Biol Evol 21:14–28.

29. Ruuskanen JO, Peitsaro N, Kaslin JV, Panula P, Scheinin M (2005) Expression and function of alpha-adrenoceptors in zebrafish: drug effects, mRNA and receptor distributions. J Neurochem 94:1559–1569.

30. Haas H, Panula P (2003) The role of histamine and the tuberomamillary nucleus in the nervous system. Nat Rev Neurosci 4:121–130.

31. Peitsaro N, Sundvik M, Anichtchik OV, Kaslin J, Panula P (2007) Identification of zebrafish histamine H1, H2 and H3 receptors and effects of histaminergic ligands on behavior. Biochem Pharmacol 73:1205–1214.

32. Candy J, Collet C (2005) Two tyrosine hydroxylase genes in teleosts. Biochim Biophys Acta 1727:35–44.

33. Chen YC, Priyadarshini M, Panula P (2009) Complementary developmental expression of the two tyrosine hydroxylase transcripts in zebrafish. Histochem Cell Biol 132:375–381.

34. Sallinen V, Kolehmainen J, Priyadarshini M, Toleikyte G, Chen YC, Panula P (2010) Dopaminergic cell damage and vulnerability to MPTP in Pink1 knockdown zebrafish. Neurobiol Dis 40:93–101.

35. Anichtchik O, Sallinen V, Peitsaro N, Panula P (2006) Distinct structure and activity of

monoamine oxidase in the brain of zebrafish (*Danio rerio*). J Comp Neurol 498:593–610.

36. Sallinen V, Sundvik M, Reenila I, Peitsaro N, Khrustalyov D, Anichtchik O, Toleikyte G, Kaslin J, Panula P (2009) Hyperserotonergic phenotype after monoamine oxidase inhibition in larval zebrafish. J Neurochem 109:403–415.

37. Anichtchik OV, Kaslin J, Peitsaro N, Scheinin M, Panula P (2004) Neurochemical and behavioural changes in zebrafish *Danio rerio* after systemic administration of 6-hydroxydopamine and 1-methyl-4-phenyl-1,2,3,6-tetrahydropyridine. J Neurochem 88:443–453.

38. Clemente D, Porteros A, Weruaga E, Alonso JR, Arenzana FJ, Aijon J, Arevalo R(2004) Cholinergic elements in the zebrafish central nervous system: Histochemical and immunohistochemical analysis. J Comp Neurol 474:75–107.

39. Kaslin J, Nystedt JM, Ostergard M, Peitsaro N, Panula P (2004) The orexin/hypocretin system in zebrafish is connected to the aminergic and cholinergic systems. J Neurosci 24:2678–2689.

40. Mueller T, Vernier P, Wullimann MF (2004) The adult central nervous cholinergic system of a neurogenetic model animal, the zebrafish *Danio rerio*. Brain Res 1011:156–169.

41. Budick SA, O'Malley DM (2000) Locomotor repertoire of the larval zebrafish: swimming, turning and prey capture. J Exp Biol 203:2565–2579.

42. Prober DA, Rihel J, Onah AA, Sung RJ, Schier AF (2006) Hypocretin/orexin overexpression induces an insomnia-like phenotype in zebrafish. J Neurosci 26:13400–13410.

43. Zhdanova IV (2006) Sleep in zebrafish. Zebrafish 3:215–226.

44. Granato M, van Eeden FJ, Schach U, Trowe T, Brand M, Furutani-Seiki M, Haffter P, Hammerschmidt M, Heisenberg CP, Jiang YJ, Kane DA, Kelsh RN, Mullins MC, Odenthal J, Nusslein-Volhard C (1996) Genes controlling and mediating locomotion behavior of the zebrafish embryo and larva. Development 123:399–413.

45. Burgess HA, Schoch H, Granato M (2010) Distinct retinal pathways drive spatial orientation behaviors in zebrafish navigation. Curr Biol 20:381–386.

46. Burgess HA, Granato M (2007) Modulation of locomotor activity in larval zebrafish during light adaptation. J Exp Biol 210:2526–2539.

47. Gahtan E, Tanger P, Baier H (2005) Visual prey capture in larval zebrafish is controlled by identified reticulospinal neurons downstream of the tectum. J Neurosci 25:9294–9303.

48. Peitsaro N, Kaslin J, Anichtchik OV, Panula P (2003) Modulation of the histaminergic system and behaviour by alpha-fluoromethylhistidine in zebrafish. J Neurochem 86:432–441.

49. Cachat J, Stewart A, Grossman L, Gaikwad S, Kadri F, Chung KM, Wu N, Wong K, Roy S, Suciu C, Goodspeed J, Elegante M, Bartels B, Elkhayat S, Tien D, Tan J, Denmark A, Gilder T, Kyzar E, Dileo J, Frank K, Chang K, Utterback E, Hart P, Kalueff AV (2010) Measuring behavioral and endocrine responses to novelty stress in adult zebrafish. Nat Protoc 5:1786–1799.

50. Stewart A, Wu N, Cachat J, Hart P, Gaikwad S, Wong K, Utterback E, Gilder T, Kyzar E, Newman A, Carlos D, Chang K, Hook M, Rhymes C, Caffery M, Greenberg M, Zadina J, Kalueff AV (2011) Pharmacological modulation of anxiety-like phenotypes in adult zebrafish behavioral models. Prog Neuropsychopharmacol Biol Psychiatry 35:1421–1431.

51. Brockerhoff SE, Hurley JB, Janssen-Bienhold U, Neuhauss SC, Driever W, Dowling JE (1995) A behavioral screen for isolating zebrafish mutants with visual system defects. Proc Natl Acad Sci USA 92:10545–10549.

52. Muto A, Orger MB, Wehman AM, Smear MC, Kay JN, Page-McCaw PS, Gahtan E, Xiao T, Nevin LM, Gosse NJ, Staub W, Finger-Baier K, Baier H (2005) Forward genetic analysis of visual behavior in zebrafish. PLoS Genet 1:e66.

53. Burgess HA, Johnson SL, Granato M (2009) Unidirectional startle responses and disrupted left-right co-ordination of motor behaviors in robo3 mutant zebrafish. Genes Brain Behav 8:500–511.

54. McNeill MS, Paulsen J, Bonde G, Burnight E, Hsu MY, Cornell RA (2007) Cell death of melanophores in zebrafish trpm7 mutant embryos depends on melanin synthesis. J Invest Dermatol 127:2020–2030.

55. Hermosura MC, Nayakanti H, Dorovkov MV, Calderon FR, Ryazanov AG, Haymer DS, Garruto RM (2005) A TRPM7 variant shows altered sensitivity to magnesium that may contribute to the pathogenesis of two Guamanian neurodegenerative disorders. Proc Natl Acad Sci USA 102:11510–11515.

56. Burns RS, Chiueh CC, Markey SP, Ebert MH, Jacobowitz DM, Kopin IJ (1983) A primate model of parkinsonism: selective destruction of dopaminergic neurons in the pars compacta of the substantia nigra by N-methyl-4-phenyl-1,2,3,6-tetrahydropyridine. Proc Natl Acad Sci USA 80:4546–4550.

57. Heikkila RE, Sonsalla PK (1987) The use of the MPTP-treated mouse as an animal model

of parkinsonism. Can J Neurol Sci 14: 436–440.

58. Pollard HB, Dhariwal K, Adeyemo OM, Markey CJ, Caohuy H, Levine M, Markey S, Youdim MB (1992) A parkinsonian syndrome induced in the goldfish by the neurotoxin MPTP. FASEB J 6:3108–3116.

59. Beal MF (2001) Experimental models of Parkinson's disease. Nat Rev Neurosci 2:325–334.

60. Panula P, Sallinen V, Sundvik M, Kolehmainen J, Torkko V, Tiittula A, Moshnyakov M, Podlasz P (2006) Modulatory neurotransmitter systems and behavior: towards zebrafish models of neurodegenerative diseases. Zebrafish 3:235–247.

61. Panula P, Yang HY, Costa E (1984) Histamine-containing neurons in the rat hypothalamus. Proc Natl Acad Sci USA 81:2572–2576.

62. Panula P, Pirvola U, Auvinen S, Airaksinen MS (1989) Histamine-immunoreactive nerve fibers in the rat brain. Neuroscience 28:585–610.

63. Eriksson KS, Peitsaro N, Karlstedt K, Kaslin J, Panula P (1998) Development of the histaminergic neurons and expression of histidine decarboxylase mRNA in the zebrafish brain in the absence of all peripheral histaminergic systems. Eur J Neurosci 10:3799–3812.

64. McKinley ET, Baranowski TC, Blavo DO, Cato C, Doan TN, Rubinstein AL (2005) Neuroprotection of MPTP-induced toxicity in zebrafish dopaminergic neurons. Brain Res Mol Brain Res 141:128–137.

65. Lam CS, Korzh V, Strahle U (2005) Zebrafish embryos are susceptible to the dopaminergic neurotoxin MPTP. Eur J Neurosci 21:1758–1762.

66. Bretaud S, Lee S, Guo S (2004) Sensitivity of zebrafish to environmental toxins implicated in Parkinson's disease. Neurotoxicol Teratol 26:857–864.

67. Thirumalai V, Cline HT (2008) Endogenous dopamine suppresses initiation of swimming in prefeeding zebrafish larvae. J Neurophysiol 100: 1635–1648.

68. Rink E, Wullimann MF (2002) Development of the catecholaminergic system in the early zebrafish brain: an immunohistochemical study. Brain Res Dev Brain Res 137:89–100.

69. Filippi A, Mahler J, Schweitzer J, Driever W (2010) Expression of the paralogous tyrosine hydroxylase encoding genes th1 and th2 reveals the full complement of dopaminergic and noradrenergic neurons in zebrafish larval and juvenile brain. J Comp Neurol 518:423–438.

70. Nayyar T, Bubser M, Ferguson MC, Neely MD, Shawn Goodwin J, Montine TJ, Deutch AY, Ansah TA (2009) Cortical serotonin and norepinephrine denervation in parkinsonism:

preferential loss of the beaded serotonin innervation. Eur J Neurosci 30:207–216.

71. Vuckovic MG, Wood RI, Holschneider DP, Abernathy A, Togasaki DM, Smith A, Petzinger GM, Jakowec MW (2008) Memory, mood, dopamine, and serotonin in the 1-methyl-4-phenyl-1,2,3,6-tetrahydropyridine-lesioned mouse model of basal ganglia injury. Neurobiol Dis 32:319–327.

72. Kish SJ, Tong J, Hornykiewicz O, Rajput A, Chang LJ, Guttman M, Furukawa Y (2008) Preferential loss of serotonin markers in caudate versus putamen in Parkinson's disease. Brain 131:120–131.

73. Wen L, Wei W, Gu W, Huang P, Ren X, Zhang Z, Zhu Z, Lin S, Zhang B (2008) Visualization of monoaminergic neurons and neurotoxicity of MPTP in live transgenic zebrafish. Dev Biol 314:84–92.

74. Bretaud S, Allen C, Ingham PW, Bandmann O (2007) p53-dependent neuronal cell death in a DJ-1-deficient zebrafish model of Parkinson's disease. J Neurochem 100:1626–1635.

75. Sheng D, Qu D, Kwok KH, Ng SS, Lim AY, Aw SS, Lee CW, Sung WK, Tan EK, Lufkin T, Jesuthasan S, Sinnakaruppan M, Liu J (2010) Deletion of the WD40 domain of LRRK2 in Zebrafish causes Parkinsonism-like loss of neurons and locomotive defect. PLoS Genet 6:e1000914.

76. Andres-Mateos E, Mejias R, Sasaki M, Li X, Lin BM, Biskup S, Zhang L, Banerjee R, Thomas B, Yang L, Liu G, Beal MF, Huso DL, Dawson TM, Dawson VL (2009) Unexpected lack of hypersensitivity in LRRK2 knock-out mice to MPTP (1-methyl-4-phenyl-1,2,3,6-tetrahydropyridine). J Neurosci 29:15846–15850.

77. Anichtchik O, Diekmann H, Fleming A, Roach A, Goldsmith P, Rubinsztein DC (2008) Loss of PINK1 function affects development and results in neurodegeneration in zebrafish. J Neurosci 28:8199–8207.

78. Xi Y, Ryan J, Noble S, Yu M, Yilbas AE, Ekker M (2010) Impaired dopaminergic neuron development and locomotor function in zebrafish with loss of pink1 function. Eur J Neurosci 31:623–633.

79. Matsui H, Taniguchi Y, Inoue H, Kobayashi Y, Sakaki Y, Toyoda A, Uemura K, Kobayashi D, Takeda S, Takahashi R (2010) Loss of PINK1 in medaka fish (Oryzias latipes) causes late-onset decrease in spontaneous movement. Neurosci Res 66:151–161.

80. Valente EM, Abou-Sleiman PM, Caputo V, Muqit MM, Harvey K, Gispert S, Ali Z, Del Turco D, Bentivoglio AR, Healy DG, Albanese A, Nussbaum R, Gonzalez-Maldonado R, Deller T, Salvi S, Cortelli P, Gilks WP, Latchman DS, Harvey RJ, Dallapiccola B, Auburger G,

Wood NW (2004) Hereditary early-onset Parkinson's disease caused by mutations in PINK1. Science 304:1158–1160.

81. Kitada T, Pisani A, Porter DR, Yamaguchi H, Tscherter A, Martella G, Bonsi P, Zhang C, Pothos EN, Shen J (2007) Impaired dopamine release and synaptic plasticity in the striatum of PINK1-deficient mice. Proc Natl Acad Sci USA 104:11441–11446.

82. Flinn L, Mortiboys H, Volkmann K, Koster RW, Ingham PW, Bandmann O (2009) Complex I deficiency and dopaminergic neuronal cell loss in parkin-deficient zebrafish (*Danio rerio*). Brain 132:1613–1623.

83. Robu ME, Larson JD, Nasevicius A, Beiraghi S, Brenner C, Farber SA, Ekker SC (2007) P53 Activation by Knockdown Technologies. PLoS Genet 3:e78.

84. Wienholds E, van Eeden F, Kosters M, Mudde J, Plasterk RH, Cuppen E (2003) Efficient target-selected mutagenesis in zebrafish. Genome Res 13:2700–2707.

85. Asakawa K, Suster ML, Mizusawa K, Nagayoshi S, Kotani T, Urasaki A, Kishimoto Y, Hibi M, Kawakami K (2008) Genetic dissection of neural circuits by Tol2 transposon-mediated Gal4 gene and enhancer trapping in zebrafish. Proc Natl Acad Sci USA 105:1255–1260.

86. Suster ML, Kikuta H, Urasaki A, Asakawa K, Kawakami K (2009) Transgenesis in zebrafish with the tol2 transposon system. Methods Mol Biol 561:41–63.

87. Halpern ME, Rhee J, Goll MG, Akitake CM, Parsons M, Leach SD (2008) Gal4/UAS transgenic tools and their application to zebrafish. Zebrafish 5:97–110.

Chapter 3

Methodological Strategies to Evaluate Functional Effectors Related to Parkinson's Disease Through Application of *Caenorhabditis elegans* Models

Kim A. Caldwell and Guy A. Caldwell

Abstract

Improvements to the diagnosis and treatment of Parkinson disease (PD) are dependent upon the identification and molecular understanding of modifiers of neuronal degeneration. Here, we describe the use of multi-factorial functional analyses to exploit the experimental attributes of the nematode, *Caenorhabditis elegans*, to accelerate the translational path toward identification and characterization of modifiers of dopaminergic neurogeneration. *C. elegans* is ideal for both screening and target validation of potential modifiers. Specific assays discussed in this technical overview include in vivo analyses using whole, intact, and living nematodes with readouts for age-dependent α-synuclein-proteotoxicity and 6-hydroxydopamine-induced neurodegeneration in dopamine (DA) neurons. These methods provide an integrated approach to target characterization and functional validation in *C. elegans* that allow researchers to prioritize lead candidates for translation toward mammalian systems.

Key words: *C. elegans*, Dopamine, Parkinson's disease, Alpha-synuclein, 6-OHDA, RNAi, Neurodegeneration

1. Introduction

Caenorhabditis elegans is a rapidly cultured organism (3 days from fertilized egg to adult) with an experimentally tenable lifespan (14–17 days), and studies can be designed to take exploratory concepts to mechanistic fruition, rapidly. Moreover, this microscopic nematode, which is a millimeter in length as an adult, is grown on agar Petri dishes with a bacterial food source, and thus, experiments are inexpensively performed. Well-designed experiments can be conducted with hundreds of animals for each data point or condition desired, therefore yielding statistical power across a variety of distinct experimental paradigms.

Emma L. Lane and Stephen B. Dunnett (eds.), *Animal Models of Movement Disorders: Volume I*, Neuromethods, vol. 61, DOI 10.1007/978-1-61779-298-4_3, © Springer Science+Business Media, LLC 2011

C. *elegans* has only 959 somatic cells as an adult hermaphrodite, yet it has hypodermis, intestine, muscle, glands, as well as reproductive and nervous systems. This small organism is also transparent and its cells can be readily correlated with gene and protein expression patterns using fluorescent marker constructs, including GFP (1). Thus, when examining neurodegenerative processes, such as those that occur in DA neurons, cell death can be readily observed and quantified within living organisms (Sect. 3).

The cells of C. *elegans* are also genetically invariant and anatomically defined, thereby allowing great accuracy when analyzing expression constructs. Each somatic cell in the nematode has been individually named according to lineage and is documented in the WormAtlas (Table 1). A serial-section, electron microscope-level map of the animal displaying the relationship of all cells and organs to each other is also available. Thus, cells can be followed from inception through final destination, and modifiers of cell survival are readily evaluated. Taken together, these resources have provided a unique platform for detailed cellular studies across many biological fields, including metabolism (insulin/daf signaling pathway), aging, sex determination, apoptosis, and neurodevelopment, among others. The relevance and contributions of C. *elegans*

Table 1
Key online resources for *Caenorhabditis elegans* research

Database	Web address	Description
WormBase	http://www.wormbase.org/	Gene summary pages containing functional, structural, and phylogenetic information. Links to other databases (i.e., interactome, microarray databases, expression patterns, etc.), related to query genes or ORFs. Links to published articles and C. *elegans* meeting abstracts as well
Caenorhabditis elegans WWW Server	http://elegans.swmed.edu/	Collection of hyperlinks to sites relevant to the study of C. *elegans* and other nematodes
WormBook	http://www.wormbook.org/	An online, open-access, site with peer-reviewed chapters describing nematode biology
WormAtlas	http://www.wormatlas.org/	An online database of the structural anatomy and its relationship to C. *elegans* behavior, with a complete nervous system wiring diagram at the EM level

research to that of higher eukaryotes, and humans, have been recognized at the highest levels of scientific achievement, as worm researchers were awarded the Nobel Prize in Physiology or Medicine for discovering programmed cell death (2002), RNA interference (RNAi) (2006), and the Nobel Prize in Chemistry for GFP (2008). In this context, *C. elegans* arguably represents the most well-understood and experimentally tractable animal on our planet.

C. elegans also has a fully sequenced genome and shares approximately 50% of its genes with humans, and there are many bioinformatics resources available to worm researchers, many of which can be accessed through links described in Table 1. Notably, ~70% of genes that are known to cause a genetic disease in humans have an ortholog in *C. elegans*. Application of *C. elegans* toward human disease research has already provided insights into the function of specific gene products linked to a variety of human movement disorders, such as dystonia (2), Huntington's disease (3), ALS (4), and Parkinson's disease (this volume).

C. elegans also exhibits a variety of behaviors that can be elicited using forward and reverse genetic strategies. As an example, Dr. Sydney Brenner, who pioneered the use of *C. elegans* as a model organism in the 1960s, initially worked with his group on a forward mutagenesis screen to identify mutants with abnormal locomotion (the uncoordinated, Unc, phenotype). Forward genetic screens, beginning with a phenotype of interest, often involve the use of a mutagen, such as EMS. This mutagen most often produces point mutations (G/C–A/T transitions) (5). Following successful screening for mutant phenotypes, identification for the genetic lesion(s) typically occurs. Their screen identified 77 *unc* genes that included both neuronal and muscular defects (6) in many genes that have since been identified as evolutionarily conserved components of muscle cells, synaptic transmission, or neurotransmitter release. Many forward genetic screens have been performed since this time, based on a variety of phenotypes (embryonic, morphological, reproductive, and neurological).

Reverse genetic screens require knowledge of the gene sequence of interest. Since the completion of the genome projects for *C. elegans*, humans, mouse, and other well-studied laboratory species, the field of comparative genomics has allowed *C. elegans* researchers to extensively use RNAi screening in their research to knockdown target genes and then screen for potential phenotypes. In nematodes, this method simply involves injecting, soaking, or feeding worms dsRNA that is complementary to the targeted gene of interest. The expression of the candidate gene will be greatly reduced, or silenced, in the next generation. Relevant to human movement disorders, whole genome-wide screens, as well as smaller, hypothesis-based screens, have been performed to examine a variety of phenotypes, such as aging (7), protein aggregation in ALS (4), and spinal muscular atrophy (8).

***1.1. Overview of the
C. elegans Nervous
System***

Despite its evolutionary distance from humans, the neurons of *C. elegans* display most of the hallmarks of mammalian neuronal function including ion channels, neurotransmitters (dopamine, serotonin, acetylcholine, GABA, etc.), vesicular transporters, receptors, and synaptic components (9, 10). Compared with the ~100 billion neurons of the human brain, or even the 10,000 neurons of *Drosophila*, *C. elegans* hermaphrodites have exactly 302 neurons. Thus, of the 959 somatic cells comprising the adult animal, approximately 1/3 of these cells are neurons. The nervous system has also been reconstructed in great detail to reveal the anatomy and complete connectivity of all the neurons (11). This is a remarkable accomplishment given that the cell bodies of worm neurons are approximately 2 μm in diameter (12). Worms have 118 different types of neurons that have been classified into groups based on differences in morphology, connectivity, and function, including mechanosensation, chemosensation, and thermosensation. Worms also have approximately 7,000 synapses.

Precisely, eight neurons within a *C. elegans* hermaphrodite are dopaminergic. Six of these neurons are located in the anterior region of the animal (Figs. 1 and 2a, b), and are identified by their lineage names as the four cephalic (CEP) neurons and the two anterior deirid (ADE) neurons, while the other two DA neurons are in the posterior of the animal are called posterior deirid (PDE) neurons (Fig. 2a). The two dorsal CEPs are post-synaptic to the

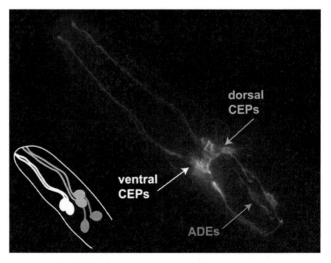

Fig. 1. A fluorescent photomicrograph depicting the anterior-most region of a *Caenorhabditis elegans* hermaphrodite. The DA neurons are illuminated using GFP driven from the DA transporter promoter (P$_{dat-1}$::GFP). The six cell bodies and neuronal processes include two pairs of cephalic (CEP) neurons and one pair of anterior deirid neurons (ADEs). The line drawing is a representation of the association of the DA neurons to each other; the dorsal CEPs *(light gray)* are post-synaptic to the ADE neurons *(darker gray)* while the ventral CEPs *(white)* are not post-synaptic to the ADE neurons.

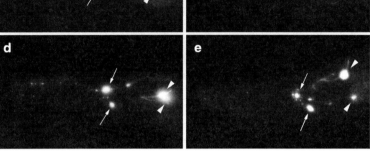

Fig. 2. Representative images of *Caenorhabditis elegans* DA neurons in normal and degenerative states. (**a**) An image depicting all eight DA neurons in a *C. elegans* hermaphrodite. The six anterior DA neurons are visible on the left (two pairs of CEP neurons (*arrows*) and one pair of ADE neurons (*large arrowheads*)), where the cell bodies and processes are highlighted. Two posterior deirid (PDE) neurons are indicated with *small arrowheads* on the right side of this image. (**b**) A magnified view of the anterior region of a worm displaying the cell bodies of the four CEP neurons (*arrows*) and the two ADE neurons (*arrowheads*). (**c**) An image displaying a 7-day-old worm that is co-expressing GFP and α-syn in DA neurons; this animal has lost three out of six anterior DA neurons. In this example, only two of the four CEP neurons (*arrows*) and one of two ADE neurons (*arrowhead*) remain. Most worms within this population are missing one or more anterior DA neurons when they are adults. (**d, e**) Exposure to 6-OHDA causes progressive degeneration of CEP neurons. (**d**) In this example, two of the four CEP neurons have already degenerated, while the remaining CEP neurites display blebbing and cell body rounding (*arrows*). The two ADE neurons in this animal are still intact (*arrowheads*). (**e**) This worm is undergoing further degeneration, whereby the two remaining CEP neurites are almost invisible and the associated cell bodies are also degenerating (*arrows*); the two ADE neurons are also beginning to degenerate (*arrowheads*).

ADE neurons, while the two ventral CEPs are not post-synaptic to the ADEs (Fig. 1).

There is evolutionary conservation of the pathways involved in the processing, packaging, and transport of DA, thus enabling researchers to utilize the worm for studying various cellular aspects of DA neuron biology. In this regard, several behavioral phenotypes have been identified as being specific for DA signaling in *C. elegans*. For example, upon exposure to exogenous DA, worms exhibit a decrease in egg laying, locomotion, and defecation. Worms also normally exhibit decreased locomotion upon entering a bacterial lawn, a DA-associated behavior called the basal slowing response (13). Researchers identified that mutations in *cat-2* resulted in no basal slowing response. This gene encodes worm tyrosine hydroxylase (TH), the rate-limiting enzyme for DA synthesis. These studies suggest that there is a role for DA in mechanosensation and food sensing (14). The basal slowing response phenotype is a potential readout for DA neuronal function and dysfunction for *C. elegans* DA models, however, it is time consuming and does not readily lend itself to screening procedures. Notably, a recently published article described an automated worm tracking system that could potentially be modified to quantify the basal slowing response in multiple individual worms in parallel (15).

1.2. C. elegans as a Model for Parkinson's Disease

Orthologs of six familial Parkinson disease (PD) genes have been identified in *C. elegans* and mutant alleles are available for these *C. elegans* PD gene orthologs (Table 2), thus enabling genetic analyses. For example, in a recent paper, *C. elegans* researchers examined double mutants in LRRK2 and PINK1 (*lrk1* and *pink1*, respectively), and determined that the gene products have an antagonist role in cellular stress response and regulation of neurite outgrowth (16). In this manner, single and double mutants of *C. elegans* PD gene orthologs provide an opportunity to evaluate pathways and interrelationships, as well as screen for potential modifiers of PD. Notably, the *C. elegans* genome does not have a homolog for human α-synuclein (α-syn) (Table 2). This exclusion has allowed researchers to overexpress wildtype (WT) or mutant α-syn in the nematode α-syn null genetic background without concerns for endogenous α-syn or the risk of eliciting a dominant negative effect (17–22).

1.2.1. Use of C. elegans as a Screening Tool to Identify Putative PD Susceptibility Genes

An expeditious route toward discerning genetic contributors to PD involves the use of animal systems amenable to screening methods to identify potential genetic modifiers. In this regard, the use of *C. elegans* has the potential to greatly accelerate the discovery of neuroprotective factors for PD (23) and other diseases of protein misfolding. These assays consist of forward and reverse genetic screens, as well as chemical screens. An example of a relevant forward genetic screen in *C. elegans* was performed for enhancers of polyglutamine aggregation in muscle cells, whereby it was determined that a general imbalance in protein homeostasis in post-synaptic muscle cells can occur if there is an increase in acetylcholine (ACh) signaling, or defective GABA signaling (24). Another example of a forward genetic analysis involved screening *C. elegans* DA neurons for insensitivity to 6-hydroxydopamine (6-OHDA), whereby new alleles of the DA transporter (DAT-1) were identified (25).

Table 2
Summary of human familial PD genes, corresponding *Caenorhabditis elegans* orthologs and mutants

PD Gene	PD Protein	*C. elegans* ortholog	E-Value	*C. elegans* chromosome location	*C. elegans* allele name	Type of mutation
PARK1	SNCA/α-syn	n/a				
PARK2	PRKN/parkin	*pdr-1*	3.4e-38	III	gk448	Knockout allele; superficially wildtype
PARK5	UCHL-1	*ubh-1*	1.2e-33	V	n/a	
PARK6	PINK1	*pink-1*	7.8e-53	II	ok3538	Knockout allele
PARK7	DJ-1	*djr-1.1* *djr-1.2*	1.6e-45 8.9e-36	II V	tm918 tm1346	Both are knockout alleles; double mutant created
PARK8	LRRK2	*lrk-1*	5.5e-66	I	tm1898	Deletion; homozygous viable
PARK9	ATP13A2	*catp-6*	2.5e-180	IV	ok3473	Knockout allele; homozygous viable
PARK11	GIGYF2	n/a				
PARK13	HTRA2	n/a				

There are many examples of reverse genetic, RNAi, screens using the full genome in *C. elegans*. Several of these screens scored phenotypes related to protein misfolding. For example, worms expressing α-syn::YFP in body wall muscle cells were screened for a modulation in protein misfolding and 80 candidates were identified (18). The candidates clustered into a few biological categories. Notably, quality control and ER/Golgi vesicle trafficking gene products were identified as those that, when knocked down, increased protein aggregation. Another category of gene products identified were aging related, but these were suppressors of protein aggregation.

Recently, candidate RNAi screens, based on the mature bioinformatics available within the *C. elegans* field, have become more common. Our laboratory performed a hypothesis-based screen for genetic modifiers of α-syn::GFP misfolding in body wall muscle cells and identified 20 candidate genes. Five of these candidates were subsequently validated as having a protective role in DA neurons as well (17). The candidates identified from this screen included orthologs of known recessive PD genes, DJ-1 and PINK1, VPS41, a vesicular protein necessary for lysosomal trafficking and biogenesis, ATG7, an autophagy-associated regulatory gene, and ULK2, a conserved serine-threonine kinase also related to the yeast autophagy protein, Atg1p.

The outcome of this screen represented an exciting group of gene products with implications for PD, both as potential susceptibility markers and novel targets for therapeutic development. For example, knockout of the lysosomal protein ATG7 has been reported to produce a neurodegenerative phenotype in mice (26). Another neuroprotective gene product identified from our screen, VPS-41, is the nematode homolog of mammalian and yeast VPS41, a protein implicated in trafficking from the Golgi to the vacuole/lysosome in yeast (27). Little was known about the function of VPS41 in mammalian systems except that it is strongly expressed within DA neurons of the *substantia nigra pars compacta*. However, subsequent mammalian studies have since revealed that VPS41 over-expression is protective against rotenone- and 6-OHDA-induced toxicity in SHSY5Y cells (28). Finally, *ULK2* was one of six genes significantly associated with single nucleotide polymorphisms in a genome-wide association study (GWAS) of PD patients (29). Therefore, it is clear that establishing a functional screening paradigm for modifiers of PD-related phenotypes in *C. elegans* has successfully yielded an intriguing collection of effectors that demonstrate the predictability of the nematode model to identify targets with high translational potential.

1.2.2. Use of C. elegans to Validate Therapeutic Targets and Chemicals

Our ability to successfully predict the probability of PD among individuals is dependent upon knowledge about genetic susceptibility factors that render certain populations at risk. In attempting to discern genetic factors associated with PD, scientists working with a variety of different organisms have generated many lists of candidate genes and proteins, but most of these leads remain mechanistically undefined. In this regard, application of *C. elegans* in a directed manner can facilitate the functional evaluation of leads originally identified in other species (yeast, cell culture, mouse, and human genomic studies). As described previously, a distinct advantage of using *C. elegans* for functional investigation of gene activity and validation is that large populations of isogenic animals can be propagated and analyzed, leading to an unequivocal level of accuracy when evaluating neurodegeneration. As shown in Fig. 2, we have established isogenic lines of transgenic *C. elegans* that overexpress human α-syn in DA neurons (P_{dat-1}::α-syn + P_{dat-1}::GFP) that enable rapid evaluation of factors (drugs and genes) that protect or enhance DA neurodegeneration. This worm strain has been successful in rapidly predicting genes that have significant consequences for neuronal survival in mammalian systems. For example, we investigated three Rab gene products, involved in regulating vesicular trafficking, that were originally identified from yeast screens as effectors of α-syn-dependent toxicity (30, 31). Suppression of cytotoxicity was recapitulated in yeast cells, *Drosophila*, rat neuronal cell cultures, and in *C. elegans*, where we showed that elevated expression of specific Rab GTPases rescue DA degeneration induced by α-syn overexpression. For example, overexpression of mammalian Rab1 successfully by-passed an α-syn-associated block of ER to Golgi trafficking (30). New molecules that impact this cellular mechanism have since been discovered in yeast and subsequently validated across multiple organisms, including *C. elegans* (32).

There is a growing body of literature demonstrating the utility of *C. elegans* for pharmacological research. A variety of compounds have been uncovered, ranging from those associated with modulating neurotransmitter activity (33, 34), anesthetics (35), lifespan extension (36), and a new calcium channel antagonist (37). Our laboratory has used *C. elegans* to conduct a small molecule screen to identify compounds that positively modulate the activity of the neuroprotective chaperone-like protein, torsinA, in vivo (38). Deficit in torsinA function is responsible for a human movement disorder termed early-onset torsion dystonia, and is linked to an in-frame 3-bp deletion in the human *DYT1* gene. These same molecules were later shown to restore functional capacity of torsinA activity in human DYT1 dystonia patient fibroblasts and reverse behavioral abnormalities in a DYT1 knock-in mouse model of early-onset torsion dystonia (38). Of relevance to PD, worms are being exploited as a system for therapeutic drug screening (39) and functional validation of drugs with a therapeutic potential for PD that have been identified in other systems (32, 40, 41).

2. Large-Scale RNAi Screening Using an α-syn Misfolding Phenotype in Non-neuronal Cells

Central to the formation of Lewy bodies, a primary pathological hallmark of PD, is α-syn (42, 43). Importantly, it was demonstrated that genomic multiplication of the WT α-syn gene results in PD, indicating that overexpression of this protein alone could lead to the disease (44). A recent GWAS on approximately 800 familial PD cases further supports that the *SCNA* (α-syn) locus is a major susceptibility factor (45). The effects of overexpression, mutation, misfolding of α-syn has led to the hypothesis that the cellular clearance of this small, aggregation-prone protein is critical to avoiding the neurodegenerative state.

As displayed in Table 3, there are two models of α-syn misfolding in *C. elegans* body wall muscle cells that can be evaluated for the consequences of α-syn overexpression and misfolding in vivo. In both models, α-syn is fused at the C-terminus to either GFP or YFP, and inclusions are visualized in body wall muscle cells. These two transgenic models were used in independent RNAi screens for gene products that modulate protein misfolding (17, 18). In both transgenic models, the aggregates become more abundant as the animals get older and can be scored over time as worms develop and age (Fig. 3).

At first, it might appear to be a daunting task to assess differences in α-syn misfolding and aggregation in body wall muscle cells visually and objectively using a compound fluorescent microscope since these aggregates are quite small. However, researchers have successfully obtained data from RNAi screens using these transgenic worm strains. In this regard, the strategies utilized were different, but equally effective, as described below. One strategy that can be used is a qualitative assessment of aggregates to compare treatment with controls. This can become second nature over time using the following criteria. The general number (none/few/many) and size of the aggregates (small/medium/large) can be scored relative to control animals. This analysis should be performed for ~20 worms per RNAi treatment (46). A worm is scored as having significant aggregation if it has multiple muscle cells with increased quantity and size of aggregates (Fig. 3a vs. Fig. 3c, d). With experience, reproducible data will consistently result from this qualitative measure of α-syn misfolding analysis. An alternative strategy is to perform a quantitative analysis of the number of aggregates, whereby only a few anterior muscle cells in each of ten animals are scored for inclusions; positive animals will have significantly more inclusions than controls (18). In both of these strategies, the level of protein misfolding in the control, α-syn, transgenic animals is directly compared to animals treated with RNAi, whereby the observer is attempting to uncover factors that, when depleted, cause an enhancement in aggregation.

Table 3
***Caenorhabditis elegans* models of α-syn expression and their corresponding phenotypes**

α-syn expression pattern	Transgenic construct(s)	Phenotypes	Reference
WT α-syn expression patterns			
Body wall muscle cells	P_{unc-54}::α-*syn*::*GFP*	α-syn accumulation visualized by misfolded GFP; used in RNAi screen	(17)
	P_{unc-54}::α-*syn*::*YFP*	α-syn accumulation visualized by misfolded YFP; used in RNAi screen	(18)
Pan neuronal	P_{aex-3}::α-*syn*	Reduced motor movement; DA neurodegeneration visualized by GFP	(19)
	P_{snb-1}::α-*syn*	Mitochondrial stress	(20)
	P_{unc-51}::α-*syn*	Endocytosis defects; motor/developmental defects	(21)
DA neuron specific	P_{dat-1}::α-*syn*	DA neurodegeneration visualized by GFP	(19)
	P_{dat-1}::α-*syn*	DA neurodegeneration visualized by GFP	(22)
	P_{dat-1}::α-*syn*	Slightly reduced DA levels; α-syn accumulation in DA neurons	(21)
Mutant α-syn expression patterns			
Pan neuronal	P_{aex-3}::α-*syn A53T*	Reduced motor movement; DA neurodegeneration visualized by GFP	(19)
	$P_{unc-119}$::α-*syn A53T*	Mitochondrial stress	(20)
	P_{unc-51}::α-*syn A30P* and *A53T*	Endocytosis defects; motor/developmental defects	(21)
DA neuron specific	P_{dat-1}::α-*syn A53T*	DA neurodegeneration visualized by GFP	(19)

Our laboratory utilized a third strategy for RNAi screening (Fig. 4a). We reasoned that the misfolding in the control α-syn genetic background might not be sensitive enough to detect moderate aggregate changes because of their normal abundance (Fig. 3a). Therefore, we co-expressed a chaperone (TOR-2, a worm ortholog of torsinA) that ameliorated the formation of the fluorescent misfolded α-syn::GFP proteins (Fig. 3b), which extended prior observations on torsinA chaperone activity (47, 48). These α-syn + TOR-2 animals were then used as the starting genetic background in an RNAi screen, which provided for a much easier qualitative analysis of aggregate levels versus controls (Fig. 3b vs. Fig. 3c, d).

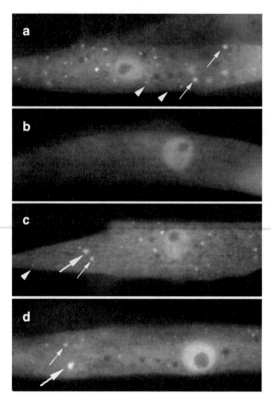

Fig. 3. α-syn misfolding and aggregation within *Caenorhabditis elegans* body wall muscle cells. (**a**) A single body wall muscle cell expressing α-syn::GFP where a moderate number of aggregates are visible following expression from the P$_{unc-54}$::α-syn::GFP construct. The aggregate sizes vary, and are typically small (*arrowhead*) or medium (*small arrow*). (**b**) A body wall muscle cell from a worm co-expressing the chaperone protein, TOR-2, and α-syn::GFP (P$_{unc-54}$::*tor-2* + P$_{unc-54}$::α-syn::GFP). α-syn misfolding is attenuated in the presence of TOR-2. (**c, d**) Single muscle cells wherein misfolded α-syn::GFP is revealed in the TOR-2 + α-syn::GFP transgenic background following RNAi knockdown of specific gene targets. The cells exhibit α-syn::GFP aggregates of varying sizes. (**c**) In this example, the aggregates are mostly small (*arrowhead*) and medium (*small arrow*) and abundant (**c**), yet there is a large aggregate in this cell (*arrow*). (**d**) In this example, the cell has only a few sparsely distributed aggregates of medium (*small arrow*) and large (*arrow*) aggregates.

Furthermore, the presence of the chaperone, which likely maintained α-syn nearer a threshold of protein misfolding, potentially assisted with the identification of genetic factors that affect the formation of misfolded oligomers versus more mature aggregates.

Notably, these *C. elegans* models of α-syn protein misfolding display an age-associated increase of inclusion formation. Since PD is a disease of aging, an age-related aggregation phenotype provided an opportunity to further select positive candidates based on candidate aggregation-inducing potency. Indeed, we were able to differentially identify modifiers of α-syn misfolding by scoring aggregation at distinct time points in the worm lifespan.

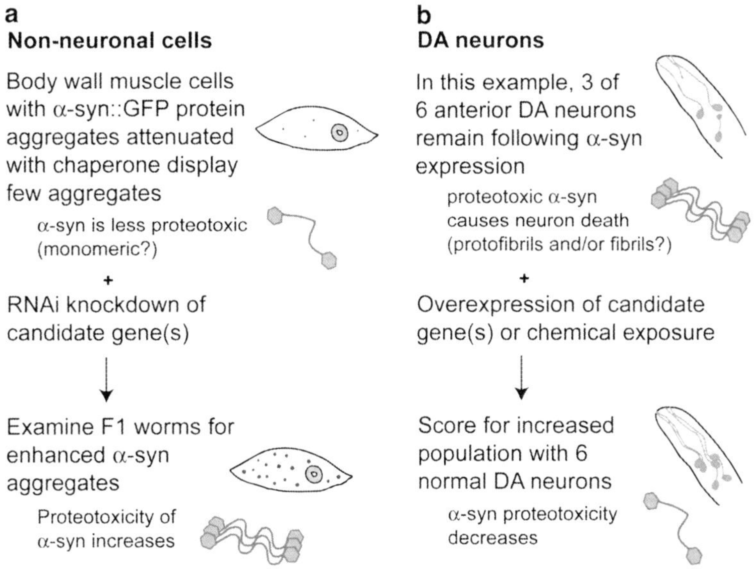

Fig. 4. Identification of factors influencing α-syn proteotoxicity in *Caenorhabditis elegans*. (**a**) RNAi screening in body wall muscle cells. A chaperone protein, TOR-2, is co-expressed with α-syn to attenuate the proteotoxicity of α-syn. The resulting expression pattern of α-syn::GFP in the muscle cells is diffuse. Following RNAi knockdown of candidate genes that affect protein misfolding, a return of the oligomeric α-syn::GFP will be visible in the muscle cells. (**b**) Target gene or chemical validation in DA neurons. Expression of α-syn in DA neurons is toxic and results in an age- and dose-dependent degeneration of the DA neurons. Over-expressed candidate genes or compounds can be evaluated for therapeutic value in this model to determine, if exposure decreases the proteotoxicity associated with α-syn.

In this regard, following RNAi knockdown, 20 genes strongly induced α-syn misfolding at an early (larval) stage of development, while there was a larger set of genes (125) that were effectors at a later chronological age (17).

3. Induction of DA Neurodegeneration in *C. elegans* and Assessment of PD Susceptibility Genes

In the era of modern "omics"-based science, technologies such as RNAi screens typically uncover many more candidates than can be readily studied. Therefore, it is useful to combine such datasets with secondary assays that exploit distinct criteria to further delineate and define functional modifiers. With respect to modeling PD, acute and chronic methods for inducing DA neurodegeneration *C. elegans* using the neurotoxin 6-OHDA or through overexpression of α-syn, respectively, are well studied. While the mechanism of action of these two neurotoxic insults is different, there are common therapeutic interventions that can be examined (Fig. 5).

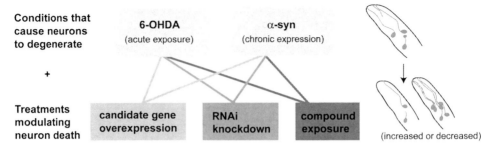

Fig. 5. Common exposure paradigms in *Caenorhabditis elegans* for examining DA neurodegeneration. 6-OHDA is an acute exposure that will cause neurodegeneration within 24 h, while transgenic overexpression of α-syn is chronic and will not result in significant degeneration until adulthood. Regardless of the condition used to induce degeneration, analysis of the neurons, and potential downstream applications and treatments, are similar. These include candidate gene overexpression, RNAi knockdown, and compound exposure. In all cases, *C. elegans* DA neurons can be examined for a decrease in neurodegeneration as a means of discovering therapeutic modalities or an increase in degeneration, which might provide information on cellular mechanisms of action.

The analysis of DA neurons in *C. elegans* is more time-consuming than protein aggregation studies, however; and therefore, researchers usually perform small-scale screening for neuroprotective agents, or validate genes or chemicals identified by other, often more high-throughput, means.

3.1. 6-OHDA-Induced Dopaminergic Toxicity

Although the precise mechanisms underlying selective neuronal vulnerability remain to be fully elucidated, the tendency of DA to induce oxidative damage in neurons renders this neurotransmitter a potential contributory factor to the degenerative process in PD. The toxicity of 6-OHDA is mediated through the formation of ROS by the generation of hydrogen peroxide and hydroxide radicals via a non-enzymatic auto-oxidation process (49, 50). After exposure to 6-OHDA, *C. elegans* DA neurons exhibit a characteristic dose-dependent pattern of apoptotic cell death that was confirmed by ultrastructural analysis (51). 6-OHDA is selectively taken up by the *C. elegans* DA transporter, DAT-1, and *dat-1* mutant worms are resistant to degeneration. Degeneration can be monitored in living animals by co-expression of GFP and occurs within a few hours of exposure (22, 51). Detailed methods for successfully performing this procedure can be found in several publications (22, 51, 52).

6-OHDA solutions (6-OHDA plus ascorbic acid, which will stabilize the 6-OHDA) must be made immediately prior to worm exposure. If the solution immediately turns pink, it has oxidized, and it will not cause DA neurodegeneration. For experimental consistency, it is important to use the same chemical supplier and lot number for 6-OHDA, as there can be vast changes in the amount of degeneration observed. Thus, for each new lot number, the 6-OHDA should be tested for DA neurodegenerative efficiency before commencing with experimental procedures. In general, since 6-OHDA oxidizes quickly, normal oxidation will usually occur after the 1-h incubation. Exposed worms should also be developmentally synchronized to the late L3–L4 staged; treatment of younger and older worms with 6-OHDA increases lethality or resistance to the toxin, respectively.

As described previously, the hermaphrodite has six anterior DA neurons (Fig. 2a, b). Most *C. elegans* laboratories analyze the six anterior DA neurons for neurodegeneration following 6-OHDA treatment (Fig. 2d, e). Worms are evaluated by scoring these anterior dopaminergic neurons as either "normal" [i.e., WT] or "degenerative" for ~50 worms per round of toxin exposure (repeat 2×). Following 6-OHDA treatment, degenerating neurons may exhibit several morphological defects, including neurite blebbing (Fig. 2d), cell body rounding (Fig. 2e), and complete cell loss.

A primary application of this strategy has been to investigate whether the toxicity associated with 6-OHDA exposure can be rescued via transgene expression, gene depletion (RNAi or mutant analysis), or chemical exposure (Fig. 5). Enhanced survival of DA neurons would indicate a mechanism, whereby DA-associated toxicity can be attenuated. For example, the 6-OHDA assay was used to screen through 11 mammalian DAergic, GABAergic, and glutamatergic receptor agonists or antagonists. Two of the 11 compounds, bromocriptine and quinpirole, protected against 6-OHDA toxicity in a dose-dependent manner (39). Likewise, we previously reported that overexpression of a chaperone-like protein (torsinA) in *C. elegans* DA neurons results in dramatic suppression of neurodegeneration following 6-OHDA treatment (22). Further investigations into this neuroprotective mechanism of action indicated that torsinA, an ER luminal protein, was influencing the turnover of DAT-1, which is a polytopic membrane protein, thus torsinA indirectly attenuated cell death by limiting the access of 6-OHDA to the DA neurons.

6-OHDA-induced DA neurodegeneration is both rapid and acute and may represent an effect that is too strong to accurately evaluate the neuroprotective influence of certain gene products or chemical modifiers. As an alternative to 6-OHDA exposure, a more subtle means of inducing neurodegeneration involves the intracellular overproduction of DA. We have also generated transgenic lines of animals that exhibit neurodegeneration by selective overexpression of TH, the *cat-2* gene product in DA neurons (22). Given the complexities of DA signaling, it may be more mechanistically challenging to interpret modifiers in this experimental context; however, this may be a suitable alternative for some lines of experimentation.

3.2. α-syn Proteotoxicity in DA Neurons

As discussed in Sect. 2, *C. elegans* transgenic lines expressing α-syn in the large body wall muscle cells display readily observable protein aggregation (using a compound microscope). Notably, however, the neuronal α-syn models do not (Table 3). The difference between the body wall muscle and neuronal models described herein is that in the former, α-syn is directly fused, at the C-terminus, with a fluorescent protein, while in the latter α-syn is not fused to a fluorescent marker protein, and is thus undetectable in live animals. Rather, the DA neurons are separately marked with GFP (independently expressed) as a means of examining neurodegeneration. Because the cell bodies of *C. elegans* neurons are very small (2 μm), the likelihood of accurately scoring differences in α-syn aggregation in these cells is minimal; therefore, neurodegeneration is a more robust phenotype. Notably, there are several *C. elegans* models of neuronal α-syn proteotoxicity, and they all exhibit a quantifiable phenotype (Table 3). Some of the models overexpress α-syn pan-neuronally, while others are DA specific. Most of the models overexpress WT α-syn and not the mutant forms. This situation correlates with human PD, where multiplication of the WT α-syn locus leads to familial PD (44), and only exceedingly rare PD cases are associated with intrinsic α-syn mutation. It should be noted that a common phenotype noted among all the transgenic lines expressing α-syn specifically in DA neurons is neurodegeneration. Thus, these models can be used to examine factors that enhance or reduce the basal level of degeneration in vivo (Fig. 4b).

In our laboratory, we have established that overexpression of WT human α-syn under control of a DA-specific promoter [P_{dat-1}::α-syn] results in age- and dose-dependent neurodegeneration (17, 22). These transgenic worms reproducibly demonstrate a high level of DA degeneration as the animals age; by day 5 of adulthood only 15% of the population has a normal complement of anterior DA neurons (Fig. 2b, c). We have used this well-characterized invertebrate model of neurodegeneration to assess gene products and chemicals for neuroprotective properties because it reproduces an important characteristic of PD – progressive dopaminergic degeneration. In this manner, large populations of isogenic animals can be propagated and analyzed, leading to an unequivocal level of accuracy in evaluating factors influencing neurodegeneration.

When analyzing transgenic animals expressing a target gene candidate, DA neuron analysis should be scheduled when synchronized worm populations reach specific ages, because degeneration increases over time. An example timeline for use with our α-syn transgenic animals follows. If the candidate gene is hypothesized to decrease neurodegeneration, worms should be analyzed at later stages, allowing for increased neurotoxicity in the control, α-syn only lines. This correlates to 7 and 10 days post-hatching, if the worms are grown at 20°C (i.e., 3- and 6-day adults). When examining candidate genes that might enhance α-syn-induced toxicity, younger worms should be analyzed [6 and 8 days post-hatching (i.e., 2- and 4-day adults, respectively)]. By judiciously considering time points for analysis, and modifying them empirically through several rounds of analysis, potential effects for candidate gene products can be uncovered (53).

The analysis of DA neurons for α-syn neurodegeneration is similar to the procedure described for 6-OHDA-induced degeneration, whereby the six anterior DA neurons are analyzed. However, the phenotypes of degenerating neurons are somewhat different following this proteotoxic insult (Fig. 2c vs. Fig. 2d, e). Degenerating α-syn neurons can vary slightly and range from neurite retraction, cell rounding, and cell loss (Fig. 2c); in our experience, we have never observed the neurite blebbing that is commonly seen following 6-OHDA exposure (Fig. 2d). Each experiment should be repeated in triplicate (30 worms/screen × 3 rounds of analysis = 90 total worms analyzed/transgenic strain; with three separate transgenic lines/experiment for a total of 270 worms analyzed per gene product).

Detailed protocols for creating expression vectors and transgenic animals are available from other sources (46, 52, 54). Two sources of online videos describing the production of transgenic *C. elegans* and scoring of DA neurons are available from the *Journal of Visualized Experiments* (JoVE) (53, 54). In considering other useful tips in the application of worm PD models, it is important to consider that α-syn-induced DA neurodegeneration experiments require that adult worms stay alive for many days. Since these hermaphrodites self-propagate frequently and worm plates can become very crowded with offspring, transfer ~100 young adult stage transgenic hermaphrodites to fresh plates containing 0.04 mg/ml 5-fluoro-2′-deoxyuridine (FUDR) (Sigma-Aldrich; F-0503). FUDR will prevent the need to continuously transfer adult worms, because it is a nucleotide analog that inhibits DNA synthesis and self-reproduction. Importantly, we have determined that the presence of progeny and over-crowding can lead to stressed worms and inconsistent DA neurodegeneration data. Additionally, depending on the nature of the study, it might be useful to score ventral and dorsal CEPs separately, because the dorsal CEP neurons synapse onto the ADE DA neurons while the ventral CEP neurons do not (Fig. 1). In our P_{dat-1}::α-syn worms, dorsal CEPs consistently degenerate significantly more often than the ventral CEPs. ADEs also degenerate significantly more than either dorsal or ventral CEP neurons (52).

4. RNAi in Neuronal Cells

Until recently, targeted knockdown in *C. elegans* neurons, and especially DA neurons, has been notoriously difficult due to selective resistance to RNAi (55) (Fig. 6a), even in supersensitive backgrounds. We have had some success with knockdown of genes in pharmacologically-sensitized backgrounds with neurons (i.e., GABAergic, cholinergic) anatomically positioned near the worm intestine and body wall, where exposure to dsRNA from feeding or soaking would be expected to be in greater concentration (56, 57). Overall, the limitation in neuronal efficacy of *C. elegans* RNAi has led most researchers to rely on the availability of worm strains carrying desired mutant alleles of specific genes to examine cellular processes within neuronal backgrounds. While many mutations are available, targeted knockdown of specific genes would greatly expand the repertoire of targets and/or provide stronger effects when the strength of existing alleles is either unknown or weak.

A significant exception to this limitation includes a study by Kuwahara et al. (21), wherein these researchers generated worm strains overexpressing WT and mutant α-syn pan-neuronally in a mutant background reported to be supersensitive to RNAi in worm neurons (*eri-1*). These animals did not exhibit neurodegeneration or other neuronal dysfunction, thereby indicating sub-pathological threshold levels of α-syn were expressed; moreover, DA neurons were not being directly evaluated in this study. However, differential phenotypic effects from a WT strain expressing the same α-syn constructs could be revealed following RNAi against established neuronal targets. Therefore, through an RNAi screen that targeted knockdown of 1,673 candidate genes implicated in nervous system function, ten positives were identified, primarily consisting of components from the endocytic pathway that exhibited growth and motor abnormalities. These data provided an interesting link between α-syn overexpression with defects in synaptic vesicles trafficking.

Notably, a recently published paper described a new method that results in substantially enhanced neuronal RNAi in worms (58). We have investigated application of this procedure for a single dopaminergic target gene and observed robust knockdown for this initial "proof of concept" target. To generate worms that are hypersensitive to RNAi pan-neuronally, the Chalfie lab expressed a gene encoding the ATP-independent double-stranded RNA channel, SID-1, necessary for systemic RNAi, in all neurons under control of *unc-119*, a gene required for neuronal development ($P_{unc-119}$::SID-1), in *sid-1* mutant animals, which are resistant to RNAi (Fig. 6b, c). This allowed for cell-specific targeted knockdown of genetic modifiers in neurons, while leaving other cells in the nematode resistant to RNAi (Fig. 6c). In addition to enhancing neuronal selective RNAi, this new method enables evaluation and screening of otherwise lethal genes.

We examined this strain for sensitivity to RNAi in DA neurons by knocking down the endogenous DA transporter (*dat-1*). As described in Sect. 3.1, DAT-1 is responsible for the selective re-uptake of DA, as well as neurotoxic DA analogs, such as 6-OHDA. Following RNAi knockdown of *dat-1* in *sid-1* mutant animals, as well as in *sid-1* mutant animals over-expressing $P_{unc-119}$::SID-1, we treated worms with 6-OHDA. We utilized two different worm

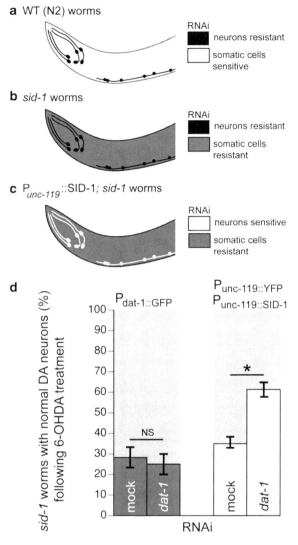

Fig. 6. Neuronal-specific RNAi in *Caenorhabditis elegans*. (**a–c**) Illustration depicting the mechanism of action for this method. (**a**) In normal, WT (N2) worms, non-neuronal somatic cells are sensitive to RNAi while neurons are typically resistant to RNAi. (**b**) *sid-1* mutant worms are resistant to RNAi in all cells, because the normal function of the *sid-1* gene product is to facilitate the uptake of double-stranded RNA. (**c**) In transgenic *sid-1* mutant worms over-expressing the SID-1 protein pan-neuronally ($P_{unc-119}$::*sid-1; sid-1*), the neurons are selectively sensitive to RNAi. Notably, in this transgenic strain, the non-neuronal somatic cells are still resistant to RNAi, because SID-1 is required cell autonomously for RNAi. (**d**) Proof of principle experiment demonstrating DA neuron-specific RNAi in P_{unc119}::YFP + $P_{unc-119}$::SID-1; *sid-1* worms using 6-OHDA exposure. The left side of the graph shows data collected from *sid-1* mutant worms expressing GFP specifically in DA neurons (P_{dat-1}::GFP; *sid-1*), while the right side of the graph shows data from *sid-1* mutant worms expressing YFP and SID-1 pan-neuronally. DA neurons were examined 24 h after 6-OHDA treatment. All worms were exposed to mock RNAi (empty vector) or *dat-1* RNAi to knockdown the endogenous DA transporter. Following knockdown, worms were exposed to 6-OHDA and examined for resistance to this toxin 24 h after treatment. Notably, only P_{unc119}::YFP + $P_{unc-119}$::SID-1; *sid-1* worms exposed to *dat-1* dsRNA (*far right column*) displayed resistance to 6-OHDA, demonstrating that DAT had been knocked down in these animals. *$P < 0.05$.

strains for this experiment, one that was not sensitive to RNAi and simply expressed GFP in DA neurons, as well as the neuronally supersensitive strain (that over-expressed SID-1 pan neuronally). Since both worm strains expressed GFP in DA neurons, we were able to assess DA neurodegeneration. Following *dat-1* RNAi, we treated L4-staged worms with 30 mM 6-ODHA and scored DA neurons 24 h later. As shown in Fig. 6d, over-expression of SID-1 pan-neuronally (right columns) resulted in robust resistance of DA neurons to 6-OHDA, while the non-sensitive strain exhibited extensive degeneration from 6-OHDA exposure. Thus, this experiment validates that the $P_{unc-119}$::SID-1 strain is sensitive to RNAi in the DA neurons. In this regard, use of this supersensitive RNAi strain alone, or in combination with worms expressing α-syn, exposure to neurotoxins, or with existing PD mutant alleles, will be a valuable tool for continued PD research in the *C. elegans* field.

5. Summary

Here, we have highlighted genetic approaches to investigate cellular mechanisms underlying neurodegeneration associated with PD in the nematode model organism, *C. elegans*. The techniques discussed included the use of transgenic *C. elegans* for assessing α-syn proteotoxicity and DA neurodegeneration resulting from α-syn overexpression, as well as 6-OHDA toxicity. Advantages and disadvantages of these models center upon differential readouts and means by which cellular stress and neurodegeneration are induced. In particular, the α-syn protein aggregation assay (in body wall muscle) is well suited to large-scale RNAi screening methodologies. Conversely, assays performed in DA neurons (RNAi, 6-OHDA, or α-syn-induced neurodegeneration) are more tedious, but are an appropriate platform for target validation or hypothesis-based screening for smaller numbers of genes or compounds. As the pathological basis for many neurodegenerative diseases is complex, the alternate screening strategies described herein may or may not target the same cellular mechanisms involved in PD. Nevertheless, the identification of multiple gene and/or drug candidates through such assays may provide previously unreported avenues for therapeutic intervention. Validation across assays and systems is equally essential for optimal translational value. Continued application of *C. elegans* for PD research represents an outstanding opportunity to conduct mechanistic studies using mutants (or RNAi knockdown) in gene and cellular pathways, as well as accelerate positive clinical outcomes for PD.

Acknowledgements

We would like to thank our collaborators and all members of the Caldwell laboratory for their collegiality and contributions to the research presented here. Special thanks to Adam Harrington for generating the data presented on neuronal RNAi and Michelle Tucci for *C. elegans* DA neuron images.

References

1. Chalfie M, Tu Y, Euskirchen G et al (1994) Green fluorescent protein as a marker for gene expression. Science 263:802–805.

2. Chen P, Burdette AJ, Porter JC et al (2010) The early-onset torsion dystonia-associated protein, torsinA, is a homeostatic regulator of endoplasmic reticulum stress response. Hum Mol Genet 19:3502–3515.

3. Voisine C, Varma H, Walker N (2005) Identification of potential therapeutic drugs for Huntington's disease using *Caenorhabditis elegans*. PLoS One. 2:e504.

4. Wang J, Farr GW, Hall DH et al (2009) An ALS-linked mutant SOD1 produces a locomotor defect associated with aggregation and synaptic dysfunction when expressed in neurons of *Caenorhabditis elegans*. PLoS Genet 5:e1000350.

5. Anderson P (1995) Mutagenesis. Meth Cell Biol 48:31–58.

6. Brenner S (1974) The genetics of *Caenorhabditis elegans*. Genetics 77:71–94.

7. Samuelson AV, Carr CE, Ruvkun G (2007) Gene activities that mediate increased life span of *C. elegans* insulin-like signaling mutants. 21:2976–2994.

8. Dimitriadi M, Sleigh JN, Walker A et al (2009) Conserved genes act as modifiers of invertebrate SMN loss of function defects. PLoS Genet 6:e1001172.

9. Chalfie M, White J (1988) The nervous system. In: Wood WB (ed) The Nematode *Caenorhabditis elegans*, Cold Spring Harbor Laboratory Press, Cold Spring Harbor, NY.

10. Bargmann CI (1998) Neurobiology of the *C. elegans* Genome. Science. 282:2028–2033.

11. White JG, Southgate E, Thomson JN et al (1986) The structure of the nervous system of *Caenorhabditis elegans*. Phil Trans R Soc Lond [Biol] 275:327–348.

12. Goodman MB, Hall DH, Avery L et al (1998) Active currents regulate sensitivity and dynamic range in *C. elegans* neurons. Neuron 20:763–772.

13. Sawin ER, Ranganathan R, Horvitz HR (2000) *C. elegans* locomotory rate is modulated by the environment through a dopaminergic pathway and by experience through a serotonergic pathway. Neuron 26:619–633.

14. McDonald PW, Jessen T, Field JR et al (2006) Dopamine signaling architecture in *Caenorhabditis elegans*. Cell Mol Neurobiol 26:593–618.

15. Ramot D, Johnson BE, Berry TL et al (2008) The Parallel Worm Tracker: a platform for measuring average speed and drug-induced paralysis in nematodes. PLoS One 3:e2208.

16. Sämann J, Hegermann SJ, Gromoff EV et al (2009) *Caenorhabditits elegans* LRK-1 and PINK-1 act antagonistically in stress response and neurite outgrowth. J Biol Chem 284: 16482–16491.

17. Hamamichi S, Rivas RN, Knight AL et al (2008) Hypothesis-based RNAi screening identifies neuroprotective genes in a Parkinson's disease model. Proc Nat Acad Sci USA 105:728–733.

18. van Ham TJ, Thijssen KL, Breitling R et al (2008) *C. elegans* model identifies genetic modifiers of alpha-synuclein inclusion formation during aging. PLoS Genet 4:e1000027.

19. Lakso M, Vartiainen S, Moilanen AM et al (2003) Dopaminergic neuronal loss and motor defects in *Caenorhabditis elegans* overexpressing human alpha-synuclein. J Neurochem 86:165–172.

20. Ved R, Saha S, Westlund B et al (2005) Similar patterns of mitochondrial vulnerability and rescue induced by genetic modification of α-synuclein, parkin, and DJ-1 in *Caenorhabditis elegans*. J Biol Chem 280:42655–42668.

21. Kuwahara T, Koyama A, Koyama S et al (2008) A systematic RNAi screen reveals involvement of endocytic pathway in neuronal dysfunction in alpha-synuclein transgenic *C. elegans*. Hum Mol Genet 17:2997–3009.

22. Cao S, Gelwix CC, Caldwell KA et al (2005) Torsin-mediated neuroprotection from cellular stresses to dopaminergic neurons of *C. elegans*. J Neurosci 25:3801–3812.

23. Caldwell GA, Caldwell KA (2008) Traversing a wormhole to combat Parkinson's disease. Dis Mod Mech 1:32–36.

24. Garcia SM, Casanueva MO, Silva MC et al (2007) Neuronal signaling modulates protein homeostasis in *Caenorhabditis elegans* post-synaptic muscle cells. Genes Dev 21: 3006–3016.

25. Nass R, Hahn MK, Jessen T et al (2005) Genetic screen in *Caenorhabditis elegans* for dopamine neuron insensitivity to 6-hydroxydopamine identifies dopamine transporter mutants impacting transporter biosynthesis and trafficking. J Neurochem 94:774–785.

26. Komatsu M, Waguri S, Chiba T et al (2006) Loss of autophagy in the central nervous system causes neurodegeneration in mice. Nature 2006 441:880–884.

27. Bowers K, Stevens TH (2005) Protein transport from the late Golgi to the vacuole in the yeast *Saccharomyces cerevisiae*. Biochim Biophys Acta 1744:438–454.

28. Ruan Q, Harrington AJ, Caldwell KA et al (2010) VPS41, a protein involved in lysosomal trafficking, is protective in *Caenorhabditis elegans* and mammalian cellular models of Parkinson's disease. Neurobiol Dis 37: 330–338.

29. Fung HC, Scholz S, Matarin M et al (2006) Genome-wide genotyping in Parkinson's disease and neurologically normal controls: first stage analysis and public release of data. Lancet Neurol 5:911–916.

30. Cooper AA, Gitler AD, Cashikar A et al (2006) Alpha-synuclein blocks ER-Golgi traffic and Rab1 rescues neuron loss in Parkinson's models. Science 313:324–328.

31. Gitler AD, Bevis BJ, Shorter J et al (2008) The Parkinson's disease protein alpha-synuclein disrupts cellular Rab homeostasis. Proc Natl Acad Sci USA 105:145–150.

32. Su LJ, Auluck PK, Outeiro TF et al (2010) Compounds from an unbiased chemical screen reverse both ER-to-Golgi trafficking defects and mitochondrial dysfunction in Parkinson's disease models. Dis Mod Mech 3:194–208.

33. Avery L, Horvitz RH (1990) Effects of starvation and neuroactive drugs on feeding in *Caenorhabditis elegans*. J Exp Zool 253: 263–270.

34. McIntire SL, Jorgensen E, Horvitz HR (1993) Genes required for GABA function in *Caenorhabditis elegans*. Nature 364:334–337.

35. Crowder CM, Shebester LD, Schedl T (1996) Behavioral Effects of Volatile Anesthetics in *Caenorhabditis elegans*. Anesthesiology 85: 901–912.

36. Evason K, Huang C, Yamben I et al (2005) Anticonvulsant medications extend worm lifespan. Science 307:258–262.

37. Kwok TC, Ricker N, Fraser R et al (2006) A small-molecule screen in *C. elegans* yields a new calcium channel antagonist. Nature 441:91–95.

38. Cao S, Hewett JW, Yokoi F et al (2010) Chemical enhancement of torsinA function in cell and animal models of torsion dystonia. Dis Mod Mech 3:386–396.

39. Marvanova M, Nichols CD (2007) Identification of neuroprotective compounds of *caenorhabditis elegans* dopaminergic neurons against 6-OHDA. J Mol Neurosci 31:127–137.

40. Wang Y, Branicky R, Stepanyan Z et al (2009) The anti-neurodegeneration drug clioquinol inhibits the aging-associated protein CLK-1. J Biol Chem 284:314–323.

41. Locke CJ, Fox SA, Caldwell GA et al (2008) Acetaminophen attenuates dopamine neuron degeneration in animal models of Parkinson's disease. Neurosci Lett 439:129–133.

42. Polymeropoulos MH, Lavedan C, Leroy E et al (1997) Mutation in the alpha-synuclein gene identified in families with Parkinson's disease. Science 276:2045–2047.

43. Conway KA, Lee SJ, Rochet JC et al (2000) Acceleration of oligomerization, not fibrillization, is a shared property of both α-synuclein mutations linked to early-onset Parkinson's disease: Implications for pathogenesis and therapy. Proc Natl Acad Sci USA 97:571–576.

44. Singleton AB, Farrer M, Johnson J et al (2003) alpha-synuclein locus triplication causes Parkinson's disease. Science 302:841.

45. Pankratz N, Wilk JB, Latourelle JC et al (2009) Genome-wide association study for susceptibility genes contributing to familial Parkinson disease. Hum Genet 124:593–605.

46. Harrington AJ, Knight AL, Caldwell GA et al (2011) *C. elegans* as a model system for identifying effectors of a-synuclein misfolding and dopaminergic cell death associated with Parkinson's disease. Methods 53:220–225.

47. McLean PJ, Kawamata H, Shariff S et al (2002) TorsinA and heat shock proteins act as molecular chaperones: suppression of alpha-synuclein aggregation. J Neurochem 83:846–854.

48. Caldwell GA, Cao S, Sexton EG et al (2003) Suppression of polyglutamine-induced protein aggregation in *Caenorhabditis elegans* by torsin proteins. Hum Mol Genet 12:307–319.

49. Kumar R, Agarwal AK, Seth PK (1995) Free radical-generated neurotoxicity of 6-hydroxydopamine. J Neurochem 64:1703–1707.

50. Folcy P, Riederer P (2000) Influence of neurotoxins and oxidative stress on the onset and

progression of Parkinson's disease. J Neurol 247 Suppl 2:II82–94.

51. Nass R, Hall DH, Miller DM 3rd et al (2002) Neurotoxin-induced degeneration of dopamine neurons in *Caenorhabditis elegans*. Proc Natl Acad Sci USA 99:3264–3269.

52. Tucci ML, Harrington AJ, Caldwell GA et al (2011) Modeling dopamine neuron degeneration in *C. elegans*. In: Manfredi G, Kawamata, H (eds) Methods in Molecular Biology: Neurodegeneration.

53. Berkowitz LA, Hamamichi S, Knight AL et al (2008) Application of a *C. elegans* dopamine neurons degeneration assay for validation of potential Parkinson's disease gene candidates. J Vis Exp doi:10.3791/835.

54. Berkowitz LA, Knight AL, Caldwell GA et al (2008) Microinjection and selection of transgenic animals in *C. elegans*. J Vis Exp doi:10.3791/833.

55. Asikainen S, Vartiainen S, Lasko M et al (2005) Selective sensitivity of *Caenorhabditis elegans* neurons to RNA inteference. Neuroreport 16:1995–1999.

56. Locke CJ, Williams SN, Schwarz EM et al (2006) Genetic interactions among cortical malformation genes that influence susceptibility to convulsions in *C. elegans*.

57. Locke CJ, Kautu BB, Berry KP et al (2009) Pharmacogenetic analysis reveals a post-developmental role for Rac GTPases in *Caenorhabditis elegans* GABAergic neurotransmission. Genetics 183(4):1357–72. Brain Res 1120:23–34.

58. Calixto A, Chelur D, Topalidou I et al (2010) Enhanced neuronal RNAi in *C. elegans* using SID-1. Nat Meth 7:554–561.

Chapter 4

Effects of Alpha-Synuclein Expression on Behavioral Activity in *Drosophila*: A Simple Model of Parkinson's Disease

Robert G. Pendleton, Xiaoyun C. Yang, Natalie Jerome, Ornela Dervisha, and Ralph Hillman

Abstract

Adult transgenic fruit flies (*Drosophila melanogaster*) carrying the human gene coding for the protein α-synuclein were tested for geotactic, locomotor, and phototactic behaviors as well as α-synuclein expression. Specific assays in adult conscious flies can be used to determine geotactic, phototactic, and locomotor behaviors. The presence of endogenous α-synuclein in the fly brain was assessed via immunological blot assays. The expression of monomeric human α-synuclein (19 kDa) in the brain of flies containing the homozygous human UAS wild-type α-synuclein transgene was shown by specific immunological blotting procedures. Behavioral testing indicated that motor rather than sensory functions were principally affected by the presence of α-synuclein, mimicking its effects in Parkinson's disease. These data illustrate that the human gene coding for α-synuclein can be expressed in the fly and produce behavioral effects in accord with behaviors seen in patients with Parkinson's disease.

Key words: Alpha-synuclein expression, *Drosophila melanogaster*, Behavioral effects, Parkinson's disease, Model organisms

1. Introduction

Parkinson's disease (PD) is a neurodegenerative disorder characterized behaviorally by abnormalities mainly involving motor function, such as tremor, rigidity, bradykinesia, and postural instability. Sensory function is largely unimpaired (1). Histopathologically, the hallmark of the disease is the presence of Lewy bodies in the substantia nigra, which are eosinophilic inclusions containing neurofilaments of which α-synuclein is a major component (2). α-Synuclein is normally present in the human brain, but is physically altered in PD, usually in normal or wild-type form, although abnormal (mutant) molecular forms are present in some patients (3). In contrast,

Emma L. Lane and Stephen B. Dunnett (eds.), *Animal Models of Movement Disorders: Volume I*, Neuromethods, vol. 61,
DOI 10.1007/978-1-61779-298-4_4, © Springer Science+Business Media, LLC 2011

wild-type fruit flies, *Drosophila melanogaster*, normally do not contain this protein. However, the human coding gene has been inserted into the w^{1118} fly genome via plasmid insertion during early development, along with a binding sequence (UAS) in its promoter region for the yeast protein GAL4 (4).This transgene is thought to be silent in flies unless they are crossed with organisms homozygous for a GAL4 coding "driver" gene (10). However, as there is evidence indicating that GAL4 is toxic in *Drosophila* (5), we measured certain behavioral activities in flies obtained from Feany and Bender, which were homozygous for the UAS wild-type α-synuclein gene (4). This report describes the results of these studies.

2. Materials and Methods

2.1. Drosophila Stocks

Dr. M. Feany provided the authors with the UAS α-synuclein flies containing the transgene for human α-synuclein that had been inserted into recipient flies carrying the w^{1181} mutation. A mutant α-synuclein construct had been placed downstream of a Gal-4 binding site. The behavioral assays in the reported experiments were accomplished using this construct in the homozygous condition. The standard control stocks, wild-type Canton-S which was used in some experiments as well as the original w^{1118} stock, are both available from FLYBASE, the *Drosophila* Stock Center at Indiana University. Both control and experimental flies were stored and maintained at constant temperature (25°C) and controlled humidity on a 12-h light dark cycle.

2.2. Media Preparation

Standard *Drosophila* medium was prepared according to a modified recipe first used at the California Institute of Technology (Table 1). The change involved the replacement of sulfur-free molasses with corn syrup. Care should be taken that only dried dead yeast is used in medium preparation. Fifty milliliters of cooked media was poured into plastic flasks, which were stored for no more than 5 days before being used. Both control and experimental flasks were seeded with several grains of live yeast, immediately prior to each experiment.

The first set of ingredients, corn syrup, agar, and water, should be brought to a boil to ensure that the agar is completely dissolved. If molasses is used, it must be sulfur free. Care should be taken that the food is not overcooked. After the addition of the cornmeal and yeast, the food should be brought to a rolling boil and then removed from the heat. It should be allowed to stand until the boiling is over before adding propionic acid. If acid is added too soon, it will boil away and the food will become moldy. If too much propionic acid is used, the flies will die. It is important to use

Table 1
Recipe for *Drosophila* medium [a]

Ingredients	200 Flasks	100 Flasks	50 Flasks	25 Flasks	10 Flasks	100 Vials
Water	8400 ml	4200 ml	2100 ml	1050 ml	420 ml	1470 ml
Agar	90 g	45 g	23 g	12 g	5 g	17g
Karo	1200 ml	600 ml	300 ml	150 ml	60 ml	210 ml
Bring water, agar, and Karo to a boil. Insure that the agar is completely dissolved						
Corn meal	980 g	490 g	245 g	17g	49 g	180 g
Brewer's yeast [b]	130 g	65 g	33 g	17g	7 g	24 g
Water	2900 ml	1450 ml	725 ml	363 ml	145 ml	500 ml
Add corn meal, yeast and water. Bring to a rolling boil to cook the corn meal						
Propionic acid	68 ml	34 ml	17 ml	9 ml	3 ml	12 ml
Add propionic acid when boiling is complete						

[a] Modified from the original California Institute of Technologies Recipe
[b] Do not use live yeast. If you do, the food will explode in your flasks

dead yeast in the food preparation and live yeast before placing flies on the food. Do not over yeast the food because the flies will not lay eggs. If desired, a moderate solution of yeast in water can be used instead of yeast grains. In this case, use only one or two drops of the suspension in each vial immediately before adding flies.

2.3. Immunoblot Assays

For α-synuclein dot blot assays, 1,000 heads were removed from CO_2 anesthetized 14-day-old w^{1118} and transgenic α-synuclein (UAS) flies on glass plates at 4°C. The heads were transferred into Eppendorf tubes on ice, homogenized in 10 mM PMSF and then centrifuged at 13,000 rpm for 15 min at 4°C. The supernatants were then blotted onto nitrocellulose paper, which was subsequently immersed in blocking solution for 16 h. This solution was discarded and the membrane was washed in TTBS for 15 min. Primary antibody (mouse anti-synuclein; Zymed) diluted in buffer was incubated with the membrane for 90 min followed by secondary anti-mouse antibody. The secondary antibody was decanted and the membrane was washed with TTBS for 15 min. The coloring agent, BCIP/TNBT, was added for 10 min and the membranes were then washed in water for 5 min to stop the reaction.

2.4. Western Blot Alpha-Synuclein Assays

Transgenic α-synuclein (UAS) and w^{1118} flies were anesthetized in CO_2, placed in plastic vials, and quick frozen in a dry ice-ethanol bath. The frozen flies were placed on tiles that had been kept in a freezer and heads were collected from 400 flies of each genotype. Heads were then separately homogenized in 40 μL of PMSF and 100 μL of lysis buffer (Cell Signaling) containing protease inhibitor. It is important that the grinding of heads and the homogenizing process take place in an ice bath to prevent denaturation of proteins. The grinding process must be vigorous because of the very small size of the brain cells. The heads were homogenized, freeze–thawed, and sonicated three times creating three extracts which were numbered in consecutive order (see Fig. 1, top). After centrifugation, 32 μL of each extract supernatant was added to 8 μL of sample (loading) buffer (Sigma) containing beta-mercaptoethanol and Laemmli buffer. Into another Eppendorf tube 1 μg/μL of synthetic human α-synuclein (Sigma) was placed in 9 μL of sample buffer. The samples were heated for 10 min at 95°C and then subjected to gel electrophoresis at 145 V for 90 min. The gels were then transferred onto a nitrocellulose membrane at 25 V for 30 min and then incubated in 5% condensed milk solution for 60 min at 5°C. After washing for 30 min, the membrane was incubated with primary mouse antibody diluted in TBST (1:1,000) at 5°C for 16 h. After washing for 30 min, the membrane was incubated with secondary antibody diluted with TBST (1:1,000) at 5°C. The membrane was washed again and stained with BCIP/NBT overnight, washed, and blow dried. A protein ladder (Bio-Rad) was also run on each membrane.

2.5. Locomotor Assay

Individual control and transgenic flies were inserted into a 100×15 mm² Petri dish that had 13 mm² grids etched on the bottom. The fly was inserted into the dish through a small hole in

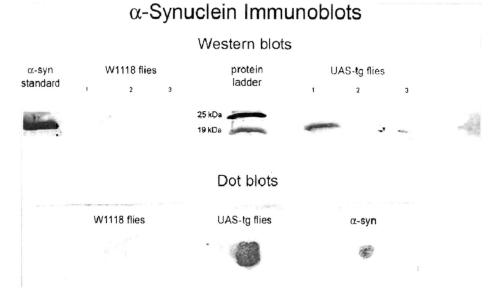

Fig. 1. Western blots (*top*) in the absence of GAL4 drivers show the presence of α-synuclein in the heads of homozygous UAS-transgenic flies. The head samples in the western protocol were from three successive extracts of 400 heads. All the α-synuclein proteins were found in the initial extract. The dot blot samples (*bottom*) were extracted once from 1,000 heads each and transferred to a nitrocellulose membrane.

the lid, which was then sealed with a piece of tape. After allowing the fly to acclimatize for 30 s, the number of gridlines the fly crossed in the next 30 s was recorded. This procedure was repeated alternating between the Canton-S and the homozygous transgenic flies, until each group was assayed ten times.

2.6. Phototaxis Assay

Under conditions of low light, control Canton-S and experimental α-synuclein flies were inserted into the closed middle section of an apparatus that had two opposing tubes radiating from each side. One tube was illuminated with white light and the other darkened with black tape. The middle section was then pushed down to place the fly in a chamber between the two tubes. The length of time it took for the flies to enter into either of the tubes as well as which tube the fly chose was recorded. The process was repeated ten times for each group. This group of behavioral assays was described in detail by Ford et al. (6) and recently reviewed (7, 8).

2.7. Geotaxis Assay

Ten flies each, first of the control flies and second of the experimental flies, were transferred into empty 95 mm × 27 mm vials on which was drawn a horizontal line 8 cm above the bottom. They were then gently tapped to the bottom of the vial three times. After 10 s, the number of flies crossing the 8-cm line was recorded. The process was randomly repeated ten times until five replicate groups of wild-type Canton-S and transgenic flies had been assayed. All flies in this and subsequent assays were 14 days old except where noted and were of both sexes.

2.8. Statistics

The statistical analyses between groups were done based on unpaired Student t-tests. Variability terms in the behavioral assays are standard error bars. The lowest level of significant difference between corresponding groups was a "p" value less than 0.05.

3. Notes

It is important that the investigator maintain both the stock cultures and the experimental flies at a constant temperature and humidity. Flies should be maintained during both development and behavioral testing at $25 \pm 1°C$ and between 40% and 60% humidity. Moreover, care should be taken that both control and experimental flies are the same age when running the behavioral assays. In our experiments, all flies were between 13 and 15 days old. If the experimentalist is using incandescent bulbs as a light source, heat can be a problem. This can be avoided either by putting a water barrier between the light source and the assay procedure or by placing the light at a suitable distance from the assay apparatus. Wild-type flies are normally positively phototactic and negatively geotactic. Therefore, the locomotor effects should be assayed with the light source either above or below the scored Petri dish and the geotactic and phototactic experiments should be assayed with the light source at the side of the apparatus.

4. Results

The presence of α-synuclein in the homozygous transgenic flies together with its absence in w^{1118} flies is shown by immunological western blot analysis (Fig. 1). This result was subsequently confirmed by dot blot analysis (data not shown). The two assays were used together for confirmatory purposes.

Feany and Bender have reported reduced climbing behavior in bipartite GAL4 driven UAS wild-type α-synuclein flies (4). Based on these observations, we examined the effects of the UAS transgene in homozygous transgenic flies compared to Canton-S controls in a similar geotaxis assay (Fig. 2a). These results were then expanded in a time course study (Fig. 2b). Canton-S flies are wild type, while the w^{1118} control flies were from the original stock into which the transgenic α-synuclein gene was transferred. The results indicate that the geotactic activity of both of these control stocks was stable and not significantly different over the 14-day period following eclosion. In contrast, the geotactic ability of the transgenic flies steadily declined over this period. The decreased geotactic activity of the transgenic flies can also be markedly reduced by crossing them with w^{1118} flies (Fig. 2c). This indicates that there is a dosage effect associated with the expression of the transgene in UAS flies lacking the driving force of the GAL-4 protein. The heterozygous human gene in *Drosophila* does not elicit aberrant behavior as opposed to the rare human response, where α-synuclein has been linked to development of autosomal dominant Parkinson's disease (9).

Two additional types of behavior were analyzed and compared in the wild-type Canton-S and UAS-transgenic flies. In each case, 14-day-old flies were used. In the first assay, the random locomotor activities of the two groups were compared. The activity of the UAS flies was substantially reduced relative to the Canton-S controls (Fig. 2d). In a phototaxis assay, however, mixed results were obtained. When the flies were placed in a central position and required to move into either a dark or lighted tube, the time required to make a decision was prolonged in the transgenic flies as compared to the Canton-S controls (Fig. 2e). However, the frequency of the decision which was made (light vs. dark) was unchanged.

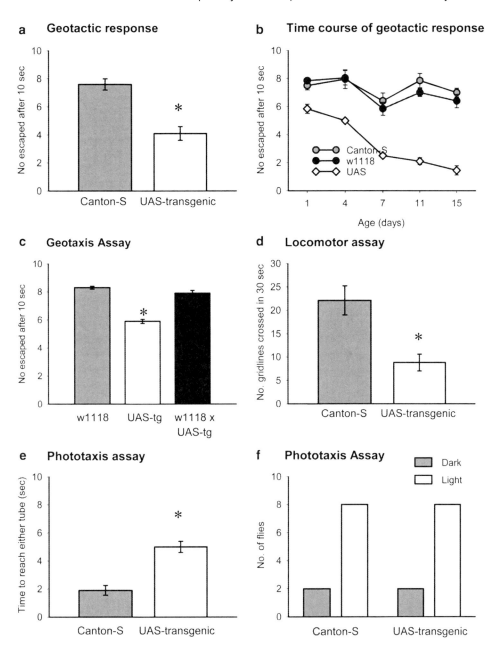

Fig. 2. (**a**) The geotactic response is inhibited in transgenic flies. Each *bar* is the mean and standard error for five groups of ten flies each, each of which was tested ten times in succession. The flies were 14 days old at the time of testing (*$p < 0.05$). (**b**) The loss of geotactic activity is progressive in transgenic flies over 14 days after eclosion. In comparison, the responses of both the wild-type Canton-S controls and flies from the stock (w^{1118}) into which the UAS transgene had been inserted were stable and not significantly different (*$p > 0.05$) during the 15 days of the study. Each *bar* is the mean and standard error of five groups of ten flies each. (**c**) The loss of geotactic activity in homozygous transgenic flies can be reversed by crossing with homozygous controls. There was no significant difference between the w^{1118} controls and the heterozygous w^{1118}/UAS-transgenic flies. Each *bar* is the mean and standard error of five groups of ten flies each (*$p < 0.05$). (**d**) Locomotor activity is reduced in UAS-transgenic flies relative to wild-type Canton-S controls. Each *bar* is the mean and standard error of ten flies tested individually. Mean activity level in the two groups of 14-day-old flies was significantly different (*$p < 0.05$). (**e**) The time-line response to light is delayed in UAS-transgenic flies compared to wild-type Canton-S controls (*$p < 0.05$). Each *bar* is the mean and standard error of 10, 14-day-old flies. (**f**) The overall response pattern to light was identical in the ten flies from the Canton-S and UAS-transgenic groups shown in (**e**).

5. Discussion

Our results indicate that homozygous UAS wild-type α-synuclein flies express α-synuclein in the absence of a GAL4 activator protein normally found in yeast. Thus, the α-synuclein gene is not silent in this model, as generally assumed (10). This point is particularly important because results obtained by others show that GAL4 may itself have biological activity, specifically, toxic activity in the eye development in *Drosophila* (5). Thus, omission of this gene from model organisms, in which novel biological effects are to be ascribed to its companion human gene, is desirable. The level of expression of α-synuclein in non-GAL4 containing flies is sufficient to elicit decreases in both geotactic and locomotor activities in the adult fly. In the phototaxis assay, where response time was slowed in the UAS flies, no effects were noted in the frequency of the responses obtained to the competing sensory stimuli presented. These data make this model more attractive as a model of Parkinson's disease, as it is a disease affecting primarily motor performance, while sensory function is not markedly altered (see above).

Moreover, previously reported results from this laboratory (8) showed that this decline in climbing activity can be prevented by addition of the clinically active anti-Parkinson drugs l-DOPA, DA1 and DA2 agonists and anticholinergic drugs to the fly food. Also, Haywood and Snavely have shown that expression of *parkin*, an enzyme which targets synuclein for proteasomal destruction is protective in this model (11). All this information supports the importance of the transgenic fly as a simple, inexpensive, and useful model for studying Parkinson's disease and as a tool for the investigation of the mechanisms by which α-synuclein results in its development. It may also be used as a test system for drug discovery.

References

1. Damjanov I (2000) Pathology for the Health Related Professions. Saunders Company
2. Wakabayashi K, Tanji K, Mori F, Takahashi H (2007) The Lewy body in Parkinson's disease: molecules implicated in the formation and degradation of a-synuclein aggregates. Neuropathology 27: 494–506
3. Polymeropoulos MH, Lavedan C, Leroy E, Ide SE, Dehejia A, Dutra A, Pike B, Root H, Rubenstein J, Boyer R, Stenroos ES, Chandrasekharappa S, Athanassiadou A, Papapetropoulos T, Johnson WG, Lazzarini AM, Duvoisin RC, Di Iorio G, Golbe LI, Nussbaum RL (1997) Mutation in the a-synuclein gene identified in families with Parkinson's disease. Science 276: 2045–2047
4. Feany MB, Bender WW (2000) A *Drosophila* model of Parkinson's disease. Nature 404: 394–398
5. Kramer JM, Staveley BE (2003) GAL4 causes developmental defects and apoptosis when expressed in the developing eye of *Drosophila melanogaster*. Genet Mol Res 2: 43–47
6. Ford SC, Napolitano LM, McRobert SP, Tomkins L (1989) Development of behavioursal competence in young *Drosophila melanogaster*

adults. Journal of Insect Behaviour 2: 575–590

7. Hillman R, Pendleton RG (2007) Animal Models of Movement Disorders. Elsevier Academic Press

8. Pendleton RG, Parvez F, Sayed M, Hillman R (2002) Effects of pharmacological agents upon a transgenic model of Parkinson's disease in *Drosophila melanogaster*. J Pharmacol Exp Ther 300: 91–96

9. Gasser T (2007) Update on the genetics of Parkinson's disease. Mov Disord 22 Suppl 17: S343–350

10. Brand AH, Dormand EL (1995) The GAL4 system as a tool for unravelling the mysteries of the *Drosophila* nervous system. Curr Opin Neurobiol 5: 572–578

11. Haywood AF, Staveley BE (2004) Parkin counteracts symptoms in a *Drosophila* model of Parkinson's disease. BMC Neurosci 5: 14

Chapter 5

Neurological Evaluation of Movement Disorders in Mice

Simon P. Brooks

Abstract

To better understand the mouse models we are working with, it is desirable to be able to probe for underlying functional deficits, which not only provide clues to the neuropathology of the animal, but also provide functional targets for therapeutic interventions. This chapter describes four methods for identifying and assessing neurological deficits in mice. These methods have been chosen because of their sensitivity in identifying deficits, their reliability in that they have been tried and tested and produce high quality reproducible data, and their accessibility as they require very little equipment and are therefore an inexpensive option. The first approach to be described is the modified SHIRPA primary screen, which is a battery of simple manual manipulations and observations assessed using rating scales. The primary screen is a high throughput method of detecting neurological and physiological abnormalities, and the general health status of the mouse. For the purposes of the present chapter, only the neurological aspects of the screen have been described. Also described is the rotarod apparatus and method. The rotarod can be used in either a stepped (using individual speeds), or accelerating method with both measuring general motor coordination. To measure gait, the footprint test is described, where mice simply have to run along a corridor with painted feet, such that the pattern of the footfall is recorded on paper that lines the floor of the apparatus. Different measurements between footfalls provide a description of the gait of the mouse. A measure of balance and fine limb control was also included in the form of the balance beam (or elevated bridge) test. Our version of the test assesses how well mice can run up a tapered beam, with measures of latency to cross and the number of footslips used as the principal readouts. Together, these simple tests can provide a detailed analysis of neurological deficits in mouse models of disease.

Key words: SHIRPA, Rotarod, Balance beam, Elevated bridge, Gait, Footprint, Mouse

1. General Introduction

Mouse models of disease are at the vanguard of medical science. They are an essential research tool that provides the researcher with the opportunity to explore disease pathogenesis and development and the opportunity to systematically evaluate novel therapeutic interventions. However, regardless of how well the mouse

Emma L. Lane and Stephen B. Dunnett (eds.), *Animal Models of Movement Disorders: Volume I*, Neuromethods, vol. 61,
DOI 10.1007/978-1-61779-298-4_5, © Springer Science+Business Media, LLC 2011

model may recapitulate the disease, the value of the model can be severely negated if the behavioural tests to measure the symptoms in the animals are not present, or if present, not optimised. Of the domains of behavioural testing, the measurement of motor output is the most accessible as the behaviours to be evaluated are generally readily observable and fall within the animals' normal behavioural repertoire. Nevertheless, the evaluation of these behaviours can be problematic if inappropriate or erroneous test protocols are followed. A peculiarity associated with mouse testing, is that almost all but the most recent of behavioural tests were designed with the rat in mind, and require adaptations of apparatus and protocols for the mouse that do not relate purely to the relative size differences between the species. Mice are generally a little more "troublesome" to work with than rats, and as such minor adaptations to the apparatus and the procedure that focus specifically on the normal behavioural repertoire of the mouse, can pay dividends in terms of the quality of the behavioural readout and experimental efficiency. It should also be noted that although these tests have been in existence for many years, in many instances the protocols are still undergoing the process of refinement in mice, and although there is variation between rat strains, due to the great number of mouse strains available and in common usage, the diversity between these lines is vast and requires considerable consideration from the outset. Whereas one mouse line may comply fully with the test requirements another may not, or can not do so, perhaps due to some physiological peculiarities of the line such as for example, retinal degeneration. With all this in mind, this review aims to describe tried and tested protocols for measuring movement and motor function in mice, with a focus specifically on four of the most widely used tests that have demonstrated operational sensitivity in the mouse. The chapter is not designed to be related to a specific class or classes of mouse model of motor disorder, but rather a resource for those readers that seek to probe their mice for motor abnormalities regardless of the cause of underlying neuropathology, and as such, little reference to specific disorders will be made, except where necessary to highlight a point.

The tests have also been chosen as being those that provide good operational sensitivity for relatively little financial cost, and are therefore accessible to all. Most behavioural tests measure more than a single facet of function and motor tests are no different in this respect. The tests chosen are the rotarod as a test of general motor coordination that may also measure fatigue, the footprint test for gait analyses and the balance beam or elevated bridge which is principally a test of balance that requires good motor coordination. However, the opening section looks not at a specific test, but at a neurological screen that is comprised of many tests organised as a primary, secondary and tertiary screen. This battery of tests is known as the SHIRPA (1) and was designed for high throughput screening of multiple mouse lines on an industrial scale. The original primary screen of the SHIRPA was a rapid assessment of neurological deficits, more recently a modified version of the screen (the modified SHIRPA (2)) has been developed and is described below.

2. The SHIRPA (Primary Screen)

2.1. Introduction

The SHIRPA screen (1) named after the institutes that developed it [SmithKline Beecham Pharmaceuticals, Harwell MRC Mouse Genome Centre and Mammalian Genetics Unit, Imperial College School of Medicine (St Mary's), Royal London Hospital, St Bartholemew's and the Royal London School of Medicine phenotype assessment] was originally a behavioural mutagenesis screen consisting of three batteries of behavioural tests. The primary, secondary and tertiary screens have been found to be sensitive in identifying neurological abnormalities in a range of mutant mouse lines (3–8). Each successive screen probed for more complex behavioural phenomena. The primary screen was designed to uncover rudimentary and superficial neurological and physiological deficits, through the use of simple observational methods and rating scales. The secondary screen was designed to probe motor deficits with tests such as the rotarod and spontaneous motor activity, with the final tertiary screen probing for cognitive and behavioural (psychiatric) deficits. The full SHIRPA screen was an extensive battery of tests, which in its entirety was well beyond the means of all but the most well funded of institutions, as it required an extensive range of equipment and numerous personnel to run the tests. More recently, a modified version of the SHIRPA (2) was developed (the modified SHIRPA), which has focused and developed the original primary screen. The following section will focus on describing the procedures used in the modified SHIRPA protocol for identifying neurological abnormalities in mice, the details of which can be found online at the EMPReSS (European Mouse Phenotyping Resource of Standardised Screens) website (http://empress.har.mrc.ac.uk/).

The modified SHIRPA is divided into four discrete sections (Sets 1–4) distinguished by the arena that is being used (although Set 4 supine restraint is used rather than an arena). Within each section, several test protocols are used that are primarily observational or superficially invasive, for example simple handling assessments. Each of the tests utilises a simple observational rating scale, which in each case has a minimal value of 0 and a maximal value of up to 6, on which the severity, presence or absence of a behaviour is rated. As the modified SHIRPA also assess other physiological functions and general health [salivation, provoked biting, lacrimation (tear production), piloerection; tail elevation; body tone; heart and respiration rate, skin colour, pinna and corneal reflex and eyelid closure], these are excluded from the present chapter. Described below are the overtly neurological assessments contained within the modified SHIRPA, which are arranged in the text such that the scores for each observation appear in brackets after the description of the behavioural readout. As most of the ratings work on a theoretical sliding scale from severe to less severe (or vice versa), only the scores for the least and the greatest points on the scale are described. In some instances, the scores on the scale vary based on slightly different manipulations of the animals, or slight variations in the scoring of the behaviour being observed, and therefore are not intuitive, consequently the scoring of these scales are described in greater detail or in full.

2.2. Methods

The arena used for the Set 1 assessments is a simple cylindrical viewing jar (15 cm × 11 cm diameter), in which the animal is placed and observed for three neurological measures: body position which is rated from completely flat (0) to continuous vertical jumping (5), with sitting or standing (3) being rated mid-way between the two extremes; spontaneous activity is rated for no activity (0) to extremely vigorous activity (4); and tremor was scored as 0 for "important" to 2 which equates to no tremor observed.

For the Set 2 tests, the mouse is transferred from the cylinder to a rectangular test arena (55 cm × 33 cm × 18 cm) without being directly handled. The test arena floor is a grid of 15, 11-cm squares with which to measure horizontal motor activity. On depositing the mouse on to the floor of the new test arena, a stop watch is started and locomotor activity is recorded over a 30-s time period. However, the initial observation determines levels of transfer arousal (arousal caused by the transfer of the mouse between the arenas) measured on a scale from "coma" (0) to "manic" (6) with 3 being "a brief freeze followed by active movement" as would be expected in normal mice. The locomotor activity count is simultaneously measured by the number of squares entered with all four feet within the 30-s time frame. An assessment of gait is scored on a scale from "incapacitated" (0) to normal (3). Pelvic elevation is measure of hind limb integrity and is frequently an early sign of neurological dysfunction in mice, and is rated on a scale from "markedly flattened" (0) to "elevated over 3 mm" (3). Touch escape (Fig. 1a) is an escape reaction in response to a touch of the animal's back and can be viewed as a measure of reactivity. It is scored from no response (0) to a vigorous escape response prior to being touched (3), with a score of 1 or 2, respectively, if the escape response occurred for a light or heavy touch. Positional passivity determines the reactivity in the mice with scores being defined by whether the mouse struggles whilst being restrained sequentially with different methods (by the tail, followed by a finger grip on the

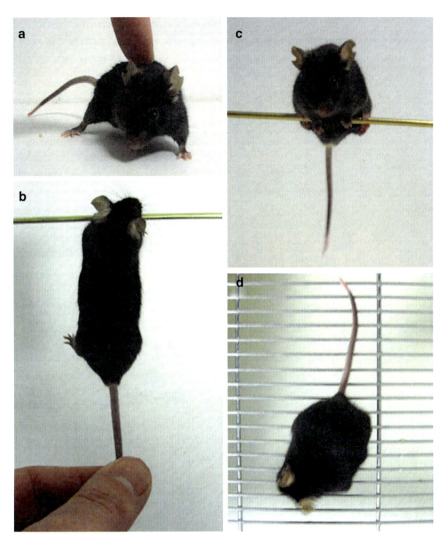

Fig. 1. Tests from the modified SHIRPA screen. Force touch assesses the reactivity of the mice to a touch on the back from above (**a**). The use of a beam or a grid can be used to determine the grip strength of the mouse (**b**) when pulled gently by the tail on the horizontal plane (picture taken from above), and the same position is the starting point for the wire climb task (**c**). Negative geotaxis assesses the ability of the mouse to rotate through 180° from a head down position (**d**). In this instance, a cage lid was used as a climbing mesh.

neck, followed by being laid out in its back and then finally when held by the hind legs, scored from 4 to 1, respectively, with "no struggling" scored as 0).

For Set 3, the assessment of the mice is made either on or above the arena whilst being handled by tail grip between the thumb and forefinger. Trunk curl and limb clasping are scored as either absent (0) or present (1). Visual placing determines whether the mice display the normal forelimb reaching reflex when being lowered by the tail on to a flat surface and is scored from absent (0) to early extension of the forelimb from 25 cm away or more (4).

Grip strength (Fig. 1b) is measured by allowing the mouse to grip a metal grid (12-mm mesh) that covers a section of the arena top, and is assessed as being absent (0) through to unusually strong (4). Toe pinch is the gentle compression of the mid digit of a hind foot with a pair of forceps and measures the pulling away reflex. Toe pinch is measured on a 5-point scale from "no reflex" (0) to "strong repeated pulling motion" (4). The final manipulation with Set 3 uses a horizontal wire (~3 mm diameter) to determine the ability of the animal to grip with its hindlimbs (Fig. 1b, c). The mouse is lowered onto the wire whilst holding the tail, and then with constant but gentle force is manipulated into the horizontal position whilst gripping the wire with the forepaws. The mouse is then released from the grip and is scored as "falls immediately" (0) to "active grip with hindlegs" (4), for its ability to grip with the hindlegs.

The measures that comprise the complete set of tests for Set 4, are less overtly neurological in nature and record parameters of general health such as skin colour and heart rate. Nevertheless, righting reflex, contact righting reflex and negative geotaxis are all measured. Righting reflex examines the ability of the mouse to land on its feet after being flicked in the air by the tail such that it undergoes a backward somersault. The height of the manoeuvre should be the minimum required to complete the assessment. Prior to flipping the mouse in the air, the mouse should be placed on its back to determine whether it is able to right itself under normal circumstances, with a failure to do so producing a score of 0. In this instance, the mouse should not be flipped. If the mouse has an intact righting reflex and is subsequently flipped, scoring is dependent on whether it lands on its back, side or feet resulting in respective scores of 1, 2 or 3. For the contact righting reflex, the mouse is placed in a narrow horizontally positioned perspex tube (30 cm × 3 cm diameter), which allows the animal to have contact with a surface to aid its recovery to a normal position. The tube is rotated through 180° such that the mouse is upside down. The contact righting reflex is either absent (0) or present (1). Finally, negative geotaxis (Fig. 1d) refers to the ability of the mouse to turn on a vertically positioned grid from being in the head down position to the head up position by rotating the body through 180°. Some mice may not be able to maintain their position on the grid at all (scored as 0), or may freeze for the duration of the 30 s test (1), move slightly but not turn (2), turn and freeze (3) or turn and climb the grid (4).

The scores from the SHIRPA allow the researcher to produce an overview of a mouse line, with the greatest value of this approach being that it provides the researcher with a clear indication towards the nature of the deficits, which may then be assessed in greater detail with additional tests. However, the data from the SHIRPA can also stand alone as a valid assessment of neurological deficits in a mouse line.

2.3. Notes

The modified SHIRPA provides a rapid assessment of any underlying deficits that a mouse model may have, but the cost of the observational rating scale approach is that it lacks sensitivity and the scientific rigour associated with more objective approaches. Whereas the mice do not require and indeed should not be trained on any aspect of the screen, it is crucial that the research scientist is trained to produce consistent scores between the mice on each aspect of the assessment procedure. The lack of objectivity is more problematic when more than a single individual scientist is assessing the animals, which then requires a high degree of rating reliability between the individuals. This can be difficult to achieve and can drift with time requiring the retraining of staff.

Whilst there are several tests that comprise the SHIRPA, it is not necessary to run them all and specific probes of function can be extracted to suit ones needs. Likewise, variations in the design of the apparatus are also acceptable, but it should be considered that the SHIRPA is designed to be a standardised test procedure, consequently modifications of the test equipment could exclude the results from being regarded as a SHIRPA screen. Nevertheless, most laboratories are likely to have similar but not exact pieces of equipment that would suffice for the tests required, and as such would provide valuable readouts without additional expense.

As the SHIRPA uses rating scales, the statistical approach for the analyses has to be non-parametric in nature, and therefore has less statistical power than tests that generate interval or ratio data sets. As a consequence, the use of statistical tests other than the standard analysis of variance is required, such as the Mann–Whitney U-test or Kruskal–Wallis non-parametric analysis of variance, where the computations are based on median values rather than means. This general lack of sensitivity may account for the reported discrepancies in the literature, where conventional tests of motor function return significant differences between synphilin-1 parkinsonian mice and their wildtype littermates that were not detected by the modified SHIRPA (9).

3. The Rotarod (Rotorod)

3.1. Introduction

The rotarod test was originally developed in the 1950s (10), specifically for the evaluation of neurological deficits in rats and mice and is now one of the most widely used tests of motor function in the mouse. The test measures motor coordination by assessing how long the animals are able to remain on a horizontal, rotating beam. The original design of the apparatus employed a number of fixed speed rotations, but an accelerating beam method (11), where a small number of runs on a beam that increases in speed from 0 rpm for a predefined time, is now more commonly used.

Whilst the apparatuses may be identical in design for the two methods, the tests measure different aspects of motor coordination as the accelerating version of the task requires the mouse to continually adapt its gait in response to the changing speed of the rod. Consequently, the tests may be differentially sensitive to the same neurological insult, or be able to differentiate between subtly different lesions caused by the administration of the same toxin (12). In mice, the rotarod has been demonstrated to be a sensitive method of assessing disease progression in acute or longitudinal studies of neurodegeneration, for example in genetically- or chemical-induced neurodegeneration in mouse models of HD, PD, cerebellar ataxia, Niemann-Pick disease and amyotrophic lateral sclerosis (13–28), but can also discriminate lesioned from non-lesioned mice (29, 30), and between different intact mouse lines (31–33). It is a simple and rapid test to execute, promoting a higher throughput of animals and thus increasing readout sensitivity further, and provides instant behavioural data that require little expertise to interpret.

3.2. Methods

Rotarod apparatuses can be made in-house, but more commonly are acquired commercially. One of the most commonly used commercially available mouse rotarod apparatuses is the Ugo Basile apparatus (Fig. 2) that comes with a 3 cm diameter rod that has ~1 mm horizontal grooves on the surface, which are designed to provide the mouse with better grip whilst running the task. The rod length is 30 cm, sub-divided with four partitions to create separate berths such that five mice may be run simultaneously. Below the beam in each of the five berths is a trip switch linked to a timer. The switch is raised slightly at the onset of each trial which initiates

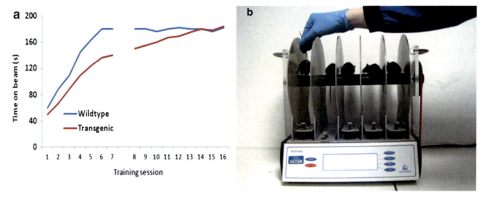

Fig. 2. Extensive training schedules that permit the mice to reach asymptotic performance are essential for optimising behavioural readouts and for valid interpretation of the data (**a**). Mice are sensitive to distractions when on the rotarod. Simple precautions such as loading the machine in an appropriate manner, avoiding the casting of a shadow or placing a moving arm above the mice (**b**), can help prevent erroneous results from being returned.

a timer for the specific birth, which is subsequently stopped when the mouse falls from the beam and lands on the switch, forcing it into the inactive position.

Commercial rotarod apparatuses have several electronically controlled settings that permit the researcher to produce customised trial protocols. Trial duration, rod speed and rotation direction can all be specified, as can the rate of acceleration of the rod for accelerating rod trials. The ability to customise the trials can greatly increase the sensitivity of the test, in contrast to trials on in-house built equipment that may only function at a single set speed.

The key to providing consistently high quality, low variance data sets, is in the execution of an extensive training regime that permits the animals to attain stable asymptotic performance prior to testing. The amount of training required varies between mouse lines and disease models, but typically will take in excess of eight trials beyond the preliminary beam exposure, to assure that peak performance has been attained. For the initial exposure to the beam, the mouse is placed on the rod turning at a low speed, for example 5 rotations per minute (rpm), where it is permitted to remain for 1 min. In this period, the mouse will explore the apparatus whilst maintaining its balance on the beam. If the animal falls or jumps from the beam, it must be replaced immediately. The beam is then accelerated with the mouse in situ up to 10 rpm, where the pace of the beam is maintained for a further 1 min. At this speed, the mouse becomes more focused on remaining on the beam. After the 1 min has elapsed, subsequent increases in speed should take into account the specific features of the mouse lines being tested, for example age and level of disability in the cohort. For normal C57BLK/6j mice or mice at an age without an overt phenotype, 3 min at speeds of over 30 rpm are advisable. On achieving a level of competence on this initial exposure to the beam, the proper training trials can begin. These training trials are defined by whether the accelerating or stepped version of the task is to be used. Training for the stepped trials does not require the animal to be trained on each of the steps that are to be used in the test, but ideally would require the animal to perform consistently at the highest speed they are able to attain, and for the test duration that the experimenter wishes to use when collecting data. One additional option here is to train the animals on each of the speeds that are to be used in the study and collect the acquisition data as a measure of motor learning, but this slows the training process. For the accelerating version of the test, the animal is placed on the rod rotating at the lowest speed, and the trip switch is set to record task acquisition (this is optional, but provides a readout of motor learning), with the training trial starting with the onset of the acceleration procedure. If the mouse falls from the beam, it can be replaced as long as the beam is not travelling too quickly for the safe placement of the mouse. Replacing the animal is also not possible if the acquisition data are being collected for either of the test protocols. These extensive training schedules ultimately provide a high baseline providing the most sensitive assessment of performance degradation as the disease develops, for example.

The aim of this training regime then, is to allow the mouse to learn the operation of the task. Whether the mouse is running on the accelerated or the stepped protocol, the test trials for each animal are terminated when the animal falls from the beam landing on and triggering, a trip switch situated directly below each of the five berths of the rod. When triggered, the trip switch stops the timer associated with that particular berth, providing an accurate measure of how long the animal was able to remain on the rod. For the test procedure, it is advisable to use more than a single run per test session to produce consistent results. Typically in our laboratory, three runs per session are used for the accelerating rotarod with the first run being a practise run followed by the test runs at 15-min intervals. The data for the test runs can be averaged for each animal to reduce the effect of outliers. To avoid the effects of fatigue, at least 15 min should be allowed between the runs for each animal. On the Ugo Basile accelerating rotarod, the beam starts at 0 rpm and accelerates to 40 rpm over a 5-min period. For the stepped speed protocol, any number of speeds can be selected within this range and run for any duration, with the trade-off being the increased number of steps taking longer to run for each mouse which can make for prolonged testing sessions. Typically, five speeds of 1 min duration would produce good sensitivity.

3.3. Notes

One of the major issues with the rotarod is the length or number of training sessions required to attain a level of performance that represents asymptotic motor performance. Typically in the literature, researchers fail to train their animals or give them minimal training prior to data collection. This raises questions regarding the validity of their interpretations of the data, do they (transgenic or lesioned mice for example) have a motor deficit as reported, or a motor learning deficit? In the latter case, they may be able to perform the task as competently as their wildtype or sham lesioned counterparts, but due to the lack of training are not permitted the opportunity to express this, as they fail to acquire the task as rapidly as the comparator group (Fig. 2a). With continued training, the transgenic or lesioned mice may be able to achieve their asymptotic level of performance that may not differ from that of their wildtype littermates. Each cohort of mice or disease model must undergo a training regime based on their own abilities, and the researcher should refrain from attempting to use a standardised protocol across different groups.

Although the rotarod is a simple and effective measure of motor function, performance on the task can be affected by a number of factors. As with many behavioural tests, distracting stimuli within the environment are able to cause the mice to fall from the beam, consequently the test must be run in a suitable testing room that is quiet and free from distracting stimuli. Potentially, the greatest distracting influence on the performance of the mice is the research scientist, particularly when placing new mice on the rod when other mice have just started their test. The experimenter should ensure that if they are right handed they load the five berths of the rotarod from left to right (vice versa for left handed people), thereby insuring that mice that have already started the test are not exposed to the experimenters' hand/arm moving around directly above their heads (Fig. 2b). Whilst at the onset of trials, the trials can be restarted if it is clear that the performance deficit was due to extraneous causes; however, this may cause additional problems if the animal that was upset by the distraction has to be placed back on to the rod, whilst other mice are running. Often it is advisable to utilise four or less of the berths rather than all five if the mice being used are particularly sensitive to distractions, and it should be recognised that the more difficult the task requirement, the more susceptible the mice will be to falling from the beam.

In some instances, the mice will lose their footing on the beam, but fail to fall from it, instead clinging to the beam surface and rotating with it. In this instance, the trial should still be classed as a fail as the mouse has lost its footing and is simply preventing itself from falling from the rod. Where possible, the time should be recorded as soon as this event takes place. In order to prevent this from happening, the ridged surface of the rod can be covered with a smoother material that makes it difficult for the mice to grip when they lose their footing. In our laboratory, we use rubber from a bicycle innertube which is not slippery to the touch even when wet, but allows the animals to fall and trip the timer when they lose their footing.

Weight differences between the control group and the experimental group, whether drug-, lesion-or transgene-induced, can seriously affect rotarod performance in adult mice; heavier mice typically demonstrate impairment on the task when compared with their lighter littermates. Unfortunately, there are few measures that can be taken to overcome this, other than to food restrict the heavier mice to produce comparable weights between the groups. However, if the underlying cause of the weight differences is not known, weight restriction may have unknown effects on the other physiological systems in the mice, for example energy metabolism that may also produce deleterious effects on performance.

4. The Footprint Test and Gait Analysis

4.1. Introduction

The analyses of the footfall patterns in mice can provide a highly sensitive functional readout of abnormality in the animals' gait when running along a corridor or on a treadmill (16, 34–36). The test has been found to be sufficiently sensitive to distinguish between mouse background strains (31), and the longitudinal development of movement disorders in a number of neurological disease states including PD, HD, amyotrophic lateral sclerosis, cerebellar ataxia, and ataxia telangiectasia, plus SOD1 and complexin 1 knock-outs (16, 20, 23, 31, 34, 37, 38). It is also able to determine the effects of neurological lesions and spinal chord contusions on movement (39, 40). This section will focus on footprint analysis as measured by the manual footprint method rather than the digital tracking, treadmill systems, as the manual footprint method is accessible to all due to the minimal financial investment required, in contrast to the digital imaging systems, the cost of which can be prohibitive to some. Regardless of the approach, the assessments of gait are provided by measuring different aspects of the footfall pattern, for example stride length (39, 41), and fore-paw/hind-paw overlap, base width between the forepaws and hind-paws and movement time (16, 20). Digital systems are able to provide further measures (see below). These measures imbue the gait analysis tests with a high translational value, in that the motor output that is measured has direct correlates with measures in people. This coupled with the simplicity of the test, make the footprint test a user-friendly, informative, sensitive and inexpensive test of motor function with high translational value.

4.2. Materials and Methods

The footprint test in its simplest form requires nothing more than paint, paper and a corridor type of arena in which to test the animals. Typically, corridors are less than 1 m in length, but can be anything from a length of ~10 cm (34), or greater. In our laboratory until recently, we use a 60 cm apparatus (31), which allows multiple strides to be assessed thereby providing greater reliability in the scoring (Fig. 3a–b). Typically, corridors are 10 cm in width, but again there is some variation. In some instances corridors are not used at all, with freely moving mice being filmed from below through a clear flooring (35, 42), however collecting consistent footfall patterns with this method may be difficult as the mice are frequently changing direction. Some corridors are fitted with dark goal boxes at one end or a tunnel to provide the mouse with an incentive to run from a lit area to a dark area where they feel secure.

The mice should be habituated to the apparatus prior to training. During the habituation phase, the mice are permitted to explore the whole of the apparatus, including any goal boxes that

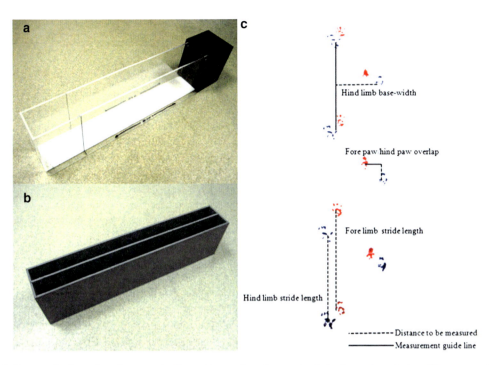

Fig. 3. Runways for the gait analysis are simple corridors and can be constructed using various designs. The runway at the top (**a**) has the advantage of a goal box, but has the disadvantage of being wide (10 cm) which permits the mice to deviate on their paths resulting in increased variability in the footprint pattern. Having clear perspex sides may cause distraction. The lower runway (**b**) is made of a dark plastic and has two 5 cm runways divided by a removable partition wall. The narrow construct maintains the consistent direction of the mouse. The three measures of gait (**c**) are stride length, base width and overlap, the measurement of which are depicted.

may be present. Typically, two habituation sessions of 15 min each would be sufficient, but this may vary by mouse strain. Mice will usually run a corridor with little encouragement, but some mouse strains may prove to be more resistant, in which case a brief training regime may be required. Training consists of teaching the mice the location of the goal box and encouraging them to run the corridor to find it. Such training regimes tend to be rapid as the task is simple and consists of encouraging the mouse to move into the goal box, starting from a very short distance away and gradually increasing this distance until the mouse readily runs the full length of the corridor. If this is not successful, there are a number of other methods available that may aid the training procedure (see Notes below).

For the mouse to provide foot steps, the researcher is required to paint the fore and hind feet of the mice, typically in two different colours such that the fore and hind-paw prints are distinguishable. The paint to be used is the non-toxic, water-based paint that is readily available. The mouse then has to walk or run along a narrow corridor that has paper lining the floor, such that painted footprints are left as the mouse travels the length of the corridor.

Once the mouse has run the corridor, the footfall patterns are analysed through the measurement of several parameters (Fig. 3c). Stride length is assessed by measuring the distance between the centre of the plantar of the fore foot, to the centre of the hind foot plantar on the same side of the body, within the same stride. Base width is the lateral (across the body of the mouse) measure of the distance between the front paws or the hind-paws and essentially measures the degree of lateral displacement of the limbs when the feet are planted. However, as the paw prints are produced in a moving animal, the corresponding front and hind limb prints are always out of alignment. The measurement of the base width distance is created by drawing a perpendicular line from the centre of one fore or hind limb across what would be the body of the animal, such that is bisects at a right angle, a second line from the plantar of the corresponding limb from the opposite side of the body. The base width is then the cross-body distance between the initial plantar and the bisection point. The final gait measurement is the degree of overlap between the fore and hind foot falls. When rodents walk in a straight line, the hind foot lands close to the same place as the fore foot of the same side of the body. Mice with motor deficits, commonly demonstrate a short fall in the overlap pattern. This measure may demonstrate a specific loss of function, for example rigidity, typically in the hind limbs. To measure overlap, the distance between the centre of the plantar of the front paw print and the corresponding hind footprint plantar of successive strides are measured. In addition to these measures, the mice are timed to run the corridor to provide a general measure of motor function. Ideally, at least two measures for each of the parameters should be taken.

4.3. Notes

The first gait analysis techniques used the available video technology to film and analyse stride length (41). The modern digital camera and tracking systems capture gait abnormalities whilst the mice are walking or running on a treadmill or on a transparent surface (35, 37, 38, 42, 43). There are clear advantages to this type of system as the many different parameters can be analysed as video imaging is not restricted to analysing foot falls, but can determine joint angles and limb placements and in some instances, the force required to perform the movement can be detected (35, 37, 38, 41–43). However, these systems can be extremely expensive with some systems requiring considerable computing power to analyse a single trial due to the very high quality of the digital image.

In some instances, it is not possible to identify specific parts of the foot as the animal may be dragging or not placing weight on one or more limbs. In such instances, measures from the centre of the foot fall should be used where possible and where this is not possible the mouse should be excluded from the study. It should be considered whether a mouse in such a deteriorated state is informative and the researcher should always err on the humanitarian side of caution in such conditions.

The width of the corridor can affect the quality of the data that is collected. Narrow corridors of around ~5 cm in width are ideal, as the mouse has little space to deviate from a straight line, thus the quality of the data is greatly improved (see Fig. 3a, b). Whereas a broader width can introduce greater variability across the measures as the mice can have a tendency to meander along the corridor rather than run in a straight line. Consequently, when choosing which footprints to use from the data set, regular stride patterns should be used that are deemed to be representative of the animals' performance. Typically, two complete stride sequences should be used for each animal. The additional bonus to using a narrow corridor is that the mice require very little training.

If the animals are proving difficult to train there may be ways with which to enhance the training. The obvious way is to increase the contrast between the desirable areas of the apparatus (the goal box) and the rest of the corridor. This can be done quite simply by running the experiments in a dimly lit room and positioning a bright lamp over the start point of the corridor to create a bright area which the mouse finds aversive, thereby encouraging it to run to the dark goal box. Raising the corridor by a few centimetres (10–15 cm) at the goal box end by placing something beneath that end of the apparatus can also induce the mice to run. For reasons unknown, mice are sometimes happy to run uphill despite not being willing to run on a level surface.

5. Balance Beam (Elevated Bridge)

5.1. Introduction

The balance beam or elevated bridge test was designed to assess the balance and motor coordination of rodents by making them cross a narrow elevated bridge (16). Balance and motor coordination are assessed by the time taken for the animals to cross the bridge and whether they fell from it, and the number of fore and hind limb slips that the mice produce during the crossing. Hind limb footslips are arguably the most sensitive measure of performance deficits. The technique has been substantially refined by generally increasing test sensitivity through the introduction of beams of different shapes and widths and an expansion in the number of measures used to quantify the movement of the mouse. The test is sensitive in a number of models of neurodegeneration including HD, Alzheimer's disease, Niemann-Pick disease and cerebellar ataxia mouse models (13, 16, 18, 44–48). The use of video technology and playback facilities are also beneficial for the recording of each trial and more recently video systems that produce detailed digital images that when coupled with analyses software, provide a high-tech approach to gait analyses. Whilst the technological approach provides more detail across a greater number of parameters, it is also the expensive option. Consequently, this section of the review will focus on the simple low-tech approach that is accessible to all.

5.2. Materials and Methods

Balance beams can be constructed in a variety of shapes and sizes. Typically, they are of wooden construction and around a metre in length and of a width of less than 5 cm (16). To increase the sensitivity of the test several beams of different shapes and widths/diameters can be used (16), although this can greatly increase the time taken to run the tests. A good compromise in this instance is to use a tapered beam such that the beam narrows as the mouse progresses along it (49). Our current balance beam (Fig. 4) is a modified version of a previously designed beam for rats (49), and has a tapered width that narrows from 1.5 cm at the lower end (start end) of the beam to the goal box. However, the measuring of each run is taken from a start and stop point (see below), where the widths of the beam are 1.2 cm at 15 cm from the lower end of the beam, and 0.5 at 10 cm from the goal box. A ledge runs along either side of the beam 2 cm from its top and of 0.5 cm width. The use of a ledge permits the accurate assessment of footslips as a measure of limb coordination/balance. To further increase sensitivity of the task and to encourage the mice to run the beam reliably the

Fig. 4. Our current balance beam with goal box and start and stop point marked with *white dots* on the side. Along the length of the beam is a ledge with enables the accurate assessment of footslips (*inset*).

beam is angled at 17° from the horizontal such that the mouse is able to run uphill (the highest point of the beam is 58 cm from the ground), which as started previously, seems to encourage them to perform tests, but may in some instances also increase test sensitivity. A further refinement which reduces the variance in the data, is the use of start and stop areas at the beginning and end of the beam as typically mice take some time to start the task and typically slow down to a stop prior to entering the goal box. These start and stop areas are defined by a start line (15 cm from the lower beam end) and end line (10 cm from the goal box) between which the traverse time and footslips are measured. Our apparatus also has a goal box, as the use of a goal box can also be beneficial for expediting training and increasing the reliability of the test performance. As the beam is elevated above ground level, a soft landing area is essential beneath the beam, which in our laboratory consists of numerous folder towels.

Different methods have been used to measure motor coordination and balance on the beam. The most commonly used method is to place the animal at one end of the beam such that it has to traverse the whole length of the beam to find the goal box. This method requires some training of the animals prior to testing and the use of a goal box is advisable. A second method places the animals in the middle of the beam and takes the distance the animals travel, regardless of direction, as the measure of performance (45, 47, 48, 50). This method does not require any pretraining or the use of a goal box.

The following protocol is the protocol used in our laboratory, which has been frequently refined over the years to produce the current method. Goal box training is as described for the gait analysis test; animals are habituated to the goal box for 15 min on 2 consecutive days prior to training, and then trained to enter the goal box initially from a position directly adjacent to the box. This distance is then gradually increased along the beam, until the animals are running the full length of the beam to reach the box reliably. The onset of each trial begins with the mouse being placed on the extreme end of the beam in the start area and facing away from the beam. The time it takes the mice to turn towards the beam is used as a somewhat crude measure of motor coordination, which nevertheless can be informative. On turning, the mouse will then begin to run the beam, where recording of the traverse time begins when the head of the mouse crosses the start line and ends when the head of the mouse crosses the finish line. Between these two points, the experimenter records the number of fore and hind limb footslips from one side of the body, whilst a video camera records the animal from the opposite side, allowing footslips from the opposite side of the body to be assessed post hoc. Videoing of the mice also provides a permanent record for the performance of each mouse. Each mouse should ideally undergo three runs at each

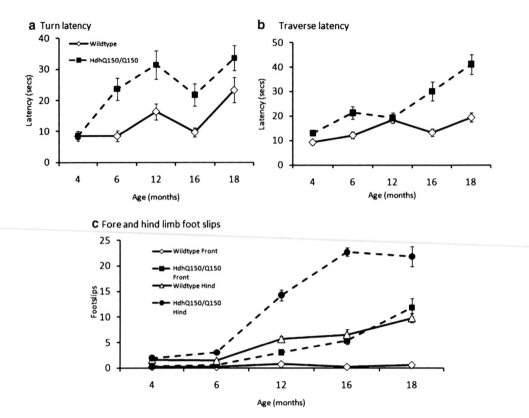

Fig. 5. The progressive nature of motor deficits in the Hdh[(CAG)Q150/Q150] mouse captured by the turn (**a**) and traverse (**b**) latencies and the footslip (**c**) measures of the balance beam test.

test session, the first being a refresher run designed to remind the mouse of the task requirements, with the remaining two runs being for data collection. Some studies have used four or five trials per session (51, 52). Each mouse should take less than 120 s to run, and all but the most disabled mice should be able to traverse the beam in 30 s. An example of the data from the balance beam is provided in Fig. 5.

5.3. Notes

Success with the balance beam, as with all behavioural tests, is largely determined by the quality and extent of the training procedure. A well-trained mouse will traverse the beam quickly and efficiently, whereas a mouse that is not well trained may stop frequently mid-beam. These problems must be overcome during the training phase of the experiments, but in longitudinal studies with genetically modified animals, the possibility exists that non-motor deficits may become a confounding influence on the performance of the animal, for example an increase in apathy or anxiety. In such cases, the animals must be excluded from the study if they fail to run the beam, despite having an otherwise normal appearance and perhaps performing well on other motor tests. Prior to the exclusion of the animals, however they should be placed on the beam and encouraged to run the beam with a pre-test trial using gentle nudges. This procedure allows the experimenter to reliably asses if the animal is likely to participate in the task, and highlights the problem with the balance beam test, which is non-compliance.

Tests such as the rotarod require the animals to comply with the test requirements or experience the negative consequence of falling from the beam. If on the balance beam the animal is happy to sit motionless there is nothing that the experimenter is able to do other than remove the animal from that particular trial or even the study, for repeat offenders. This can to an extent be overcome during the training phase of the protocol where animals should not be permitted to settle on the beam and should be manually dissuaded from doing so. Nevertheless, some mouse strains are more susceptible to inactivity (31), and should be avoided as background strains for genetically modified mice that are to undergo behavioural testing.

In studies that are using animals that have an advanced disease state, the animals may not be able to run the beam in the conventional manner. Typically, mice become dysfunctional in their hind limbs first which may result in relatively subtle motor deficits, which are not readily observed from observations of their normal daily activities, but become manifest on the balance beam which requires a high degree of motor function. Typically, the mice may rely on their forelimbs for propulsion and simply use their hind limbs for extra grip resulting in the hindlimbs not stepping in the normal way and thus not producing footslips. This tends to not be a problem if the cohort of animals to be tested is large, but in small groups, this could skew the footslip data considerably and it may be advisable in such circumstances to omit the footslip counts from the data for these animals. Of course, consideration should be made that the animal in such a state of disease does not repeatedly fall from the beam on successive testing sessions. Typically, if an animal falls from the beam on two consecutive trials in our laboratory, it is scored the maximum traverse time permitted (120 s) and removed from the study.

6. Summary

The aim of this review was to provide an overview of three commonly used tests of motor function that are sensitive for probing different motor functions in the mouse, but are also highly accessible to scientists who have little behavioural experience and limited funds to spend on behavioural equipment. As such these tests are ideal, but it should be considered that no behavioural test is problem free, whether those problems be in the execution of the test or in the theoretical interpretation of the data. As a consequence the researcher should ideally choose to collect data from more than a single test where possible, as data from similar tests are able to validate results, and results from dissimilar tests further ones knowledge of the mouse and/or disease. Finally, as with all behavioural tests, the general good health and welfare of the mouse is paramount to the collection of reliable data.

Acknowledgements

Our work in the field has been supported by grants provided by the Cure Huntington's Disease Initiative Foundation Inc. and the European Huntington's Disease Network.

References

1. Rogers DC, Fisher EM, Brown SD, Peters J, Hunter AJ, Martin JE (1997) Behavioral and functional analysis of mouse phenotype: SHIRPA, a proposed protocol for comprehensive phenotype assessment. Mamm. Genome 8: 711–713

2. Masuya H, Inoue M, Wada Y, Shimizu A, Nagano J, Kawai A, et al. (2005) Implementation of the modified-SHIRPA protocol for screening of dominant phenotypes in a large-scale ENU mutagenesis program. Mamm. Genome 16: 829–837

3. Ferdinandusse S, Zomer AW, Komen JC, van den Brink CE, Thanos M, Hamers FP, Wanders RJ, van der Saag PT, Poll-The BT, Brites P (2008) Ataxia with loss of Purkinje cells in a mouse model for Refsum disease. Proc. Natl. Acad. Sci. USA 105: 17712–17717

4. Hatcher JP, Jones DN, Rogers DC, Hatcher PD, Reavill C, Hagan JJ, Hunter AJ (2001) Development of SHIRPA to characterise the phenotype of gene-targeted mice. Behav. Brain Res. 125: 43–47

5. Lackner P, Beer R, Heussler V, Goebel G, Rudzki D, Helbok R, Tannich E, Schmutzhard E (2006) Behavioural and histopathological alterations in mice with cerebral malaria. Neuropathol. Appl. Neurobiol. 32: 177–188

6. Lalonde R, Dumont M, Paly E, London J, Strazielle C (2004) Characterization of hemizygous SOD1/wild-type transgenic mice with the SHIRPA primary screen and tests of sensorimotor function and anxiety. Brain Res. Bull. 64: 251–258

7. Rafael JA, Nitta Y, Peters J, Davies KE (2000) Testing of SHIRPA, a mouse phenotypic assessment protocol, on Dmd(mdx) and Dmd(mdx3cv) dystrophin-deficient mice. Mamm. Genome 11: 725–728

8. Rogers DC, Jones DN, Nelson PR, Jones CM, Quilter CA, Robinson TL, Hagan JJ (1999) Use of SHIRPA and discriminant analysis to characterise marked differences in the behavioural phenotype of six inbred mouse strains. Behav. Brain Res. 105: 207–217

9. Jin HG, Yamashita H, Nakamura T, Fukuba H, Takahashi T, Hiji M, Kohriyama T, Matsumoto M (2008) Synphilin-1 transgenic mice exhibit mild motor impairments. Neurosci. Lett. 445: 12–17

10. Dunham NW, Miya T (1957) A note on a simple apparatus for detecting neurological deficit in rats and mice. J. Am. Pharm. Assoc. (Baltim.) 46(3): 208–209

11. Jones BJ, Roberts DJ (1968) The quantiative measurement of motor inco-ordination in naive mice using an accelerating rotarod. J. Pharm. Pharmacol. 20: 302–304

12. Monville C, Torres EM, Dunnett SB (2006) Comparison of incremental and accelerating protocols of the rotarod test for the assessment of motor deficits in the 6-OHDA model. J. Neurosci. Methods 158: 219–223

13. Brooks S, Higgs G, Janghra N, Jones L, Dunnett SB (2010) Longitudinal analysis of the behavioural phenotype in YAC128 (C57BL/6J) Huntington's disease transgenic mice. Brain Res. Bull

14. Brooks S, Higgs G, Jones L, Dunnett SB (2010) Longitudinal analysis of the behavioural phenotype in Hdh((CAG)150) Huntington's disease knock-in mice. Brain Res. Bull

15. Brooks S, Higgs G, Jones L, Dunnett SB (2010) Longitudinal analysis of the behavioural phenotype in Hdh(Q92) Huntington's disease knock-in mice. Brain Res. Bull

16. Carter RJ, Lione LA, Humby T, Mangiarini L, Mahal A, Bates GP, Dunnett SB, Morton AJ (1999) Characterization of progressive motor deficits in mice transgenic for the human Huntington's disease mutation. J. Neurosci. 19: 3248–3257

17. Caston J, Hilber P, Chianale C, Mariani J (2003) Effect of training on motor abilities of heterozygous staggerer mutant (Rora(+)/Rora(sg)) mice during aging. Behav. Brain Res. 141: 35–42

18. Elrick MJ, Pacheco CD, Yu T, Dadgar N, Shakkottai VG, Ware C, Paulson HL, Lieberman AP (2010) Conditional Niemann-Pick C mice demonstrate cell autonomous Purkinje cell neurodegeneration. Hum. Mol. Genet. 19: 837–847

19. Glynn D, Bortnick RA, Morton AJ (2003) Complexin II is essential for normal neurological function in mice. Hum. Mol. Genet. 12: 2431–2448

20. Glynn D, Drew CJ, Reim K, Brose N, Morton AJ (2005) Profound ataxia in complexin I knockout mice masks a complex phenotype that includes exploratory and habituation deficits. Hum. Mol. Genet. 14: 2369–2385

21. Ikeda M, Kawarabayashi T, Harigaya Y, Sasaki A, Yamada S, Matsubara E et al. (2009) Motor impairment and aberrant production of neurochemicals in human alpha-synuclein A30P+-A53T transgenic mice with alpha-synuclein pathology. Brain Res. 1250: 232–241

22. Kennel PF, Fonteneau P, Martin E, Schmidt JM, Azzouz M, Borg J, Guenet JL, Schmalbruch H, Warter JM, Poindron P (1996) Electromyographical and motor performance studies in the pmn mouse model of neurodegenerative disease. Neurobiol. Dis. 3: 137–147

23. Knippenberg S, Thau N, Dengler R, Petri S (2010) Significance of behavioural tests in a transgenic mouse model of amyotrophic lateral sclerosis (ALS). Behav. Brain Res. 213: 82–87

24. L'Episcopo F, Tirolo C, Caniglia S, Testa N, Serra PA, Impagnatiello F, Morale MC, Marchetti B (2010) Combining nitric oxide release with anti-inflammatory activity preserves nigrostriatal dopaminergic innervation and prevents motor impairment in a 1-methyl-4-phenyl-1,2,3,6-tetrahydropyridine model of Parkinson's disease. J. Neuroinflammation. 7: 83

25. Lalonde R, Hayzoun K, Selimi F, Mariani J, Strazielle C (2003) Motor coordination in mice with hotfoot, Lurcher, and double mutations of the Grid2 gene encoding the delta-2 excitatory amino acid receptor. Physiol Behav. 80: 333–339

26. Lin CH, Tallaksen-Greene S, Chien WM, Cearley JA, Jackson WS, Crouse AB, Ren S, Li XJ, Albin RL, Detloff PJ (2001) Neurological abnormalities in a knock-in mouse model of Huntington's disease. Hum. Mol. Genet. 10: 137–144

27. Stam NC, Nithianantharajah J, Howard ML, Atkin JD, Cheema SS, Hannan AJ (2008) Sex-specific behavioural effects of environmental enrichment in a transgenic mouse model of amyotrophic lateral sclerosis. Eur. J. Neurosci. 28: 717–723

28. Yang EJ, Jiang JH, Lee SM, Yang SC, Hwang HS, Lee MS, Choi SM (2010) Bee venom attenuates neuroinflammatory events and extends survival in amyotrophic lateral sclerosis models. J Neuroinflammation. 7: 69

29. Haelewyn B, Freret T, Pacary E, Schumann-Bard P, Boulouard M, Bernaudin M, Bouet V (2007) Long-term evaluation of sensorimotor and mnesic behaviour following striatal NMDA-induced unilateral excitotoxic lesion in the mouse. Behav. Brain Res. 178: 235–243

30. Iancu R, Mohapel P, Brundin P, Paul G (2005) Behavioral characterization of a unilateral 6-OHDA-lesion model of Parkinson's disease in mice. Behav. Brain Res. 162: 1–10

31. Brooks SP, Pask T, Jones L, Dunnett SB (2004) Behavioural profiles of inbred mouse strains used as transgenic backgrounds. I: motor tests. Genes Brain Behav. 3: 206–215

32. McFadyen MP, Kusek G, Bolivar VJ, Flaherty L (2003) Differences among eight inbred strains of mice in motor ability and motor learning on a rotorod. Genes Brain Behav. 2: 214–219

33. Tarantino LM, Gould TJ, Druhan JP, Bucan M (2000) Behavior and mutagenesis screens: the importance of baseline analysis of inbred strains. Mamm. Genome 11: 555–564

34. Barlow C, Hirotsune S, Paylor R, Liyanage M, Eckhaus M, Collins F, Shiloh Y, Crawley JN, Ried T, Tagle D, Wynshaw-Boris A (1996) Atm-deficient mice: a paradigm of ataxia telangiectasia. Cell 86: 159–171

35. Clarke KA, Still J (1999) Gait analysis in the mouse. Physiol Behav. 66: 723–729

36. D'Hooge R, Hartmann D, Manil J, Colin F, Gieselmann V, De Deyn PP (1999) Neuromotor alterations and cerebellar deficits in aged arylsulfatase A-deficient transgenic mice. Neurosci. Lett. 273: 93–96

37. Amende I, Kale A, McCue S, Glazier S, Morgan JP, Hampton TG (2005) Gait dynamics in mouse models of Parkinson's disease and Huntington's disease. J. Neuroeng. Rehabil. 2: 20

38. Wooley CM, Sher RB, Kale A, Frankel WN, Cox GA, Seburn KL (2005) Gait analysis detects early changes in transgenic SOD1(G93A) mice. Muscle Nerve 32: 43–50

39. Fernagut PO, Diguet E, Labattu B, Tison F (2002) A simple method to measure stride length as an index of nigrostriatal dysfunction in mice. J. Neurosci. Methods 113: 123–130

40. Ma M, Basso DM, Walters P, Stokes BT, Jakeman LB (2001) Behavioral and histological outcomes following graded spinal cord contusion injury in the C57Bl/6 mouse. Exp. Neurol. 169: 239–254

41. Heglund NC, Taylor CR, McMahon TA (1974) Scaling stride frequency and gait to animal size: mice to horses. Science 186: 1112–1113

42. Clarke KA, Still J (2001) Development and consistency of gait in the mouse. Physiol. Behav. 73: 159–164

43. Kale A, Amende I, Meyer GP, Crabbe JC, Hampton TG (2004) Ethanol's effects on gait dynamics in mice investigated by ventral plane videography. Alcohol Clin. Exp. Res. 28: 1839–1848

44. Grusser C, Grusser-Cornehls U (1998) Improvement in motor performance of Weaver mutant mice following lesions of the cerebellum. Behav. Brain Res. 97: 189–194

45. Hilber P, Caston J (2001) Motor skills and motor learning in Lurcher mutant mice during aging. Neuroscience 102: 615–623

46. Lalonde R, Joyal CC, Thifault S (1996) Beam sensorimotor learning and habituation to motor activity in lurcher mutant mice. Behav. Brain Res. 74: 213–216

47. Le Cudennec C, Faure A, Ly M, Delatour B (2008) One-year longitudinal evaluation of sensorimotor functions in APP751SL transgenic mice. Genes Brain Behav. 7 Suppl 1: 83–91

48. Lorivel T, Hilber P (2007) Motor effects of delta 9 THC in cerebellar Lurcher mutant mice. Behav. Brain Res. 181: 248–253

49. Schallert T, Woodlee MT, Fleming SM (2003) Experimental focal ischemic injury: behavior-brain interactions and issues of animal handling and housing. ILAR J. 44: 130–143

50. Joyal CC, Beaudin S, Lalonde R (2000) Longitudinal age-related changes in motor activities and spatial orientation in CD-1 mice. Arch. Physiol Biochem. 108: 248–256

51. Kashiwabuchi N, Ikeda K, Araki K, Hirano T, Shibuki K, Takayama C, Inoue Y, Kutsuwada T, Yagi T, Kang Y, . (1995) Impairment of motor coordination, Purkinje cell synapse formation, and cerebellar long-term depression in GluR delta 2 mutant mice. Cell 81: 245–252

52. Takahashi T, Kobayashi T, Ozaki M, Takamatsu Y, Ogai Y, Ohta M, Yamamoto H, Ikeda K (2006) G protein-activated inwardly rectifying K+ channel inhibition and rescue of weaver mouse motor functions by antidepressants. Neurosci. Res. 54: 104–111

Chapter 6

Rodent Skilled Reaching for Modeling Pathological Conditions of the Human Motor System

Jenni M. Karl and Ian Q. Whishaw

Abstract

Almost all nervous system motor disorders result in impaired use of the upper limb. Skilled reaching, including the ability to reach for, grasp, and eat a piece of food (reach-to-eat), when impaired, limits patient independence and quality of life. The present paper describes a rat preclinical model of skilled reaching that is useful for investigating human motor system pathology and potential therapies. Rat skilled reaching is described using end-point measures of success, scoring of the movement, and biometric measures. The chapter also presents arguments justifying the model in relation to nervous system organization and shows how the model applies to a number of human motor system pathologies, including stroke, Parkinson's disease, Huntington's disease, and spinal cord injury.

Key words: Skilled reaching, Rat reaching, Single pellet reaching, Hand shaping, Evolution of skilled movement, Movement elements, Eshkol-Wachmann movement notation, Rat vs. human homology, Rodent models of stroke, Rodent models of Parkinson's disease, Rodent models of Huntington's disease, Rodent models of spinal cord injury

1. Introduction

Skilled reaching, also called reach-to-eat, is an act in which an animal reaches with a forelimb for a food item that it then places in its mouth for eating. It is evolutionarily ancient and is well-developed in animal species that are widely used for neurobiological research, including mice, rats, and nonhuman primates (1). For this reason, it is a useful behavior for studying the motor system and modeling neurobiological diseases in humans. The present paper describes general methods for studying skilled reaching in the rat. Rats are a sister clade to primates and, like primates, are especially adept at skilled reaching (2). Almost identical procedures can be used for

Emma L. Lane and Stephen B. Dunnett (eds.), *Animal Models of Movement Disorders: Volume I*, Neuromethods, vol. 61,
DOI 10.1007/978-1-61779-298-4_6, © Springer Science+Business Media, LLC 2011

studying skilled reaching in other rodent species, nonhuman primates, and humans. The procedures described here include end-point measures of success, measures that describe the act in terms of component gestures, and measures that describe movement of body segments and their kinematics. Combined, these various assessment measures provide a comprehensive description of the act and how it is changed by damage to the motor system.

Four types of reaching tasks have been developed to assess skilled forelimb movements in rats. These include the Montoya staircase task (3), the tray reaching task (4, 5), the pasta matrix reaching task (6), and the single pellet reaching task (2, 7). In addition, a number of studies have described the structure and impairments in forelimb use during spontaneous use of the forelimb in different types of behaviors, including spontaneous food handling during eating (8). The present paper describes the very formal structure of the single pellet task, provides a number of scoring systems for its analysis, and provides a list of justifications for its use. The following sections describe the method, assessments, and interpretation of skilled reaching.

2. Methods

2.1. Apparatus

The single pellet reaching task allows for the same movement to be elicited repeatedly, filming of the movement from any angle, and partition of the behavior into discrete trials. The single pellet reaching box is made of clear Plexiglas. Its appearance and dimensions are illustrated in Fig. 1 (2). In the center of the front wall is a vertical slot 1 cm wide. On the outside of the wall, in front of the slot, mounted 3 cm above the floor is a 4-cm-deep shelf. Two indentations on the surface of the shelf are located 2 cm from the inside of the wall to hold the food. The indentations are aligned with the edges of the slit through which the rats can reach. Because the rat pronates the hand medially to contact and grasp food, grasping is most easily performed with the hand contralateral to the indentation, and so the baited indentation can dictate which hand is used and so is useful because the task provides a condition of restraint useful for constraint-induced movement therapy.

2.2. Feeding and Food Familiarization

Prior to and during skilled reach training, animals are gradually food restricted to 90–95% of their normal body weight. This is accomplished by feeding 20 g of Purina Rat Chow once in the morning. One week prior to training, rats are familiarized with 45 mg dustless precision banana-flavored pellets (product #F0021) that are the food targets for reaching, before their single morning feeding. Because the banana pellets are a favored food, only a short duration of food deprivation is required to motivate performance.

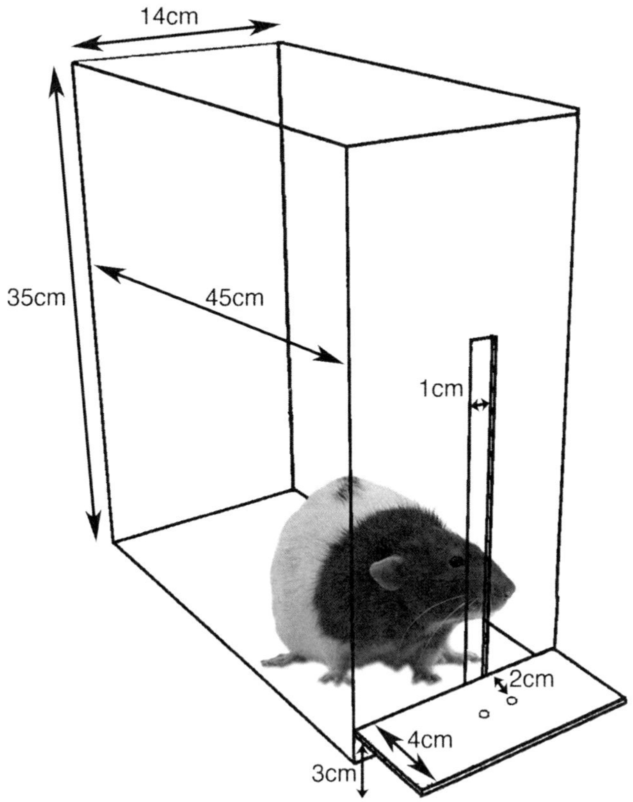

Fig. 1. The single pellet reaching task and apparatus dimensions. An animal must walk from the back of the apparatus to the front, sniff to identify the presence of a food pellet in one of the indentations, and then retrieve that pellet with the contralateral hand and bring the food to the mouth for eating.

2.3. Video Recording

Performance during single pellet reaching is video recorded using a camcorder. A wide variety of commercially available camcorders capable of conventional and high-speed recording are available. Filming should be done with at least 1,000th of a second shutter speed to avoid image blurring. The preferred camera position is head on to the reaching box such that the animal's behavior is filmed from the front, but views from the side, above, or below are useful. A cool fluorescent (Lowel Caselight 4, Brooklyn, NY, USA) studio light source, in addition to the overhead lights of the testing room, provides the illumination necessary for filming. Representative video clips and still frames are captured from digital video recordings with the video-editing software Final Cut Pro (http://www.apple.com) for subsequent archiving and analysis. Pictures can be cropped and adjusted for color and brightness contrast in Adobe Photoshop V.7.0.

2.4. Apparatus Habituation

The aim of the first 10-min training session is to habituate a rat to the reaching box and task. Initially, a rat spends most of its time exploring the box. A number of banana pellets are placed on the reaching tray to promote a rat's interest. It is acceptable for the rat to start to reach for pellets with the tongue. Once sufficiently interested, pellets may be moved further from the box opening to encourage reaching with the hands. It is important that early in the training process multiple pellets are made available on the tray to encourage reaching and reinforce use of the hands.

2.5. Establishment of Hand Dominance

Once a rat consistently uses the hands to reach for banana pellets, only two banana pellets are presented at a time, one in each indentation in the tray of the reaching box. The animal will likely favor a single hand, which is then reinforced, by baiting the contralateral indentation only. Subsequent reaching sessions promote successful reaching with what is now designated as the dominant hand. Nevertheless, the protocol can be easily modified in order to have an animal use either or both hands.

After each reach, the subsequent food pellet is withheld until the rat walks to the far end of the box. With continued training, a reaching "trial" consists of walking to the front of the box, a reach, and a return to the back of the box. To encourage targeted reaching, the food is periodically withheld on a trial in order to teach the rat to sniff to determine whether food is present before making a reach. Rats use olfaction and not vision to detect the presence of food (9).

Once rats are performing discrete trials and successfully identifying food presence, scoring can begin. Typically, a rat is given 20 trials each day, but the number of trials can be varied depending upon the objectives of the experiment. There are wide individual differences in the success achieved by rats and so the objective of baseline training is to ensure that each rat's performance is optimal (10).

2.6. Performance Can Be Represented Using End-Point Measures

The utility of skilled reaching behavior is that it can be evaluated by a number of objective measures, including end-point measures of success. The following scoring terminology describes the scores that are applied in order to establish end-point performance on the task.

1. *Trial*: A trial consists of the approach of an animal from the rear of the cage to the front slot, the movement of the forelimb into the slot in an attempt to grasp the food, the consumption of the food if successful, and the return to the back of the box.

2. *Reach*: A reach is any advance of the hand through the slot.

3. *Hit*: A hit is a trial in which the animal grasps the food pellet and brings it to its mouth for eating. In a trial, an animal may make a number of reaches before achieving a hit (or knocking a food pellet off the shelf).

4. *Single-reach-hit*: A single-reach-hit (or single-reach-success) is a trial in which the animal makes a single reach movement, successfully grasps the food pellet, and transports it to its mouth.

5. *Miss*: A miss is any reach action that fails to grasp the food and transport it to the mouth. The food may be considered a miss if it is ignored, knocked off the shelf, or dropped after being grasped.

Performance can be analyzed in terms of individual animals or groups of animals and can be averaged across the 20 trials of each day's testing or over multiple days of testing. Figure 2 illustrates end-point measures from a group of rats pretrained to reach and then tested for 25 days following a stroke to the forelimb region of the motor cortex contralateral to the preferred hand. The dotted lines on each graph represent baseline prestroke performance. Figure 2a illustrates total success and it can be seen that the stroke initially impairs total success, but by about 24 days poststroke, animals recover to near baseline performance. Following this, performance deteriorates somewhat with continued training. It is proposed that this decline (*), termed "learned bad use," signifies a shift in behavior to a "habit" strategy that is less sensitive to the constraints of reinforcement (11). Figure 2b illustrates single reach successes and shows that recovery on this measure is minimal. Figure 2c illustrates total number of reaches and shows that in terms of reach activity there is a marked increase in activity. The change in success and reach number between the first and second day poststroke is interesting. Animals may initially try reaching poststroke and then quit, a behavior termed "learned nonuse." It is proposed that they need to attempt a reach poststroke in order to learn that they are behaviorally impaired (12). The most important lesson provided by Fig. 2 is that the choice of end-point measure can influence the result obtained (recovery, no recovery, or increased activity). The most accurate profile into the effects of an experimental manipulation can be obtained with multiple end-point measures.

2.7. A Reaching Trial Is Composed of Three Oppositions

A reach is more than the insertion of the forelimb into the slot to grasp a food pellet. It consists of all of the preceding postural adjustments, the support from the triad of nonreaching limbs, and the adjustments of the body associated with bringing the food to the mouth. A reach can be described by all of the body events associated with limb advancement and withdrawal (13). A reaching trial can also be described in terms of three body–pellet relationships or "oppositions" (14). An opposition is a body-to-food target relationship that is independent of variations in an animal's movement. The three oppositions of a reach are illustrated in Fig. 3 and are defined as follows:

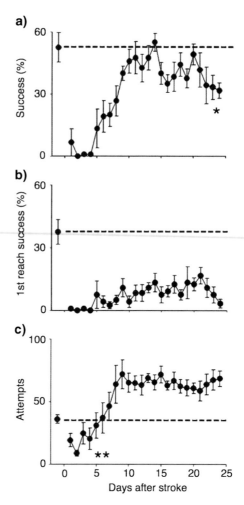

Fig. 2. End-point measures of skilled reaching behavior (mean ± SEM) before and after forelimb motor cortex stroke. (**a**) Total number of successes, (**b**) total number of successes on first reach attempt, and (**c**) total number of reach attempts. Note: Each end-point measure gives a different view of recovery. On total successes, recovery approaches presurgical performance (*dotted line*); on first attempt successes, performance remains severely impaired; and on total reach attempts, performance is enhanced in terms of overall limb use over recovery time. Other features of performance include: (1) learned nonuse, in which performance declines from the first postoperative day to the second postoperative day. (2) Restraint induced therapy, in which success increases over the first postoperative week as the rat is forced by food location to use its affected hand. (3) Performance variability at asymptote. (4) Learned baduse in which success declines with continued rehabilitation.

1. Snout-pellet opposition: The first component of a reach is a snout-pellet opposition that is represented by the trajectory of the rat's snout from the back of the box to the point that it sniffs the food pellet.

2. Hand-pellet opposition: The second component of a reach is a hand-pellet opposition that is represented by the movement of

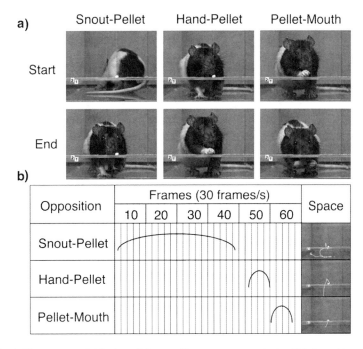

Fig. 3. Three sequential body-pellet oppositions compose a single skilled reaching act. (**a**) Photos illustrating the start and end of each opposition; snout-pellet, hand-pellet, and pellet-mouth. (**b**) *Left panel*: The proportional duration of each opposition in frames per second. Right panel: Movement trajectory of the snout (orientation), hand (transport), and hand (withdraw) during the performance of each opposition.

the animal's hand from the point of its last step to the point that it grasps the food pellet.

3. Pellet-mouth opposition: The third component of a reach is a pellet-mouth opposition that is represented by the trajectory of the food held in a rat's hand to the point that it is grasped by the mouth.

The utility in describing a reach in terms of three oppositions is that oppositions are invariant even while the movements used to achieve each opposition may vary between trials, individual animals, and test sessions. Each opposition can be considered a learned stimulus-response (S-R). The analysis of reach learning and the analysis of recovery of reaching following nervous system injury suggest that reach learning involves both the acquisition of each opposition as well as the acquisition of the sequence of three oppositions. Thus, oppositions are scored across training days such that after successful task acquisition, each trial should consist of the three oppositions performed in order.

2.8. A Reaching Act Is Composed of Four Gestures

A gesture is defined as a single, continuous, nonweight bearing movement of a body part or simultaneous combination of body parts. Gestures may occur smoothly to achieve a single successful

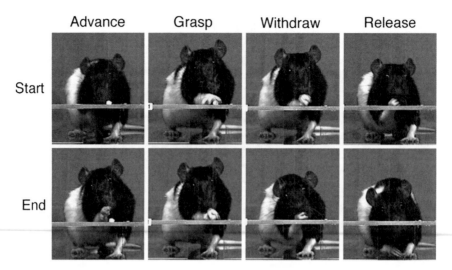

Fig. 4. Four gestures (nonweight-bearing movements) compose a single skilled reaching act. Photos illustrating the start and end of each gesture; Advance, Grasp, Withdraw, Release.

reach, but an animal may also introduce pauses between gestures, repeat gestures, or fail to perform a gesture, suggesting that they are discrete components of the reaching act. An analysis of reach learning suggests that an animal must learn to link successive gestures to perform a reaching act. As is illustrated in Fig. 4 and defined below, a successful reach comprises four gestures (11, 15).

1. *Advance.* The advance gesture begins with the hand lifted from the floor. The forelimb moves through the slot to approach the food.

2. *Grasp.* The grasp gesture begins with the hand pronating over the food target and is completed as the digits flex in an attempt to close around the pellet.

3. *Withdraw.* The withdraw gesture begins with a movement of the forelimb away from the food location to the mouth.

4. *Release.* The release gesture consists of a movement of the hand away from the mouth or the shelf to either the floor or the starting position for the advance.

Table 1 illustrates the scoring procedure for gestures. Note that in a trial a rat may make many different numbers and combinations of gestures on both miss and hit reaches. For a normal rat, the withdraw gesture is very unlikely to occur unless the reach is a hit, but after brain injury to sensory cortex resulting in loss of haptic information from the hand, these gestures may occur on both hit and miss reaches (16).

Table 1
Gesture sheet and exemplar scores

Scenario	Gestures				Total	Outcome
	Advance	Grasp	Withdraw	Release		
1	/	/	/	/	4	Hit
2	/	/		/	3	Miss
3	//	//	/	/	6	Hit
4	/	//		/	4	Miss

"/" indicates one gesture and the frequency with which that gesture is observed per trial
Note: An exemplary reach is composed of only four gestures resulting in a hit

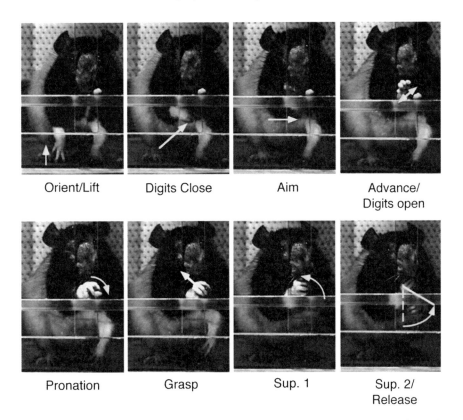

Orient/Lift	Digits Close	Aim	Advance/ Digits open
Pronation	Grasp	Sup. 1	Sup. 2/ Release

Fig. 5. Eleven sequential movement elements compose a single skilled reaching act. Photos illustrating each movement element; Orient, Lift, Digits Close, Aim, Advance, Digits Open, Pronation, Grasp, Supination 1, Supination 2, and Release.

2.9. A Reaching Act Is Composed of 11 Movement Elements

Movement elements are the individual movements of a limb segment around one or more joints. Figure 5 illustrates the 11 movements comprising a reach, which are derived from a conceptual framework obtained from Eshkol-Wachmann Movement Notation (EWMN) (2). Briefly, EWMN describes the relation, and changes in relation, between the parts of the body. The body is treated as a system of jointed axes (i.e., body and limb segments). A limb is

defined as any body part that either lies between two joints or has a
joint and a free extremity. These are imagined as straight lines (axes)
of constant length which move with one end fixed to the center of
a sphere. Using EWMN, the same movements can be notated in
several polar coordinate systems. The coordinates of each system are
determined in reference to the next proximal or distal limb or body
segment, to the animal's body midline axis, and to the environment. By considering the same behavior in multiple coordinate systems, invariances in that behavior may emerge in some coordinate
systems but not others. Thus, the behavior may be invariant in relation to some or all of the following: the animal's longitudinal axis,
gravity, or the next proximal or distal body segment.

The movement elements are difficult to determine directly
from the video records because of a pliable fur coat and musculature of the forelimb, which obscure the exact location and movement of skeletal segments. Nevertheless, a close approximation of
segmental movements can be made from the video record as has
been confirmed by cineradiographic visualization of the skilled
reaching act (17). Analysis of the movement elements of successive
reaches across both individual and groups of normal rats suggests
that the movement is remarkably stable within a rat strain. For this
reason, it is typical to score only the first three successful reaches of
a test session when scoring the movement elements. The scoring
method is illustrated in Table 2, and the movement elements are
determined by the following characteristics:

Table 2
Movement element rating scale

	Normal	Impaired	Absent[a]
1. Orient	0	0.5	1
2. Lift	0	0.5	1
3. Digits close	0	0.5	1
4. Aim	0	0.5	1
5. Advance	0	0.5	1
6. Digits open	0	0.5	1
7. Pronation	0	0.5	1
8. Grasp	0	0.5	1
9. Supination 1	0	0.5	1
10. Supination 2	0	0.5	1
11. Release	0	0.5	1

Note: An exemplary reach would achieve a score of "0" on the
movement elements rating scale.
[a]A mild impairment is indicated by a score of 0.5

1. *Orient*. The rat locates and fixates on the pellet using the nose and usually completes the movement by sniffing the food.

2. *Lift*. The limb is lifted from the floor with the upper arm and the digits are swung to the midline of the body.

3. *Digits close*. As the limb is lifted, the hand takes a collected position in which the digits are semiflexed and the hand is supinated so that the palm faces the midline of the body.

4. *Aim*. Using the upper arm, the elbow is adducted so that the forearm is aligned along the midline of the body with the hand located just under the mouth. This movement involves fixation of the distal portion of the limb so that the digits remain aligned with the midline of the body. This is likely produced by a movement around the elbow that reverses the direction of movement of the hand to compensate for the adduction of the elbow.

5. *Advance*. The head is lifted and the limb is advanced directly forward, above, and beyond the food pellet.

6. *Digits open*. As the limb is advanced, the digits are extended and slightly opened.

7. *Pronation*. Using a movement of the upper arm, the elbow is abducted, pronating the hand over the food. Full pronation of the hand onto the food is aided by a rotation of the hand around the wrist and opening of the digits.

8. *Grasp*. As the pads of the palm touch the food, the food is grasped by closure of the digits. This typically occurs as an independent movement before the hand is withdrawn.

9. *Supination 1*. As the limb is withdrawn, the hand is dorsiflexed and is supinated by 90° by a movement around the wrist and by adduction of the elbow. These movements can occur as soon as the food is grasped and are complete as the hand is withdrawn from the slot.

10. *Supination 2*. As the rat sits back with the food held in the hand, the hand is supinated by a further 90° and ventroflexed to present the food to the mouth.

11. *Food release*. The digits are opened and the food is transferred to the mouth. The food is usually grasped with the teeth or the tongue.

Each of the movements is rated on a 0 to 1 point scale. If the movement appears normal, the movement is given a score of "0," if the movement appears slightly abnormal but is recognizable, the movement is given a score of "0.5," and if the movement is absent or compensated for entirely by movement of a different body part, the movement is given a score of "1."

It is important to note that although a brain injury may impair or abolish any of the movement elements, any element can be substituted for by compensatory movements of other body parts and so scoring of movement elements can provide insights into functional loss, functional recovery, as well as compensatory behaviors (18).

2.10. There Are Six Hand-Shaping Transitions

The trajectory of the hand from its initial position on the floor to the food target, then to the mouth, and finally back to the floor is associated with a number of hand rotations and hand and digit shapes (19). A hand-shaping rating scale is used to describe the changes in hand position and digit shape during the act of reaching (20). During the reach, the palm of the hand moves through rotations of supinate-pronate-supinate-supinate-prontate while at the same time the digits are extended and flexed and opened and closed in various configurations. The sequence of hand-shaping movement is illustrated in Fig. 6, and the details of the shaping movements are described below.

1. *Flex–close.* As the hand is lifted from the floor to the aiming position, the digits flex and close so that the hand and digits are in a "collected" configuration. During the movement, the palm of the hand is supinated by about 90°.

2. *Extend.* As the hand is extended through the slot, the digits are extended. The hand palm remains in a mainly supinated position.

3. *Open.* As the hand pronates over the food pellet in an arpeggio movement with digit 5 followed by 4,3,2 contacting the shelf surface, the digits are opened.

4. *Flex–close.* As the digits or palm of the hand contact the food, the digits are flexed and closed in a grasp to purchase the food. Following the grasp, the palm of the hand is partially supinated.

5. *Extend.* As the hand approaches the mouth, the palm almost fully supinates and the digits are partially extended to release the food from the hand to the mouth, which grasps the food with the incisors or laps the food with the tongue.

6. *Extend–open.* As the hand is lowered toward the surface of the floor, the hand and digits move from a collected position to a manipulatory movement that takes up a stationed position on the floor with the digits extended and opened.

The six hand shape transitions are scored using a 0 to 1 point scale. If the movement is normal a score of "0" is given, if the movement is absent a score of "1" is given, and if the movement is present but abnormal a score of "0.5" is given. Hand rotation and hand shaping are under at least partially independent control because damage to corticospinal pathways can largely abolish hand rotation while leaving some aspects of hand shaping intact (21).

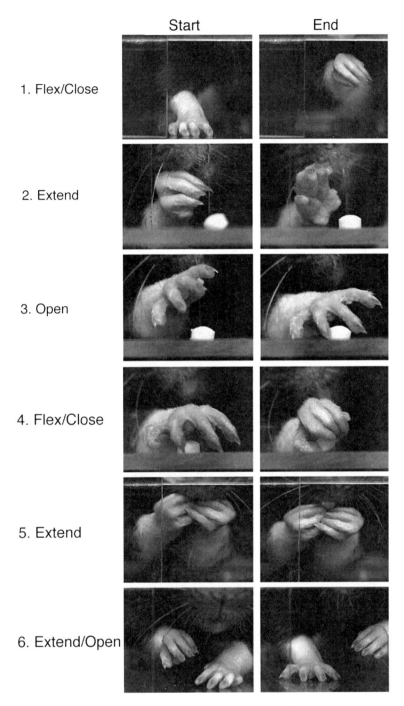

Fig. 6. Six sequential hand shape transitions compose a single skilled reaching act. Photos illustrating the start and end of each hand shape transition. Note: The digits can be Flexed (digits bent toward the palm), Extended (digits straightened away from the palm), Closed (digits adjacent to each other), or Open (digits spread apart from each other).

***2.11. Rats Make
Independent Digit
Movements
in Grasping***

During food grasping, rats do not perform digit movements in which digit 1 (thumb) is placed in opposition to the other digits, as is characteristic of primate precision grips. Nevertheless, the digits are moved somewhat independently in the grasping movement. Studies have not examined whether they similarly move independently in releasing food to the mouth.

Kinematic digit analysis describes the independent movement of individual digits during grasping (22). To capture the individual digit movements, reaching is filmed using a high-speed camera (Photron, Fastcam-ultima APX, Rev. 1.08; High Speed Imaging, http://www.hsi.ca) capable of capturing 1,000 frames/s with the assistance of a cold light source. Images are captured from either the frontal or lateral perspective. To freeze the movements without compromising the quality of the image, the shutter speed is set to 1/1,000 s. Representative reaching movements from high-speed video are captured with Final Cut Express HD (V.3.5; http://www.apple.com) and analyzed frame by frame by motion measurement software, Peak Motus (Version 8; http://www.peakperform.com). The data is acquired in manual mode using the cursor to digitize moving points (i.e., the tips of individual digits).

Figure 7 illustrates the trajectories of each digit (2–5) in the horizontal plane and vertical plane during a normal reach. Trajectories from the horizontal plane characterize the abduction/

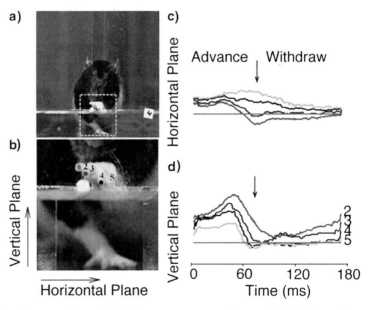

Fig. 7. Rats use independent digit movements to complete a single skilled reaching act. (**a**) Frontal view of a rat completing the single pellet reaching task. (**b**) Magnified view of the reaching hand. The tips of digits 2–5 are digitized and tracked during the reaching action. (**c**) Trajectories from the horizontal plane characterize the abduction/adduction of the digits. (**d**) Trajectories from the vertical plane characterize rotary movements of the digits. Note: The trajectory of each digit is unique and somewhat independent of the other digits.

adduction movements of the digits. Trajectories from the vertical plane characterize rotary movements of the digits. Independent digit movements are quantified using digit coupling and digit individuation indices.

Digit coupling is a measure of how each digit moves in relation to other digits. The coupling index is quantified from the trajectory measures. Spearmen correlation coefficients are calculated between pairs of digits (with one digit as a dependent variable and another as an independent variable, e.g., digits 2,3 or digits 2,5, etc.) for individual reaches. The coupling index for a given digit pair approaches "+1" when the digits move synergistically and approaches "−1" when the digits move independently.

Digit individuation is a measure of how independent a given digit moves relative to another moving digit. Correlation coefficients for pairs of digits are grouped and averaged [(rdigits 2,3 + rdigits 2,4 + rdigits 2,5)/3]. An individuation index approaching "−1" expresses higher independence, whereas an individuation index approaching "+1" indicates lower independence.

3. Discussion and Justifications

As noted in the introduction of this paper, there are a number of ways of measuring the reach-to-eat movement in the rat and in addition there are many other tests for measuring rat forelimb function, including tests of placing, tests of ground reaction forces, and tests of gait in locomotion (23, 24). The single pellet reaching task cannot supplement all of these other tests, still it has a number of experimentally useful features that recommend its use for neurobiological research. The following include a number of justifications that highlight the strengths of the single pellet reaching task.

3.1. Skilled Reaching Is a Member of a Family of Allied Gestures

In both the human and the rat, skilled reaching features a small number of conserved forelimb gestures that are featured in many limb movements, including walking, crawling, crossing a grid, and exploring a vertical surface (25). For all of these movements, the digits take a collected shape as they are lifted from a substrate; this collection position is represented by flexed and closed digits featured in the "lift," "aim," and "release" postures of skilled reaching. For all of these movements, the collected position is maintained as the limb is advanced for action, as is featured by the "advance" in skilled reaching. In all of these movements, the digits are shaped prior to making a manipulatory action as occurs for placing the hand on the floor or the wall, grasping a rung of a ladder, or grasping a target, as occurs for the "grasping" of food in the single pellet task. A distinguishing feature of skilled reaching is that the hand makes the additional movement of withdrawal in which the hand is brought to the mouth to release the food. Finally, the manipulatory movement of replacing the hand on the floor after a reach is a movement that is also featured in tasks of placing and walking. Many brain injuries that impair hand use in grasping food also impair hand placing at the completion of the reaching act (26).

3.2. Skilled Reaching Features a Number of Forelimb Movement Primitives

A number of recent studies support the idea that the motor system is organized into a number of primitive gestures or synergies. Using relatively long trains of electrical stimulation, Graziano (27) in primates and Ramanathan (28) in rats find that motor cortex is organized into a small number of ethologically relevant movements. These are likely linked together and adapted in various ways to perform a wide range of unskilled and skilled movement actions. Skilled reaching by the rat features a number of these movements, including reaching to a target, grasping, withdrawing the limb to the mouth, and sometimes holding and manipulating the food at the midline of the body with both hands for eating. Thus, the skilled reaching task and its various analytical methods also assess a number of anatomically relevant gestures and their integration into a complex act.

3.3. Rat Skilled Reaching Can Be Generalized to Humans

A wide range of animal species in different animal orders use a forelimb for grasping food and bringing it to the mouth for eating. It is not surprising that, as sister clades, rodent and primate reaching might be quite similar. Indeed, the similarities are such that very similar analytical procedures can be used with rodents, nonhuman primates, and humans. Nevertheless, it is important to note a number of differences between rodent reaching and human reaching (29). Humans use vision to locate a target and guide a hand to it, whereas rats use olfaction. Humans and rats both make extensive use of proximal movements in order to move the hand to a target, but freedom of movement of the upper arm of humans is enabled by a ball and socket shoulder while freedom of movement in the rat is enabled by a freely moving scapula tethered only by muscles. Humans pronate the hand by relative movement of the ulnar and radius bones of the lower arm, whereas these bones are fused in the rat. To compensate, the rat can make rotary movements at the wrist, whereas humans cannot. As has been noted, humans also feature advantages in the use of precision grips. These differences aside, the similarities in the movements in rodents and primates justify the rat as an inexpensive and easily used model of human skilled movement.

3.4. Reaching Depends upon a Number of Aspects of Motor Learning

It is well-known that human skilled reaching is dependent upon both maturation and learning. Learning is featured in the acquisition of the skilled reaching task in the rat, as animals must learn the three oppositions of locomoting to the food, advancing a limb to grasp, and bringing the food to the mouth. During task learning, the three oppositions are acquired sequentially and after brain damage they may be unequally affected and may require relearning (14). In addition, in order to account for individual differences in performance by rats, Gholamrezaei and Whishaw (10) have suggested that rats may differentially utilize one of two reaching strategies, a goal strategy in which they reach to a particular location and a habit strategy in which they learn a movement sequence. The importance of recognizing the relative differences in learning strategies is that following brain injury there may be systematic shifts in strategy use. For example, following injury, an animal may become less reliant on goal learning and more reliant on habit learning, and may thus display a deterioration in performance as featured in the "learned baduse" of animals that are overtrained following motor cortex stroke. Learning also influences the performance of animals following nervous system injury. An animal returning to the task for the first time following injury may "learn" that it now has difficulty performing the behavior and so may quit, a change referred to as "learned nonuse." Unless forced to use the affected limb by the constraint of having the food located contralateral to its affected hand or by having a bracelet placed on the wrist of its affected hand (5), the rat may completely neglect the use of its affected limb. The learning component of constraint-induced movement therapy can be documented in the recovery of success over the first two post-operative weeks.

3.5. Skilled Reaching Can Assess Recovery of Function

One of the goals in using animal models is to develop procedures that can restore lost functions following brain injury. A variety of procedures are directed toward that end, including constraint-induced therapies, enriched environments, pharmacological treatments, brain grafts, and stimulation of endogenous neuronal regrowth. One of the strengths of the skilled reaching task is that it allows for multiple measures of performance and thus, minimizes the risk of obtaining false positive results. In order to demonstrate true recovery of function after nervous system injury, it is required that all of the behavioral measures described in the present paper be normalized. No therapeutic manipulation has yet achieved this effect. Nevertheless, recovery after an injury may also be partial, and the numerous measures described here provide sufficient sensitivity to detect partial recovery.

3.6. Skilled Reaching Can Assess Compensation

A finding of the many studies that have developed models of the injured nervous system is that skilled reaching performance and success are remarkably resistant to injury. The analytical procedures described here show that the restitution of performance that is seen after many different types of injury is due to the development of compensatory movements. For example, impairments in hand rotation can be substituted by body rotation. In the absence of

Table 3
Studies investigating skilled reaching impairments in motor conditions of the relevant rat analogues

Condition	Human	Rat
Stroke	(15)	(11, 18, 20, 22, 30–32)
Parkinson's disease	(33, 34)	(35, 36)
Huntington's disease	(37, 38)	(39, 40)
Spinal cord injury	(41–43)	(44–47)

strategies to restore function, strategies to enhance compensation can be very useful in therapy. The measures described here, by emphasizing the normal movements used in reaching, also allow sensitive measurement of compensatory movements and provide insights into how compensation can be enhanced.

3.7. Skilled Reaching Can Model Specific Human Disorders

The methods of studying skilled reaching developed for the rat can be applied almost directly to humans. Using a hand to pick up a piece of food to place it in the mouth for eating is an everyday human act. By having a human subject perform the task in an analogous paradigm allows almost identical analytical procedures for scoring. This can prove useful because preclinical procedures developed in nonhuman animal models can be tested almost identically in human subjects. Table 3 lists four conditions, stroke, Parkinson's disease (PD), Huntington's disease (HD), and spinal cord injury (SCI), in which similar measures have been used to identify similar impairments in the reach-to-eat movement in human subjects and in rat models of the same conditions. What is noteworthy about these studies is that the similarities in behavioral impairments are easy to identify. For this reason, these impairments can be used to index the effectiveness of behavioral treatments.

4. Conclusion

Details of the skilled reach-to-eat movement have been systematically described using end-point, qualitative, and kinematic measures for the rat. The same behaviors and measurement procedures can be used with nonhuman primates and with humans. Similarities in end-point measures, oppositions, movement elements, and hand shaping support the conclusion that skilled reaching is a homologous behavior in rats and humans. For this reason, skilled reaching is a sensitive measure of motor system damage and can be used to determine the validity of rodent models of movement disorders and the efficacy of potential therapeutics for the human conditions. When qualitative movement analysis techniques are combined with end-point measures, the test can dissociate impairment, recovery, and compensation as well as identify the learned adjustments associated with these processes.

Acknowledgments

Supported by the Natural Sciences and Engineering Research Council of Canada.

References

1. Iwaniuk AN & Whishaw IQ (2000) On the origin of skilled forelimb movements. Trends in Neuroscience, 23:327–376.
2. Whishaw IQ & Pellis SM (1990) The structure of skilled forelimb reaching in the rat: a proximally driven movement with a single distal rotary component. Behavioural Brain Research, 41:49–59.
3. Montoya CP, Campbell-Hope LJ, Pemberton KD, Dunnett SB (1991) The "staircase test": a measure of independent forelimb reaching and grasping abilities in rats. Journal of Neuroscience Methods, 36:219–228.
4. Peterson GM (1932) Mechanisms of handedness in the rat. Comparative Psychology Monographs, 9:21–43.
5. Whishaw IQ, O'Connor RB, Dunnett SB (1986) The contributions of motor cortex, nigrostriatal dopamine and caudate-putamen to skilled forelimb use in the rat. Brain, 109:805–843.
6. Ballermann M, Metz GA, McKenna JE, Klassen F, Whishaw IQ (2001) The pasta matrix reaching task: a simple test for measuring skilled reaching distance, direction, and dexterity in rats. European Journal of Neuroscience, 14:27–37.
7. Castro AJ (1972) The effects of cortical ablations on digital usage in the rat. Brain Research, 37:173–185.
8. Whishaw IQ (2005) Prehension. In: Whishaw IQ, Kolb B (Eds.) The behaviour of the laboratory rat: a handbook with tests. Oxford: Oxford University Press, pp. 162–170.
9. Whishaw IQ & Tomie JA (1989) Olfaction directs skilled forelimb reaching in the rat. Behavioural Brain Research, 32:11–21.
10. Gholamrezaei G & Whishaw IQ (2009) Individual differences in skilled reaching for food related to increased number of gestures: evidence for goal and habit learning of skilled reaching. Behavioral Neuroscience, 123:863–874.
11. Alaverdashvili M, Foroud A, Lim DH, Whishaw IQ (2008b) "Learned baduse" limits recovery of skilled reaching for food after forelimb motor cortex stroke in rats: A new analysis of the effect of gestures on success. Behavioural Brain Research, 188:281–290.
12. Erickson CA, Gharbawie OA, Whishaw IQ (2007) Attempt-dependent decrease in skilled reaching characterizes the acute postsurgical period following a forelimb motor cortex lesion: an experimental demonstration of learned nonuse in the rat. Behavioural Brain Research, 179:208–218.
13. Whishaw IQ, Gorny B, Tran-Nguyen LT, Castaneda E, Miklyaeva EI, Pellis SM (1994) Making two movements at once: impairments of movement, posture, and their integration underlie the adult skilled reaching deficit of neonatally dopamine-depleted rats. Behavioural Brain Research, 61:65–77.
14. Gharbawie OA & Whishaw IQ (2006) Parallel stages of learning and recovery of skilled reaching after motor cortex stroke: "oppositions" organize normal and compensatory movements. Behavioural Brain Research, 175:249–262.
15. Foroud A & Whishaw IQ (2006) Changes in the kinematic structure and non-kinematic features of movements during skilled reaching after stroke: a Laban Movement Analysis in two case studies. Journal of Neuroscience Methods, 158:137–149.

16. Gharbawie OA, Gonzalez CLR, Whishaw IQ (2005) Skilled reaching impairments from the lateral frontal cortex component of middle cerebral artery stroke: a qualitative and quantitative comparison to focal motor cortex lesions in rats. Behavioural Brain Research, 156:125–137.

17. Alaverdashvili M, Leblond H, Rossignol S, Whishaw IQ (2008a) Cineradiographic (video X-ray) analysis of skilled reaching in a single pellet reaching task provides insight into relative contribution of body, head, oral, and forelimb movements in rats. Behavioural Brain Research, 192:232–247.

18. Whishaw IQ (2000) Loss of the innate cortical engram for action patterns used in skilled reaching and the development of behavioral compensation following motor cortex lesions in the rat. Neuropharmacology, 39:788–805.

19. Sacrey LA, Alaverdashvili M, Whishaw IQ (2009) Similar hand shaping in reaching-for-food (skilled reaching) in rats and humans provide evidence of homology in release, collection, and manipulation movements. Behavioural Brain Research, 204:153–161.

20. Gharbawie OA, Auer RN, Whishaw IQ (2006) Subcortical middle cerebral artery ischemia abolishes the digit flexion and closing used for grasping in rat skilled reaching. Neuroscience, 137:1107–1118.

21. Whishaw IQ, Sarna JR, Pellis SM (1998) Evidence for rodent-common and species-typical limb and digit use in eating, derived from a comparative analysis of ten rodent species. Behavioural Brain Research, 96:79–91.

22. Alaverdashvili M & Whishaw IQ (2008c) Motor cortex stroke impairs individual digit movement in skilled reaching by the rat. European Journal of Neuroscience, 28:311–322.

23. Woodlee MT, Asseo-Garcia AM, Zhao X, Liu SJ, Jones TA, Schallert T (2005) Testing forelimb placing "across the midline" reveals distinct, lesion-dependent patterns of recovery in rats. Experimental Neurology, 191:310–317.

24. Muir GD, Whishaw IQ (1999) Ground reaction forces in locomoting hemi-parkinsonian rats: a definitive test for impairments and compensation. Experimental Brain Research, 126:307–314.

25. Whishaw IQ, Travis SG, Koppe SW, Sacrey LA, Gholamrezaei G, Gorny B (2010a) Hand shaping in the rat: Conserved release and collection vs. flexible manipulation in overground walking, cylinder exploration, and skilled reaching. Behavioural Brain Research, 206:21–31.

26. Schallert T, Fleming SM, Leasure JL, Tillerson JL, Bland ST (2000) CNS plasticity and assessment of forelimb sensorimotor outcome in unilateral rat models of stroke, cortical ablation, parkinsonism and spinal cord injury. Neuropharmacology, 39:777–787.

27. Graziano M (2006) The organization of behavioral repertoire in motor cortex. Annual Review of Neuroscience, 29:105–134.

28. Ramanathan D, Conner JM, Tuszynski MH (2006) A form of motor cortical plasticity that correlates with recovery of function after brain injury. Proceeding from the National Academy of Sciences, USA. 103:11370–11375.

29. Whishaw IQ, Pellis SM, Gorny BP (1992) Skilled reaching in rats and humans: evidence for parallel development or homology. Behavioural Brain Research, 47:59–70.

30. Alaverdashvili M & Whishaw IQ (2010) Compensation aids skilled reaching in aging and in recovery from forelimb motor cortex stroke in the rat. Neuroscience, 167:21–30.

31. Gharbawie OA, Karl JM, Whishaw IQ (2007) Recovery of skilled reaching following motor cortex stroke: do residual corticofugal fibers mediate compensatory recovery? European Journal of Neuroscience, 26:3309–3327.

32. Whishaw IQ, Pellis SM, Gorny BP, Pellis VC (1991) The impairments in reaching and the movements of compensation in rats with motor cortex lesions: an endpoint, videorecording, and movement notation analysis. Behavioural Brain Research, 42:77–91.

33. Whishaw IQ, Suchowersky O, Davis L, Sarna J, Metz G, Pellis SM (2002) Impairment of pronation, supination, and body co-ordination in reach-to-grasp tasks in human Parkinson's disease (PD) reveals homology to deficits in animal models. Behavioural Brain Research, 133:165–175.

34. Doan J, Melvin KG, Whishaw IQ, Suchowersky O (2008) Bilateral impairments of skilled reach-to-eat in early Parkinson's disease patients presenting with unilateral or asymmetrical symptoms. Behavioural Brain Research, 194:207–213.

35. Metz GA & Whishaw IQ (2002) Drug-induced rotation intensity in unilateral dopamine-depleted rats is not correlated with end point or qualitative measures of forelimb or hindlimb motor performance. Neuroscience, 111:325–336.

36. Miklyaeva EI, Castaneda E, Whishaw IQ (1994) Skilled reaching deficits in unilateral dopamine-depleted rats: impairments in movement and posture and compensatory adjustments. The Journal of Neuroscience, 14:7148–7158.

37. Ho AK, Manly T, Nestor PJ, Sahakian BJ, Bak TH, Robbins TW, Rosser AE, Barker RA (2003) A case of unilateral neglect in Huntington's disease. Neurocase, 9:261–273.

38. Klein A, Sacrey LR, Dunnett SB, Whishaw IQ, Nikkhah G (2010) Proximal movements compensate for distal movement impairments in a reach-to-eat task in Huntington's disease: New insights into motor impairments in a real-world skill. Neurobiology of Disease, (accepted with revisions).

39. Whishaw IQ, Zeeb F, Erickson C, McDonald RJ (2007) Neurotoxic lesions of the caudate-putamen on a reaching for food task in the rat: acute sensorimotor neglect and chronic qualitative motor impairment follow lateral lesions and improved success follows medial lesions. Neuroscience, 146:86–97.

40. Fricker-Gates RA, Smith R, Muhith J, Dunnett SB (2003) The role of pretraining on skilled forelimb use in an animal model of Huntington's disease. Cell Transplantation, 12:257–264.

41. Koshland GF, Galloway JC, Farley B (2005) Novel muscle patterns for reaching after cervical spinal cord injury: a case for motor redundancy. Experimental Brain Research, 133–147.

42. Laffont I, Briand E, Dizien O, Combeaud M, Bussel B, Revol M, Roby-Brami A (2000) Kinematics of prehension and pointing movements in C6 quadriplegic patients. Spinal Cord, 38:354–362.

43. Treanor WJ, Moberg E, Buncke HJ (1992) The hyperflexed seemingly useless tetraplegic hand: a method of surgical amelioration. Paraplegia, 30:457–466.

44. Kanagal SG, Muir GD (2007) Bilateral dorsal funicular lesions alter sensorimotor behaviour in rats. Experimental Neurology, 205: 513–524.

45. Krajacic A, Ghosh M, PUentes R, Pearse DD, Fouad K (2009) Advantages of delaying the onset of rehabilitative reaching training in rats with incomplete spinal cord injury. European Journal of Neuroscience, 29:641–651.

46. Krajacic A, Weishaupt N, Girgis J, Tetzlaff W, Fouad K (2010) Training-induced plasticity in rats with cervical spinal cord injury: effects and side effects. Behavioural Brain Research, 214:323–331.

47. McKenna JE & Whishaw IQ (1999) Complete compensation in skilled reaching success with associated impairments in limb synergies, after dorsal column lesion in the rat. The Journal of Neuroscience, 1885–1894.

Chapter 7

High-Throughput Mouse Phenotyping

Sabine M. Hölter and Lisa Glasl

Abstract

Here, we describe the systematic mouse phenotyping approach of the German Mouse Clinic (GMC), that works as an open-access phenotyping platform, and of the European Mouse Disease Clinic, which is an EU-funded multi-centre project characterising mutants generated by the large-scale mouse mutagenesis project European Conditional Mouse Mutagenesis Program. We explain the aims and the general framework of these large-scale projects and the resulting consequences for the phenotyping strategies. Then, we focus on the description of the behavioural tests used in the GMC to detect motor and non-motor symptoms in mouse mutants that are genetic models of human movement disorders or neurodegenerative diseases.

Key words: Behavioural phenotyping, Mouse, Functional genomics, Comprehensive screening, Motor abilities, Parkinson's disease, Gait, Non-motor symptoms

1. Introduction

1.1. Why High Throughput?

With the completion of the genome sequences of human, mouse, rat and several other species, the mouse became the most used model organism in the endeavour to develop a complete functional annotation of the human genome and to employ this information to better understand human disease and its underlying physiological and pathological basis (1–3). Large-scale systematic mutagenesis efforts are underway to generate null and conditional mutations for every coding gene of the mouse genome, e.g. in the framework of the European Conditional Mouse Mutagenesis Program (EUCOMM; (4) http://www.eucomm.org) and the International Knockout Mouse Consortium (IKMC; (5) http://www.knockoutmouse.org).

As a consequence, the demand for phenotyping facilities for the systematic analysis of mouse mutants is increasing (6, 7).

Emma L. Lane and Stephen B. Dunnett (eds.), *Animal Models of Movement Disorders: Volume I*, Neuromethods, vol. 61, DOI 10.1007/978-1-61779-298-4_7, © Springer Science+Business Media, LLC 2011

"Systematic" and "high throughput" means in the context of large-scale phenotyping: comprehensive and at the same time cost- and time-efficient "screening" for phenotypes. Thus, large-scale mouse phenotyping facilities, the so-called mouse clinics, should provide analysis of, ideally, all organ systems of the mouse and must have a standardised and efficient workflow. There are two major arguments for the comprehensive analysis approach: first, most genes have pleiotropic effects and affect more than one body system, which might be overlooked in focused investigations of specific biological processes of interest (8). Second, both from an economical and reduction of animal numbers point of view, one might as well get as much information as possible out of a cohort of mouse mutants, once all the efforts of generating and breeding them have been invested.

1.2. German Mouse Clinic

In 2001, the German National Genome Research Network (NGFN) funded the first infrastructure for large-scale mouse phenotyping at the Helmholtz Zentrum Munich (formerly named GSF) to accelerate the identification and analysis of new mouse models for human diseases. This phenotype assessment centre was called "German Mouse Clinic" (GMC). The director, Martin Hrabé de Angelis, got together clinical and experimental experts from the fields of allergy, behaviour, cardiovascular, clinical chemistry, dysmorphology, energy metabolism, eye, immunology, lung function, molecular phenotyping, neurology, nociception, pathology and steroid metabolism to set up the different GMC screens under one roof (http://www.mouseclinic.de). The strongly interacting team of screeners decided on a hierarchical phenotyping strategy to save on time and costs: the first step is the so-called primary screen through which all mice of a mutant line, comprising equal numbers of mutants and littermate controls of both sexes, are run. Each GMC screen performs one or more fast and effective tests in the primary screen, together providing a comprehensive overview of the phenotypes of a mutant line. Each screen chose its test(s) for the primary screen with the goal of being sensitive enough to not miss a phenotype relevant to its disease area (9). At this time, the GMC behavioural screen had chosen the modified Hole-Board Test (10) and the GMC neurological screen the SHIRPA protocol (11) as its primary screen tests.

The second step are the so-called secondary or tertiary investigations, which are more specific, but also more costly and time-consuming (i.e. the term "screen" is not really appropriate in this context). If an interesting phenotype is detected in the primary screen or if there are other scientific reasons to investigate more in a particular field, each GMC screen offers more in-depth analysis with specialised methods (8, 12). For such secondary screens, a new cohort of a mutant line must be provided, and this cohort then runs only through the specific secondary screen(s) agreed with the mouse provider.

The GMC runs as an open-access phenotyping platform on a collaborative basis (9). This means that any scientist worldwide who is willing to breed an appropriate cohort of mice, ship it to the GMC and publish potential results together with the GMC screeners generating these results can request a "slot" for his or her mutant mouse line(s) through the GMC Web page. This has been used by many scientists so far, and the waiting list is still exceeding capacity (7).

1.3. EUMORPHIA and EUMODIC

In 2002, the European Commission started the project European Union Mouse Research for Public Health and Industrial Applications (EUMORPHIA; http://www.eumorphia.org) under framework program 5 (FP5) which comprised 18 research centres in 8 European countries, the GMC being one of them. The main focus of EUMORPHIA was to incorporate the scientific expertise and resources of many European mouse genetics centres in the development of novel approaches in phenotyping, mutagenesis and informatics to improve the characterisation of mouse models for understanding human molecular physiology and pathology. Similar to the modular GMC screen structure, EUMORPHIA was organised in work packages for each body system or disease field. In each work package, the members discussed and worked together on finding the best and most efficient methods for mouse phenotyping with respect to reproducibility of results and streamlining of testing procedures, and developed Standard Operating Procedures (SOPs) (13). When this project ended in 2006, the consortium had created:

- European Mouse Phenotyping Resource of Standardised Screens (EMPReSS), a public database of the SOPs generated by the EUMORPHIA scientists that can be used to phenotype a mouse (http://www.empress.har.mrc.ac.uk). ·

- EuroPhenome (http://www.europhenome.org), an open-source project to develop a software system for capturing, storing and analysing raw phenotyping data from SOPs contained in EMPReSS (14, 15).

- European Mouse Disease Clinic (EUMODIC, http://www.eumodic.org), a new project funded by the European Commission under FP6 to generate phenome data on 500–650 mutant mouse lines using the EMPReSS SOPs.

As a result of this European cooperation, the GMC adapted its primary screen workflow and tests according to the EMPReSS SOPs (16, 17). The GMC behavioural screen now performs the Open Field (OF) test and Prepulse Inhibition of the Acoustic Startle Reflex in the primary screen, and the GMC neurological screen the modified SHIRPA protocol, the Grip Strength and the Accelerating Rotarod test.

1.4. The Future

The next step is the International Mouse Phenotyping Consortium (IMPC) (18, 19). This consortium is currently forming and includes the European Mouse Clinics involved in EUMODIC as well as Asian, Australian and North American Mouse Clinics. Its aim is the primary phenotyping of mutant mouse lines on identical genetic background for all of the approximately 20,000 genes in the mouse genome, which are generated by the partners of the IKMC. In the course of the IMPC formation, the choice of tests, procedural details and also the primary screen workflow of a mouse clinic will be discussed and optimised again, probably largely based on the experiences gained in the EUMODIC project.

2. GMC Behavioural Screen Choice of Tests for Animal Models of Movement Disorders

2.1. General Issues

Due to the high-throughput assignment of the GMC, the choice of the GMC behavioural screen for tests to analyse animal models of movement disorders follows the rationale of efficiency and efficacy: only as much as necessary to clearly characterise a phenotype. The first priority for the choice of a test is the availability of a protocol that is known to reliably detect the expected phenotype(s). Second priorities are cost and space requirement of the equipment and time needed to test a cohort of mice. Ease of performance, extent of necessary experimenter training and ease of cleaning equipment to maintain the hygiene conditions necessary in a large-scale animal facility are third-level priorities.

To analyse a mouse line specifically for movement disorders, which would be considered a secondary investigation in the GMC, the behavioural screen starts with the Open Field test (if a primary screen was not performed before on this mouse line) to first get a baseline measure of spontaneous locomotor and exploratory activity in a novel environment. Then, we continue with an automated gait analysis because this is also a short test that does not involve much handling of the animals, followed by beam walking, ladder, vertical pole test and accelerating Rotarod to check motor coordination and balance. We also test the animals' muscular strength using a grip strength meter. Since many motor-disabling diseases are neurodegenerative and accompanied by non-motor symptoms, like impairment of olfactory function (in particular Parkinson's disease, but also Huntington's and Alzheimer's diseases (20, 21)) and cognitive decline, on mouse models of such diseases, we also apply tests for olfactory function and a social discrimination and an object recognition test for the assessment of social and object recognition memory. Our test protocols for all of these tests are described below.

In general, we choose the order of the tests such that the least stressful baseline tests, that involve a minimal amount of handling, are done first because it is known that such tests can be affected by handling and experimental experience (22). Thus, the more handling of the mice is involved in an experimental protocol, the later it is performed in the series of tests; i.e. we usually perform our olfactory paradigm last.

The temperature, humidity, ventilation, noise intensity and lighting intensity in the animal rooms must be maintained at levels appropriate for mice. Environmental factors, like diet, animal husbandry, housing conditions or cage structure, can also affect test results, therefore keeping these factors constant during a series of tests is recommended. In particular, changing cages shortly before testing animals should be avoided because cage changes arouse the animals. To allow for acclimatisation, mice should be brought to the testing room at least 10 min, better 30 min, before the start of an experiment. Importantly, for each test, mutants and controls should always be tested concurrently or alternatively to control for circadian rhythms of many physiological and biochemical parameters or any other potentially confounding effects of testing day (e.g. experimenter performance). After each run, all equipment exposed to a mouse must be cleaned carefully with a disinfectant not only to erase olfactory traces, but also to maintain hygiene conditions.

The majority of behavioural parameters vary according to age, sex and strain, i.e. genetic background. It is important to keep this in mind when designing experiments and interpreting results.

2.2. Open Field

2.2.1. Rationale

The Open Field test, used once as a novel environment, creates a non-conditioned conflict situation between the natural tendencies of mice to explore on the one hand and to avoid open spaces, in this case the centre of the Open Field, on the other hand. Hence, it allows an evaluation of exploratory drive, reactivity to novelty and emotionality (23–26).

2.2.2. Method

The Open Field test is carried out according to the standardised phenotyping screens developed by the EUMORPHIA partners and available at http://www.empress.har.mrc.ac.uk.

The ActiMot system test apparatus from TSE is a square-shaped frame with two pairs of light beam strips, each pair consisting of one transmitter strip and one receiver strip (Fig. 1). These basic light barrier strips are arranged at right angles to each other in the same plane to determine the X and Y coordinates of the animal, and thus its location (XY frame). Each strip is equipped with 16 infrared sensors with a distance between adjacent sensors of 28 mm. With two further pairs of unidimensional light barrier strips (Z1 and Z2), rearing can also be detected. The light barriers are scanned with a frequency of 100 Hz each on fast computer platforms. Whenever an even number of light beams are interrupted, the centre of gravity is calculated to lie between adjacent sensors.

Fig. 1. Test arena for open field test.

The test apparatus, where the mouse is placed, consists of a transparent and infrared light-permeable acrylic test arena (inner measures: $45.5 \times 45.5 \times 39.5$ cm) with a smooth floor, coloured light grey. The illumination levels are set at approximately 150 lux in the corners and 200 lux in the middle of the test arena.

At the beginning of the experiment, all animals are transported to the test room and left undisturbed for at least 30 min before the testing starts. Then, each animal is placed individually into the middle of one side of the arena facing the wall and allowed to explore it freely for 20 min. After each trial, the test arena is cleaned carefully with a disinfectant.

For data analysis, the arena is virtually divided by the software into two areas, the periphery defined as a corridor of 8 cm width along the walls and the remaining area representing the centre of the arena (41% of the total arena in our TSE system). The following parameters are calculated by the software: distance travelled, resting and permanence time as well as speed of movement for the whole arena, the periphery and the centre. Additionally, rearing frequency, percentage distance travelled and percentage time spent in the centre as well as the latency to first entry in centre and centre entry frequency are calculated. The time course of distance travelled, rearing frequencies as well as percentage distance travelled and percentage time spent in the centre are additionally analysed in 5-min intervals.

Alterations in the exploration of the exposed, central area are considered to reflect alterations in emotionality/anxiety. Reductions in spontaneous rearing activity can indicate balance and/or motor problems, and in such cases usually coincide with deficits in performance on the Accelerating Rotarod (see below). Reductions in distance travelled or speed of movement can indicate motor problems, but could also reflect other alterations, e.g. in motivation or in sensory modalities.

2.2.3. Pitfalls

Activity levels and exploratory behaviour in the Open Field depend on many factors, e.g. the shape, size and illumination of the arena, presence or absence of walls or floor texture (23, 27). Absolute values measured can also depend on the sampling rate of the equipment devices. Any stress before, as well as noise or any other disturbances during the test session, must be avoided to really get a baseline measurement of spontaneous activity. This usually requires tight control of activities outside of the testing room.

Most labs run several mice in the Open Field test in parallel, with multiple Open Fields situated in one room. In the GMC, we currently run two Open Field tests in parallel, with each Open Field apparatus situated in an approximately 1×1-m small room so that each Open Field apparatus is individually enclosed, i.e. the animal is isolated, and thus more sheltered from any visual, acoustic and olfactory stimuli during testing. We found that increases in Open Field activity induced by restraint stress are more reproducible if parallel tests are run in individualised Open Field rooms, rather than in one big room (Annemarie Wolff-Muscate, unpublished observations).

2.3. Gait Analysis

2.3.1. Rationale

Disturbances or problems with gait are characteristic features of several neurodegenerative disorders, like Parkinson's disease, Huntington's disease or amyotrophic lateral sclerosis. Walking includes important aspects, like coordination, balance and proprioception and there are some (genetic and pharmacological) mouse models of neurodegenerative disorders that replicate particular aspects of the disease, including gait disturbances (28–31). For a detailed analysis of motor coordination, mice are tested in the automated gait analysis system "CatWalk" (Fig. 2; Noldus, Wageningen).

2.3.2. Method

For gait analysis, the mice ambulate in a dark room over an elevated glass walkway (plastic walls, width 8 cm, length 100 cm), which is illuminated along the long edge with a fluorescent tube. The light is completely internally reflected until the mouse paw makes contact with the glass. The light escapes, is scattered and the contact

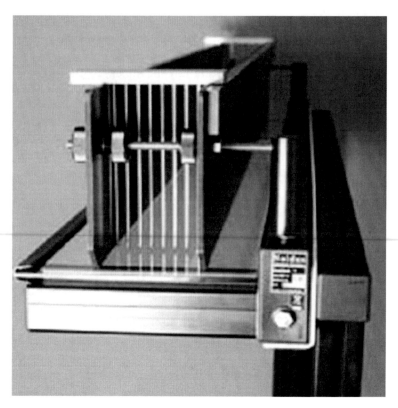

Fig. 2. CatWalk apparatus (see also http://www.noldus.com).

area becomes visible. The brightness of a print represents the pressure exerted on the glass floor by this paw. A camera (Pulnix Camera RM-765) situated below the middle of the walkway tracks the footprint pattern that is converted by a frame grabber into a digital image, which can be analysed with the "CatWalk" software Version 7.1.

Data are collected without knowledge of group assignment in at least three uninterrupted runs per animal. To circumvent variability based on the acceleration and deceleration caused by the initiation and ending of walking, we record a field of view of 40 cm in the middle part of the walkway to get four- to eight-step cycles during the middle of each pass. The recorded runs can be analysed in different ways. For a qualitative description, the walking pattern can be visualised in a manner similar to the print marks a mouse would leave by running over a paper with its paws dipped in ink. Moreover, a time-based gait diagram, representing the duration of contacts with the glass plate (stand), is available. The footfall patterns window shows the order in which the paws are placed and indicates the corresponding step sequence categories. Additionally, the print intensity curves for each single paw placement are shown in a graphical feature.

For a quantitative analysis, the CatWalk software automatically calculates a wide number of parameters in several categories: parameters related to individual footprints, for example the width and length of a paw print, the duration of contact and the pressure exerted by a paw. In order to correct for body weight differences, values are also expressed as ratios between contralateral paws. Parameters related to the position of footprints are, for example, stride length and base of support (average width between either the front paws or the hind paws). Parameters related to time-based relationships between footprints are, for example, footfall patterns and phase dispersions. Max contact at (%) reflects the time of maximum contact at which the largest part of a paw contacts the glass plate during the total time of the contact, relative to the duration of the stand phase of an individual paw placement and can be seen as the point where the braking phase turns into the propulsion phase.

To quantify the time-based relationships between footprints, the CatWalk software uses the correspondence of the initial contacts between the paws to calculate the interpaw coordination with phase dispersions (PhDs) and couplings. To describe the temporal relationship between the placement of two paws, the timing between the initial contacts of an anchor to a target paw is expressed as the percentage of the total step cycle time, which is the time between two initial contacts of a paw. For diagonal (right fore paw–left hind paw (RF–LH) and LF–RH) and ipsilateral (RF–RH; LF–LH) pairs, the forepaw is designated as the anchor of the reference paw, and for girdle (LF–RF; LH–RH) pairs, the right paw is designated as the anchor. In rodents, for the diagonal pairs, the target paw normally runs synchronously with the anchor paw, so the initial contacts occur at the same time resulting in a PhD value around 0%. The PhD value for girdle pairs, paws that alternate, usually results in approximately 50% when walking with a moderate speed. A positive PhD value indicates that the target paw makes initial contact after that of the anchor paw. A negative PhD value indicates that the initial contact of the target paw preceded that of the anchor paw. When PhD exceeds 75%, the target paw is more closely associated with the next anchor paw. Therefore, in that case, the target is assigned to the next anchor, leaving the previous anchor without a target, resulting in a PhD of –25%. Couplings are computed in the same way as PhD. However, in the case of couplings, a target paw can never precede an anchor paw. Therefore, the value of couplings ranges between 0 and 100.

For a more detailed description of the Catwalk method, see (32).

2.3.3. Pitfalls

To get nice footprints without disturbing signals produced by dirt, it is essential to clean the glass plate accurately with a good glass cleaner. Since urine and faeces produce also signals, it is sometimes necessary to clean the glass floor several times while testing the same mouse. We do not train the mice in the CatWalk before testing, since we want to see their spontaneous walking behaviour, and in our experience this is not necessary to get appropriate runs. Normally, mice start to run repeatedly from one side to the other without any interruptions after they have explored the gangway with a few slower crossings. Mice that are less active, very old or anxious often need a bit more time to acclimate and perform good runs. If it is not possible to get enough good runs, the mouse can be tested again at another time point. Testing within a more active phase of the day is normally sufficient. Very active or aroused mice tend to run very fast in the early crossings. These trials should not be taken for analysis.

2.4. Beam Walking

2.4.1. Rationale

The Beam Walking test is widely used to examine sensorimotor function in rodents. In this test, the animals have to traverse an elevated beam and their performance is observed and rated. The method is well-established to assess and follow up sensorimotor coordination after neuronal injuries, like stroke (33, 34), ablation (35, 36) or pharmacological treatment (37). But also the influence of training or aging on sensorimotor function has been studied in rodents with or without injury, as well as in genetic and pharmacological disease models (30, 38–40).

2.4.2. Method

Mice have to traverse a 1-m wooden round beam (15 mm diameter) that is elevated 14 cm above the bench attached with tape to a neutral cage as the starting point and the mouse's home cage as the goal box (Fig. 3). The Beam Walking test starts with a short training period, in which the animal is repeatedly (about five times) placed on the beam to learn to traverse the beam at a single blow. Initially, the animal is allowed to start on a position on the beam directly next to the home cage so that the mouse finds the way home quite easily. In the successive training trials, the distance from the start position to the home cage is increased stepwise until the full length is achieved. During the first training trials, some mice have to be encouraged by carefully nudging and pushing, but normally they rapidly learn to traverse the beam. After a recovery time of about 5 min, three to five testing trials are conducted with 5-min intertrial intervals (ITIs). The mice are placed at one end of the beam and have to walk over the full length of the beam to their home cage. The time needed to traverse the beam and the number of foot slips (one or both hind or front limbs slipped from the beam) and number of falls off the beam are measured. If mice fall, the timer is paused, the mice returned to the position on the beam they fell from and the timer restarted. In addition, in particular when video recording the beam crossings, other aspects like the number of steps or body posture can be analysed as well.

2.4.3. Pitfalls

In the case that a mouse does not start walking on the beam, it sometimes helps to pull it slightly on the tail (backwards). Some animals (very fat or very old) are not able to walk over the beam

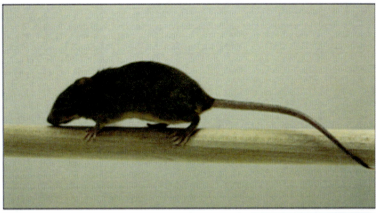

Fig. 3. Beam walking. The *upper picture* shows normal performance. The *lower picture* depicts a hind paw slip.

properly, making an excessive number of slips or showing an ungainly kind of gait, lying flat on the beam with its hind legs clasped around it, stretching and drawing the body forwards just by the front limbs. Such a trial would pass with a maximum number of slips recorded, calculated by the mean number of normally executed steps on the beam with the hind paws (i.e. 20). But while training, you should try to show/teach such a mouse how to walk over the beam with all four paws by bringing the body in the right position and supporting the walk by holding it by the tail. The amount of training trials should be the same for all mice.

2.5. Ladder

2.5.1. Rationale

Animals are required to walk along a horizontal ladder on which the spacing of the rungs is variable to assess skilled walking, limb placement and limb co-ordination (Fig. 4). Walking across the ladder needs precise paw placement and grasping movements. The irregular rung arrangement demands adjustment of paw placement, stride length and also of the regular limb co-ordination. Additionally, each footfall requires control of the body weight support to immediately correct for any paw misplacement. Animals with

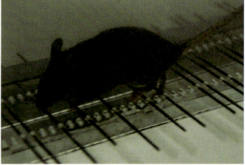

Fig. 4. Ladder with irregular rungs.

deficits in the motor system have problems to adjust to the irregular spacings, whereas normal animals are able to do this within a few trials. The ladder walking test has been shown to be sensitive to chronic movement deficits after adult and neonate lesions to the motor system, including rat models of stroke (41–44), Parkinson's disease (45, 46) and spinal cord injury (47, 48).

2.5.2. Method

The ladder apparatus is a 1-m long horizontal ladder (5 cm wide), enclosed with transparent walls (20 cm high) with irregular spacings (0.5–2 cm) between the rungs that become wider at the end of the gangway. All animals have to cross the ladder in the same direction. For the analysis, the time the animal needs to traverse the ladder is recorded and the number of front and hind paw misplacements is counted. The number of stops and the time an animal is stationary can be included in the measurement too.

2.5.3. Pitfalls

For the ladder test, training is not usually necessary, as mice normally start to run over the ladder more or less uninterrupted when placed in the start area. If a mouse does not start walking over the ladder, it sometimes helps to pull it *slightly* on the tail (backwards) before it is allowed to start. After a few runs, many mice start to explore the apparatus by poking their nose through the rungs or turning around midway (which should be prohibited), making it impossible to get valid data. Some mice show this behaviour in the first trials and afterwards begin to cross the ladder uninterrupted. Therefore, we collect data from the first exposure onwards. Afterwards, we decide for the entire cohort of mice if these first trials count as training or testing trials.

2.6. Vertical Pole Test

2.6.1. Rationale

The Vertical Pole test is used to test the motor co-ordination during turning behaviour in mice (Fig. 5). The test was developed to assess basal ganglia-related movement disorders and is regarded to be highly sensitive to nigrostriatal dysfunction (28, 29, 49–52). MPTP-treated mice display slower times compared to controls, and the impairments can be reversed by L-DOPA (49–51).

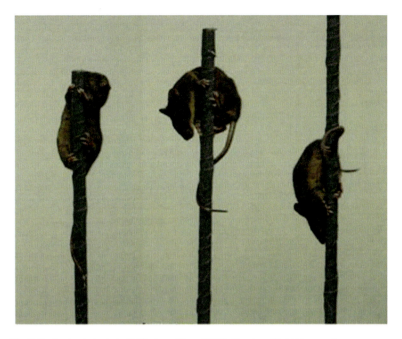

Fig. 5. Vertical pole test. *Left*: Start position. *Middle*: Turning. *Right*: Descending.

2.6.2. Method

The apparatus consists of a 50-cm high, taped pole (diameter 1 cm). When the animals are placed head upwards on the pole, they orient themselves downward and descend the length of the pole to the ground. To bring the mouse in the right start position, the pole is tilt over nearly horizontally. While putting it up, the mouse's tail has to be held fixed on the pole by the experimenters' fingers; otherwise, the mouse starts turning before the vertical position is reached.

The time to turn downwards and the time to descend are measured and according to the performance of the mouse, each trial is scored on a scale of 5–0: 5 = turning on top, 4 = turning below midway, 3 = descending sideways, 2 = backwards, 1 = falls/springs, 0 = stays over 60 s on top. With ITIs of 5–10 min, the mice receive two training trials and three to five test trials.

2.6.3. Pitfalls

Since we frequently test older mice, we fixed at the base of the pole a foam rubber to prevent the mice from injuries by falling down. To keep up the mouse's motivation for the task, it should be brought back to the home cage directly after it has reached the base. Normally, the mice start to orient themselves downwards immediately after the experimenter's hand is taken off the tail, so the experimenter has to be very quick starting the timer. When the mouse does not start immediately, the experimenter has to be aware of the moment when the mouse starts to turn and to note this time point so that the exact duration of the turning movement is measured as well as the latency to turn, which can be an indication for an akinetic phenotype.

If the mice, especially older or overweight ones, do not start with the downward orientation by themselves, they sometimes need to be shown how to do this and the amount of training trials can be increased. For this, the experimenter moves the tail of the mouse to one side (try left and right!), whereupon the majority of mice start to turn their body to the opposite side and learn after few trials to do the task. If this is not sufficient or the mouse cannot hold on tight to the pole while turning, it can be further supported by guiding slightly on the tail.

2.7. Accelerating Rotarod

2.7.1. Rationale

The Rotarod test is used to assess motor co-ordination and balance in rodents (16, 53, 54). Mice have to keep their balance on a rotating rod, and that requires a variety of proprioceptive, vestibular and fine-tuned motor abilities. The time (latency) it takes the mouse to fall off the rod rotating at different speeds or under continuous acceleration (e.g. from 4 to 40 rpm) can be measured.

2.7.2. Method

Motor coordination and balance are assessed using the Rotarod apparatus from Bioseb (Fig. 6; Letica LE 8200) with an accelerating speed of the rotating rod from 4 to 40 rpm in 300 s. The rod diameter

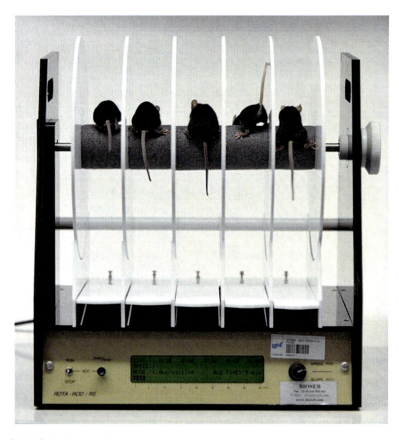

Fig. 6. Rotarod apparatus.

is approximately 4.5 cm made of hard plastic material covered by soft rubber foam (cut from insulation material to cover water pipes) with a lane width of approximately 5 cm.

The test phase consists of three trials separated by 15-min ITIs. On each trial, three mice are placed on the rod leaving an empty lane between two mice. The rod is initially rotating at 4 rpm constant speed to allow positioning of all mice in their respective lanes. Once all mice are positioned, the trial is started and the rod accelerated from 4 to 40 rpm in 300 s. Latency and rpm at which each mouse falls off the rod are measured. Passive rotations are counted as a fall and the mouse is immediately carefully removed from the rod. After each trial, the apparatus is disinfected and let dry. The accelerating Rotarod protocol is carried out according to the standardised phenotyping screens developed by the EUMORPHIA partners and available at http://www.empress.har.mrc.ac.uk.

2.7.3. Pitfalls

We perform the Rotarod test without training trials, but sometimes it is necessary to show or motivate the mice (especially old mice) in the beginning to stay on the rotating rod. This can be done by slightly pushing or lifting up the mouse while the rod rotates with constant speed before starting the increasing speed modus.

2.8. Grip Strength

2.8.1. Rationale

The Grip Strength test measures the muscle strength of forelimbs and combined forelimb–hind limb grip strength of mice (55).

2.8.2. Method

To measure the grip strength of mice, we are using the commercially available Grip Strength Meter apparatus (Fig. 7; Bioseb or TSE). The system is supplied with a single grid ($400 \times 180 \times 200$ mm) which connects to the sensor.

Fig. 7. Grip strength meter, forelimb measurement.

This protocol is carried out according to the standardised phenotyping screens developed by the EUMORPHIA partners and available at http://www.empress.har.mrc.ac.uk. For the forelimb measurement, the mouse is lowered gently over the top of the grid so that only its front paws can grip the grid. Keep the torso horizontal parallel to the grid and pull the mouse back steadily (not jerking) until the grip is released down the complete length of the grid. When the animal releases the grid, the maximal grip strength value of the animal is displayed on the screen. Record the value manually/automatically. Forelimb and hind limb measurement: Gently lower the mouse over the top of the grid so that *both* its front paws and hind paws can grip the grid. Keep the torso parallel to the grid and pull the mouse back steadily (not jerking) until the grip is released down the complete length of the grid. Record the value manually/automatically. Three trials per condition are carried out, alternating forelimb strength only, and the combined forelimb/hind limb grip strength.

2.8.3. Pitfalls

For the evaluation of the results, it is essential to weigh the animal, since body weight influences the grip force (56). Mice with missing digits (e.g. after toe clipping) or other injuries on the claws should be omitted from this test. When testing the fore paw grip strength only, take care that the mouse cannot reach the grip with its hind paws.

2.9. Social Discrimination

2.9.1. Rationale

The social discrimination test allows the assessment of social recognition memory formation by exposing a test animal to conspecifics and measuring the time of active investigation of the stimulus animals, relying mainly on olfactory function. It makes use of the natural tendency of mice to investigate novel stimuli more than known ones. It is a discriminatory task, since the novel and the familiar stimulus animals are presented together during the test session.

2.9.2. Method

We use a modified version of the social discrimination procedure previously described by Engelmann and co-workers (57, 58). Adult ovariectomised 129S1/SvlmJ female mice are used as stimulus animals. The test animals are separated by transferring them to fresh cages 2 h before starting the session, and the illumination level in the testing room is set at approximately 200 lux. The social discrimination procedure consists of two 4-min exposures of stimulus animals to the test animal in the test animal's cage (Fig. 8). During the first exposure ("sampling"), one stimulus animal is exposed to the test animal, and after a retention interval of 2 h this stimulus animal is re-exposed to the test animal together with an additional, previously not presented stimulus animal during this second ("test") exposure. During each exposure, the duration of investigatory behaviour of the test animal towards the stimulus animal(s)

Fig. 8. Social discrimination, test session. The two dark-coloured mice are the stimulus animals, one is familiar, the other one novel.

is recorded by a trained observer blind to the genotype with a handheld computer. Data are analysed by using the Observer 4.1 Software (Noldus, Wageningen). A significantly longer investigation duration of the unfamiliar stimulus animal compared to the familiar one (i.e. the conspecific previously presented during the sampling phase) is taken as an evidence for an intact recognition memory. A social recognition index is calculated as time spent investigating the unfamiliar subject/time spent investigating both subjects.

2.9.3. Pitfalls

This protocol was established using C57BL/6J wild-type mice, i.e. in our hands these mice reliably show social recognition memory with this testing protocol after a retention interval of 2 h. But learning and memory abilities depend strongly on the genetic background, therefore it is essential for this paradigm to also test littermate controls to see if the control mice properly form recognition memory with this protocol. If they do not, it might be necessary to test a new cohort of this mouse line with a slightly altered protocol, e.g. a different retention interval (probably shorter) or longer exposure times to the stimulus animal (e.g. 6 or 8 min). With such modifications, we could convincingly demonstrate an improvement in social recognition memory in urocortin 3-deficient mice, reproducing results in several cohorts and in different labs (59).

2.10. Object Recognition

2.10.1. Rationale

Also the Object Recognition task is based on the natural tendency of rodents to spend more time exploring an unfamiliar stimulus than a familiar one, in this case the stimulus being an object, as described by Ennaceur and Delacour (60). As with social discrimination, also when testing object recognition, we use a discriminatory paradigm. The following procedure with three sample phases and ITIs of 15 min between the sample phases were chosen as used by Genoux and co-workers (61).

2.10.2. Method

The arena consists of a type II mouse cage ($20 \times 14 \times 26$ cm) made of plastic. The small objects to be discriminated are made of either metal or plastic. They are fixed to the bottom of the apparatus to ensure that they cannot be moved by the mice. As far as could be ascertained, the objects have no natural significance for the mice and they have never been associated with a reinforcer. Any natural preference for one of the objects was excluded by a natural preference test. The illumination level in the testing room is set at approximately 200 lux.

All mice are given two habituation days before testing in which they are placed into the empty arena and allowed to freely explore it for 10 min. On the testing day, two identical objects are placed in the arena and the test mouse is allowed to explore the two objects for 5 min (Fig. 9). Exploration of an object is defined as touching the object with the nose. The mouse is then kept individually in a cage until the next sample phase is introduced. The time spent in between the sample phases (ITIs) is 15 min and in total the two identical objects are presented three times for 5 min

Fig. 9. Object recognition, test session with one familiar and one novel object.

to each mouse. After retention intervals of 3, 24, 48 h and 1 week, during which the test mouse is kept singly in a cage, one of the previously encountered familiar objects is substituted for a new, unfamiliar one. The mouse is put back into the arena for another 5 min for a retention test and object exploration times are scored again. To exclude any positional effects, the position of the new object (left or right) is counterbalanced in each experimental group. Exploration times are always recorded by a trained observer blind to the genotype with a handheld computer, and data are analysed by using the Observer 4.1 Software (Noldus, Wageningen). As in the social discrimination paradigm, a significantly longer investigation duration of the unfamiliar object compared to the familiar one is taken as an evidence for an intact recognition memory. An object recognition index is calculated as time spent investigating the unfamiliar object/time spent investigating both the familiar and unfamiliar object. If retention is tested repeatedly, e.g. the first time after 3 h, and then a second time either after 24, 48 h or 1 week, a new unfamiliar object is chosen for the second retention test, i.e. a different one from the unfamiliar object used during the first retention test. The same kind of object is always used for the familiar object.

After each session, the cages and objects are cleaned carefully with a disinfectant. To run this test efficiently, multiple copies of the objects are produced and all fresh cages equipped with objects needed for a testing day are prepared beforehand, used once, and cleaned at the end of the testing day.

Using this procedure with three sample sessions and a first retention test after 3 h, male and female C57BL/6J mice still show significantly higher exploration times of the novel object during a second retention test after 1 week (Karin Weindl, unpublished observations). With this protocol, we could identify deficits in object recognition after a retention interval of 24 h in $Prkg1^{BKO}$ mutants, which are deficient of the cGMP-dependent protein kinase type I (62), and in mutants of the DJ-1 gene, which is associated with Parkinson's disease (63).

2.10.3. Pitfalls

It is crucial to always use fresh, clean cages and objects for each session and each mouse to avoid any confounding effects of olfactory traces. It is also crucial to keep the mice single-housed during the sample and the retention intervals when using this protocol. If the mice are group-housed between trials, C57BL/6J mice at least do not even show intact object recognition after 3 h (Karin Weindl, unpublished observations). This may be different in other mouse strains or with other protocols using different settings, e.g. longer sampling phases. When choosing objects, it is also important to ensure in preference tests that none of the objects are more attractive to the mice than the other ones used. We strongly recommend avoiding objects that mice like to gnaw at, e.g. plastic cubes with protruding vertices, because prolonged gnawing at an object renders the test non-analysable. Metal cubes do not seem to be problematic because the mice do not like to gnaw on metal.

2.11. Testing Olfactory Abilities

2.11.1. Rationale

With a "Simultaneous Smell Discrimination Task", we can detect graded differences in the olfactory abilities of mice. The procedure was adapted from Mihalick and colleagues (64).

2.11.2. Method

For the conditioning procedure, mice are placed on a restricted feeding regime to maintain their body weight at around 90% body weight of their free-feeding level for the duration of training and testing. Experimental sessions are conducted 5 days per week, one session with 12–20 trials per day. The testing is conducted using standard mouse cages (18.5 cm wide, 29.5 cm long, 13 cm high), two-thirds are covered with a plastic lid, and at the far end of the cage a small amount of bedding is placed (Fig. 10). Odorants are presented on fresh bedding shavings (ratio: 1 ml per 3 g shavings) in two circular, plastic dishes (3 cm diameter) that are mounted horizontally on a carrier, separated by a vertical barrier between them. For each trial, the dishes are filled with fresh odorised bedding shavings, and the carrier is inserted in the front part of the test cage. Between the trials, the mouse is separated from the front part by a barrier inserted by the experimenter.

During pretraining, a single dish is presented on each trial, with equivalent frequency in the left or right location. First, mice are trained for 3 days to dig in that dish filled with clean, unscented bedding shavings to retrieve a small piece of chocolate. For the discrimination tasks, two identical dishes are presented simultaneously. In the simple condition, one dish is scented with the odorant Phenethylacetate (SIGMA, Schnelldorf, Germany; smelling like apple; diluted to a concentration of 10%) that is designated [S+]

Fig. 10. Olfactory test set-up. *Left*: Test cage during an intertrial interval with inserted barrier. The carrier with the dishes is inserted on the right side of the barrier. *Right*: Mouse digging in one of the two dishes, which are mounted on a carrier and separated by a transparent vertical barrier.

and the other with the same amount of solvent (Diethyl phthalate; SIGMA, Schnelldorf, Germany). On the first day, digging in the dish scented with [S+] is consistently rewarded with chocolate for up to 18 successful trials. To train the mice to associate [S+] smell and not the chocolate with the reward, on the second day the chocolate is buried in both dishes but animals are just allowed to dig in [S+]. Training is completed after 18 successful trials. On the next 2 days, the mice have to discriminate between [S+] and another odorant Methyl *trans*-cinnamate (SIGMA, Schnelldorf, Germany; diluted to a concentration of 10%; smelling like strawberry) that is designated [S−]. A correct choice is defined as digging first in the dish with [S+]-scented shavings. The number of errors (digs) in the unrewarded dish and the dish of first choice (choice accuracy in %) is recorded. In a more difficult condition, the mice then have to discriminate between different binary mixtures of [S+] and [S−] that make the odorants increasingly similar ([S+]:[S−] vs. [S+]:[S−] 70:30 vs. 30:70; 55:45 vs. 45:55; 53:47 vs. 47:53; 51:49 vs. 49:51; 50:50 vs. 50:50). A correct choice is defined as digging first in the dish with the higher amount of [S+] and this is followed by the delivery of a small piece of chocolate by the experimenter with forceps. In a failed discrimination, if the mice chose the [S−] rather than the [S+], both dishes are immediately removed from the cage, terminating the trial. One concentration is tested per day in 15–20 trials.

After this phase, a smell sensitivity test is conducted. For this, mice are tested in binary steps of dilution, starting from a concentration of 10%. Mice have to discriminate between a dish, scented with [S+] and another dish with the solvent. To determine the threshold, a mouse will be tested on the next, lower dilution binary step (three trials per each concentration step, three to eight steps per day), if it responded correctly. If the mouse responded incorrectly, it will be retested with the previous, higher concentrated binary step.

In all test conditions, the positions of the dishes are alternated in a pseudorandom fashion to prevent the mice from making selection decisions based on location.

2.11.3. Pitfalls

In the olfactory test, several mice can be trained and tested in parallel by successively presenting the dishes. The exact number depends on the experimenters' experience and behaviour of the mice. In the digging training phase, the dishes can be presented to more mice at the same time. In the discrimination tasks, when the experimenter must react immediately to the mouse's behaviour, just a few dishes can be presented at the same time. If a mouse is very anxious and does not dig, more handling often helps. If group-housed males fight in their home cage and this affects training performance, separating mice might be of benefit. It is not recommended to test males and females in the same room at the same time. The training phase for the olfactory test can be prolonged, if the mice or some of them need more training to learn to dig or to associate the reward with the odorant. To be sure that the mice learn to associate the reward only with the "right" odorant, it is crucial that the mice never get the chocolate that is buried in the "wrong" dish. Otherwise, the following trials the mouse would continue to dig in the wrong dish if it can smell the chocolate, ignoring the added odorant. Therefore, the experimenter has to be very observant, reacting immediately when the mouse starts to dig in the wrong dish and preventing it from getting the chocolate by removing the dishes. In the sensitivity test, when the chocolate is given by hand, the experimenter has to take care to reward the mouse during or directly after digging in the right dish. But also take care that the mouse is actually digging because if the reward is presented too late or too early, the mouse will reduce or even stop the digging movements after a few trials.

3. Concluding Remarks

To avoid misinterpretation, behavioural phenotyping results should be considered in the context of analyses of other body systems and neuropathological and pathological analyses. This is particularly important when screening for phenotypes in mutants of genes with unknown functions. Locomotor or exploratory phenotypes may be confounded or even caused by, for example, sensory deficits, skeletal malformations or metabolic alterations. Likewise, apparent cognitive phenotypes may not be centrally mediated but may be caused by alterations in sensory perception. In the GMC, we rely on the optokinetic drum test performed by the Eye Screen (http://www.mouseclinic.de/research/screens/eye) to test if the mice can see, and on the startle response curve generated during measurement of prepulse inhibition of the acoustic startle reflex, to detect hearing impairments. The GMC nociceptive screen informs us of alterations in pain perception. Olfactory abilities can be tested with our olfactory test battery and with the social discrimination test.

To summarise, when detecting phenotypes indicative of movement disorders, we strongly recommend ensuring that sensory functions are intact, checking for differences in body weight or growth curves and the exclusion of skeletal malformations.

Acknowledgements

This work was partially supported by the Bundesministerium für Bildung und Forschung within the framework of the NGFN-Plus (FKZ: 01GS0850 and 01GS08174), the European Commission (EUMODIC: LSHG-2006-037188) and the "Helmholtz Alliance for Mental Health in an Ageing Society" (HELMA).

References

1. Nadeau JH, Balling R, Barsh G, Beier D, Brown SD, Bucan M, et al. (2001) Sequence interpretation. Functional annotation of mouse genome sequences. Science 291: 1251–1255

2. O'Brien T, Woychik R (2003) Our small relative. Nat Genet 33: 3–4

3. Nobrega MA, Pennacchio LA (2004) Comparative genomic analysis as a tool for biological discovery. J Physiol 554: 31–39

4. Friedel RH, Seisenberger C, Kaloff C, Wurst W (2007) EUCOMM--the European conditional mouse mutagenesis program. Brief Funct Genomic Proteomic 6: 180–185

5. Collins FS, Rossant J, Wurst W (2007) A mouse for all reasons. Cell 128: 9–13

6. Auwerx J, Avner P, Baldock R, Ballabio A, Balling R, Barbacid M, et al. (2004) The European dimension for the mouse genome mutagenesis program. Nat Genet 36: 925–927

7. Abbott A (2009) Mouse genetics: The check-up. Nature 460: 947–948

8. Fuchs H, Gailus-Durner V, Adler T, Aguilar-Pimentel JA, Becker L, Calzada-Wack J, et al. (2010) Mouse Phenotyping. Methods

9. Gailus-Durner V, Fuchs H, Becker L, Bolle I, Brielmeier M, Calzada-Wack J, et al. (2005) Introducing the German Mouse Clinic: open access platform for standardized phenotyping. Nat Methods 2: 403–404

10. Ohl F, Holsboer F, Landgraf R (2001) The modified hole board as a differential screen for behavior in rodents. Behav Res Methods Instrum Comput 33: 392–397

11. Rogers DC, Fisher EM, Brown SD, Peters J, Hunter AJ, Martin JE (1997) Behavioral and functional analysis of mouse phenotype: SHIRPA, a proposed protocol for comprehensive phenotype assessment. Mamm Genome 8: 711–713

12. Fuchs H, Gailus-Durner V, Adler T, Pimentel JA, Becker L, Bolle I, et al. (2009) The German Mouse Clinic: a platform for systemic phenotype analysis of mouse models. Curr Pharm Biotechnol 10: 236–243

13. Brown SD, Chambon P, de Angelis MH (2005) EMPReSS: standardized phenotype screens for functional annotation of the mouse genome. Nat Genet 37: 1155

14. Mallon AM, Blake A, Hancock JM (2008) EuroPhenome and EMPReSS: online mouse phenotyping resource. Nucleic Acids Res 36: D715–718

15. Morgan H, Beck T, Blake A, Gates H, Adams N, Debouzy G, et al. (2010) EuroPhenome: a repository for high-throughput mouse phenotyping data. Nucleic Acids Res 38: D577–585

16. Mandillo S, Tucci V, Holter SM, Meziane H, Banchaabouchi MA, Kallnik M, et al. (2008) Reliability, robustness, and reproducibility in mouse behavioral phenotyping: a cross-laboratory study. Physiol Genomics 34: 243–255

17. Gailus-Durner V, Fuchs H, Adler T, Aguilar Pimentel A, Becker L, et al. (2009) Systemic first-line phenotyping. Methods Mol Biol 530: 463–509

18. Abbott A (2010) Mouse project to find each gene's role. Nature 465: 410

19. Wurst W, de Angelis MH (2010) Systematic phenotyping of mouse mutants. Nat Biotechnol 28: 684–685

20. Moberg PJ, Doty RL, Mahr RN, Mesholam RI, Arnold SE, Turetsky BI, Gur RE (1997) Olfactory identification in elderly schizophrenia and Alzheimer's disease. Neurobiol Aging 18: 163–167

21. Moberg PJ, Doty RL (1997) Olfactory function in Huntington's disease patients and at-risk offspring. Int J Neurosci 89: 133–139

22. McIlwain KL, Merriweather MY, Yuva-Paylor LA, Paylor R (2001) The use of behavioral test batteries: effects of training history. Physiol Behav 73: 705–717

23. Archer J (1973) Tests for emotionality in rats and mice: a review. Anim Behav 21: 205–235

24. Crawley JN (1989) Animal models of anxiety. Current Opinion in Psychiatry 2: 773–776

25. Weiss SM, Lightowler S, Stanhope KJ, Kennett GA, Dourish CT (2000) Measurement of anxiety in transgenic mice. Rev Neurosci 11: 59–74

26. Choleris E, Thomas AW, Kavaliers M, Prato FS (2001) A detailed ethological analysis of the mouse open field test: effects of diazepam, chlordiazepoxide and an extremely low frequency pulsed magnetic field. Neurosci Biobehav Rev 25: 235–260

27. Walsh RN, Cummins RA (1976) The Open-Field Test: a critical review. Psychol Bull 83: 482–504

28. Sedelis M, Schwarting RK, Huston JP (2001) Behavioral phenotyping of the MPTP mouse model of Parkinson's disease. Behav Brain Res 125: 109–125

29. Fleming SM, Salcedo J, Fernagut PO, Rockenstein E, Masliah E, Levine MS, Chesselet MF (2004) Early and progressive sensorimotor anomalies in mice overexpressing wild-type human alpha-synuclein. J Neurosci 24: 9434–9440

30. Carter RJ, Lione LA, Humby T, Mangiarini L, Mahal A, Bates GP, Dunnett SB, Morton AJ (1999) Characterization of progressive motor deficits in mice transgenic for the human

Huntington's disease mutation. J Neurosci 19: 3248–3257

31. Fischer LR, Culver DG, Tennant P, Davis AA, Wang M, Castellano-Sanchez A, Khan J, Polak MA, Glass JD (2004) Amyotrophic lateral sclerosis is a distal axonopathy: evidence in mice and man. Exp Neurol 185: 232–240

32. Hamers FP, Lankhorst AJ, van Laar TJ, Veldhuis WB, Gispen WH (2001) Automated quantitative gait analysis during overground locomotion in the rat: its application to spinal cord contusion and transection injuries. J Neurotrauma 18: 187–201

33. Alexis NE, Dietrich WD, Green EJ, Prado R, Watson BD (1995) Nonocclusive common carotid artery thrombosis in the rat results in reversible sensorimotor and cognitive behavioral deficits. Stroke 26: 2338–2346

34. Ohlsson AL, Johansson BB (1995) Environment influences functional outcome of cerebral infarction in rats. Stroke 26: 644–649

35. Shelton SB, Pettigrew DB, Hermann AD, Zhou W, Sullivan PM, Crutcher KA, Strauss KI (2008) A simple, efficient tool for assessment of mice after unilateral cortex injury. J Neurosci Methods 168: 431–442

36. Goldstein LB, Davis JN (1990) Beam-walking in rats: studies towards developing an animal model of functional recovery after brain injury. J Neurosci Methods 31: 101–107

37. Majchrzak M, Brailowsky S, Will B (1992) Chronic infusion of GABA into the nucleus basalis magnocellularis or frontal cortex of rats: a behavioral and histological study. Exp Brain Res 88: 531–540

38. Plaas M, Karis A, Innos J, Rebane E, Baekelandt V, Vaarmann A, Luuk H, Vasar E, Koks S (2008) Alpha-synuclein A30P point-mutation generates age-dependent nigrostriatal deficiency in mice. J Physiol Pharmacol 59: 205–216

39. Kline AE, Massucci JL, Marion DW, Dixon CE (2002) Attenuation of working memory and spatial acquisition deficits after a delayed and chronic bromocriptine treatment regimen in rats subjected to traumatic brain injury by controlled cortical impact. J Neurotrauma 19: 415–425

40. Dluzen DE, Liu B, Chen CY, DiCarlo SE (1995) Daily spontaneous running alters behavioral and neurochemical indexes of nigrostriatal function. J Appl Physiol 78: 1219–1224

41. Emerick AJ, Kartje GL (2004) Behavioral recovery and anatomical plasticity in adult rats after cortical lesion and treatment with monoclonal antibody IN-1. Behav Brain Res 152: 315–325

42. Riek-Burchardt M, Henrich-Noack P, Metz GA, Reymann KG (2004) Detection of chronic sensorimotor impairments in the ladder rung walking task in rats with endothelin-1-induced mild focal ischemia. J Neurosci Methods 137: 227–233

43. Farr TD, Liu L, Colwell KL, Whishaw IQ, Metz GA (2006) Bilateral alteration in stepping pattern after unilateral motor cortex injury: a new test strategy for analysis of skilled limb movements in neurological mouse models. J Neurosci Methods 153: 104–113

44. Ploughman M, Attwood Z, White N, Dore JJ, Corbett D (2007) Endurance exercise facilitates relearning of forelimb motor skill after focal ischemia. Eur J Neurosci 25: 3453–3460

45. Metz GA, Whishaw IQ (2002) Cortical and subcortical lesions impair skilled walking in the ladder rung walking test: a new task to evaluate fore- and hindlimb stepping, placing, and coordination. J Neurosci Methods 115: 169–179

46. Faraji J, Metz GA (2007) Sequential bilateral striatal lesions have additive effects on single skilled limb use in rats. Behav Brain Res 177: 195–204

47. Z'Graggen WJ, Metz GA, Kartje GL, Thallmair M, Schwab ME (1998) Functional recovery and enhanced corticofugal plasticity after unilateral pyramidal tract lesion and blockade of myelin-associated neurite growth inhibitors in adult rats. J Neurosci 18: 4744–4757

48. Metz GA, Merkler D, Dietz V, Schwab ME, Fouad K (2000) Efficient testing of motor function in spinal cord injured rats. Brain Res 883: 165–177

49. Ogawa N, Hirose Y, Ohara S, Ono T, Watanabe Y (1985) A simple quantitative bradykinesia test in MPTP-treated mice. Res Commun Chem Pathol Pharmacol 50: 435–441

50. Ogawa N, Mizukawa K, Hirose Y, Kajita S, Ohara S, Watanabe Y (1987) MPTP-induced parkinsonian model in mice: biochemistry, pharmacology and behavior. Eur Neurol 26 Suppl 1: 16–23

51. Matsuura K, Kabuto H, Makino H, Ogawa N (1997) Pole test is a useful method for evaluating the mouse movement disorder caused by striatal dopamine depletion. J Neurosci Methods 73: 45–48

52. Fernagut PO, Chalon S, Diguet E, Guilloteau D, Tison F, Jaber M (2003) Motor behaviour deficits and their histopathological and functional correlates in the nigrostriatal system of dopamine transporter knockout mice. Neuroscience 116: 1123–1130

53. Wahlsten D, Metten P, Phillips TJ, Boehm SL, Burkhart-Kasch S, Dorow J, et al. (2003)

Different data from different labs: lessons from studies of gene-environment interaction. J Neurobiol 54: 283–311

54. Capacio BR, Harris LW, Anderson DR, Lennox WJ, Gales V, Dawson JS (1992) Use of the accelerating rotarod for assessment of motor performance decrement induced by potential anticonvulsant compounds in nerve agent poisoning. Drug Chem Toxicol 15: 177–201

55. Tilson HA (1990) Behavioral indices of neurotoxicity. Toxicol Pathol 18: 96–104

56. Maurissen JP, Marable BR, Andrus AK, Stebbins KE (2003) Factors affecting grip strength testing. Neurotoxicol Teratol 25: 543–553

57. Engelmann M, Wotjak CT, Landgraf R (1995) Social discrimination procedure: an alternative method to investigate juvenile recognition abilities in rats. Physiol Behav 58: 315–321

58. Richter K, Wolf G, Engelmann M (2005) Social recognition memory requires two stages of protein synthesis in mice. Learn Mem 12: 407–413

59. Deussing JM, Breu J, Kuhne C, Kallnik M, Bunck M, Glasl L, et al. (2010) Urocortin 3 modulates social discrimination abilities via corticotropin-releasing hormone receptor type 2. J Neurosci 30: 9103–9116

60. Ennaceur A, Delacour J (1988) A new one-trial test for neurobiological studies of memory in rats. 1: Behavioral data. Behav Brain Res 31: 47–59

61. Genoux D, Haditsch U, Knobloch M, Michalon A, Storm D, Mansuy IM (2002) Protein phosphatase 1 is a molecular constraint on learning and memory. Nature 418: 970–975

62. Feil R, Holter SM, Weindl K, Wurst W, Langmesser S, Gerling A, Feil S, Albrecht U (2009) cGMP-dependent protein kinase I, the circadian clock, sleep and learning. Commun Integr Biol 2: 298–301

63. Pham TT, Giesert F, Rothig A, Floss T, Kallnik M, Weindl K, et al. (2010) DJ-1-deficient mice show less TH-positive neurons in the ventral tegmental area and exhibit non-motoric behavioural impairments. Genes Brain Behav 9: 305–317

64. Mihalick SM, Langlois JC, Krienke JD, Dube WV (2000) An olfactory discrimination procedure for mice. J Exp Anal Behav 73: 305–318

Chapter 8

MRI of Neurological Damage in Rats and Mice

Mathias Hoehn

Abstract

The present chapter introduces the potential of magnetic resonance imaging (MRI) for in vivo investigations of experimental stroke in rodents. Aspects in setting up the experiment, particular for this technique, are presented as well as considerations for the choice of anaesthesia protocol. Structural changes induced by the evolution of the pathophysiological processes are demonstrated. Furthermore, a broad range of MRI options for the investigations of functional deficits and outcome improvement are pointed out. Emphasis is set on the multimodal approach capability of MRI to obtain complementary structural, haemodynamic, metabolic and, finally also, functional information in longitudinal studies of the complex pathophysiological evolution of stroke.

Key words: Anaesthesia, Animal fixation, Stroke, Structural damage, Functional damage, Fibre tracking, Functional connectivity

1. Introduction

The application of non-invasive imaging is usually sought when diseases are studied, where the underlying pathophysiology develops over a longer period. During longitudinal observation, information becomes accessible about the evolution of damage and, potentially, also of recovery processes. This is also of high interest when therapeutic strategies are investigated and their efficiency evaluated.

During such longitudinal studies, not only structural reorganizations caused by the primary damage are of interest, but also functional deficits and improvements, respectively. In consequence, several modifications on or additions to the basic requirements of the imaging technique are needed to provide the requested information. In the following chapter, we use stroke as the disease with which to characterize the information content obtainable with

Emma L. Lane and Stephen B. Dunnett (eds.), *Animal Models of Movement Disorders: Volume I*, Neuromethods, vol. 61,
DOI 10.1007/978-1-61779-298-4_8, © Springer Science+Business Media, LLC 2011

magnetic resonance imaging (MRI) and present the MRI modalities according to the evolution of stroke (1).

In general, rats and mice generate quite similar demands on the technical realization of producing state-of-the-art in vivo imaging. Therefore, all aspects of imaging the lesion evolution over time are discussed in reference to rat experiments. Where appropriate, deviations from the protocol for adjustments to mice are explained.

2. Methods

2.1. General Remarks

The principle of MRI depends on the irradiation of the sample under investigation with a radio frequency (RF) pulse while the corresponding signal is registered by an antenna operating in the same frequency range. To avoid disturbance of the environment by the RF pulse from the MRI and, on the other hand, to exclude the pick-up of external RF signals by the detector, the whole magnet system is usually installed inside a Faraday cage to "seal" the MRI detection system from the outside world. Moreover, the strong cryo-magnet has a stray field which can seriously affect sensitive electronics and disturb reliable measurements. Therefore, electronic equipment, e.g. for the recording of physiological variables, must remain outside the Faraday cage, typically at a distance of few meters from the magnet. As the sensors need to reach the animal in the magnet, it is important to make sure that such cables or tubes do not act as RF antennas, importing the RF noise inside the Faraday cage.

2.2. Anaesthesia

Animals must be anaesthetized during the whole MRI experiment because some of the measurement sequences produce noise easily at 90 dB or beyond. An awake animal not accustomed to this situation would be stressed, consequently influencing the physiological condition. Moreover, movement artefacts must be avoided during the measurement to assure good-quality images. Injection-type anaesthesia protocols have the disadvantage of having to reinject new dose after certain period, often requiring movement of the animal to the outside of the magnet or preparing a catheter. The best applicable are inhalation anaesthetics, such as halothane or isoflurane. They are easy to dose, are applied through a nose mask on the animal and permit to keep the animal physiologically stable and quiet for several hours with no artefact concerns during the measurement. For functional activation studies, the deep anaesthesia of fluorinated inhalation anaesthetics is not useful as the stimulus is not well-processed in the cortex any longer. In the past, functional activation studies used α-chloralose because it preserves best the functional-metabolic coupling (2), but the severe side effects (3, 4) demanded those experiments to be terminal. Recently, we developed a protocol suitable for longitudinal investigations using the well-tolerated medetomidine for sedation of the animals (5, 6). Medetomidine is an α_2-adrenoreceptor agonist which is quickly compensated by its antidote, atipamezole.

2.3. Animal Fixation

In order to avoid movement artefacts and to keep the animal in a physiologically appropriate position, it is laid prone in a polycarbonate half-tube (Fig. 1). The bottom of this cradle contains a meander of circulating warm water which is adjusted through a feedback system via a rectal temperature probe. The head of the animal is fixed in a head holder with ear bars and a tooth bar. An air balloon positioned under the thorax and connected to a pressure transducer monitors amplitude and frequency of the thorax expansion which can be used for analysis of anaesthesia depth. Depending on the particular experimental requirements, further physiological monitoring devices can be added or even special electrodes can be used for recording of EEG and evoked potentials (7, 8).

2.4. Quantitative MRI

MRI allows the measurement of various physical variables (Table 1). These are used – after evaluation and calibration – to depict a range of aspects from structural to haemodynamic and metabolic to functional parameters. Conventional experiments emphasize one particular parameter without excluding the others necessarily, thus weighting the contrast for a certain variable. This has the advantage of being fast, but the disadvantage that a non-weighted parameter can still generate a contrast, which was expected from the selected, originally weighted one, or that two parameters lead to low contrast because of competitive contrast mechanisms. Therefore, it may often be favourable to invest slightly more experimental scan time to record quantitative data of a single variable of

Fig. 1. Animal cradle for fixation in the MRI system. The half-pipe cradle (centre of the figure) holds the animal laid in prone position over the bed with the circulating warm water to keep the animal at physiological temperature. At left end of the half-pipe in the photograph, the fixation of the ear bars can be seen with the mask cone in between used for application of the breathing gases and the inhalation anaesthesia (the tooth bar is hidden in the mask cone). At the right end of the cradle, the connectors for the gas tubes and the warm water are positioned. Reproduced with permission from Medres company.

Table 1
Physical MRI variables and their use of in vivo imaging

MRI variable	In vivo application
Structural information	
T1 relaxometry	Blood-brain barrier intactness (+ Gd chelate contrast agent)
T2 relaxometry	Anatomic contrast
	Vasogenic edema
	Chronic ischemic lesion
Apparent diffusion coefficient (ADC)	Acute ischemic lesion
Diffusion tensor imaging (DTI)	Fibre orientations
Haemodynamic information	
Perfusion-weighted imaging	Tissue perfusion
T2*-weighted imaging	CBV, CBF, MTT (+contrast agent bolus)
Metabolic information	
ADC	Ion and water homeostasis (ATP dependent)
Functional information	
Manganese-enhanced MRI (MEMRI)	Axonal tracing
DTI	Connections of brain nuclei by fibres
Resting-state functional MRI (rs-fMRI)	Brain connectivity network
fMRI	Stimulus-specific brain activation

CBV cerebral blood volume, *CBF* cerebral blood flow, *MTT* mean transit time

choice: in these parameter maps, the contrast between tissue areas is generated from pure difference in the selected parameter in these tissues (9, 10). This makes interpretation of the pathophysiological cause of the contrast unambiguous.

2.5. Structural Damage Registration

In order to choose the best measurement sequence, it is necessary to have prior knowledge of the pathophysiological cascade of events during the window of observation (11).

2.5.1. Early Phase of Stroke Evolution

Rapidly after the onset of ischemia, cell respiration fails and the ATP pool is depleted. The cells depolarize, as they no longer can maintain ionic gradients across their membranes. In consequence, osmotically driven inflow of water into the cells leads to a reduction of the apparent diffusion coefficient (ADC) (12). ADC measurements, thus, show a significant reduction by approximately 30–40% in the ischemic core, depicting the ischemic territory very early and with sensitive contrast (13). Quantitative analysis of the degree of ADC reduction and correlation with the tissue area of energy breakdown has led to an ADC threshold for a severely damaged ischemic core region and a further threshold for the

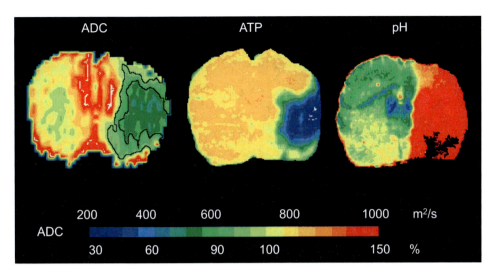

Fig. 2. ADC thresholds for ischemic core and penumbra. Coronal images of ADC (*left*), distribution of ATP (*centre*) and tissue pH (*right*) after 2 h of MCA occlusion are depicted. The area of tissue acidosis is clearly larger than the region of ATP depletion. For comparison, the corresponding area sizes are outlined in the ADC image, thus indicating the two different ADC thresholds (77% of normal for the ischemic core, 90% of normal for the penumbra). Reproduced with permission from (14).

metabolic penumbra characterized by anaerobic glycolysis and tissue acidosis (Fig. 2) (14).

In a permanent occlusion model, the ADC reduction persists for a few days, then increases again to produce a pseudonormalization, followed by an ADC value above normal (15). These changes reflect the pronounced expression of the vasogenic edema during the first week post-stroke, followed by the cystic transformation. In an early reperfusion situation, the ADC may normalize very quickly in relation to rapid recovery of the energy metabolism (16, 17). With several hours delay, a secondary decline of the ADC occurs in the same territory that was characterized by ADC reduction at the end of the primary occlusion period. Thus, within the first few hours after successful reperfusion, the ADC may severely underestimate the damage in the ADC maps (Fig. 3).

Another parameter commonly applied is the T2 relaxation time. T2-weighted images show the early vasogenic edema as hyperintensity after a few hours (18). This hyperintensity persists and increases to its maximum at approximately 1 week after stroke onset. T2-weighted imaging, however, is not sensitive enough to detect the beginnings of T2 increases during the very early phase of the vasogenic edema with sufficient contrast (16). Quantitative T2 maps, on the other hand, allow the reliable detection of the ischemic territory even within an hour of stroke onset (16). This T2 increase is not reversible during early reperfusion when ADC transiently normalizes again. Moreover, the further development of T2 during

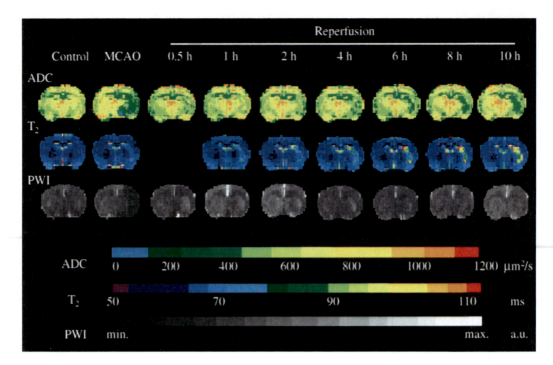

Fig. 3. Representative time course of apparent diffusion coefficient (ADC), perfusion-weighted signal intensity and T2 maps before ischemia (Control), at the end of middle cerebral artery occlusion (MCAO) and at different time points of reperfusion. Note the transient recovery of ADC at the early phase of reperfusion, followed by secondary deterioration. The increase of T2 value was continued in the lesion area during early recirculation. Perfusion-weighted signal intensity images did not reveal significant delayed hypoperfusion after retraction of the occluder. Reproduced with permission from (16).

this reperfusion phase – a further increase or stagnation – has been described as a good indicator for the severity of damage (16).

MR angiography provides information about the vascular occlusion and the position of a clot in thromboembolic stroke models (19). This contrast agent-free angiographic method allows repetitive screening for success of lysis. Together with perfusion MRI, it offers a detailed haemodynamic characterization of the occlusion period: vascular occlusion and reopening, tissue perfusion and collateral supply (despite occlusion of the main vessel).

T1-weighted MRI can be used together with iv injection of a gadolinium contrast agent to depict disturbances of the blood-brain barrier (BBB) (20). In the regions of open BBB, the contrast agent leaks out of the vascular bed into the parenchyma, inducing a T1 reduction in the tissue and thus resulting in a diagnostic hyperintensity.

2.5.2. Chronic Phase of Stroke Evolution

In the chronic phase, ADC and T1-weighted imaging are of little help, particularly as the pseudonormalization phase may veil the pathophysiological processes, thus leading to false negative interpretation. T2-weighted imaging is the most widely used parameter for presentation of the lesion extent in the chronic phase. It should be noted, however, that even T2 maps carry the risk of false interpretation due to the development of T2 contrast over time: After the first wave of T2 increase during the pronounced expression of the vasogenic edema, T2 normalizes again in parallel with the resolution of the edema (21). This T2 normalization will continue if the lesioned tissue develops only selective neuronal death. In the case of pannecrosis with following cystic transformation, however, a second T2 increase occurs at this time. Thus, the time profile of T2 during the chronic phase is indicative of pannecrosis while a lack of T2 increase may be interpreted either as tissue recovery or selective neuronal death (21, 22).

Finally, during the late phase, several weeks after stroke induction, a hypointensity may appear in T2*-weighted images in some areas of the ischemic territory. The reason for this T2*-sensitive hypointensity was found to be iron accumulation in macrophages around vessels with delayed degradation. Iron ions of erythrocytes, entering the parenchyma through vascular leaks and incorporated by the macrophages, are the source of the image contrast, thus serving as sensitive monitor of delayed vessel degradation and ensuing macrophage activity (23, 24).

2.6. Functional Connectivity

The study of a focal lesion, such as stroke, may also inflict only local functional deficits. However, it is also conceivable that areas distant to the primary lesion site are affected or that plastic reorganization processes point to interactions between the lesion and close or even distant brain areas. Thus, the focal lesion may have repercussions on a whole network of interacting areas. In order to better understand the whole consequences of the damage and also the underlying mechanisms of any observed functional improvements, it is of great interest to investigate the connectivity of the brain as a whole, before and at times after lesion induction. Here, several recently developed methods of in vivo MRI contribute to these investigations. The following three paragraphs introduce axonal tracing, fibre tracking and resting-state connectivity measurements as important tools.

2.6.1. Axonal Tracing

Manganese ions are taken up by cells in a similar way to calcium because of their same charge and van der Waals radius. After incorporation into vesicles, they are transported anterogradely along the axons and are even passed on transsynaptically (25, 26). Due to their electron configuration, manganese 2+ ions act as efficient T1 contrast agent – this effect is exploited for *M*anganese-*E*nhanced *MRI*, MEMRI. Thus, transport of manganese ions from the local injection site of manganese chloride can be followed in vivo. This can be used to trace connections and assess the intactness of a connection (27). Appearance of manganese-induced T1 contrast in the thalamus after local injection into the somatosensory cortex, thus, demonstrates the intact cortico-thalamic connection while lack of hyperintense signal in the thalamic nucleus would indicate loss of the axonal pathway.

2.6.2. Fibre Tracking

MEMRI allows the probing of the intactness of specific connection of two nuclei (28), but due to rather hazy, weak contrast along the axons, the specific pathway cannot be depicted with sufficient accuracy for analysis. Here, the method of fibre tracking, based on *Diffusion Tensor Imaging* (DTI) or its more complex variant DSI (diffusion spectrum imaging), can act as a complementary technique (29–31). Using DTI, the myelinated fibre bundles are presented between defined points. By this method, the orientation of fibres between connecting points can be determined, the fibre location relative to the extent of the lesion assessed and, potentially, even the degree of damage of these connecting fibres estimated. A further highly interesting aspect of DTI fibre tracking is the monitoring of deterioration of existing fibres and, in parallel, the sprouting of new fibre connections, intra-hemispherically as well as transhemispherically (32). Thus, DTI can be combined with other techniques providing functional connectivity information or structural correlates thereof. This, then, allows the determination of structural connections and the functional assessment of these connections (33, 34). An example is given in Fig. 4 for the fibre bundle changes after stroke in connection with fMRI-based localization of the corresponding activated S1 area.

2.6.3. Functional Connectivity of Brain Networks

Functional MRI of the brain's resting-state (rs-fMRI) has gained considerable interest during the past few years, first in human

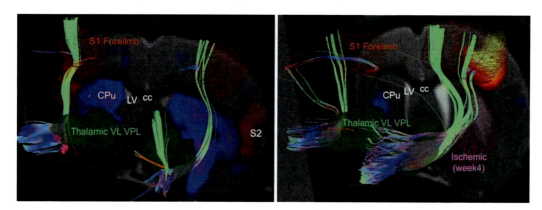

Fig. 4. Changes of the thalamo-cortical connectivity after stroke: combination of fibre tracking with functional connectivity and with activation of the forepaw representation area in the S1 somatosensory cortex in the rat brain. *Left*: Diffusion spectrum imaging (DSI) was used for thalamo-cortical fibre tracking. Fibre bundles on both hemispheres were overlaid on a functional connectivity map obtained in a healthy control animal. The connectivity map reflects interhemispheric connectivity between bilateral S2 (*orange*) and caudate putamen (CPu; *blue*) obtained from resting-state fMRI. Representation volumes of S1 of forelimb and of thalamic nuclei VPL were determined from histological brain atlas. *Right*: Results of DSI fibre tracking in the same animal, but 4 weeks after stroke induction on the right hemisphere. The thalamo-cortical fibre bundles are overlaid on a BOLD functional activation map showing S1 response, recorded in a separate session at 5 weeks after stroke. The ischemic lesion volume at this time point was determined from T2 images and is illustrated as wireframe (*magenta*) on the right hemisphere. Comparison with the pre-stroke situation shows appearance of a new fibre bundle, bypassing the lesion volume and entering the S1 more frontally than the original fibre bundle before stroke induction (reprinted with permission: YB Kim, D. Kalthoff, C. Po; unpublished results).

studies (35) but more recently also in small animal investigations (36). It is based on the analysis of synchronous noise fluctuations in situations of complete rest, i.e. without specific stimulations, as they are typically applied in functional activation studies. Extensive image post-processing of a series of images under resting condition allows the determination of correlation between defined regions, both intra- and transhemispherically (37). This information is commonly interpreted as reflecting the existence of a network of functional connectivity in the brain, whereby the technique is sometimes also called functional connectivity MRI (fcMRI). While the application to pathophysiological models in small animals is still in its beginning, there are already first studies dealing with changes of the functional connectivity after stroke in rats (38–40). Expectations about the future potential contribution of this technique are high as this approach may elucidate normal networks as well as disease-induced alterations and deficits of this network. Compensations and reorganizations may be better investigated and understood in the future using fcMRI, especially in combination with MEMRI-based axonal tracing (28) and fibre tracking by DTI.

2.7. Functional Activation: Deficit and Recovery

In contrast to the above-discussed functional connectivity approaches, functional brain activation imaging (fMRI) relies on blood oxygenation level-dependent (BOLD) signal changes in specific brain areas responding to specific external stimuli. Using a well-established paradigm of electrical forepaw stimulation (2), the activation of the contralateral somatosensory cortex can be reliably monitored. This protocol has been used already in several studies for the investigation of stroke-induced functional alterations. Based on an α-chloralose anaesthesia protocol, Dijkhuizen et al. studied the functional S1 activation at different times after stroke in separate animal groups and found transient transhemispheric reorganization during the early ischemic phase, followed by relocation to the original representation area at later times (41, 42). To avoid the need for group comparisons, we developed a new anaesthesia protocol based on the α_2-adrenoreceptor agonist medetomidine (6) that is well-tolerated and permits repetitive anaesthesia sessions during longitudinal studies. In a stroke investigation, fMRI was combined with electrophysiological recording of somatosensory-evoked potentials (SSEPs) during a 7-week period after stroke induction (43). The combination with the SSEPs allowed the conclusion that the neurovascular coupling remains preserved during the whole observation period, thus confirming the validity of fMRI as a reliable indicator of functional deficit and recovery (Fig. 5). In this study, no reorganization processes, transient or permanent, were noted while only the original representation field was activated in cases of functional improvement. The fMRI results of the spontaneous recovery at approximately 3 weeks after stroke were in full correspondence with parallel sensorimotor behaviour

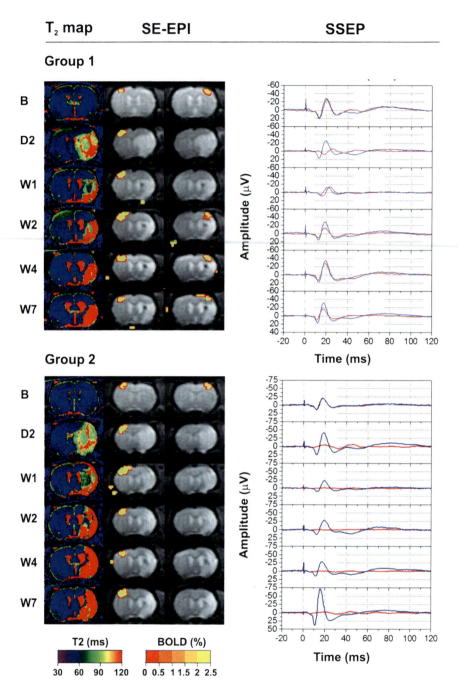

Fig. 5. Structural and functional imaging and SSEP recording following transient focal cerebral ischemia in the rat. T2 maps (first column), spin-echo echo-planar fMRI images (SE-EPI) upon right (second column) and left (third column) forepaw stimulation. Corresponding SSEP signals from S1FL (fourth column; left-hemispheric SSEP in *blue*; right-hemispheric SSEP in *red*) from animal with transient loss (Group 1) and permanent loss (Group 2) of BOLD activation in the right somatosensory cortex. BOLD activation and normal SSEP signal are observed in both somatosensory cortices during baseline measurements (B), i.e. before transient MCAO. In animals with transient loss of BOLD activation in the right S1FL area (Group 1), the right SSEP signal is delayed and diminished 2 days after MCAO (D2) and delayed 1 week after MCAO (W1). In parallel with re-emergence of BOLD activation 2 weeks after MCAO (W2), a normal SSEP signal is observed again. In animals with permanent loss of BOLD activation in the right S1FL area (Group 2), no restoration of the SSEP signal is observed.

tests (43). Finally, the fMRI signal in the S1 cortex was found to be a reliable prognostic indicator at 3 weeks: animals with the chance for spontaneous recovery showed a positive activation signal at this time, whereas animals without this activation at 3 weeks did not show functional improvement during the whole 6 months after stroke (44).

In a follow-up study, the same authors were able to show a therapeutic effect of stem cell implantation 3 weeks after stroke induction (44). Animals with spontaneous recovery were excluded at 3 weeks after stroke. The others were divided into two groups at this time point, untreated and stem cell-treated, and were monitored for functional deficit and improvement during a 6-month period. Functional improvement was beginning at approximately 20 weeks, but was seen only in those animals receiving stem cells.

fMRI acts as the specific read-out parameter probing the activation capability in response to a specific stimulus. Thus, localization and reorganization of representation fields for certain stimuli are monitored. Furthermore, functionality of selected representation fields can be assessed under diseased conditions with this method. Using this "focal brain activation" approach in combination with the connectivity investigations discussed above, the components of the cerebral functional network can be assessed with MRI for their contribution to observed functional deficit and functional improvement, respectively.

3. Notes

In the following section, a few typical examples of errors or technical problems are described together with possible solutions.

3.1. Movement Artefacts

Strong thorax expansion during breathing movements may lead to nick movements of the head rotating around the horizontal fixation axis of the ear bars. To minimize or avoid this nick movement, the body should be stretched slightly to reduce the normal hunching position of the rodents. For this purpose, the head of the animal is first fixed in the head holder; then the animal is held on the neck and the body is slightly pulled backwards to reduce the curved spine position. Then, the shoulder area can be fixed additionally with tape to further restrain movements. This precaution is particularly useful for diffusion-weighted imaging (ADC imaging or DTI imaging) as the diffusion-weighted MR sequence is intrinsically particularly sensitive to movements.

Further movement artefacts of a shifting centre of gravity of images in a long series of images, such as for rs-fMRI or, even worse, for DTI, may be due to a high duty cycle of the gradients, leading to heating of the gradient coils and resulting in a shift of the images along the phase axis. Here, technical care must be taken to reduce the speed of the measurement sequence to allow for better heat dissipation in the gradient sets. Alternatively, extensive data post-processing can be performed co-registering the images to a common template.

3.2. Shine-Through Effect in Diffusion-Weighted Imaging

As described above, during acute stroke, the ischemic territory is characterized by a decrease in ADC. This lower diffusivity leads to a lower signal damping in diffusion-weighted images for ischemic tissue than for normal tissue (higher ADC). The ischemic area is, in consequence, depicted as a hyperintense region in diffusion-weighted images. However, in a time window when the vasogenic edema has developed already, this appears as hyperintense in T2-weighted images. On the other hand, diffusion-weighted images are also always T2-weighted to some extent. That means that under conditions of existing vasogenic edema, the diffusion-weighted images always show some hyperintensity, independent of diffusion weighting or not, i.e. the hyperintensity of the T2-weighting effect "shines through" the diffusion weighting. To separate these two superimposing effects, quantitative ADC maps and T2 maps must be recorded at this point. Only thus can the cytotoxic edema (ADC reduction on ADC maps) be differentiated reliably from the vasogenic edema (T2 increase on T2 maps).

3.3. Anaesthesia Conditions for fMRI in Rodents

Basically, α-chloralose and medetomidine are described as anaesthesia protocols compatible with the requirement that the functional activation remains preserved. While α-chloralose is not compatible with longitudinal studies (cf above), medetomidine must sensitively be adapted to the particular experimental conditions. Mice and rats require different doses (45). We have found that between rat or mouse strains or even between genders there exists sometimes the need to adapt the dose of medetomidine to assure fully sedated animals.

4. Summary

Using stroke as one of the well-studied cerebral lesions with MRI, we have tried to show the broad range of structural, haemodynamic and even functional information obtainable with this versatile technique. It is MRI's intrinsic strength that one technique allows the recording of complementary information, thus providing a much more complete picture about a rather complex pathophysiological development.

Acknowledgement

Support by the Nederlands Organisatie voor Wetenschappelijk Onderzoek and Alexander-von-Humboldt Stiftung (NWO-AvH) is gratefully acknowledged.

References

1. Weber R, Ramos-Cabrer P, Hoehn M. Present status of MRI and MRS in animal stroke models. Journal of Cerebral Blood Flow & Metabolism 2006;26:591–604.

2. Ueki M, Linn F, Hossmann KA. Functional activation of cerebral blood flow and metabolism before and after global ischemia of rat brain. J Cereb Blood Flow Metab 1988;8(4):486–494.

3. Hedenquist P, Hellebrekers LJ. Laboratory animal analgesia, anesthesia and euthanasia. In: Hau J, Van Hoosier GLJ, editors. Handbook of laboratory animal science. Volume Vol.1: essential principles and practices. Florida: CRC Press; 2003. pp. 413–455.

4. Silverman J, Muir WW, 3rd. A review of laboratory animal anesthesia with chloral hydrate and chloralose. Lab Anim Sci 1993;43(3):210–216.

5. Ramos-Cabrer P, Weber R, Wiedermann D, Hoehn M. Continuous noninvasive monitoring of transcutaneous blood gases for a stable and persistent BOLD contrast in fMRI studies in the rat. NMR in Biomedicine 2005;18(7):440–446.

6. Weber R, Ramos-Cabrer P, Wiedermann D, van Camp N, Hoehn M. A fully noninvasive and robust experimental protocol for longitudinal fMRI studies in the rat. Neuroimage 2006;29(4):1303–1310.

7. Brinker G, Bock C, Busch E, Krep H, Hossmann K-A, Hoehn-Berlage M. Simultaneous recording of evoked potentials and T2*-weighted MR images during somatosensory stimulation of rat. Magnetic Resonance in Medicine 1999;41(3):469–473.

8. Busch E, Hoehn-Berlage M, Eis M, Gyngell ML, Hossmann K-A. Simultaneous recording of EEG, DC potential and diffusion-weighted NMR imaging during potassium induced cortical spreading depression in rats. NMR in Biomedicine 1995;8(2):59–64.

9. Eis M, Hoehn-Berlage M. A time-efficient method for combined T_1 and T_2 measurement in magnetic resonance imaging: evaluation for multiparameter tissue characterization. MAGMA 1994;2:79–89.

10. Eis M, Hoehn-Berlage M. Quantitative diffusion MR imaging: Optimization of measurement parameters for improved accuracy and reproducibility. Plesser T, Wittenburg T, editors; 1995. p. 39–61.

11. Dirnagl U, Iadecola C, Moskowitz MA. Pathophysiology of ischaemic stroke: an integrated view. Trends in Neuroscience 1999;22:391–397.

12. Hoehn-Berlage M. Diffusion-weighted NMR imaging: application to experimental focal cerebral ischemia. NMR in Biomedicine 1995;8(7–8):345–358.

13. Back T, Hoehn-Berlage M, Kohno K, Hossmann KA. Diffusion nuclear magnetic resonance imaging in experimental stroke. Correlation with cerebral metabolites. Stroke 1994;25(2):494–500.

14. Hoehn-Berlage M, Norris DG, Kohno K, Mies G, Leibfritz D, Hossmann K-A. Evolution of regional changes in apparent diffusion coefficient during focal ischemia of rat brain: The relationship of quantitative diffusion NMR imaging to reduction in cerebral blood flow and metabolic disturbances. Journal of Cerebral Blood Flow & Metabolism 1995;15:1002–1011.

15. Knight RA, Dereski MO, Helpern JA, Ordidge RJ, Chopp M. Magnetic resonance imaging assessment of evolving focal cerebral ischemia. Comparison with histopathology in rats. Stroke 1994;25:1252–1261.

16. Olah L, Wecker S, Hoehn M. Secondary deterioration of apparent diffusion coefficient after 1-hour transient focal cerebral ischemia in rats. Journal of Cerebral Blood Flow & Metabolism 2000;20(10):1474–1482.

17. Olah L, Wecker S, Hoehn M. Relation of apparent diffusion coefficient changes and metabolic disturbances after 1 hour of focal cerebral ischemia and at different reperfusion phases in rats. Journal of Cerebral Blood Flow & Metabolism 2001;21(4):430–439.

18. Hoehn-Berlage M, Eis M, Back T, Kohno K, Yamashita K. Changes of relaxation times (T1, T2) and apparent diffusion coefficient after

permanent middle cerebral artery occlusion in the rat: temporal evolution, regional extent, and comparison with histology. Magnetic Resonance in Medicine 1995;34(6):824–834.

19. Hilger T, Niessen F, Diedenhofen M, Hossmann KA, Hoehn M. Magnetic resonance angiography of thromboembolic stroke in rats: indicator of recanalization probability and tissue survival after recombinant tissue plasminogen activator treatment. Journal of Cerebral Blood Flow & Metabolism 2002;22(6):652–662.

20. Busch E, Krueger K, Fritze K, Allegrini PR, Hoehn-Berlage M, Hossmann K-A. Blood-brain barrier disturbances after rt-PA treatment of thromboembolic stroke in the rat. Acta Neurochirurgica Supplementum 1997;70:206–208.

21. Wegener S, Weber R, Ramos-Cabrer P, Uhlenkueken U, Sprenger C, Wiedermann D, Villringer A, Hoehn M. Temporal profile of T2-weighted MRI distinguishes between pannecrosis and selective neuronal death after transient focal cerebral ischemia in the rat. Journal of Cerebral Blood Flow & Metabolism 2006;26(1):38–47.

22. Wegener S, Weber R, Ramos-Cabrer P, Uhlenkueken U, Wiedermann D, Kandal K, Villringer A, Hoehn M. Subcortical lesions after transient thread occlusion in the rat: T2-weighted magnetic resonance imaging findings without corresponding sensorimotor deficits. Journal of Magnetic Resonance Imaging 2005;21(4):340–346.

23. Weber R, Wegener S, Ramos-Cabrer P, Uhlenküken U, Hoehn M. Detection of chronic macrophage activity in response to degrading blood vessels in experimental cerebral ischemia in rats: a high resolution MRI study. Symposium "Neuro-Visionen 2 - Perspektiven in NRW", Ferdinand Schöningh GmbH & Co, Paderborn 2005:230–232.

24. Justicia C, Ramos-Cabrer P, Hoehn M. MRI detection of secondary damage after stroke - Chronic iron accumulation in the thalamus of the rat brain. Stroke 2008;39(5):1541–1547.

25. Pautler RG, Koretsky AP. Tracing odor-induced activation in the olfactory bulbs of mice using manganese-enhanced magnetic resonance imaging. Neuroimage 2002;16(2):441–448.

26. Pautler RG, Silva AC, Koretsky AP. In vivo neuronal tract tracing using manganese-enhanced magnetic resonance imaging Magnetic Resonance in Medicine 1998;40(5):740–748.

27. van der Zijden JP, Wu O, van der Toorn A, Roeling TP, Bleys RL, Dijkhuizen RM. Changes in neuronal connectivity after stroke in rats as studied by serial manganese-enhanced MRI. Neuroimage 2007;34(4):1650–1657.

28. Soria G, Wiedermann D, Justicia C, Ramos-Cabrer P, Hoehn M. Reproducible imaging of rat corticothalamic pathway by longitudinal manganese-enhanced MRI (L-MEMRI). Neuroimage 2008;41:668–674.

29. Boska MD, Hasan KM, Kibuule D, Banerjee R, McIntyre E, Nelson JA, Hahn T, Gendelman HE, Mosley RL. Quantitative diffusion tensor imaging detects dopaminergic neuronal degeneration in a murine model of Parkinson's disease. Neurobiology of Disease 2007;26:590–596.

30. Song S-K, Yoshino J, Le TQ, Lin S-J, Sun S-W, Cross AH, Armstrong RC. Demyelination increases radial diffusivity in corpus callosum of mouse brain. NeuroImage 2005;26:132–140.

31. Sun S-W, Liang H-F, Le TQ, Armstrong RC, Cross AH, Song S-K. Differential sensitivity of in vivo and ex vivo diffusion tensor imaging to evolving optic nerve injury in mice with retinal ischemia. NeuroImage 2006;32:1195–1204.

32. Granziera C, D'Arceuil HD, Zai L, Magistretti PJ, Sorensen AG, de Crespigny AJ. Long-term monitoring of post-stroke plasticity after transient cerebral ischemia in mice using in vivo and ex vivo diffusion tensor MRI. The Open Neuroimaging Journal 2007;1:10–17.

33. Thuen M, Olsen O, Berry M, Pedersen TB, Kristoffersen A, Haraldseth O, Sandvig A, Brekken C. Combination of Mn2+−enhanced and diffusion tensor MR imaging gives complementary information about injury and regeneration in the adult rat optic nerve. Jounral of Magnetic Resonance Imaging 2009;29:39–51.

34. Obenaus A, Jacobs RE. Magnetic resoannce imaging of functional anatomy: use for small animal epilepsy models. Epilepsia 2007;48 Supp4:11–17.

35. Biswal B, Yetkin FZ, Haughton VM, Hyde JS. Functional connectivity in the motor cortex of resting human brain using echo-planar MRI. Magnetic Resonance in Medicine 1995;34:537–541.

36. Zhao FQ, Zhao TJ, Zhou L, Wu QL, Hu XP. BOLD study of stimulation-induced neural activity and resting-state connectivity in medetomidine-sedated rat. Neuroimage 2008;39:249–260.

37. Kalthoff D, Seehafer JU, Po C, Wiedermann D, Hoehn M. Functional connectivity in the rat at 11.7T: impact of physiological noise in resting state fMRI. Neuroimage 2011;54:2828–2839.

38. Kim YR, Biswal BB, Rosen BR. Causality Analysis using Resting State BOLD fMRI in Normal and Ischemic Rat Brains. 2007; Berlin.

39. van der Marel K, van Meer M, Wang K, Otte W, Berkelbach van der Sprenkel J, Dijkhuizen R. Analysis of changes in functional connectivity patterns with serial resting state fMRI after transient ischemic stroke in rat brain. 2009 April; Honolulu. p 697.

40. van Meer MP, van der Marel K, Wang K, Otte WM, El Bouazati S, Roeling TA, Viergever MA, Berkelbach van der Sprenkel JW, Dijkhuizen RM. Recovery of sensorimotor function after experimental stroke correlates with restoration of resting-state interhemispheric functional connectivity. J Neurosci 2010;30(11):3964–3972.

41. Dijkhuizen RM, Ren J, Mandeville JB, Wu O, Ozdag FM, Moskowitz MA, Rosen BR, Finklestein SP. Functional magnetic resonance imaging of reorganization in rat brain after stroke. Proc Natl Acad Sci USA 2001; 98(22):12766–12771.

42. Dijkhuizen RM, Singhal AB, Mandeville JB, Wu O, Halpern EF, Finklestein SP, Rosen BR, Lo EH. Correlation between brain reorganization, ischemic damage, and neurologic status after transient focal cerebral ischemia in rats: a functional magnetic resonance imaging study. J Neurosci 2003;23(2): 510–517.

43. Weber R, Ramos-Cabrer P, Justicia C, Wiedermann D, Strecker C, Sprenger C, Hoehn M. Early prediction of functional recovery after experimental stroke: Functional magnetic resonance imaging, electrophysiology, and behavioral testing in rats. Journal of Neuroscience 2008;28(5):1022–1029.

44. Ramos-Cabrer P, Justicia C, Wiedermann D, Hoehn M. Stem cell mediation of functional recovery after stroke in the rat. PLoS One 2010;5:e12779.

45. Adamczak J, Farr TD, Seehafer JU, Kalthoff D, Hoehn M. High field BOLD response to forepaw stimulation in the mouse. Neuroimage 2010;51:704–712.

Chapter 9

Quantification of Brain Function and Neurotransmission System In Vivo by Positron Emission Tomography: A Review of Technical Aspects and Practical Considerations in Preclinical Research

Nadja Van Camp, Yann Bramoullé, and Philippe Hantraye

Abstract

Unlike many other imaging techniques, positron emission tomography, PET, necessitates gathering a broad array of competences: biologists/physicians have to interact with physicists, chemists and mathematicians to acquire and analyze PET data. The ensemble of a PET imaging experiment, from the creation of the isotope to the interpretation of the imaging data, requires a unique combination of high-tech equipment to be installed preferentially on the same site, such as: a cyclotron, a radiochemistry laboratory, a PET camera, an animal facility and a data treatment and storage facility.

In the material and methods chapter for each process, we discuss the requirements of the equipment and the usual procedures, from the creation of the isotope to the modelling of the PET data.

Depending on the animal model (mouse, rat, non-human primate, …) or the isotope (^{11}C, ^{18}F, …) used, the challenges and requisites for setting up a PET imaging experiment will be different. The notes section discusses some important considerations on animal handling in PET imaging and the basic experimental set-up to evaluate the characteristics of a radiotracer.

This chapter is concluded with some practical examples related to Parkinson disease and neuroinflammation.

Key words: Positron emission tomography, PET, Radiochemistry, Radiotracer, Brain imaging, Macaque, Rodents

1. Introduction

Preclinical and clinical researches in neurodegenerative diseases have focused recently on the development of interventions that aimed at either halting or slowing down the neurodegenerative processes involved in these disorders. These new therapeutic approaches have stimulated the search for objective markers that

Emma L. Lane and Stephen B. Dunnett (eds.), *Animal Models of Movement Disorders: Volume I*, Neuromethods, vol. 61, DOI 10.1007/978-1-61779-298-4_9, © Springer Science+Business Media, LLC 2011

would help to assess (ideally to quantify) changes in disease progression potentially triggered by these innovative treatments. In this respect, the contribution of positron emission tomography (PET) imaging, a nuclear imaging technique that provides a versatile means of examining a large variety of molecular/cellular processes involved in normal and pathological cerebral function, has been essential.

Relying on the use of radiolabelled molecules (radiotracers) that after systemic administration diffuse freely in the body to selectively bind to specific biological targets (neuroreceptors, enzymes, false transmitters, transporters, intracellular molecules, etc. …), the technique is currently used in various aspects of neuroscience, neuropharmacology and the validation of new therapies (including drug candidates) by academic research institutions, clinical units or pharmaceutical companies. The enthusiasm around the use of this technique is probably due to the fact that PET bears several advantages over other imaging techniques: in addition to its great versatility, PET imaging benefits from a remarkable sensitivity enabling the detection of nanomolar concentrations of a radiotracer. Nevertheless, the technique also requires heavy infrastructures to be brought into play, such as cyclotron facilities, radiochemistry laboratories equipped with shielded hot cells and automated chemical units for radioligand preparation, PET tomographs and a complete multidisciplinary array of human competences, at least including chemists, physicists, biologists, physicians and image processing specialists.

Despite the fact that PET carries radio-security issues both for the subject and the experimenter limiting its ease of use and increasing the cost of the technique, there are currently hundreds of radiolabelled compounds of pharmacological and clinical interest ready to be used on the shelf. This makes PET the most "translational" imaging method in neurosciences with numerous successful cases of radiotracer translation from preclinical validation studies to clinical research applications. For many years reserved for non-human primate use because of the relatively low intrinsic spatial resolution of the available PET detectors, recent technological advances in the physics of PET and the design of dedicated high-resolution microbe tomographs for rodent use have given scientists access to a vast array of new applications. What is now referred to as "in vivo molecular imaging" paves the way for a myriad of new applications of the technique not only in biology and preclinical studies, but more importantly for translational applications in medicine and industry.

The execution of PET imaging experiments, from the production of the radioisotope to the interpretation of the biological data, requires the teamwork and coordination of scientists with different but complementary skills in order to orchestrate the succession of protocols applied on a unique combination of the high-tech equipment.

This chapter deals, thus, with the general principles and instrumentation issues of PET imaging as well as with the general methods to produce a radiotracer to set up an imaging experiment and extract biologically relevant measures (receptor density, receptor affinity, enzymatic reaction rate, etc.) from the raw radioactivity events acquired in list mode by the PET scanner.

2. Materials

2.1. Radiochemistry

2.1.1. Properties of a Radiotracer

A major challenge in the production of radiotracers for brain imaging studies is the ability of the radioligand to pass the blood brain barrier (BBB), but not its radiometabolites, which is related to different factors. First, the lipophilicity of the radiotracer should be moderate: it should be sufficiently high to ensure a good cerebral uptake, but not too high to ensure a low non-specific binding. A high lipophilicity also favours binding to blood proteins and, thus, reduces the fraction of radiotracer freely available in the plasma (1). The most commonly used index of compound lipophilicity is $logP$, where P is the n-octanol/water partition coefficient of the unionized species. At a pH of 7.4, $logP$ should be between +0.5 and +3.0 and optimally equally to +2.0. To ensure a negligible brain uptake of the radioactive metabolites, radiometabolites should preferentially be hydrophilic. Second, passive entry into the brain is promoted by low molecular weight (500 Da), a small cross-sectional area (<80 Å^2), a low hydrogen-bonding capacity and a lack of formal charge. Third, the radioligand should be a poor substrate for drug efflux pumps, like P-glycoprotein, which restrict the entry of (radio)pharmaceuticals to the brain.

In view of the metabolic pathways, the choice of the labelling position of the isotope is essential to guarantee the in vivo stability of the radiotracer. A well-known example is the 5-HT$_{1A}$ antagonist, [^{11}C]-WAY-100635 (Fig. 1). This molecule can be labelled in the 2-methoxy position or in the carbonyl of the amide group. As the liver metabolism cleaves at the level of the amide bond, this generates, respectively, the amine [^{11}C]-WAY-100634 which crosses the BBB or the [^{11}C]-hexanecarboxylic acid metabolite which is too polar to enter into the brain. The labelling of the precursor needs, thus, to be performed on the carbonyl group.

In addition, the pharmacokinetics of the radiotracer and the biological process under study need to be compatible with the half-life of the isotope. For example, for blood flow measurements, it is preferable to use [^{15}O]-water (half-life $t_{1/2} = 2$ min) which enables repetitive injections within a short period of time. In contrast, when studying slower biological processes like receptor binding or metabolism, it is preferable to select a radionuclide with a longer half-life, such as fluorine-18 ($t_{1/2} = 110$ min), allowing longer

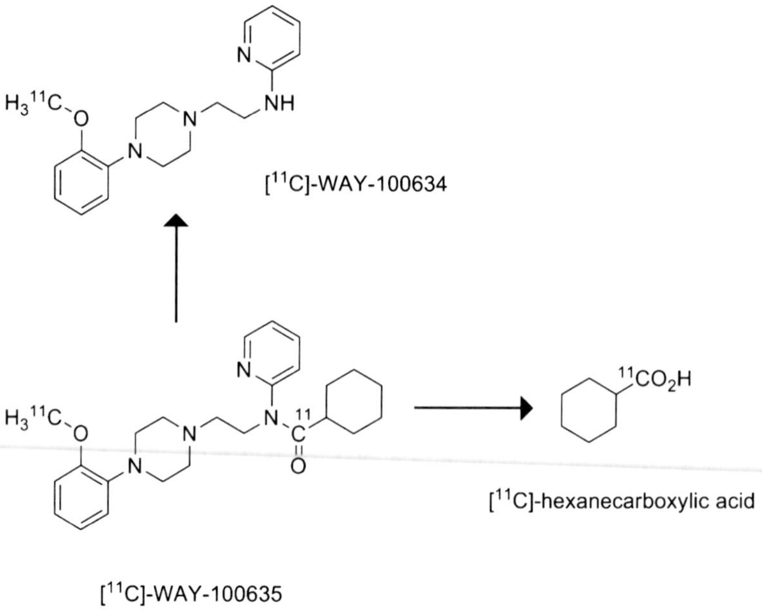

Fig. 1. Radiometabolites of [^{11}C]-WAY-100635 after labelling in the 2-methoxy position or in the carbonyl of the amide group.

acquisition times. In vivo follow-up studies of larger biomolecules, such as peptides and/or antibodies with slow kinetics, need radio-isotopes with very long half-lives, such as copper-64 ($t_{1/2} = 12.7$ h) or iodine-124 ($t_{1/2} = 4.2$ day).

2.1.2. Production of the Radionuclides

Table 1 lists some of the positron emitters and their half-lives, main mode of production and positron energies.

A major constraint of positron-emitting nuclides is the very short half-life of most isotopes which necessitates an onsite production. An exception to this is fluorine-18 which, with a half-life of almost 2 h, can be distributed to centres up to several hundred kilometres away. In addition, this relatively long half life allows a more sophisticated radiochemistry and its position weak energy (635 keV maximum) represent a good point in terms of desimetry to clinical applications.

The radionuclides are produced in a cyclotron consisting of five major components: an ion source, a magnet, a radiofrequency switching circuit, an ion beam extractor and a target. The ion source provides an electrical discharge across a volume of gas-generating ions, e.g. negative hydrogen ions, which are constrained into a spiral orbit by a uniform magnetic field. The ion beam is subsequently accelerated by an alternating electric field in order to increase the energy of the ions. Once the beam reaches sufficient energy, it passes through a thin carbon foil that strips the electrons from the ions, yielding an energized beam of protons that is used to irradiate a target of non-radioactive atoms. Depending on the target, various radionuclides can be produced, as outlined in Table 1.

Table 1
Main characteristics of some position emitting isotopes

Isotope	Half-life (min)	Production mode	Maximum β^+ energy (MeV)	Maximal free path in water (mm)
[^{18}F]	110	$^{18}O(p,n)^{18}F$	0.64	2.3
[^{11}C]	20	$^{14}N(p,\alpha)^{11}C$	0.96	3.9
[^{15}O]	2	$^{15}N(p,n)^{15}O$	1.72	8.0
[^{13}N]	10	$^{16}O(p,\alpha)^{13}N$	1.20	5.1
[^{76}Br]	960	$^{75}As(^3He,2n)^{76}Br$	3.70	19.0
[^{124}I]	6,048	$^{124}Te(p,n)^{124}I$	1.50	7.0

2.1.3. Radiosynthesis

Four main considerations need to be taken into account when using short-lived β^+-emitting isotopes: the reaction time factor, the dilution factor, the radiation protection and the labelling precursor.

The reaction time factor. First, it is preferable to choose a radiosynthesis scheme with the shortest reaction time possible, such as the one-step labelling. If this is not feasible, the isotope has to be introduced during the very last steps of the preparation of the radioligand. Second, the purification method should be as efficient as possible using ideally solid-phase extraction (SPE) cartridges and/or a semi-preparative high-performance liquid chromatography (HPLC). Finally, the quality control system, which assures the radionuclidic and chemical purities, as well as the sterility and pyrogenicity of the radiotracer, should be robust and rapid. Overall, the maximum preparation time for the radiotracer (including quality control and syringe preparation) should not exceed 2–3 times the radioactive half-life of the positron emitter selected for the study.

The dilution factor. The quantity of the primary precursors produced by the cyclotron is very low (order of nanomolar) in contrast to the quantity of labelling precursor (order of micromolar). As a consequence of the stoichiometry, the conversion rate is high and the radioactive reagent is consumed very rapidly producing high radiochemical yields within a short time. The small amounts of precursors require, thus, the use of high-quality reagents and a high clean status of equipment. For example, atmospheric carbon dioxide contamination should be avoided when recovering [^{11}C]-carbon dioxide from the cyclotron to reach the highest possible specific activity (Bq/mol). On the other hand, the advantage from a technical point of view is the simplified technical handling, e.g. by using smaller chemical units and reactors.

Radiation protection. Radiochemical syntheses are performed in fully automated chemical-processing units, in which the radionuclide is

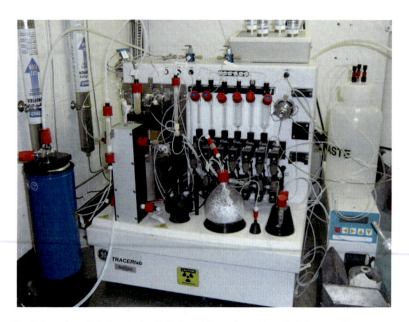

Fig. 2. An automated chemical unit designed for carbon-11 methylation reactions.

brought together with the appropriate reagents and with the labelling precursor to produce the radiotracer. In addition to an automated radiosynthesis, the chemical units minimize the radiation dose received by the radiochemists (Fig. 2).

The labelling precursor. This is chosen in function of its reactivity with the primary or secondary precursors, and having access to various labelling precursors is crucial for radiotracer synthesis.

2.1.4. Reaction Schemes

Carbon-11

The ubiquitous presence of carbon makes carbon-11 an attractive and important positron-emitting isotope. However, it is beyond the scope of this chapter to describe extensively carbon-11 radiochemistry and for review the reader is referred to Allard et al. (2). The most commonly applied production method to obtain carbon-11 is the $^{14}N(p,\alpha)^{11}C$ nuclear reaction. If the target of the cyclotron is filled with a nitrogen/oxygen gas mixture, [^{11}C]-carbon dioxide is produced (blue box, Fig. 3), whereas when a nitrogen/hydrogen gas mixture is present, [^{11}C]-methane is recovered (green box, Fig. 3). [^{11}C]-carbon dioxide and [^{11}C]-methane are referred to as primary precursors. As [^{11}C]-methane is unable to react directly with the labelling precursor, subsequent rapid reactions are needed to produce more reactive molecules, referred to as the secondary precursors. These react with the labelling precursor yielding the final radiotracer. Even though [^{11}C]-carbon dioxide has very reactive properties, it can be transformed in secondary precursors for some chemical reactions, e.g. the methylation and carbonylation reaction (Fig. 3).

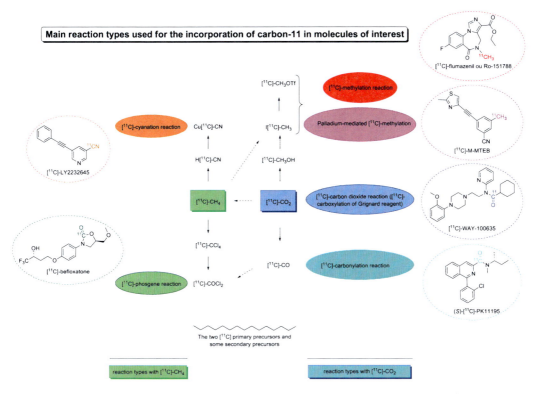

Fig. 3. Representation of the main reaction types used for the incorporation of carbon-11 in molecules of interest, starting from the two [^{11}C] primary precursors and some of the main secondary precursors.

[^{11}C]-carbon dioxide as the primary precursor:

1. The most frequently used reaction is [^{11}C]-methylation (red box, Fig. 3) using [^{11}C]-carbon dioxide as primary precursor and [^{11}C]-methyl iodide (I[^{11}C]-CH$_3$) and [^{11}C]-methyl triflate ([^{11}C]-CH$_3$OTf) as secondary precursors (3). For example, [^{11}C]-flumazenil (4) and [^{11}C]-raclopride (5) can be generated through this chemical pathway.

2. Palladium-catalyzed reactions can be considered as an indirect [^{11}C]-methylation reaction (pink box, Fig. 3). The secondary precursor [^{11}C]-methyl iodide allows for example, the synthesis of the two radioligands of the metabotropic glutamate receptor (mGluR5) antagonists, the [^{11}C]-M-MTEB via a Suzuki cross-coupling reaction (6) and the [^{11}C]-MPEP via a Stille reaction (7).

3. Reaction of [^{11}C]-carbon dioxide with Grignard reagents (dark blue box, Fig. 3) can form [^{11}C]-carboxylic acids or the more reactive acid chloride species. A further treatment of this acid chloride with amines leads to [^{11}C]-amides. This route is one of the two synthesis pathways used to label the [^{11}C]-WAY-100635 on the carbonyl group (8).

4. Finally, the [^{11}C]-carbonylation reaction using [^{11}C]-carbon monoxide as the secondary precursor (light blue box, Fig. 3) (9) allows access to imides, ketones, amides and acrylamides. For example, (S)-[^{11}C]-PK11195, the reference ligand for peripheral benzodiazepine receptor (PBR), can be obtained via this reaction (10).

[^{11}C]-methane as the primary precursor:

1. [^{11}C]-methane allows the production of the highly reactive secondary precursor, the [^{11}C]-phosgene molecule ([^{11}C]-COCl$_2$) (green box, Fig. 3). The potent, reversible and competitive mono-amine oxidase-A (MAO-A) inhibitor, [^{11}C]-befloxatone, is for example, obtained via this route (11). Recently, a simplified [^{11}C]-phosgene preparation method has been proposed (12).

2. [^{11}C]-methane reacts with ammoniac over platinum to produce the [^{11}C]-hydrogen cyanide ([^{11}C]-HCN) which is the starting material for [^{11}C]-cyanation reactions (orange box, Fig. 3) (13). This secondary precursor can be used directly to synthesize amino acids (14) or can be transformed in another secondary precursor, the [^{11}C]-CuCN, to obtain, for example, the mGluR5 antagonist [^{11}C]-LY2232645 (15).

Fluorine-18

An important difference between carbon-11 and fluorine-18 radiochemistry follows from the limited chemical methods that are available to incorporate fluorine-18 into the target molecule, which is in contrast to carbon-11 radiolabelling. It is beyond the scope of this chapter to describe extensively fluorine-18 radiochemistry and for a review of this domain the reader is referred to the recent special issues 2 and 3 of the volume 3, edited in 2010 by Current Radiopharmaceuticals (16–18).

Briefly, two main reactions are used: *electrophilic* and *nucleophilic radiofluorination*. The first method uses molecular [^{18}F]-fluorine ([^{18}F]-F$_2$) generally obtained via the ^{20}Ne(d,α)^{18}F nuclear reaction or more recently via the ^{18}O(p,n)^{18}F reaction, where the cyclotron target is filled with ^{18}O gas. These two nuclear reactions involve an exchange step between carrier non-radioactive fluorine and [^{18}F]-fluorine, inducing a poor specific radioactivity. The second radiofluorination uses the anion [^{18}F]-fluoride. This method is more selective and provides significantly higher yields and specific radioactivities than the electrophilic radiofluorination. Therefore, nucleophilic radiofluorination is often preferred, particularly for brain imaging studies, where high-specific radioactivities are required.

Electrophilic radiofluorination. [^{18}F]-fluorine is not the only electrophilic-labelling reagent used for this reaction. It is sometimes suitable to involve less-reactive fluorinating reagents, such as trifluoromethyl-[^{18}F]-hypofluorite (CF$_3$O[^{18}F]-F) (**19**) or acetyl-[^{18}F]-hypofluorite (CH$_3$CO$_2$[^{18}F]-F) (**20**). The labelling precursor, consisting of a nucleophile entity such as an alkene, an enol or an organostannyl compound, is then radiolabelled with one of these [^{18}F] electrophilic reagents to produce the [^{18}F]-radiotracer. Starting from [^{18}F]-fluorine and the 6-trimethyl-stannyl-labelling precursor, the 6-[^{18}F]-fluoro-L-DOPA can be obtained using a fully automated computer-controlled module (TRACERlab FX$_{FDOPA}$™, GE Medical Systems) (**21**).

Nucleophilic radiofluorination. As previously mentioned, this method is the most efficient and allows significantly higher yields and specific radioactivity than the electrophilic radiofluorination reaction. The anion [^{18}F]-fluoride, recovered after an ^{18}O(p,n)^{18}F nuclear reaction, is highly soluble in liquid [^{18}O]-water, which served as a the target for the nuclear reaction. However, the [^{18}F]-fluoride is a strong base, but a poor nucleophilic reagent. The most efficient and commonly used procedure to increase the nucleophilic force of [^{18}F]-fluoride consists of fixing [^{18}F]-fluoride on an anion exchange resin before eluting it with a diluted aqueous potassium carbonate solution. This creates a chelation between the [^{18}F]-fluoride anion and the potassium cation. This chelation is then weakened using a kryptand, the polyaminoether Kryptofix-222® (4,7,13,16,21,24-hexaoxa-1,10-diazabicyclo[8.8.8]hexacosane). After concentration to dryness, the complex (K-[^{18}F]-F-K$_{222}$) is often further dried by azeotropic evaporation with acetonitrile. The reactive nucleophile [^{18}F]-fluoride reagent is then dissolved in polar solvents, such as DMSO, DMF, acetonitrile or bulky alcohols (**22**). Typical reaction conditions involve high reaction temperatures, between 80 and 200°C, and even microwave irradiations. In general, the reaction time lasts a few minutes to half an hour. Depending on the target molecule, being an aliphatic or aromatic entity, the nucleophilic radiofluorination occurs by direct or indirect nucleophilic fluorination.

Direct nucleophilic fluorination (left side, Fig. 4) is the method of choice to radiolabel a target molecule with a fluorine-18 because it consists of a one-step process by direct reaction of the radioactive precursor, [^{18}F]-fluoride, with the labelling precursor. Within direct fluorination, short reaction steps (one to two reactions) can be considered as well, e.g. to remove a protective or to remove/transform a strong electron-withdrawing group.

1. Direct nucleophilic *aliphatic* substitution (pink box, Fig. 4) with [^{18}F]-fluoride, which is essentially an S$_N$2-type reaction, is used, e.g. to produce the high-affinity dopaminergic D$_2$ receptor radiotracer, [^{18}F]-fallypride (synthesized from the corresponding tosylate-labelling precursor) (**23**).

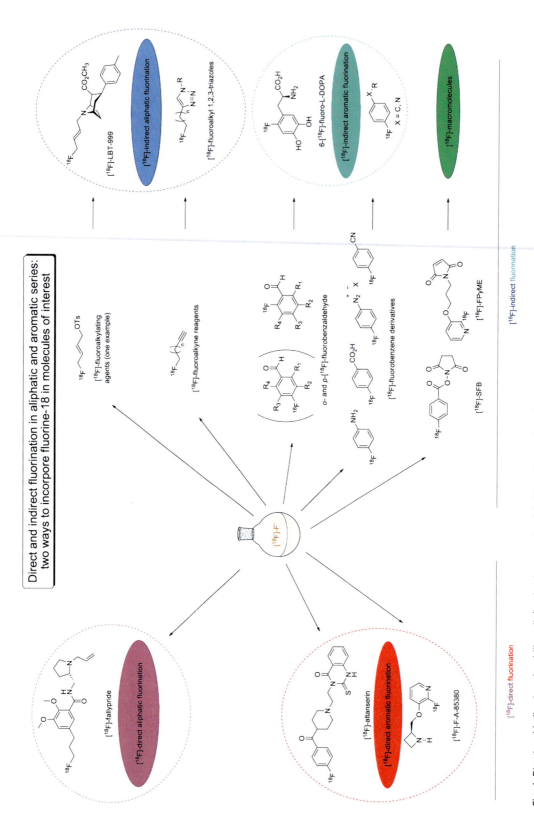

Fig. 4. Direct and indirect nucleophilic radiofluorination ways in aliphatic and aromatic series.

2. Direct nucleophilic fluorination is also possible in the *aromatic* series: The substitution reaction in homoaromatic series (red box, Fig. 4) requires aromatic nuclei carrying both a good leaving group and a strong electron-withdrawing substituent placed *para* or *ortho* to the leaving group. The serotoninergic $5HT_{2A}$ receptor ligand [^{18}F]-altanserin has been obtained via this route (24). Unlike the homoaromatic counterpart, heteroaromatic substitution reaction requires only a good leaving group. The nicotinic cholinergic $\alpha_4\beta_2$ receptor radioligand [^{18}F]-F-A-85380 has been synthesized at a position *ortho* to the ring nitrogen of the pyridine (25).

Indirect nucleophilic fluorination (right side, Fig. 4) is used when direct nucleophilic fluorination is not feasible due to the relatively harsh reaction conditions, such as high pH and high reaction temperatures, not supported by many labelling precursors.

1. To overcome this problem, [^{18}F]-indirect *aliphatic* fluorination reactions (dark blue box, Fig. 4) use prosthetic groups, such as [^{18}F]-alkylating agents (26) and [^{18}F]-alkyne reagents (also developed to radiolabel macromolecules by "click chemistry" (27)). A Radioligands obtained by this reaction scheme is [^{18}F]-LBT-999, a dopamine transporter ligand starting from the (E)-4-[^{18}F]-fluorobut-2-enyl tosylate and the tropane derivative as the labelling precursor (28).

2. Many other prosthetic groups have been developed to succeed in the indirect *aromatic* fluorination (light blue box, Fig. 4). Especially, [^{18}F]-fluorobenzene and [^{18}F]-fluorobenzaldehyde derivatives are very convenient prosthetic groups. Notably, the latter is used to produce the 6-[^{18}F]-fluoro-L-DOPA with a specific radioactivity higher than 74GBG/μmol in contrast to the specific radioactivity less than 1GBG/μmol obtained via the electrophilic fluorination route (29).

The increasing interest of macromolecules (green box, Fig. 4) for treatment and diagnosis of disease has contributed to the development of [^{18}F]-fluorine radiolabelling of oligonucleotides, peptides, proteins and antibodies. Many other [^{18}F]-prosthetic groups reacting in a chemoselective way with the macromolecule have, thus, been designed. Two of these most commonly used prosthetic groups are the *N*-succinimidyl-4-[^{18}F]-fluorobenzoate ([^{18}F]-SFB) (30) and the maleimide 1-[3-(2-[^{18}F]-fluoropyridin-3-yloxy)propyl]pyrrole-2,5-dione ([^{18}F]-FPyME) (31) which react selectively with thiol group of cysteine residues.

2.2. PET Instrumentation

2.2.1. The Physics of PET

The vast majority of annihilation events between positions and surrounding electrons, give rise to the emission of two 511keV photons that are emitted in almost exactly opposite directions (180°), albeit with a directional uncertainty of about ± 0.5° (Fig. 5). This physical particularity yields some uncertainty in the exact determination of the site of annihilation, thereby limiting the spatial resolution of the technique.

The distance that the positron travels prior to annihilation (free path) is dependent on the initial emission energy and the medium through which it is travelling (Table 1). Depending on the isotope, the mean distance travelled by the positron before annihilation varies, affecting the actual spatial resolution in the images. As a consequence, even if the point of annihilation is very close to the point of emission, the image generated from the distribution of annihilation radiation only approximates the actual distribution of the radionuclide itself.

2.2.2. Coincidence Detection of the Annihilation Events

The detection of the positron emission is achieved by the simultaneous (coincident) detection of the two collinear 511 keV photons resulting from the annihilation process by two opposite detectors surrounding the positron-emitting object. Surrounding the subject's head by a large number of detectors (up to 20,000 for the most recent tomographs) permits a multi-coincidence detection and a spatial reconstruction of the various annihilation events (Fig. 6).

Typically, in a circular set-up, each detector is placed in coincidence with about half of the total detectors in the ring (Fig. 6a, b), whereas in a polygonal array set-up the detector is in coincidence

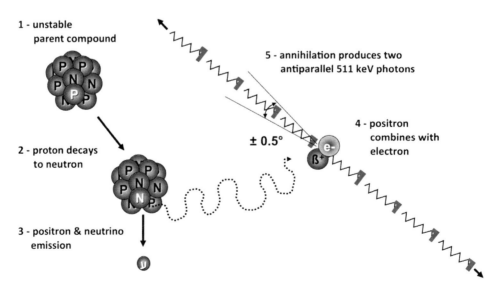

Fig. 5. Annihilation of the positron, resulting in two 511 keV photons that are emitted in almost exactly opposite directions (180°), albeit with a directional uncertainty of about ± 0.5°.

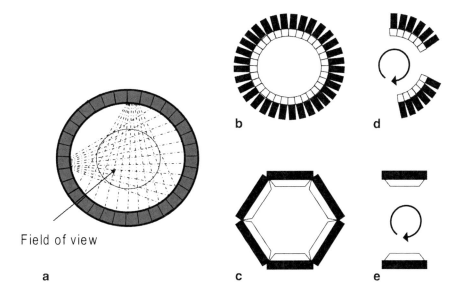

Field of view

Fig. 6. Detector configurations: Circular set-up (**a**, **b**), polygonal array set-up (**c**), partial detector rings (**d**), Anger camera systems (**e**).

with the opposed detector bank (Fig. 6c). PET systems with only partial detector rings are less expensive, but require rotation of the detector assembly round the longitudinal axis of the patient to complete the acquisition of the projection data (Fig. 6d). Continuous, large-area detectors, such as those found in multi-head Anger camera systems and used for single-photon emission-computed tomography (SPECT), have now been appropriately modified and are as well used as for coincidence imaging of positron emitters (Fig. 6e).

Each coincidence event, i.e. simultaneous detection of a photon pair within a time window of a few nanoseconds, represents a line in space, the Line of Response (LOR), connecting the two detectors along which the annihilation process occurred. The raw data collected by a PET scanner are, thus a list of "coincidence events" representing near-simultaneous detection of annihilation photons by a pair of opposite detectors. It takes approximately 3 ns for photons to traverse the field of view of a PET scanner, and in addition to measure the position of arriving annihilation photons, the detectors placed in coincidence also measure their time of arrival. Each detected photon is, thus, tagged with a detector position and a detection time: if the detection time difference between two photons is smaller than the coincidence window (traditionally 5–10 ns), the two events are considered to be physically correlated to the same annihilation event (Fig. 7). Time-of-flight PET uses this time-of-flight difference between the detection of the two photons to better locate the

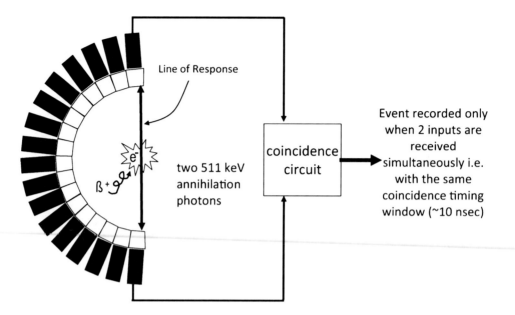

Fig. 7. Line of response and representation of a true coincidence.

annihilation position of the emitted positron along the LOR (for review, see Conti et al. (32)).

Besides true coincidence events arising from single positron annihilation followed by detection of the photon pair without any precedent interaction of the photon pair with the medium (solid line in Fig. 8a), two other major events might cause coincidences. Scattered coincidence (solid line in Fig. 8c) results from a single annihilation with one of the photons (dashed line) undergoing a change in direction (due to the Compton scatter effect) in the medium before detection. Though a true incidence, the LOR is displaced and thus degrading image contrast and resolution. In addition to true coincidences, random or accidental coincidence events (Fig. 8b) occur when two photons not arising from the same annihilation event (dashed lines) are detected by two opposite detectors within the coincidence time window of the system. These *falsely* detected coincidences (full lines) are uniformly distributed in time and will cause isotope concentrations to be overestimated if not corrected for. Photons are inherent to attenuation as well, reducing the number of coincidences and thus underestimating the amount of radioligand within a region of interest, if not corrected for. Although the magnitude of this correction for small animal subjects is much smaller than for humans (1.3 for a 3-cm diameter set-up in mouse versus 1.6 for a 5-cm diameter in rat versus 45 for a 40-cm diameter in human), it is important to

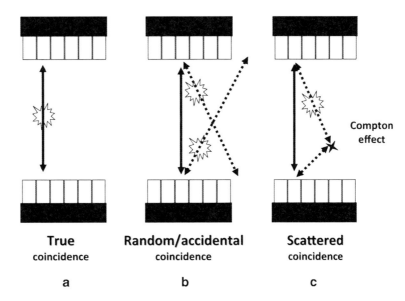

True coincidence   **Random/accidental** coincidence   **Scattered** coincidence

a   b   c

Fig. 8. Different types of coincidences: True (**a**), random (**b**) and scattered (**c**).

correct the data for attenuation in order to obtain accurate quantitative images of the tracer distribution (33).

In conclusion, in order to achieve quantitative images, the acquired data need to be corrected for random, scattered and attenuation effects during the image reconstruction process. Random coincidences are usually subtracted by the delayed coincidence method while scatter subtraction is based on model calculation. Correction for photon attenuation is by far the most important correction, especially on large animal brains.

2.2.3. Attenuation Correction

A significant fraction of the photons arising from annihilation events fail to reach the detectors due to their auto-attenuation in the body. In addition, at the surface of the body, the mean attenuation is less, as the LOR passes mostly through air. These differences in attenuation will result in distortions in the reconstructed image if not corrected for. If the anatomical properties of the subject/object are known, the measurement along each LOR can be corrected for the attenuation effect. Several methods of correcting for attenuation in PET exists: the material properties of the object/subject can be obtained using an external rotating pin single photon-emitting source, e.g. [^{68}Ge] or [^{57}Co], or by using an X-ray-computed tomography scanner (for more information, the reader is referred to Fahey et al. (33)). In this case, the attenuation coefficients are scaled to the appropriate energy

(511 keV) using tabulated values. The attenuation map is then projected along each LOR to give the attenuation coefficient of the coincidences.

2.2.4. Detector Design

Detectors for annihilation radiation should have good stopping power for 511 keV photons and the ability to resolve the time of arrival of incident photons within a few nanoseconds at most. Additionally, because these detectors are configured without collimators, they must be capable of operating in high photon flux environments. These extreme conditions have led to the development of specialized detectors. The most important practical features of scintillation detectors include a high mass density and a high effective atomic number to maximize the crystal stopping power. In addition, a fast crystal (short scintillation decay time) allows the use of a narrow coincidence window and therefore a notable reduction of the random coincidences count rate. Finally, the detector system should have a good energy resolution to reduce the fraction of detected scattered gamma rays as much as possible. Typical choices are bismuth Germinate (BGO), which has excellent stopping power, or lutetium oxyorthosilicate (LSO), which has good stopping power and excellent signal speed. Another choice is gadolinium silicate (GSO), which has intermediate stopping power but excellent signal speed and also offers improved energy resolution (Table 2).

2.2.5. Commercially Available Preclinical PET Scanners

While the first generations of PET tomographs were characterized by limited resolution and sensitivity, the latest generation of PET scanners combines a relatively high spatial resolution, a high sensitivity with a large field of view resulting in better counting statistics and improved region-of-interest definition due to decreased partial volume effects. Table 3 gives a brief overview of the main characteristics of several commercially available PET imaging systems.

Table 2
Properties of different solid scintillation detectors

Material	Density (g/cm³)	Atomic number	Decay time (ns)
NaI (TI)	3.76	11, 53	230
BGO	7.13	83, 32, 8	300
LSO	7.4	71, 32, 8	40
GSO	6.71	64, 32, 8	60
Csi (TI)	4.51	55, 53	1,000

Table 3
Properties of Preclinical PET scanners

	FOV (cm)		Resolution FWHM (mm)	Sensitivity (threshold)	Attenuation correction method	Crystal	Scintillation detector	Manufacturer
	Transaxial	Axial						
Triumph™	3.7/7.5	11.6	0.9	2/4%	CT	LYSO/LGSO	APD	Gamma Medica-Ideas
MOSAIC	12.8	11.6	2.7	1.3% (400 keV)	Cs-137	GSO	PMT	Philips
FOCUS 220	7.6	19	1.4	4%	Ge-68; Co-57	LSO	PSPMT	Siemens
FOCUS 120	10	7.6	1.3	6.5 (250 keV)	Ge-68	LSO	PSPMT	Siemens
INVEON	10	12.7	1.4	10% (100 keV)	Co-57; CT	LSO	PSPMT	Siemens
eXplore Vista	6	4.6	1.6	4% (250 keV)	Ge-68	LYSO-GSO	PMT	GE Healthcare
ClearPET	13/30	11	1.48	3.2% (350 keV)	ns	LYSO-LuYAP	PSPMT	Raytest
HIDAC	17	28	1.0	1.8% (200 keV)	ns	Position sensitive gas ionisation chamber		Oxford Positron systems
SHR-7700	33	16.3	2.6	ns	Ge-68	BGO	PMT	Hamamatsu Photonics KK

LSO lutetium oxyorthosilicate, *LYSO* lutetium yttrium oxyorthosilicate, *GSO* gadolinium orthosilicate; *Lu YAP* lutetium yttrium aluminium perovskite, *PSPMT* position-sensitive photomultiplier tubes, *APD* avalanche photomultiplier diode, *ns* not specified

2.3. Arterial Input Function and Metabolites

2.3.1. Arterial Input Function

The change of the radioligand concentration in function of time is called the arterial input function, AIF. The AIF is generally obtained by manual, repeated blood sampling at a rate which is at least as fast as the frame rate, followed by external counting of the plasmatic radioactivity concentration by a γ-counter. Besides its labour-intensive aspects, manual withdrawal presents several drawbacks, such as: repeated exposure of the staff to radioactivity; a limited temporal resolution and the collection of a relatively large amount of blood with possible significant effects on the physiological parameters of the animal. The latter aspect especially is the major obstacle to measure the AIF in small animals. Automatic sampling and the radioactivity counting of small blood volumes resolve some of the major issues of manual sampling. Weber et al. were the first to propose an automated continuous measurement of the arterial whole blood activity of small animals using a femoral a shunt, presenting any blood loss (34).

An alternative approach has been described by Pain and colleagues (35) who developed a beta-microprobe, which has thereafter been validated by other groups to quantify the AIF (36–38). The probe is a local β^+-radioactivity counter that takes the advantage of the limited range of positrons before annihilation occurs. For [^{18}F], it was estimated that the detection volume corresponding to 90% of the detected positrons is a cylinder of approximately 0.8 mm radius centred on the probe axis. However, the dimension of the artery and the position of the probe within the artery can significantly influence the amount of signal. Pain and colleagues solved this issue by normalizing the measured input function to the activity of a late blood sample (34). Again, the measured input function represents the whole blood time-activity curve (TAC), whereas compartmental modelling requires the plasma TAC.

A third, completely non-invasive alternative is the input function derived from PET images by drawing a region of interest over an intra-arterial or intra-cardiac volume within the image. However, image-derived AIF requires a large arterial tissue compartment lying within the field of view, as image-derived AIF from small arteries is inherent to partial volume effects. In addition, the whole blood TAC, and not the plasma TAC, is obtained by this method.

2.3.2. Detection of Metabolites and Radiopharmaceuticals

Metabolite analysis plays an essential role in PET pharmacokinetic modelling: for new tracers whose in vivo behaviour has not yet been characterized; for tracers that bind to targets for which there is no reference region and for studies in which it is essential to reduce the bias introduced by the use of simplified analysis methods. The failure to correctly determine the parent tracer fraction in arterial blood leads to an overestimation in the input function which leads to an underestimation of the binding potential (39).

The gold standard method for metabolite detection and quantification is high-pressure liquid chromatography, HPLC. The traditional method for metabolite analysis using HPLC involves deproteination of the plasma sample using an organic solvent (e.g. acetonitrile or methanol) or a strong acid (e.g. trifluoro acetic acid or perchloric acid), centrifugation of the resulting denatured protein emulsion and injection of the supernatant on the HPLC column. The radioactivity concentration is mostly determined using fraction collection and counting of the fractions using a gamma counter. Depending on the amount of activity that can be administered per scan, it may also be possible to use online detection for the early samples and using the gamma counter for the later time points only. However, care has to be taken for online detection as the sensitivity limits vary greatly (39).

An introduction on good HPLC practice and an overview of new methods for metabolite analysis are provided by Passchier (39), in which methods to reduce HPLC analysis time by increasing flow and temperature or reducing column length and particle size are considered. In addition, alternatives to HPLC are presented as well, such as ultra-high PLC (UHPLC) and the column-switching method. The latter allows the direct injection of the plasma sample onto an HPLC system by passing through a suitable pre-column using a low-organic-strength solvent (e.g. 1% acetonitrile in water). The flow is directed to UV, online radiodetector and/or fraction collector. Radioactive peaks coming through the pre-column reflect polar metabolites. After a set period of time, the pre-column is switched in line with a reversed-phase HPLC column for the non-polar analytes, including the parental compound to be eluted (39).

Other techniques for radiometabolite analysis include radio-thin layer chromatography (TLC) and SPE methods.

3. Methods

The initial step in the entire process of a PET imaging experiment is the production of the radioisotope by the cyclotron. The resulting radioisotopes – [^{15}O], [^{13}N], [^{11}C], [^{18}F] – are then, at the level of the radiochemistry unit, incorporated in a molecule by a one-step chemical reaction, creating the radiotracer or radioligand, which has a particular affinity for a specific target being a receptor, an enzyme, protein aggregation, etc. This radiotracer is then transported to the PET imaging facility where it is, most often intravenously, injected into the subject already present under the scanner. Data acquisition is most often started at the moment of tracer injection for a time period, depending on the radioactive decay of the selected isotope, the tracer kinetics itself and the aim of the experimental study. The PET camera then continuously register the coincidence

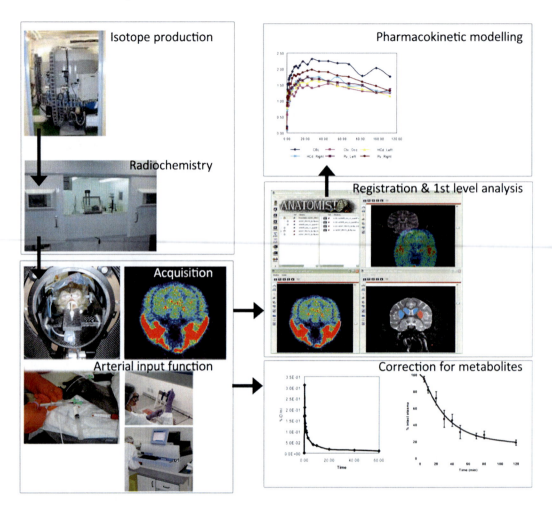

Fig. 9. The entire process of a PET imaging experiment.

of photon pairs in opposite detectors. Data are, thus, acquired in list mode and reconstructed off-line into quantifiable images using reconstruction algorithms. Radioactive concentrations present in each picture element (pixel) of PET images are colour-coded in such a way that each pixel reflects the tracer concentration in different brain tissues in Bq/cc, in the three-dimensional (3D) or four-dimensional space, the fourth being time. From these images, TACs can be obtained from pre-defined regions of interest, and later be modelled to different tissue compartment models in order to quantify binding, receptor density, etc. using commercially and/or (more generally) home-made software. Figure 9 represents the entire process of a PET imaging experiment.

3.1. Animal Handling

PET imaging is performed in live animals usually placed under chemical restraint, although a handful of groups have developed ways of performing imaging studies in awake animals. The choice of the anaesthetic compounds is crucial as direct and indirect interferences with the radiotracer kinetics are likely to occur. Very often, anaesthesia interferes with brain metabolism and receptor availability. For example: following isoflurane inhalation, blood glycaemia increases and different studies showed a reduction in the number of available DA reuptake sites (40) and an important inter-action with dopamine metabolism (41). The anaesthesia protocol, thus, depends on the animal model used, the acquisition protocol and the radiotracer used. Tables 4 and 5 give a brief overview of anaesthesia protocols used in different animal models of Parkinson disease (PD) in order to study the dopaminergic system.

Besides an adapted anaesthesia protocol, physiological moni-toring and especially thermoregulation of the animal are crucial to reduce inter-experimental variability (42).

After selection of the proper anaesthetic protocol, anaesthe-tized animals are positioned in the PET tomograph with their head held in a fixed position using a specific head holder designed such that it does not affect too heavily photon attenuation in the volume of interest and that precise positioning and reposi-tioning of the brain along stereotactic-like coordinates are guaranteed.

3.2. Acquisition of Transmission and Emission Scans

When the animal is positioned in the scanner, a blank scan without the subject and a transmission scan with the subject is obtained using a single-photon external source, e.g. [^{68}Ge] (20–40 min), or CT (a minute) to correct for γ-ray attenuation. Then, the radiotracer is injected intravenously as a bolus, bolus/infusion or infusion, either manually or – preferably – using a programmable infusion pump. The scanning protocol (which depends on the radiotracer used and the type of quantification desired, see Tables 4 and 5) may include from 1 (static scan) up to 30 or more frames (dynamic scan) with the first frame starting immediately after tracer injection and the last frame ending between 120 min and up to 240 min for ^{11}C and ^{18}F experiments, respectively. The tem-poral resolution of the reconstructed dynamic data is not con-stant, but it decreases as the imaging experiment lasts. As radionuclide counting statistics follow a Poisson distribution, the number of events is inversely related to uncertainty or noise. Consequently, at the start of the acquisition, when radioactive counts are elevated, the temporal resolution of the reconstructed data is high, though at the later time points when radioactive counts lower due to radioactive decay, the temporal resolution needs to be reduced to obtain longer time frames and better counting statistics. It is important that the temporal resolution of the reconstructed data fits well to the tracer kinetics allowing an accurate quantification of the PET data.

Table 4
PET imaging protocols in rodents

Neuro-transmitter system	Radiotracer	Species	Anaesthesia	Dose	Acquisition protocol	Data processing	Camera	Reference
DA synthesis	6-[^{18}F]-fluoro-L-DOPA + carbidopa, entacapone	Mouse	Isoflurane	3.9–9.4 MBq 200 μl	15 min post-injection; scan duration: 45 min	PMOD: ROI segmentation, time-activity curves in SUV	Quad-HIDAC tomograph	(69)
	6-[^{18}F]FMT* + carbidopa		Isoflurane	5.6–8.3 MBq 200 μl	8 min after injection, scan duration 120 min		Quad-HIDAC tomograph	
DAT	[^{18}F]-FECNT	Mouse	Isoflurane	3.3–19.8 MBq 200 μl	60 s after injection, scan duration 60 min	PMOD: ROI segmentation, time-activity curves in SUV	Quad-HIDAC tomograph	(69)
DA vesicles	[^{11}C]-dihydro-tetrabenazine	Rat	Keatmine/ xylazine	3.7 Mbq 100 g	30–60 min after injection	ASIPro6.0, ROI segmentation <ROI>/<CBL> - 1	μPET R4	(70)

	Tracer	Species	Anesthesia	Dose	Scan protocol	Analysis	Scanner	Ref
D2 receptors	[18F]-Fallypride	Mouse	Isoflurane	2.8–8.6 MBq 2.5–4.4 nmol/kg Max 200 μl	Dynamic 150 after injection Different reconstructions	PMOD: Time-activity curves in SUV; comparison simplified tissue ref. model and Ichise method (without blood sampling)	Quad-HIDAC tomograph	(71)
	[11C] Raclopride	Rat	Halothane	18.5 MBq	Dynamic for 1 h Scan 60 min post-injection; scan time 60 min	(ROI-CBL)/CBL	μPET P4	(72)
	[18F]-NMB (N-methyl-benperidol)		Ketamine/ xylazine	37 MBq/ml 0.3 ml		ROI segmentation after MRI coregistration	TierPET	(73)
	[3-(2*-18F] fluoroethyl- spiperone			55–74 MBq	Static, 3 h after injection; scan time = 30 min	Ratio ROI vs. CBL; ROI lesioned vs. unlesioned	microPET	(74)
Glucose metabolism	[18F] fluordeoxy-glucose	Mouse	Halothane	7.4 MBq	Dynamic for 1 h	ROI segmentation, time-activity curves SUV calculation	μPET P4	(72)
	[18F] fluordeoxy-glucose	Rat	Isofluraan (awake)	55 MBq	Dynamic for 1 h	Time-activity curves Technique of Patlak (blood sampling)	ATLAS small animal PET scanner	(75)
			Ketamine/ xylazine	74 MBq (max volume 0.7 ml)	Static scan after 45 min; scan time 60 min	Ratio ROI vs. CBL; ROI lesioned vs. unlesioned	microPET	(74)

Table 5
PET imaging protocols in non-human primate MPTP model

Neurotransmitter system	Radiotracer	Anaesthesia	Dose	Acquisition protocol	Data processing	Camera	Reference
Monoamines	VMAT2 (DA, NE, 5HT)	[¹¹C]-DTBZ	Induction: ketamine (10 mg/kg) + midazolam (1 mg/kg) Maintenance: ketamine (5 mg/kg) + imidazolam 0.5 mg/kg	79 MBq	40 min	Multilinear Reference Tissue Model (PMOD Technologies)	Philips Mosaic tomograph	(76, 77)
			Maintenance: isoflurane 1–2%	185 MBq	60–90 min	Logan	ECAT 953-31B tomography	(77)
Dopamine	DAT	[¹¹C]-CFT	Propofol (6 mg/kg/h)	187 MBq	60 min	SRTM	ECAT EXACT HR+SHR-7700	(78)
	[¹¹C]-PE2I	Awake		61 MBq	90 min	SRTM		(79)
AADC	6-[¹⁸F]-fluoro-L-DOPA (pretreatment carbidopa)	Induction: ketamine (10 mg/kg) + imidazolam 1 mg/kg Maintenance: ketamine (5 mg/kg) + imidazolam 0.5 mg/kg	83 MBq	100 min	Patlak graphical analysis,	Philips Mosaic tomograph	(76)
		Maintenance: isoflurane 1–2%	185 MBq	90 min		ECAT 953-31B tomography	(77)

[18F]-FMT benzerazide	Induction: ketamine (15 mg/kg) Maintenance: methoxyflurane	370–555 MBq	90 min	Three-compartment, three-kinetic, rate-constant model	Siemens-CTI ECAT EXACT HR 47-slice scanner	(80, 81)
D2 [11C]-raclopride	Awake	84 MBq	90 min	Simplified reference tissue model	SHR-7700 PET camera	(79)
	1.5 mg/kg alfadolone and alfaxolone acetate (Saffan®; Arnolds Veterinary Products, Shropshire, UK) Maintenance: 9–14 mg/kg/h Saffan	Bolus + (B/I) 0.5, 13, and 80 nmol/ kg	2.5 h	Scatchard analysis	Siemens Exact ECAT HR+ whole-body tomograph in 3-D mode	(82)

3.3. Image Reconstruction

The PET data are acquired as a list of individual events, list mode, providing a high temporal with full spatial resolution. List-mode data can then be binned into sinograms with the required spatial and temporal sampling for reconstruction. In order to achieve quantitative images, the acquired data need to be corrected for random, scatter and attenuation effects, as well as radioactive decay and the dead time of the detector system. In addition, the acquired data need to be corrected for detector inhomogeneity using the normalization data. Depending on the exploitation of the reconstructed images, different reconstruction methods are preferred. A commonly used analytic reconstruction algorithm is the filtered back projection (FBP). FBP is an analytic technique that produces a representation of the object within the limit of an infinite number of projection angles and noise-free data. However, when the data are noisy, FBP produces images with streak artefacts. Today, most image reconstruction techniques are iterative and use statistical iterative reconstruction algorithms, such as the ordered subsets expectation maximization (OSEM) algorithm, which has become the method of choice in many cases. Their main advantage is that they possess different noise properties to FBP. Briefly, starting from an initial guess for an activity distribution, data are forward projected according to the scanner geometry. The resulting projections are compared to the measured projections and the error correction is used for correcting the estimate. The new estimate is then forward projected and the comparisons between estimated and measured projections yield the next correction. This loop is iterated until estimated and measured projections agree with their statistics (43).

3.4. Data Processing

3.4.1. Image Registration

If PET can provide absolute quantitative images of brain function, it is not a method that allows simple correlation of these functional findings with the underlying anatomy. This frequently necessitates both specific hardware and software developments. Recent technological developments in the field, aiming to combine various imaging modalities within the same piece of equipment (PET–CT, PET–MRI), allow a very good correlation between two image sets, one representing function (PET) and the other anatomy (CT or MRI). However, preclinical hybrid scanners only exist for small animals like rodents but not for larger animals, such as the non-human primate. In addition, many of the preclinical imaging centres do not have access to hybrid scanners and, thus, need to co-register the PET data onto MRI data sets.

Co-registration of images from different modalities or alignment of images from repeated acquisitions of the same modality is significantly facilitated when the animal positioning is independent from the imaging modality. This can be obtained by using specific head holders that are PET- and/or MRI-compatible and that position the animal's head in the same configuration (Fig. 10, top,

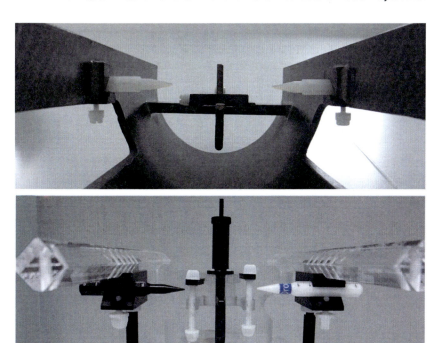

Fig. 10. Front view of two stereotactic-like positioning devices: Top, the macaque holder for SIEMENS Concorde 220 microPET and bottom, the macaque holder for the 7 tesla Varian MRI scanner, courtesy MIRCen (CEA, Fontenay-aux-Roses, France).

the macaque holder for SIEMENS Concorde 220 microPET and bottom, the macaque holder for the 7 tesla Varian MRI scanner; courtesy MIRCen (CEA, Fontenay-aux-Roses, France)). This allows an accurate, stereotactic-like positioning of the head of the animals into each of the imaging modalities, and a precise repositioning of the same animal's brain over longitudinal studies.

The second step of the co-registration process uses registration software that is commercially available, custom-made or freely available on the Internet (http://www.brainvisa.info/index_f.html developed by LNAO, Neurospin, CEA, Saclay, France). There are three basic steps in performing image registration: converting the image data sets to a common image format, determining the transformation required to register one of the image sets with the other and displaying the results. Figure 9 illustrates co-registration of PET and MRI image of a macaque using the freely available Brainvisa/Anatomist software.

Ideally, anatomical data are obtained from the same animal; however, not every PET imaging facility has an MRI system available. An acceptable compromise might be the use of population-based atlases or brain templates, many of which are freely available on the Internet.

Table 6
Feuly available MRI brain templates/atlases

Mouse C57/BL6J	http://www.loni.ucla.edu/MAP http://www.mbl.org/mbl_main/atlas.html	
Rat Sprague-dawley	http://expmr.ki.se/research/ratatlas.jsp	(83)
Macaca Mulatta	http://brainmap.wisc.edu/monkey.html http://www.bic.mni.mcgill.ca/ServicesAtlases/HomePage	(84, 85)
Fascicularis	www.nil.wustl.edu/labs/kevin/ni/cyno/cyno.html	(86)
	http://www.cima.es/labs-en/instrumental-techniques-micropet/ technologies/1	(44)
Nemestrina	http://www.nil.wustl.edu/labs/kevin/ni/n2k/ http://www.loni.ucla.edu/Atlases/Atlas_Detail.jsp?atlas_id=2	(87) (88)
Baboon Papio anubis	http://www.nil.wustl.edu/labs/kevin/ni/b2k/	(86)

Table 6 shows some of the freely available MRI brain atlases for the mouse, rat and non-human primate brain.

An alternative to MRI/PET co-registration is the registration to radiotracer-specific PET templates, as has been reported for the macaque 6-[^{18}F]-fluoro-L-DOPA, [^{11}C]-DTBZ) (44) and the rat brain ([^{18}F]-FECT, [^{11}C]-raclopride, [^{18}F]-FDG) (45). In addition, mapping in vivo imaging data to in vitro histological data has been introduced recently as a way to co-register functional images onto the actual anatomy of the studied animal (46, 47). Dauget and colleagues presented a detailed protocol to achieve a 3D reconstruction of a series of 2D stained histological sections of a primate brain and to match the reconstructed histological volume with in vivo MRI images of the same animal. They applied the same methodology for specific brain regions, such as the thalamus, and the resulting 3D digital atlas was then used to delineate and analyze PET data. The more precise segmentation of the thalamic nuclei resulting from the superior contrast obtained by histological staining allowed discerning significant differences in the kinetics of the radiotracer between different thalamic nuclei, which might be related to differences in receptor densities in these regions of interest (47).

3.4.2. First-Level Analyses   Volume of interests (VOIs) corresponding to the anatomical structures of interest are manually or automatically segmented on MR images and then projected onto the corresponding PET images applying the transformation obtained from the PET–MRI registration.

The mean radioactive concentrations (expressed in Bq/ml) of each individual VOI can then be calculated for each PET frame and plotted as a function of time, yielding the so-called TACs which constitute the first level of quantification of the PET data.

The raw data obtained from the PET images (in Bq/ml) is dependent on the injected activity and mass of the radioligand, but also on the weight of the animal. In order to compare TACs obtained from different acquisitions, the raw data need to be normalized for these variables. Knowing the specific activity of the radiotracer (defined in MBq/mole of tracer), the injected activity and the weight of the animal, TACs can be expressed, respectively, as: the mass of tracer per volume unit of brain tissue (mole/ml), the percentage of injected dose per volume unit (%ID/ml) or the standard uptake value (SUV, %ID/ml/kg) in function of the time.

3.4.3. Pharmacokinetic Modelling of the PET Data

It is important to differentiate *absolute quantification* of radioactive events, which is essentially a counting problem, from *biological quantification* which aims to derive a biologically relevant measure (receptor density, receptor affinity, enzymatic reaction rate, etc. ...) from the raw activity concentrations measured by the PET scanner. One of the challenges of in vivo PET studies is, thus, to provide accurate values of these biological parameters using experimental protocols easily applicable in routine examinations in living subjects. This step is the pharmacokinetic compartmental modelling of the TAC and the generation of biologically pertinent data, such as the uptake rate constant (K_i value) determined in caudate/putamen for 6-[^{18}F]-fluoro-L-DOPA using the multiple-time graphic analysis or the in vivo [^{11}C]-PE2I-binding kinetics (B_{max}, K_d) quantified using a more complex multi-injection modelling approach (48, 49).

The term compartment does not necessarily coincide with the boundaries of an organ or its substructures, but refers to a physiologically separated (whether in space or time) pool of radioligand. The time course of the ligand concentration in the different compartments is described by rate equations. For example, brain regions containing receptors have at least three compartments (also referred to as the two-tissue compartment). Radioligands are delivered in arterial blood, which can be considered as the first compartment. From the arterial blood, the radioligand passes the BBB and gains access to the second compartment known as the free compartment, which anatomically coincides with the interstitial fluid and intracellular cytoplasm. The third compartment is the region of specific binding that contains the high-affinity receptors. One may also consider a non-specific binding compartment exchanging with the free compartment. However, in practice, for most radioligands the non-specific binding compartment is in rapid equilibrium with the free compartment and these two compartments are treated as a single compartment, often referred to as the non-displaceable compartment (50). Full kinetic modelling, thus, requires the concentration of the available, i.e. non-metabolized and non-bound to blood cells or proteins, ligand in the arterial plasma. Given that the fraction of the non-metabolized radiotracer rapidly decreases after intravenous administration, it is important to correct the plasma input function curve for peripheral metabolism in order to avoid the overestimation of the real amount of unchanged radiotracer reaching the brain. The metabolites and the rate of the metabolism are determined by HPLC analysis using a fairly large number of arterial blood samples. The true arterial ligand concentration is then derived from the plasma concentration by multiplication with the instantaneous parent fraction, i.e. the time-dependent unmetabolized fraction of the substance (51). An example of a full compartment model is presented in Fig. 11.

However, the AIF is not sufficient to determine all rate constants of the multiple compartment, and complex experimental paradigms, such as a multi-injection approach, are needed to solve the different equations defining the model that describes the complex evolution of the radiotracer within the various compartments present in the living organism under study. Such a multi-injection/ multi-compartmental modelling approach was used by Delforge and co-workers (49) to quantify the in vivo interactions between the D_2 receptor sites and ^{11}C-FLB457 in the baboon brain. The model was composed of four compartments (plasma, free ligand in the brain, non-specific and specific binding in the brain) and seven parameters (including the D_2 receptor site density). The arterial

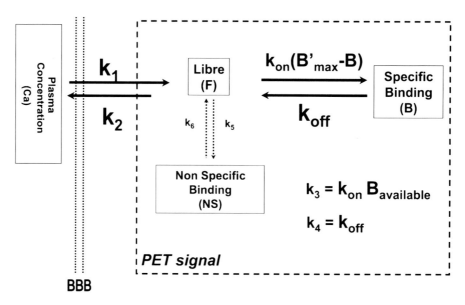

Fig. 11. An example of a full compartment model indicating the plasma compartment that exchanges with the free compartment over the blood brain barrier (BBB). The free compartment exchanges with the non-specific binding compartment and the specific binding compartment. The exchange rates between the compartments are represented by the kinetic constant, *k*.

plasma concentration, corrected for metabolites, was used as the input function describing the rate of blood-to-brain transfer of the radiotracer. The experimental protocol, which consisted of three injections of labelled and/or unlabelled ligand, allowed the evaluation of all model parameters from a single PET experiment yielding direct estimates of the receptor density (B_{max} values) and apparent affinity (K_d values) of the tracer in different brain regions.

However, such multi-injection compartmental modelling protocols may prove difficult to implement in routine research and particularly in humans as an important number of blood samples need to be withdrawn and because of the difficulties encountered in accurately measuring the arterial input function and to correctly identifying the compartment configuration. Therefore, the four compartment model is simplified to a three-compartment model under the assumption of a rapid equilibrium between free and non-specifically bound compartments resulting in a single compartment of free and non-specifically bound ligand. However, the three-compartment model still requires the plasma concentration of the unmetabolized ligand, which might be problematic for some experimental set-ups. In these situations, simplified modelling approaches might be suitable.

Simplified modelling means that there are fewer compartments, and thus less-complex parameter estimation. This reduces the calculation time and the complexity of the experimental set-up, but it might as well induce bias in the observations as the real situation has been simplified. Among the simplified modelling approaches exist: the linear approaches, the so-called graphical analyses (such as the Patlak and Logan plot), the reference tissue models and the equilibrium approaches. Before using these simplified models, some assumptions need to be made, such as the achievement of a rapid equilibrium, the lack of specific binding in the reference tissue and the absence of radiometabolites in the brain. The sacrifice of these simplified models is the bias you might introduce, the inability to quantify absolutely the micro-parameters (i.e. rate constants, such as k_2, k_3) and K_d or B_{max}. Instead, these models allow the calculation of macro-parameters, such as the distribution volume, the binding potential and K_1. For every new tracer, it is indispensable to validate the simplified approach with a higher compartment model. Reference tissue models assume that in the brain exists a region, a reference tissue, in which the target is not present but where the radioligand is assumed to display a similar pharmacokinetic profile as in the specific compartment. The reference region, thus, provides an estimate of the so-called non-specific binding of the non-displaceable compartment. The graphical methods, such as the logan plot for reversible ligands (e.g. [11C]-Raclopride) or the Patlak for irreversible ligand (e.g. 6-[18F]-fluoro-L-DOPA), generally allow a quick estimation of the biological parameter of interest by graphically fitting a straight line to experimental data using regression analysis. Other methods like the true equilibrium methods are also quite robust, but their use is largely limited to experiments with tracers that are suitable for constant infusion.

In conclusion, not a single method is applicable to all PET tracers, but rather a careful case-by-case analysis should be conducted to select the best model-based methods required for each radiotracer. It is well beyond the scope of the present chapter to detail existing modelling methods and the reader is referred to more specialized review on these subjects (52).

4. Notes: Biological Applications

4.1. Receptor–Ligand Interaction Studies

Probing in vivo the regional distribution and affinity of receptors in the brain (of rodent to non-human primates and humans) has become possible by the use of selective PET ligands. The methodology used to characterize a ligand–receptor binding in vivo is largely inspired on that used to characterize receptors in vitro.

This includes:

1. The comparison of the regional distribution of the radiotracer with the known regional distribution of the receptors under study.

2. The assessment of saturability of the tracer binding in vivo using either saturation studies by co-injection or injection of an unlabelled analogue prior to the radiotracer and/or by dose-dependent displacement paradigms.

3. The assessment of (when possible) the stereo-specificity of the ligand by displacement with the "non-active" enantiomer (when chiral analogues are available for the target receptor, only the pharmacologically active enantiomer should displace in vivo the radiotracer from its binding sites).

4. The assessment of whether or not dose-dependent displacements (reflecting increasing receptor occupancy in vivo) are correlated with increased pharmacological effects (53).

One example of such an approach was the in vivo characterization of the so-called benzodiazepine receptor central-type first published in 1984 (54). The TAC measured in a region of interest (total binding) reflects both the fraction of specific and non-specific binding of the radiotracer. Using a dose-dependent displacement of the radioligand ([^{11}C]-flumazenil) with a non-radioactive receptor analogue (Fig. 12) allows evaluation of the specificity and reversibility of the radioligand binding. The dose-related decrease in the radioactivity present in the region of interest reflects the actual competition between the radiotracer (initially bound on its receptor)

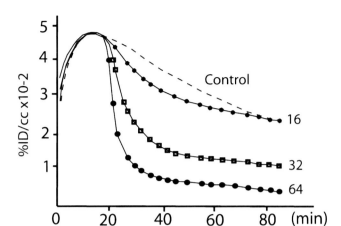

Fig. 12. Dose-dependent displacement of the radioligand ([^{11}C]-flumazenil) with a non-radioactive receptor analogue, adapted from (54).

and cold compound progressively chasing the tracer at the binding domain. This paradigm allows determination of the dose-dependent displacement which can be plotted as a function of the increasing dose of displacing compound (right panel in Fig. 13), as is in general done for in vitro studies, and from which the affinity of the displacer for the receptor can be estimated in vivo. Moreover, using the very same approach, Schmid and colleagues (55) were able to demonstrate the presence of several benzodiazepine receptor subtypes in various sub-regions of the primate brain, identifying a high- and a low-affinity binding sites in several cortical areas in contrast to a single high-affinity binding sites in the cerebellum.

These initial studies with [^{11}C]-Flumazenil (54) validated this radiotracer as a very selective benzodiazepine receptor ligand, and because of its ubiquitous distribution on neurons as one of the best biomarkers of neuronal density and viability, proven valuable in various pathologies associated with cell loss, such as amyotrophic lateral sclerosis, epilepsy, panic disorder or Machado-Joseph disease.

Interestingly, in vivo displacement of [^{11}C]-Flumazenil by several non-labelled ligands selective for the same binding site but with different pharmacological profiles (agonist, partial agonist, antagonist, partial inverse and inverse agonists) combined with electroencephalographic activity recordings established the correlation between benzodiazepine receptor occupancy and anticonvulsant/convulsant action of different drugs acting at the benzodiazepine–GABA receptor complex (54, 56). As shown in Fig. 13, agonists showed positive linear correlations, with full agonists having a steeper slope as compared to partial agonists, whereas the correlation with inverse agonists displayed negative slopes, with partial inverse agonists having a less-steep slope than full inverse agonists.

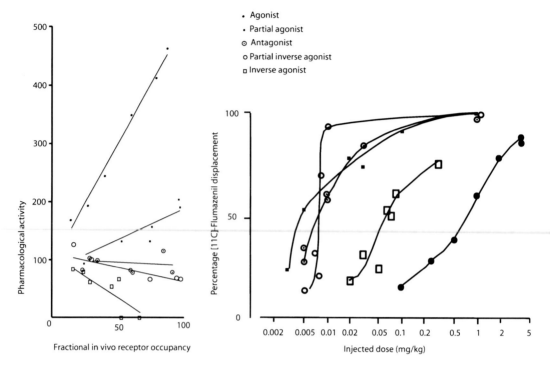

Fig. 13. *Left panel*, pharmacological activity in function of receptor occupation as defined by PET imaging: Agonists showed positive linear correlations. *Right panel*, displacement plotted as a function of the increasing dose of displacing compound, adapted from (54, 56).

4.2. Assessing Brain Diseases In Vivo

4.2.1. Example 1: Parkinson's Disease

Parkinson's disease is a neurodegenerative disorder which manifests clinically when more than 80% of the midbrain dopaminergic neurons are lost. This dopaminergic neuronal loss and dysfunction can be detected by PET imaging well before the appearance of clinical symptoms using radioligands selectively targeting (1) the dopamine reuptake system at the pre-synaptic level neuron; (2) dopamine metabolism by using the false dopamine precursor 6-[^{18}F]-fluoro-L-DOPA or (3) the post-synaptic level of the dopaminergic synapse (for review, see Cropley et al. (57)). Targeting the pre-synaptic neuron reflects the neuronal loss and results in a decreased binding as compared to the healthy brain. Targeting the post-synaptic neuron might detect the compensatory mechanisms brought into play following the loss of pre-synaptic inputs, and hence an increased binding of the radioligand as compared to the healthy brain. The increase is either due to an up-regulation of the post-synaptic receptor or an increased affinity of the binding sites resulting from the loss of endogenous ligand, i.e. dopamine. These direct methods of pre- and post-synaptic imaging of dopamine neurotransmission system have been largely used in the preclinical and clinical assessment of lesion intensity and the efficacy of different therapies, such as cell transplantation, neuroprotection and/or neurotrophic factor strategies.

Specific developments were recently made to assess changes in the concentration of the endogenous dopamine. In these experimental set-ups, the PET ligand acts as a detector of changes in the concentration of an endogenous competitor via displacement of the tracer. Morris and colleagues (58) evaluated the displacement sensitivity of several high-affinity ligands of the D_2 receptor. They demonstrated that [^{11}C]-raclopride had the highest sensitivity, followed by [^{18}F]-Fallypride, [^{18}F]-FESP, [^{18}F]-FLB, [^{11}C]-NMSP and [^{11}C]-Epidepride (lowest). Assessment of changes in the endogenous dopamine concentration is evaluated by using a dopamine-releasing agent, met-amphetamine, which induces a dose-related decrease of the [^{11}C]-raclopride post-synaptic binding in vivo. In this paradigm, decreased D_2-receptor occupancy by endogenous dopamine in the putamen of parkinsonian animal models or untreated PD patients results in an increased binding potential of the radioligand, and thus a reduction in amphetamine-induced displacement of [^{11}C]-raclopride-specific binding (59, 60). Recently, a new quantification method was presented to separately assess changes in binding affinity and D_2 receptor density in one single experiment using [^{11}C]-raclopride. This new approach showed that the increased [^{11}C]-raclopride signal in the DA-depleted striatum of parkinsonian animals was essentially attributable to an increased affinity of [^{11}C]-raclopride binding to the D_2 receptors. After gene therapy, mediated by adeno-associated viral vectors encoding tyrosine-hydroxylase (AAV5–TH) and its cofactor (rAAV5–GCH1), these changes were reversed essentially through a restoration of the endogenous pool of dopamine. Importantly, the method showed that the extent of the restoration in the [^{11}C]-raclopride binding affinity was well-correlated with the amount of DA synthesized by the viral vectors as well as the recovery in spontaneous motor behaviours (61).

4.2.2. Example 2: Neuroinflammation

In many neurodegenerative diseases, such as Parkinson's and Huntington's diseases, neuronal loss is often associated with (and likely preceded by) neuroinflammation (62). The PET imaging target of neuroinflammation is the 18 kDa Translocator Protein (TSPO) also referred to as the peripheral benzodiazepine receptor (PBR). The rationale for using TSPO as a marker of inflammatory processes is that this molecular entity, which forms part of the mitochondrial transition pore, is expressed on the outer mitochondrial membrane of activated microglia and/or astrocytes. The large increase of this protein observed during neuroinflammation makes TSPO a potential biomarker and an attractive target for in vivo assessment of microglial/astrocytic activation using PET. Despite several disadvantages (such as a high non-specific binding), the gold standard radioligand for TSPO remains [^{11}C]-PK11195 (63–65). Nevertheless, over recent years, many other radioligands have been developed and evaluated, and their evaluation in different animal models has been extensively reviewed by Chauveau and colleagues (66). In general, all these ligands have been evaluated for their in vivo or in vitro biodistribution, for their metabolism and their specificity for TSPO as determined by displacement and/or pre-saturation PET studies using fixed doses of a cold-competitive compound. However, dose-dependent displacement studies as performed for the [^{11}C]-flumazenil have, to our knowledge, not been performed for any of the candidate TSPO ligands. Nevertheless, a recent in vitro binding study by Owen and colleagues suggested that several TSPO ligands, among which [^{11}C]-PBR28, [^{11}C]-DAA116 and [^{11}C]-DPA713, display different binding patterns, also defined as high-affinity, low-affinity and medium (two-site)-affinity binding. The knowledge on the in vivo binding affinities of each TSPO radioligand is, thus, indispensable for a correct interpretation of ligand binding and for an accurate quantification of TSPO expression in vivo using PET. Indeed, due to different affinity sites, a reduction in radioligand binding might reflect a change in affinity for TSPO and not decreased density of the TSPO (67, 68).

References

1. Pike VW (2009) PET radiotracers: crossing the blood-brain barrier and surviving metabolism. Trends Pharmacol Sci 30: 431–440

2. Allard M, Fouquet E, James D, Szlosek-Pinaud M (2008) State of art in C-11 labelled radiotracers synthesis. Curr Med Chem 15: 235–277

3. Jewett DM (1992) A simple synthesis of [11C] methyl triflate. Int J Rad Appl Instrum A 43: 1383–1385

4. Maziere M, Hantraye P, Prenant C, Sastre J, Comar D (1984) Synthesis of Ethyl 8-Fluoro-5,6-Dihydro-5-[C-11]Methyl-6-Oxo-4H-Imidazo[1,5-A][1,4]BenzodiaZepine-3-Carboxylate (Ro 15.1788-11C) - A Specific Radioligand for the Invivo Study of Central Benzodiazepine Receptors by Positron Emission Tomography. Int J Appl Radiat Isot 35: 973–976

5. Langer O, Någren K, Dolle F, Lundkvist C, Sandell J, Swahn CG, Vaufrey F, Crouzel C, Maziere B, Halldin C (1999) Precursor synthesis and radiolabelling of the dopamine D2 receptor ligand [11C]raclopride from [11C] methyl triflate. J Label Compd Radiopharm 42: 1183–1193

6. Hamill TG, Krause S, Ryan C, Bonnefous C, Govek S, Seiders TJ, et al. (2005) Synthesis, characterization, and first successful monkey imaging studies of metabotropic glutamate receptor subtype 5 (mGluR5) PET radiotracers. Synapse 56: 205–216

7. Yu M, Tueckmantel W, Wang X, Zhu A, Kozikowski AP, Brownell AL (2005) Methoxyphenylethynyl, methoxypyridylethynyl and phenylethynyl derivatives of pyridine: synthesis, radiolabeling and evaluation of new PET ligands for metabotropic glutamate subtype 5 receptors. Nucl Med Biol 32: 631–640

8. Maiti DK, Chakraborty PK, Chugani DC, Muzik O, Mangner TJ, Chugani HT (2005) Synthesis procedure for routine production of [carbonyl-11C]desmethyl-WAY-100635. Appl Radiat Isot 62: 721–727

9. Långström B, Itsenko O, Rahman O (2007) [11C]Carbon monoxide, a versatile and useful precursor in labelling chemistry for PET-ligand development. J Label Compd Radiopharm 50: 794–810

10. Rahman O, Kihlberg T, Långström B (2002) Synthesis of N-methyl-N-(1-methylpropyl)-1-(2-chlorophenyl)isoquinoline-3-[C-11]carboxamide ([C-11-carbonyl]PK11195) and some analogues using [C-11]carbon monoxide and 1-(2-chlorophenyl)isoquinolin-3-yl triflate. J Chem Soc Perkin Trans 23: 2699–2703

11. Dolle F, Valette H, Bramoulle Y, Guenther I, Fuseau C, Coulon C, Lartizien C, Jegham S, George P, Curet O, Pinquier JL, Bottlaender M (2003) Synthesis and in vivo imaging properties of [C-11]befloxatone: A novel highly potent positron emission tomography ligand for mono-amine oxidase-A. Bioorg Med Chem Lett 13: 1771–5

12. Bramoulle Y, Roeda D, Dolle F (2010) A simplified [C-11]phosgene synthesis. Tetrahedron Lett 51: 313–316

13. Iwata R, Ido T, Takahashi T, Nakanishi H, Iida S (1987) Optimization of [C-11] HCN Production and No-Carrier-Added [1-C-11]Amino Acid Synthesis. Appl Radiat Isot 38: 97–102

14. Gillings NM, Gee AD (2001) Synthesis of [4-C-11]amino acids via ring-opening of aziridine-2-carboxylates. J Lab Compd Radiopharm 44: 909–920

15. Mathews WB, Monn JA, Ravert HT, Holt DP, Schoepp DD, Dannals RF (2006) Synthesis of a mGluR5 antagonist using [C-11]copper(I) cyanide. J Lab Compd Radiopharm 49: 829–834

16. Roeda D, Dolle F (2010) Aliphatic Nucleophilic Radiofluorination. Current Radiopharmaceuticals 3: 81–108

17. Ermert J, Coenen HH. Nucleophilic (2010)18F-Fluorination of Complex Molecules in Activated Carbocyclic Aromatic Position . Current Radiopharmaceuticals 3, 109–126.

18. Ermert J, Coenen HH (2010)No-Carrier-Added [18F]Fluorobenzene Derivatives as Intermediates for Built-up Radiosyntheses . Current Radiopharmaceuticals 3, 127–160.

19. Neirinckx RD, Lambrecht RM, Wolf AP (1978) Cyclotron isotopes and radiopharmaceuticals--XXV An anhydrous 18F-fluorinating intermediate: Trifluoromethyl hypofluorite. The International Journal of Applied Radiation and Isotopes 29: 323–327

20. Ehrenkaufer RE, Potocki JF, Jewett DM (1984) Simple Synthesis of F-18-Labeled 2-Fluoro-2-Deoxy-D-Glucose - Concise Communication. Journal of Nuclear Medicine 25: 333–337

21. de Vries EFJ, Luurtsema G, Brussermann M, Elsinga PH, Vaalburg W (1999) Fully automated synthesis module for the high yield one-pot preparation of 6-[F-18]fluoro-L-DOPA. Applied Radiation and Isotopes 51: 389–394

22. Lee SJ, Oh SJ, Chi DY, Lee BS, Ryu JS, Moon DH (2008) Comparison of synthesis yields of 3'-deoxy-3'-[F-18]fluorothymidine by nucleophilic fluorination in various alcohol solvents. J Lab Compd Radiopharm 51: 80–82

23. Mukherjee J, Yang ZY, Das MK, Brown T (1995) Fluorinated Benzamide Neuroleptics. 3. Development of (S)-N-[(1-Allyl-2-Pyrrolidinyl) Methyl]-5-(3-[F-18] Fluoropropyl)-2, 3-Dimethoxybenzamide As An Improved Dopamine D-2 Receptor Tracer. Nuclear Medicine and Biology 22: 283–96

24. Tan PZ, Baldwin RM, Fu T, Charney DS, Innis RB (1999) Rapid synthesis of F-18 and H-2 dual-labeled altanserin, a metabolically resistant PET ligand for 5-HT2A receptors. J Lab Compd Radiopharm 42: 457–467

25. Dolle F, Dolci L, Valette H, Hinnen F, Vaufrey F, Guenther I, Fuseau C, Coulon C, Bottlaender M, Crouzel C (1999) Synthesis and nicotinic acetylcholine receptor in vivo binding properties of 2-fluoro-3-[2(S)-2-azetidinylmethoxy]pyridine: A new positron emission tomography ligand for nicotinic receptors. Journal of Medicinal Chemistry 42: 2251–2259

26. Zhang MR, Suzuki K (2007) [F-18]Fluoroalkyl agents: Synthesis, reactivity and application for development of PET ligands in molecular imaging. Current Topics in Medicinal Chemistry 7(18): 1817–1828

27. Ross TL. (2010)The click chemistry approach applied to fluorine-18 . Current Radiopharmaceuticals 3(3), 202–223

28. Dolle F, Helfenbein J, Hinnen F, Mavel S, Mincheva Z, Saba W, et al. (2007) One-step radiosynthesis of [F-18]LBT-999: a selective radioligand for the visualization of the dopamine transporter with PET. J Lab Compd Radiopharm 50: 716–723

29. Horti A, Redmond DE, Soufer R. No-Carrier-Added (NCI) (1995) Synthesis of 6-[F-18] Fluoro-L-Dopa Using 3,5,6,7,8,8A-Hexahydro-7,7,8A-Trimethyl-[6S-(6-Alpha,8-Alpha,8-Alpha-Beta)]-6,8-Methano-2H-1,4-Benzoxazin-2-One. J Lab Compd Radiopharm 36: 409–423

30. Vaidyanathan G, Zalutsky MR (1992) Labeling Proteins with F-18 Using N-Succinimidyl 4-[F-18]Fluorobenzoate. Nucl Med Biol 19: 275–281

31. de Bruin B, Kuhnast B, Hinnen F, Yaouancq L, Amessou M, Johannes L, Samson A, Boisgard R, Tavitian B, Dolle F (2005) 1-[3-(2-[F-18] fluoropyridin-3-yloxy)propyl]pyrrole-2,5-dione: Design, synthesis, and radiosynthesis of a new [F-18]fluoropyridine-based maleimide reagent for the labeling of peptides and proteins. Bioconjug Chem 16: 406–420

32. Conti M (2009) State of the art and challenges of time-of-flight PET. Physica Medica 25: 1–11

33. Chow PL, Rannou FR, Chatziioannou AF (2005) Attenuation correction for small animal PET tomographs. Phys Med Biol 50: 1837–1850

34. Weber B, Burgee C, Bivo P, Buck A (2002) A femoral arteriovenous shunt facilitates arterial whole blood sampling in animals. Eur Jmucl med 29: 319–323

35. Pain F, Laniece PL, Mastrippolito R, Gervais P, Hantraye P, Besret L (2004) Arterial input function measurement without blood sampling using a beta-microprobe in rats. J Nucl Med 45: 1577–1582

36. Pain F, Dhenain M, Gurden H, Routier AL, Lefebvre F, Mastrippolito R, Laniece P (2008) A method based on Monte Carlo simulations and voxelized anatomical atlases to evaluate and

correct uncertainties on radiotracer accumulation quantitation in beta microprobe studies in the rat brain. Phys Med Biol 53: 5385–5404

37. Maramraju S, Stoll S, Woody C, Schlyer D, Schiffer W, Lee D, Dewey S, Vaska P (2007) A LSO beta microprobe for measuring input functions for quantitative small animal PET. IEEE 33rd Anuual Northeast Bioengineering Conference 56–57

38. Schlyer DJ, Stoll SP, Woody CL, Vaska P, Shokouhi S, Volkow ND, Fowler JS (2002) A beta microprobe for in situ measurement of the blood radioactivity curve in PET scanning of small animals. J Nucl Med 43: 211

39. Passchier J (2009) Fast high performance liquid chromatography in PET quality control and metabolite analysis. Q J Nucl Med Mol Imaging 53: 411–416

40. Votaw J, Byas-Smith M, Hua J, Voll R, Martarello L, Levey AI, Bowman FD, Goodman M (2003) Interaction of isoflurane with the dopamine transporter. Anesthesiology 98: 404–411

41. Adachi YU, Yamada S, Satomoto M, Higuchi H, Watanabe K, Kazama T (2005) Isoflurane anesthesia induces biphasic effect on dopamine release in the rat striatum. Brain Res Bull 67: 176–181

42. Fueger BJ, Germim J, Hildebrandt I, Tran C, Halpern BS, Stout D Plelfs AE, Weber WA (2006) Impact of animal handling on the results of FDG PET studies in mice. J mud med 47:999–1006

43. Ziegler SI (2005) Positron Emission Tomography: Principles, Technology, and Recent Developments. Nucl Phys A 752: 679–687

44. Collantes M, Prieto E, Penuelas I, Blesa J, Juri C, Marti-Climent JM, et al. (2009) New MRI, 18F-DOPA and 11C-(+)-alpha-dihydrotetrabenazine templates for Macaca fascicularis neuroimaging: advantages to improve PET quantification. Neuroimage 47: 533–539

45. Casteels C, Vermaelen P, Nuyts J, Van Der Linden A, Baekelandt V, Mortelmans L, Bormans G, Van LK (2006) Construction and evaluation of multitracer small-animal PET probabilistic atlases for voxel-based functional mapping of the rat brain. J Nucl Med 47: 1858–1866

46. Rubins DJ, Melega WP, Lacan G, Way B, Plenevaux A, Luxen A, Cherry SR (2003) Development and evaluation of an automated atlas-based image analysis method for micro-PET studies of the rat brain. Neuroimage 20: 2100–2118

47. Dauguet J, Condq F, Hantraye P, Frouin V, Delzescaux T (2011) Generation of a 3D atlas of the nuclear division of the thalamus based on histological sections of primate: Intra- and intersubject atlas-to-MRI warping. IRBM 30: 281–291

48. Delforge J, Loc'h C, Hantraye P, Stulzaft O, Khalili-Varasteh M, Maziere M, Syrota A,

Maziere B (1991) Kinetic analysis of central [76Br]bromolisuride binding to dopamine D2 receptors studied by PET. J Cereb Blood Flow Metab 11: 914–925

49. Delforge J, Bottlaender M, Loc'h C, Guenther I, Fuseau C, Bendriem B, Syrota A, Maziere B (1999) Quantitation of extrastriatal D2 receptors using a very high-affinity ligand (FLB 457) and the multi-injection approach. J Cereb Blood Flow Metab 19: 533–546

50. Ichise M, Meyer JH, Yonekura Y (2001) An introduction to PET and SPECT neuroreceptor quantification models. J Nucl Med 42: 755–763

51. van den Hoff J (2005) Principles of quantitative positron emission tomography. Amino Acids 29: 341–53

52. Innis RB, Cunningham VJ, Delforge J, Fujita M, Gjedde A, Gunn RN, et al. (2007) Consensus nomenclature for in vivo imaging of reversibly binding radioligands. J Cereb Blood Flow Metab 27: 1533–1539

53. Hantraye P, Kaijima M, Prenant C, Guibert B, Sastre J, Crouzel M, Naquet R, Comar D, Maziere M (1984) Central type benzodiazepine binding sites: A positron emission tomography study in the baboon's brain. Neurosci Lett 48: 115–120

54. Hantraye P, Kaijima M, Prenant C, Guibert B, Sastre J, Crouzel M, Naquet R, Comar D, Maziere M (1984) Central type benzodiazepine binding sites: A positron emission tomography study in the baboon's brain. Neurosci Lett 48: 115–120

55. Schmid L, Bottlaender M, Fuseau C, Fournier D, Brouillet E, Maziere M (1995) Zolpidem displays heterogeneity in its binding to the nonhuman primate benzodiazepine receptor in vivo. J Neurochem 65: 1880–6

56. Brouillet E, Chavoix C, Bottlaender M, Khalili-Varasteh M, Hantraye P, Fournier D, Dodd RH, Maziere M (1991) In vivo bidirectional modulatory effect of benzodiazepine receptor ligands on GABAergic transmission evaluated by positron emission tomography in non-human primates. Brain Res 557(1–2): 167–176

57. Cropley VL, Fujita M, Innis RB, Nathan PJ (2006) Molecular imaging of the dopaminergic system and its association with human cognitive function. Biol Psychiatry 59: 898–907

58. Morris E, Yoder K (2006) PET displacement sensitivity: Predicting binding potential change for PET tracers based on their kinetic characteristics. Neuroimage 31 (Suppl 2): T106

59. Doudet DJ, Holden JE, Jivan S, McGeer E, Wyatt RJ (2000) In vivo PET studies of the dopamine D2 receptors in rhesus monkeys with long-term MPTP-induced parkinsonism. Synapse 38: 105–113

60. Doudet DJ, Jivan S, Ruth TJ, Holden JE (2002) Density and affinity of the dopamine D2 receptors in aged symptomatic and asymptomatic MPTP-treated monkeys: PET studies with [11C]raclopride. Synapse Jun;44: 198–202

61. Leriche L, Björklund T, Breysse N, Besret L, Gregoire MC, Carlsson T, Dolle F, Mandel RJ, Deglon N, Hantraye P, Kirik D (2009) Positron emission tomography imaging demonstrates correlation between behavioral recovery and correction of dopamine neurotransmission after gene therapy. J Neurosci 29: 1544–1553

62. Banati RB (2002) Visualising microglial activation in vivo. Glia 40: 206–217

63. Bartels AL, Leenders KL (2007) Neuroinflammation in the pathophysiology of Parkinson's disease: evidence from animal models to human in vivo studies with [11C]-PK11195 PET. Mov Disord 22: 1852–1856

64. Cagnin A, Gerhard A, Banati RB (2002) In vivo imaging of neuroinflammation. Eur Neuropsychopharmacol 12: 581–586

65. Doorduin J, de Vries EF, Dierckx RA, Klein HC (2008) PET imaging of the peripheral benzodiazepine receptor: monitoring disease progression and therapy response in neurodegenerative disorders. Curr Pharm Des 14(31): 3297–3315

66. Chauveau F, Boutin H, Van CN, Dolle F, Tavitian B (2008) Nuclear imaging of neuroinflammation: a comprehensive review of [11C] PK11195 challengers. Eur J Nucl Med Mol Imaging 35: 2304–1239

67. Owen DR, Gunn RN, Rabiner EA, Bennacef I, Fujita M, Kreisl WC, Innis RB, Pike VW, Reynolds R, Matthews PM, Parker CA (2011) Mixed-affinity binding in humans with 18-kDa translocator protein ligands. J Nucl Med 52: 24–32

68. Owen DR, Howell OW, Tang SP, Wells LA, Bennacef I, Bergstrom M, Gunn RN, Rabiner EA, Wilkins MR, Reynolds R, Matthews PM, Parker CA (2010) Two binding sites for [3H] PBR28 in human brain: implications for TSPO PET imaging of neuroinflammation. J Cereb Blood Flow Metab 30: 1608–1618

69. Honer M, Hengerer B, Blagoev M, Hintermann S, Waldmeier P, Schubiger PA, Ametamey SM (2006) Comparison of [18F]FDOPA, [18F] FMT and [18F]FECNT for imaging dopaminergic neurotransmission in mice. Nucl Med Biol 33: 607–614

70. Strome EM, Cepeda IL, Sossi V, Doudet DJ (2006) Evaluation of the integrity of the dopamine system in a rodent model of Parkinson's disease: small animal positron emission tomography compared to behavioral assessment and autoradiography. Mol Imaging Biol 8: 292–299

71. Honer M, Bruhlmeier M, Missimer J, Schubiger AP, Ametamey SM (2004) Dynamic imaging of striatal D2 receptors in mice using quad-HIDAC PET. J Nucl Med 45: 464–470

72. Wang X, Sarkar A, Cicchetti F, Yu M, Zhu A, Jokivarsi K, Saint-Pierre M, Brownell AL (2005) Cerebral PET imaging and histological evidence of transglutaminase inhibitor cystamine induced neuroprotection in transgenic R6/2 mouse model of Huntington's disease. J Neurol Sci 231: 57–66

73. Bauer A, Zilles K, Matusch A, Holzmann C, Riess O, Von HS (2005) 94Regional and subtype selective changes of neurotransmitter receptor density in a rat transgenic for the Huntington's disease mutation. J Neurochem : 639–650

74. Araujo DM, Cherry SR, Tatsukawa KJ, Toyokuni T, Kornblum HI (2000) Deficits in striatal dopamine D receptors and energy metabolism detected by in vivo microPET imaging in a rat model of Huntington's disease. Exp Neurol 166: 287–297

75. Shimoji K, Ravasi L, Schmidt K, Soto-Montenegro ML, Esaki T, Seidel J, Jagoda E, Sokoloff L, Green MV, Eckelman WC (2004) Measurement of cerebral glucose metabolic rates in the anesthetized rat by dynamic scanning with 18F-FDG, the ATLAS small animal PET scanner, and arterial blood sampling. J Nucl Med 45: 665–672

76. Collantes M, Prieto E, Penuelas I, Blesa J, Juri C, Marti-Climent JM, et al. (2009) 18F-DOPA and 11C-(+)-alpha-dihydrotetrabenazine templates for Macaca fascicularis neuroimaging: advantages to improve PET quantification. Neuroimage 47: 533–539

77. Doudet DJ, Rosa-Neto P, Munk OL, Ruth TJ, Jivan S, Cumming P (2006) Effect of age on markers for monoaminergic neurons of normal and MPTP-lesioned rhesus monkeys: a multitracer PET study. Neuroimage 30: 26–35

78. Saiki H, Hayashi T, Takahashi R, Takahashi J (2010) Objective and quantitative evaluation of motor function in a monkey model of Parkinson's disease. J Neurosci Methods 190: 198–204

79. Nagai Y, Obayashi S, Ando K, Inaji M, Maeda J, Okauchi T, Ito H, Suhara T (2007) Progressive changes of pre- and post-synaptic dopaminergic biomarkers in conscious MPTP-treated cynomolgus monkeys measured by positron emission tomography. Synapse 61: 809–819

80. Bankiewicz KS, Forsayeth J, Eberling JL, Sanchez-Pernaute R, Pivirotto P, Bringas J, Herscovitch P, Carson RE, Eckelman W, Reutter B, Cunningham J (2006) Long-term

clinical improvement in MPTP-lesioned primates after gene therapy with AAV-hAADC. Mol Ther 14: 564–570

81. Eberling JL, Pivirotto P, Bringas J, Bankiewicz KS (2004) Comparison of two methods for the analysis of [18F]6-fluoro-L-m-tyrosine PET data. Neuroimage 23: 358–363

82. Chefer SI, Kimes AS, Matochik JA, Horti AG, Kurian V, Shumway D, Domino EF, London ED, Mukhin AG (2008) Estimation of D2-like receptor occupancy by dopamine in the putamen of hemiparkinsonian Monkeys. Neuropsychopharmacology 33: 270–278

83. Schweinhardt P, Fransson P, Olson L, Spenger C, Andersson JL (2003) A template for spatial normalisation of MR images of the rat brain. J Neurosci Methods 129: 105–113

84. McLaren DG, Kosmatka KJ, Oakes TR, Kroenke CD, Kohama SG, Matochik JA, Ingram DK, Johnson SC (2009) A population-average MRI-based atlas collection of the rhesus macaque. Neuroimage 45: 52–59

85. McLaren DG, Kosmatka KJ, Kastman EK, Bendlin BB, Johnson SC (2010) Rhesus macaque brain morphometry: a methodological comparison of voxel-wise approaches. Methods 50: 157–165

86. Black KJ, Koller JM, Snyder AZ, Perlmutter JS. Atlas template images for nonhuman primate neuroimaging: baboon and macaque. Methods Enzymol (2004) 385: 91–102

87. Black KJ, Koller JM, Snyder AZ, Perlmutter JS (2001) Template images for nonhuman primate neuroimaging: 2. Macaque. Neuroimage 14: 744–748

88. Frey S, Pandya DN, Chakravarty MM, Bailey L, Petrides M, Collins DL (2011) An MRI based average macaque monkey stereotaxic atlas and space (MNI monkey space). Neuroimage doi: 10.10.16/j.neuroimage.2011.01.040

Chapter 10

Optical Approaches to Studying the Basal Ganglia

Joshua L. Plotkin, Jaime N. Guzman, Nicholas Schwarz,
Geraldine Kress, David L. Wokosin, and D. James Surmeier

Abstract

Altered synaptic integration is a major factor for many neurological disorders involving the basal ganglia, including Parkinson's disease and Huntington's disease. Despite the fact that most synaptic integration occurs within dendrites, nearly all we know about the physiology of basal ganglia neurons comes from somatic measurements. This is particularly true of neurons in the striatum, the major input nucleus of the basal ganglia. Principal spiny projection neurons (SPNs) of the striatum have fine caliber dendrites that are inaccessible using traditional patch-clamp electrodes. Two-photon laser scanning microscopy (2PLSM) and two-photon laser uncaging (2PLU) offer alternative strategies for studying synaptic integration in basal ganglia neurons, like SPNs. These methods also allow subcellular organelles like mitochondria to be monitored in physiologically meaningful settings. Combining these approaches with electrophysiological and optogenetic methodologies builds a powerful arsenal of investigational tools. This chapter describes how optical methodologies are being applied to the study of the basal ganglia.

Key words: Two-photon laser scanning microscopy, Two-photon laser uncaging, Optogenetics, Spiny projection neurons

1. Nonlinear Fluorescence Microscopy

1.1. Background

Although most synaptic integration occurs in fine neuronal dendrites (1–3), most physiological studies are limited to recordings from somata. The refinement of visualized patch-clamp techniques has allowed large dendrites in some neurons to be directly recorded from (4–6), but fine dendrites remain elusive to traditional electrophysiological probing. Recently, optical approaches have supplemented our experimental toolbox, making fine dendrites accessible.

Emma L. Lane and Stephen B. Dunnett (eds.), *Animal Models of Movement Disorders: Volume I*, Neuromethods, vol. 61, DOI 10.1007/978-1-61779-298-4_10, © Springer Science+Business Media, LLC 2011

While ultra-short pulse lasers have been used to create high-contrast fluorescent images for over 20 years (7), it has only been recently that improvement in laser technology has allowed reliable, long wavelength, computer-controlled excitation spectra to be produced. This has made two-photon laser scanning microscopy (2PLSM) an important tool for biologists as the use of long wavelength light has allowed thick, biologically meaningful specimens to be imaged with high spatial resolution (8). The neuroscience community has been the largest group to embrace the 2PLSM technology, using it to probe neuron form and function in brain slices and in intact animals (7, 9–11).

Within the last decade, optical workstations have been created and refined to provide user-friendly, systems level solutions for specific biological imaging applications (8) (Fig. 1). These dual-galvo-pair, double 2PLSM systems are now commercially available from at least four vendors (Prairie Technologies, Inc., Nikon, Olympus and Zeiss). These systems have enabled high-resolution optical study of individual neurons and organelles in essentially any brain region. In parallel, there has been a virtual explosion in the number and diversity of fluorescent dyes and indicators, expanding the utility of 2PLSM systems.

Two-photon excitation imaging of fluorescent dyes and indicators is most often performed with laser systems having pulse durations from 70 to 140 fs. The restriction of photon delivery in time, via mode-locking, increases the peak laser power at the sample plane, increasing the probability that an individual molecule will experience two-(or more) photon excitation. The advantage of nonlinear excitation is the restriction of fluorophore excitation to the focal volume determined by the objective lens. This allows optical sectioning (12). This mode of fluorescence excitation can greatly reduce background excitation and improves efficiency by permitting the collection of any emitted photons generated by the sampled probe. Thus, it does not require spatial filtering with a confocal pinhole aperture (which can reduce in-focus signal), since the emission signal collection is decoupled from the excitation event. This permits the collection of some of the emitted photons that are scattered while leaving the focal volume deeper into the brain slice tissue, increasing the detection sensitivity (13). Large front element objective lenses have an advantage in collecting this scattered signal – as long as the detector optical path is engineered to take advantage of the larger angles leaving the back of the objective lens (14). To effectively take advantage of this, two-photon imaging requires separate detectors optimized for full field emission collection.

1.2. System Components

In its simplest form, an optical experimental system is composed of a probe, an interaction, and an observation. A generic two-photon imaging system has been well described previously (12). This chapter briefly covers the basic components: laser, modulator, scanning

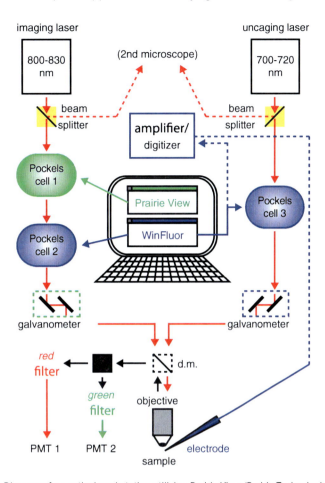

Fig. 1. Diagram of an optical workstation utilizing Prairie View (Prairie Technologies, Inc.) and Win Fluor (University of Strathclyde) software. The movement and gain of the imaging laser is controlled using the commercially provided software Prairie View. Electrophysiological recordings are made using the free software package WinFluor, which is also used to control the position, timing, and gain of the photolysis laser. Imaging and physiological data are coordinated and displayed by WinFluor.

mirrors, objective lens, optical filters, and detectors, as well as the optical workstation approach. How the first three items (laser, modulator, and mirrors) can be doubled and combined to add a second scanning channel for photo-stimulation and/or photolysis is also described.

1.2.1. Probe

The probe can be considered to be the laser and the instruments used to control the laser exposure and location. The laser most often employed is one with a Ti:Sapphire-based crystal laser with solid state 532 nm pump laser. Electro-optical modulators (Pockels cells) are used to control laser intensity since they have a very broad excitation wavelength range (700–1,200 nm) with low insertion loss (5%), rapid control (~5 μs), the laser beam is not diverted by the Pockels cell and it requires no special lenses. If possible, the average laser power should be reduced to 1.5 W before the beam reaches the Pockels cell, as this greatly reduces heating load during imaging. However, newer lasers far exceed these average powers over most of the tuning range (<900 nm). A solution to this problem (employed by Prairie Technologies and in our lab) is the addition of a manual attenuator in front of the Pockels cell, which includes a rotating achromatic half-wave plate and fixed polarizing beam splitter cube. The cube permits sharing of the laser beam (at 90°) between two different scanning systems. Systems with this manual modulator scheme should keep both arms at 50% power to avoid polarization changes at the down-stream modulator(s). For average powers higher than 1.5 W, the best strategy is to leave the laser shutter open and place a mechanical shutter after the Pockels cell to aid modulator thermal equilibrium.

Cambridge galvanometer mirror systems, which allow manipulation of the laser target location in the x and y directions, are used by most commercial laser scanning microscope companies. The microscopes used for electrophysiology in brain slice work are most often upright in nature and the objective lenses used are water-dipping style for enhanced free working distances (2–3.3 mm) permitting pipette access for patching.

1.2.2. Interaction

The interaction refers to the excitation of a biologically relevant fluorescent molecule by the laser. Fluorophores, or fluorescent proteins, have optimal excitation photon energies, defined as excitation cross-section spectra. There is an inverse linear relationship between the photon wavelength and energy. The quantum mechanical selection rules for two-photon absorption events must be different from single photon absorption, so the energy levels involved will also be different. The rule in practice is to double the single photon spectra and then investigate lower wavelengths, higher photon energies. The ability to excite tissue with deep penetrating low-energy long wavelength light is a major advantage of 2P microscopy. For large symmetric molecules (like GFP), the two-photon cross-section spectra resemble the single photon spectra doubled. This is not true for second and third harmonic imaging, a scattering technique where the photon energy and momentum are conserved. In our experience, 820 nm photons work very well for most green-emitting fluorescent calcium indicators and ~780 nm photons will optimally excite red Alexa-488, -568, and -594 dyes.

1.2.3. Observation

The strategies of dealing with the observation are the main focus of the remainder of this chapter. This includes collecting, interpreting, and displaying the experimentally collected photons in a biologically relevant way.

When working with living tissue, it is imperative to avoid inducing phototoxicity. As discussed below, in addition to frank distortions in structure, phototoxicity may also lead to more subtle changes in function, which may be overlooked during conventional physiological or anatomical experiments. Hence, great care must be taken in minimizing the stimulation exposure dose. The stimulation exposure dose is defined by the 2P excitation potential (power and space) multiplied by the duration of exposure (time). The excitation potential depends primarily on the square of the sample average power and linearly on the laser wavelength, as these are the two laser variables that historically have been the easiest to vary. Newer Ti:Sapphire lasers (Spectra Physics DeepSee and Coherent Vision) can now also control the pulse duration at the sample plane, so minimal average power can be used for a given stimulation dose. The signal obtained from 2P stimulation can be defined by the equation:

$$\langle F2 \rangle = \frac{1}{2} N t_d \frac{8n}{\pi\lambda} \frac{g_{2,p}}{R\tau} \langle P(t) \rangle^2 C \sigma_2(\lambda)\phi,$$

where N is photon counting scans, t_d is a time constant, n is the optical index of refraction of the immersion media, λ is the wavelength, $g_{2,p}$ is a constant defining the laser pulse shape, R is the laser pulse repetition rate, τ is the pulse duration, P is the power, C represents the fluorophore concentration, σ_2 is the 2PEF cross section, and ϕ represents the fraction of light collected assuming isotropic emission.

1.3. Photo-Stimulation and Photolysis

Optical workstations can be modified to permit focused laser stimulation or 2PLU during imaging. Combining two-photon laser uncaging (2PLU) with 2PLSM increases system cost and requires a second ultra-short pulse duration laser (~720 nm). In an attempt to minimize cost, lower power (10 W pump) laser systems or refurbished lasers have been used to power the 2PLU channel. Unfortunately, more affordable (compact, lower power) single-wavelength ultra-short pulse duration laser systems are not available with spectra near 700 nm (15). A more comprehensive set of criteria for selecting ultra-short pulse duration lasers for multiphoton excitation imaging can be found elsewhere (16).

The average powers reaching the sample (with a 60×, 0.9 NA objective) rarely approach 30 mW with 1 ms exposures, so very little of the full laser power (~600 mW or more) is actually used for successful 2PLU with brain slices. The average photolysis laser power is attenuated with a Pockels cell, in the same manner as the imaging laser power. Great care must be taken with the modulator bias setting to ensure the best minimum (lowest power) to the sample plane. This minimizes photolysis during the "off" times of the stimulation protocol.

Despite the fact that the doubled one-photon excitation wavelength of 4-methoxy-7-nitroindolinyl (MNI)-glutamate (caged glutamate) is 630 nm, most applications of 2PLU to date have been done with Ti:sapphire lasers, which restricts the lower wavelength to 680 nm or 720 nm (depending on the laser pump power). The wavelength we use for 2PLU of MNI-glutamate is 720 nm. This wavelength is a compromise between the older, lower power laser systems (the manually tuned, 5 W pumped two-box lasers: Verdi/Mira, Millenia/Tsunami) which often yielded low average powers below 720 nm, and the newer commercially available one-box lasers (Chameleon-210, MaiTai-XF -BB) which cannot tune below ~715 nm. The new series of pumped lasers (Chameleon-Ultra, MaiTai-HP) now permit 2P photolysis with average powers down to 680 nm or 690 nm.

Typical photolysis exposure times vary from microseconds to several milliseconds. With the Prairie Ultima system, it is possible to stimulate six distinct points within 1 ms with exposure times of only 100 μs at each point. Exposures of 0.5–1 ms at each point are much more common for experiments in slices. Spatial calibrations within the imaged field of view are verified using standard test samples to match the marked points for photolysis mirrors (galvanometers) to the voltages for the corresponding image pixel locations.

2. Imaging Applications in Brain Slices of the Basal Ganglia

2.1. The Challenge

The striatum is a complex subcortical nucleus that is populated by two major classes of spiny projection neurons (SPNs) and at least four classes of interneurons (2, 17–19). While this rich milieu undoubtedly increases the computational power of the striatum, it has traditionally posed severe constraints on the interpretation of physiological data collected in brain slices. There are three major obstacles. First, unlike other brain regions such as the cortex, hippocampus, and cerebellum, there is no obvious laminar organization in the striatum allowing positional identification of cell types (17). Moreover, with the exception of giant cholinergic interneurons, cell types are very difficult to visually distinguish using the IR-DIC optics typically used for patch-clamp recordings made in slices. Second, the two classes of SPN (which constitute ~90% of all striatal neurons) are difficult to unequivocally distinguish either anatomically or physiologically (2, 17, 18, 20). Although there are differences (21–24), they are not so pronounced that they allow cells to be readily sorted in a patch-clamp experiment. Third, although interneurons can typically be identified by their physiological properties, they are extremely rare (constituting a few percent of all the neurons), making them difficult to find using standard IR-DIC optics (25, 26).

The recent development of BAC transgenic mice expressing fluorescent proteins under the regulation of cell-type specific promoters has offered a means of identifying specific striatal neuron populations (22, 27, 28). As mentioned above, striatal SPNs come in two major flavors: those primarily expressing D_1 type dopamine receptors and projecting to the substantia nigra and internal globus pallidus (direct pathway) and those expressing D_2 type dopamine receptors and projecting to the external segment of the globus pallidus (indirect pathway) (2, 18, 20, 29). These direct pathway SPNs (dSPNs) and indirect pathway SPNs (iSPNs) have similarly sized somata and spiny dendritic arbors that extend similar radial distances. With the advent of mice expressing GFP under the control of the D_1 ($BACD_1$) or D_2 ($BACD_2$) promoters, these two cell types can readily be distinguished (23, 27, 28). BAC transgenic mice expressing GFP or tdTomato reporters under the control of appropriate promoters can also be used to readily identify interneurons in brain slices. For example, mice expressing GFP under the control of parvalbumin or other cell-type specific promoters have been used to quickly sample from cholinergic, fast-spiking GABAergic and persistent low-threshold spiking interneurons in the striatum (30, 31).

2.2. Probing SPN Dendritic Anatomy

Traditionally, the dendritic anatomy of a neuron subjected to physiological interrogation was capable of being visualized only by filling the neuron with dye or another reactive molecule (e.g., biocytin), and then fixing, processing, and sectioning the slice after the experiment. 2PLSM has changed this situation and allows visualization of dendrites during the physiological experiment. By filling the patch electrode with a fluorophore (most commonly Alexa 568 or Alexa 594; 50 μM) that readily diffuses into the soma and dendrites, the entire dendrite can be visualized (27, 32, 33). Once the fluorophore has equilibrated throughout the dendritic tree (10–20 min post patching), the dye can be imaged. Imaging can be performed simultaneously with physiological somatic recordings if desired, in part because of the lower energy exposure to the slice afforded by the reliance upon longer wavelength 2P excitation.

2.2.1. The Z-Stack

Because 2P excitation is limited to a small focal volume in the specimen, anatomical data is collected in the form of an optical section in the z-axis. These optical sections can then be assembled into a stack (z-stack) to provide a three-dimensional image of the neuron (Fig. 2a, b). The lens used to acquire a z-stack should be of sufficiently high numerical aperture (NA) to resolve fine dendritic detail, but with a field of vision capable of capturing a sufficient portion (ideally all) of an average SPN dendritic field diameter. Larger NAs allow greater photon collection and therefore increased sensitivity. However, high NA typically decreases the working distance, making patch clamping difficult. We have found that a

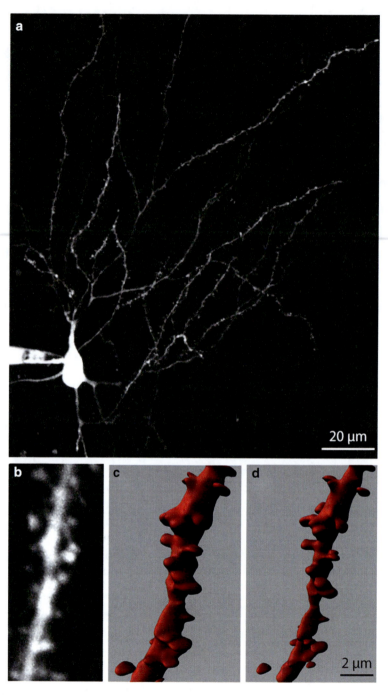

Fig. 2. Reconstructing a 3D image of an SPN. (**a**). Maximum intensity projection image of an iSPN loaded with Alexa 568. (**b**) High zoom (4.9×) MIP of an iSPN dendrite. (**c**) 3D rendering of the dendrite shown in (**b**) using Imaris software to construct an isosurface from a nondeconvolved z-series. (**d**) 3D rendering of the same dendrite in (**b**, **c**) following a 60-cycle blind deconvolution of the z-series using AutoQuant. Scale bar in (**d**) applies to (**b**, **d**).

60×/1.0 N.A. Olympus LUMPFL water-dipping lens is a reasonable compromise.

Once the lens is selected and the neuron patched and loaded with dye, there are five major variables that should be maximized to yield the highest quality z-stack.

1. The digital zoom should be adjusted to include the appropriate field of view. While digital zoom is not needed for a whole cell image, it is an important variable to adjust when taking high magnification images of dendritic regions.

2. The z-direction step size must be determined, and the z-direction limits defined. The step size should be small enough to pick up subtle anatomical features, yet not so small that the imaging time is increased unnecessarily. It should be noted that the ultimate maximal resolution is determined by the optics and laser properties given by

$$\omega_{xy} = \frac{0.325\lambda}{\sqrt{2}\mathrm{NA}^{0.91}}$$

$$\omega_{z} = \frac{0.532\lambda}{\sqrt{2}}\left[\frac{1}{n - \sqrt{n^2 - \mathrm{NA}^2}}\right]$$

where NA is the numerical aperture of the lens (and is >0.7), λ is the wavelength of the excitation laser, and n is the refraction index of the immersion medium (12) and cannot be exceeded by altering the step size. We typically adjust our step size to be approximately one-half to one-third of the estimated z-plane resolution, in accordance with standard sampling theories. In addition to the step size, the top and bottom limit planes must be defined. This is done on a cell-by-cell basis.

3. The dwell time (amount of time the laser resides on each acquired pixel) should be adjusted to provide high enough photon emission to yield a quality image without increasing imaging time more than necessary. In our experiments, a dwell time of 4–10 μs is typically a good compromise.

4. As all emitted photons are collected by the photomultiplier tubes (PMTs), it can be helpful to utilize averaging as a means of increasing signal-to-noise ratio (even though the signal-to-noise ratio is already vastly improved by 2PLSM compared to epifluorescence and confocal imaging (13)). To achieve this, multiple images can be taken at each focal plane and averaged.

5. Finally, the laser sample power should be set to achieve an optimal signal-to-noise ratio for each cell. We achieve this through the use of a Pockels cell. As the image intensity diminishes as the image plane moves deeper into the slice, it is useful to increase the laser power with focal depth. This can be easily achieved with commercially available software. With the exception

2.2.2. Deconvolution

Because the region of 2P excitation is not a single point and emitted photons are subject to diffraction by the slice, there is blurring of the acquired image. In general, 2P imaging greatly reduces blurring, as compared to confocal imaging, because of the tighter localization of the excitation volume. For instance, the optical setup in our laboratory yields a theoretical lateral resolution of 0.38 μm but an axial resolution of 1.43 μm. Blurring is most pronounced in the z-direction, because the 2P excitation region is maximally distorted in this direction. The process of removing out-of-focus signal is called deconvolution or deblurring (34, 35). The amount of blurring is directly related to the point spread function (PSF) of the imaging device. Deconvolution methods are implemented as iterative algorithms that consider an experimentally derived PSF and use a maximum likelihood estimator to minimize Poisson-distributed noise during each pass. Although the PSF can be measured and used during the deconvolution process, it is difficult to measure in real biological samples. Instead, blind deconvolution algorithms initially estimate and then refine the PSF during successive iterations (36). This function can be performed in a nearly automated fashion by commercially available software (described later). The benefits of deconvolving an image are most noticeable along the z-axis (Fig. 2c, d).

2.2.3. Direct Volume Rendering

Rendering techniques can be used to visualize the acquired and deconvolved dataset. Volume rendering is the process of creating a 2D image directly from a 3D image dataset, in this case, a series of slices that constitute a z-stack or from an intermediate 3D representation of the image data. When generating an image directly from image data, rays are cast from each point in the output image through the data (37). Samples are taken at discrete intervals along the ray usually using trilinear interpolation. The samples are used to determine a pixel's final value (Fig. 3). In the simple case of a maximum intensity projection (MIP), the largest sample value along the ray is used to determine the pixel value. More complex compositing methods consider the contribution of all samples along a ray (38).

In order to increase performance, isosurfaces or contour plots are often used as an intermediate representation (Fig. 2c, d). A 3D model is generated over a specified intensity value in the data, usually using the marching cubes algorithm (39). This computationally intensive operation is performed only once before rendering occurs, and generates a smaller, easier to render set of geometric primitives.

Most current volume rendering software takes advantage of the 3D texture mapping capabilities (40) of commonly used computer graphics cards. 3D texture mapping methods upload an

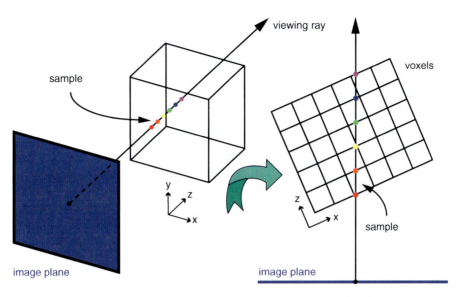

Fig. 3. Volume rendering using ray casting. A ray starting at a point on the image plane is cast through the volume. Samples are taken at evenly spaced intervals along the ray. Because the data are discrete, trilinear interpolation is used to determine the value at each sample location.

entire volume dataset into a graphics card's memory. Sampling using trilinear interpolation, and the application of color and opacity maps are performed entirely on the graphics card. This provides for great performance gains as long as the complete dataset fits in the graphics card's video RAM.

Rendering and viewing the reconstructed dataset allows for cursory visual inspection and estimates of dendrite length and spine number and types. More sophisticated software packages use image segmentation and computational geometry techniques to automate 3D dendrite tracing, branch detection, and spine counting. These programs can further categorize spines into defined shapes (i.e., mushroom, stubby, thin), which have been correlated with various states of synaptic plasticity (41–44). A number of software packages (commercial and free) exist that aid in the anatomical reconstruction and analysis of neurons. Volocity (PerkinElmer, Inc.) and Amira (Visage Imaging, Inc.) are commercial products that offer deconvolution, rendering, and image analysis features. Imaris (Bitplane, Inc.) is a rendering and analysis package aimed specifically at cellular and neuronal applications, and integrates with the AutoQuant/AutoDeblur (Media Cybernetics, Inc.) deconvolution package. NeuronStudio (Mount Sinai School of Medicine), Voxx (Indiana University), and NUPVer (Northwestern University) are free applications that have simple rendering and measurement tools. Kitware, Inc. maintains popular open source C++ software libraries for visualization, rendering, and image analysis, including the Visualization Toolkit (VTK) (http://www.vtk.org), and the National Library of Medicine's Insight Segmentation and

Registration Toolkit (ITK) (http://www.itk.org). Even with these rather advanced tools, deconvolution and rendering can take many hours or even days on current workstations. A typical z-series spanning $200 \times 200 \times 100$ μm and sampled with a pixel size of 0.36 μm and z-step of 0.2 μm can take up to half a GB. Large datasets also pose a challenge for fully automated analysis packages. Computationally complex tasks, such as spine counting, have not been successfully automated to date (at least not in our hands).

2.2.4. Limitations

While 2PLSM offers striking benefits for analyzing neuronal anatomy, there are several caveats. One caveat is that the neuronal properties might change during the time it takes to dialyze the cell and acquire an image. For SPNs, the cell must load with dye for 10–20 min before a z-stack can be made and each stack can take up to 20 min to complete. Another caveat is that if the slice physically moves, the acquired image will be distorted. Furthermore, image quality will be inversely related to their depth in the slice. This limitation is worsened by the fact that emitted photons are typically of shorter wavelength than the exciting photons, resulting in greater diffraction. As a consequence, cells chosen for imaging must not be too deep (typically no deeper than 80 μm below the surface). Dendrites descending deep into the slice, though equally likely to fill with the added fluorophore, will not be detected as efficiently. This is a real problem with SPNs, as they have spherical dendritic trees (3, 45, 46). Finally, as the optical density of brain tissue typically increases with age (45), imaging neurons in brain slices from older mice can be problematic and require adjustment of imaging conditions.

2.3. Dendritic Calcium Imaging

Compared to projection neurons in other brain regions, SPNs have very fine caliber dendrites. In fact, the distal tips of SPN dendrites have been estimated to be only about a third of a micron in diameter (3). As mentioned above, this makes direct patch clamping of distal dendrites virtually impossible with current methodologies. An alternative to direct voltage measurements is to measure calcium flux through voltage-gated calcium channels in neurons subjected to patch-clamp recording (27, 33, 47). Although there is a rough correlation between calcium entry and membrane voltage in many of the experimental situations we described, the two are not interchangeable and appropriate caution should be used in interpreting data. For example, cytosolic calcium transients may come not only from opening voltage-dependent channels in the plasma membrane, but also from intracellular sources that are not voltage sensitive (48–50). Nevertheless, the combination of electrophysiological and optical approaches has major strengths in studying neuronal properties. One of the challenges we have faced is the integration of electrophysiological and optical data acquisition modalities in the same workstation. Some of the solutions we have developed are discussed within the context of the methodology for dendritic calcium imaging.

The most common method of imaging calcium transients in living tissue involves loading the cell(s) of interest with a calcium-sensitive indicator. This fluorophore must (1) have an appropriate affinity for calcium (as such, BAPTA/EGTA are typically the starting point for the indicator's construction) and (2) be capable of changing its photon absorption or emission efficiency based on its binding state with calcium (51, 52). We only discuss imaging calcium signals using nonmembrane-permeable calcium dyes (free acids) loaded through the recording electrode. When collections of neurons are to be imaged, membrane permeable AM-ester modified dyes are a viable alternative, but these approaches are not discussed here (53–56).

The first step in designing a calcium imaging experiment is choosing an appropriate dye. Two factors should be taken into consideration when making this decision: the affinity of the dye for calcium and its absorption/emission spectrum. Calcium dyes have a wide range of affinities. Which dye is appropriate for a particular experiment will depend on the physiology of the cell type to be examined and the phenomenon to be studied. The dye should have a low enough calcium affinity that it does not disrupt basal cytoplasmic calcium signaling. Supplemental calcium buffers such as EGTA or BAPTA (which are commonly included in many internal recording solutions) should not be used in these experiments, as the calcium dye itself should be adequate to buffer calcium in the internal solution. The relationship between calcium binding and photon emission should be linear within the range of calcium concentrations being measured. In our experience, a high-affinity dye such as Fluo-4 is useful for studying calcium dynamics associated with modest physiological stimuli such as short trains of action potentials or excitatory postsynaptic potentials (EPSPs). Fluo-4 fluorescence is minimal in SPNs at resting membrane potentials (approximately –80 mV), and we have observed little evidence of dye saturation in most circumstances. The optimal concentration of Fluo-4 needed may differ between protocols. We typically find that 200 µM Fluo-4 provides a strong calcium signal without disrupting cellular function in any obvious way. It should be noted here that great care must be taken when choosing a dye and corresponding concentration when examining phenomena governed by calcium dynamics, as such events will be sensitive to exogenous buffering.

In addition to having an appropriate calcium affinity, the dye must also have 2-photon absorption/emission wavelengths appropriate for the equipment being used. Several criteria should be met by this choice:

1. The maximum 2P excitation cross section should be readily achieved using the system's laser (820 nm is fine for most green-emitting indicators);

2. The emission wavelength should be easily detected by the PMTs collecting the excited photons when used with the correct filter (i.e., avoid INDO-1);

3. If used in conjunction with an anatomical dye (such as Alexa 568) and a single stimulating laser, the excitation cross sections of the two dyes must sufficiently overlap and the emission wavelengths must be sufficiently separated to be detected by two separate PMTs using appropriate filters (Alexa-488, -568, and -594 all have 2PEFcs peaks at ~780 nm; these dyes also are sensitive to 3P photobleaching when excited in the UV energies).

2.3.2. Line Scans

Once the appropriate calcium dye is chosen, the next consideration is how to measure changes in dye emission. There are several ways to accomplish this goal. The most common methods are to measure light coming from a two-dimensional region of interest (ROI) or by performing one-dimensional line scans. We focus our attention on acquiring line scans in this section.

Line scans acquire data along a line positioned to span an area of biological interest. Limiting the laser's movement to one string of pixels greatly increases the maximum acquisition rate and temporal precision of the measurement. Line scans have proven useful for studying calcium dynamics in SPNs (27, 33, 56). While the maximum data acquisition rate of a line scan is less than can be achieved by "parking" the laser at a single spot (or even setting an ROI composed of fewer total pixels), it allows rapid imaging of multiple dendritic components simultaneously. For example, there is ample evidence that dendritic spines and shafts respond to synaptic stimuli differently (32, 33). By adjusting the orientation of a line scan, it is routinely possible to capture data from both spine heads and dendritic shafts in the same experimental trial, allowing an assessment of the change in the two compartments nearly simultaneously (Fig. 4).

The key parameters of the line scan are simply the length of the scan and the duration (time) of the scan. The *length* (number of pixels) is determined by the resolution of the image being scanned. For example, a square image scanned at 512×512 pixels (a size we typically use for dendritic imaging in SPNs) will give a line scan of at most 512 pixels. When initiated, the laser will run from some starting pixel to some final pixel in sequence and then repeat the process until the protocol is terminated. The refractory period, or the time required to move the laser beam from the end of the scan line to the beginning of the scan line is a function of the hardware. Some hardware and software allow bi-directional scanning, so that data are collected in both directions. While line scans are typically performed in straight lines, many 2PLSM software packages allow the use of complex shapes (such as curves, circles, etc.) if the experimental question requires.

The acquisition rate of a line scan is dependent upon both the length and number of pixels scanned and the pixel dwell time (amount of time the laser stimulates each pixel, usually in the 10 μs range). While the length is easily estimated by the anatomy of the

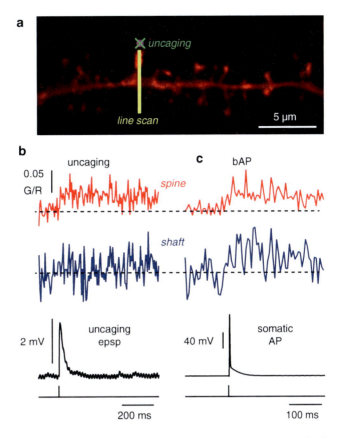

Fig. 4. Calcium imaging in a SPN dendrite using line scans. (**a**) MIP image of a distal SPN dendrite loaded with 50 μM Alexa 568 and 200 μM Fluo 4. Locations of line scan and uncaging spot are indicated (**b**, **c**). G/R fluorescent values for linescans of averaged (6) single uncaging events (**b**) or bAPs (**c**). G/R traces are shown for the indicated spine (*top*) and shaft (*middle*) and are aligned to the somatic voltage traces (*bottom*).

region being probed, the dwell time is best determined empirically, as it is sensitive to hardware. There is a trade-off between stimulating the sample long enough to detect emitted photons and briefly enough to maximize the sampling rate. Once the appropriate dwell time is determined, it can be standardized between experiments.

A major concern when performing line scans is photobleaching and/or phototoxicity. Repeated illumination of the same spot can bleach the dye, reducing its effective concentration (photobleaching), creating measurement artifacts. Repeated illumination also can damage the biological sample (phototoxicity). A common way to alleviate these problems is to shutter the laser during parts of the protocol where illumination is not required. For example, calcium transients evoked by single APs or EPSPs are small and typically require averaging multiple trials. There should be reasonably long periods between the line scans to be averaged to allow the membrane potential and cytosolic calcium concentration to return to baseline.

It is advantageous to shutter the laser during these times. The mechanical shutter on the laser is typically too slow to do this efficiently (a typical mechanical shutter has a duty cycle between 10 and 50 ms), and may induce unwanted noise. An alternative strategy is to use two Pockels cells in series. The first Pockels cell serves as the major attenuator of the laser power and is usually controlled by the 2PLSM scanning software. The second Pockels cell provides a second level of attenuation and is controlled by the electrophysiology protocol. As Pockels cells are electro-optic "shutters," and can be modulated at great speeds (2–5 ns) with simple voltage control signals (0–2 V), they are ideal for this task. Unlike the first Pockels cell in series, which is tuned to grade the laser power reaching the sample and adjust the image gain, the second Pockels cell is usually used only at two attenuation settings ("imaging baseline" and fully "closed"), with the laser only allowed to reach the sample during the times immediately flanking the physiological event to be imaged. Thus, the second modulator is used to provide fast shuttering for protocols requiring multiple stimulations. This strategy minimizes sample damage and photobleaching, while increasing the time during which reliable data can be gathered from a single neuron.

Many of the same limitations and problems raised within the context of anatomical imaging discussed above are applicable during calcium imaging. For example, simply using the intensity of the calcium fluorophore emission is problematic for between-sample comparisons. A more reliable approach is to normalize the emitted fluorescence, either to itself or to another signal. The most common method for normalizing the signal to itself is to calculate the percent change in fluorescence as $\Delta F/F_o$, where F is the peak fluorescence minus the baseline fluorescence and F_o is the average baseline fluorescence. This has many advantages, such as the ability to minimize the impact of sample depth and dye concentration. While this approach is simple to implement, the relative values are very sensitive to the F_o value, which is typically acquired during low, basal calcium conditions. Thus, small changes in the basal signal can have a big impact on the ratios. Another caveat of this method is that it does not provide a readout of drift in the basal calcium or dye concentration. An alternative method is to normalize the green calcium fluorescence signal to that of a second indicator, typically a red anatomical dye (Alexa 568 or 594) (G/R method). The Alexa dyes are insensitive to calcium concentration, so in practice they provide a strong signal when excited even when calcium concentration is low. While the green- and red-emitting dyes may not share identical photobleaching kinetics, monitoring the intensity of the red channel does provide an additional degree of confidence in the integrity of the scan (33, 57). Furthermore, as the signal is not normalized to the basal calcium concentration, this method is better suited to study manipulations that may result in altered basal calcium levels. For example, activation of G-protein-coupled receptors can result in the mobilization of intracellular calcium stores and slow changes in cytosolic calcium concentration; these changes often alter physiological outcome measures like EPSPs or synaptically evoked calcium transients. The G/R normalization will pick up this change readily.

2.3.3. Calcium Imaging in Other Basal Ganglia Neurons

One of the basal ganglia nuclei that our group has studied intensively in the last few years is the dopaminergic neurons of the substantia nigra pars compacta (SNc) (58–60). In these neurons, fluctuations in intracellular calcium concentration attend basal pacemaking, and these changes have been implicated in vulnerability of these neurons to aging and Parkinson's disease. The basics of calcium imaging in these neurons are similar to those described above for SPNs. However, there are a few differences between the cell types, both anatomical and physiological that warrant special consideration. SNc neurons have soma that are nearly twice as large as SPNs and have dendrites that are largely devoid of spines (61, 62). Data acquisition parameters must be adjusted to accommodate these differences in size (i.e., line scan length). A more difficult problem is posed by the aspiny nature of the dendrites. This makes the location of local glutamatergic inputs uncertain. While this may have little effect on the design and interpretation of many experiments, it can be an obstacle to overcome if examining the effects of excitatory synaptic inputs. Another consideration is that SNc neurons are autonomous pacemakers and have dendritic calcium oscillations at "rest." This is quite different from SPNs, which are hyperpolarized and quiescent in slices. Much attention has been paid to the mechanisms underlying these dendritic calcium oscillations in SNc neurons, both as a mechanism of pacemaking and a risk factor for disease (59, 60). These oscillations can be studied using the same approaches outlined above for SPNs. Line scans either along or across a dendrite can be performed to measure the kinetics of the dendritic calcium oscillations with great precision. In our experience, dwell times of 10–13 μs with a pixel size of 0.18 μm have been the most successful in measuring these oscillations (60). These are good starting settings, but may obviously need to be tweaked to fit to the properties of other systems. As mentioned above, another consideration is the choice of calcium indicator (and its concentration). In SNc dopaminergic neurons, the "resting" calcium concentration is modestly high and important to pacemaking. We have used low concentrations of relatively high-affinity dyes (Fluo-4, Fura 2, and Oregon green) to minimize the disruption of pacemaking while maximizing our ability to track the cytosolic calcium concentration.

2.4. Synaptic Stimulation Using 2PLU

A major function of striatal SPNs is to integrate afferent gluta-matergic synaptic inputs (primarily from the cortex and thalamus) and convey the outcome of this computation to other basal ganglia nuclei (2, 17, 20, 63). In spite of its importance, the dendritic mechanisms underlying this integration are largely unapproachable with conventional physiological tools. The reasons for this are anatomical. As mentioned above, there is no clear lamination of excitatory afferents to the striatum, making it difficult to reliably stimulate a well-defined input to SPNs. Also, cortical pyramidal neurons innervate SPN dendrites en passant, with a single pyramidal neuron forming only one or a few synapses on any given SPN dendritic tree (64). This makes the finding of dendritic spines activated by stimulation of a given pyramidal neuron extremely difficult. Recently, Higley et al. (65) have attempted to address this problem by using minimal local electrical stimulation guided by 2PLSM imaging. The stimulating electrode is placed next to an SPN dendrite of interest and the stimulation intensity adjusted to obtain an EPSP and calcium transient in a single dendritic spine within the field of view. While this is an immensely powerful technique, it cannot guarantee that a single spine is being activated. The possibility will always exist that other spines are stimulated but that they are outside the field of view or have calcium transients below the detection threshold of the experiment.

An alternative approach to electrical activation of glutamatergic afferents is to optically uncage glutamate near a synapse. Chemically "caging" glutamate renders it inactive until the chemical cage is degraded by photons (66–68). The most widely used caged glutamate is 4-methoxy-7-nitroindolinyl (MNI)-glutamate. A major advantage of this compound is that the bond between the MNI group and glutamate is readily photocleaved with 2P stimulation (68). The stimulation wavelength is sufficiently far away from that used for exciting calcium dyes, allowing simultaneous calcium imaging and glutamate uncaging. However, this requires two separate ultra-short pulse laser systems.

The relatively high price of MNI-glutamate typically makes large-scale bath application cost-prohibitive. To keep experiments cost effective, MNI-glutamate is usually applied to the recording chamber either through a small volume recirculating pump or through an injection syringe (with speed and flow-rate control) placed at the surface of the slice. We have experience with both methods and find each to have its own strengths and weaknesses. The recirculation pump allows for more even penetration of the caged compound through the depth of the slice, but limits any other pharmacological manipulations that can be made during the experiment because it is difficult to switch drugs in and out. On the other hand, while local perfusion of the caged compound to the surface of the slice may not yield equivocal concentrations of the drug with depth, it does allow an efficient and highly controlled means of performing paired pharmacological manipulations in the same cell.

As described above, the 2P laser used for uncaging can be finely positioned in the plane of focus using a pair of galvanometer mirrors. This allows glutamate to be uncaged at a single point (or a small number of rapid changing points) within this plane (Fig. 4). As spine heads are a major site of glutamatergic synaptic inputs in SPNs (46, 69, 70), these are typically the anatomical landmark to which the uncaging spot is targeted (the laser is targeted immediately adjacent to the spine head). Once MNI-glutamate is added to the slice and the target(s) selected for uncaging, the stimulus intensity must be adjusted to release the proper amount of glutamate. The power of the uncaging laser can be easily tuned with a Pockels cell modulator (68). There are two ways this has been approached: (1) by tuning the laser power to deliver a consistent photon intensity (as determined by photobleaching of the red Alexa dye) at the point of interest (32) or (2) by tuning the laser power to give a physiological response (such as somatic EPSP) of a desired amplitude. Photobleaching offers a means of calibrating laser power (and the amount of glutamate released) in a way that is independent of the sample biology (i.e., glutamate receptor density). This makes quantitative comparisons of the responses of different neuronal populations feasible (32, 65). However, as the uncaging laser is aimed squarely at the dendritic spine or shaft of interest, great care must be taken to avoid photodamage. Moreover, for reliable comparisons, this approach requires that MNI-glutamate concentration is the same throughout the depth of the slice being imaged. This limits this approach to setups with recirculation delivery of the caged compound. When applying MNI-glutamate with local perfusion, the second calibration strategy is advisable. This approach calls for adjusting the laser power to produce a given biological response, like a somatic EPSP. While this can make the interpretation of cell-to-cell comparisons complicated, it does allow for pharmacological manipulations to be made and interpreted.

In addition to glutamate, a slew of other neurotransmitters and physiologically important molecules have been conjugated to caging molecules. These include: dopamine, GABA, calcium, IP_3, and various neuropeptides. As we have limited experience with the majority of these, we refer the reader to other reviews for a more complete overview of their details (71–73). One note to keep in mind is that the photolability of the majority of caged compounds has been determined using single photon stimulation. As not all photosensitive molecules have good 2P cross sections, the utility of these compounds for 2P laser uncaging must be confirmed. Furthermore, the inertness of the caged molecule must be rigorously confirmed when applying it to a new cell type or even a new cellular compartment. For example, IP_3 caged by a single 1-(2-nitrophenyl)ethyl (NPE) group has been shown to produce calcium release in Purkinje neuron somata but not in dendrites when photo-stimulated. Closer examination revealed that IP$_3$ caged with a single NPE group actually blocks calcium release in these neurons, possibly accounting for the lack of effect seen in dendrites. The addition of a second NPE caging group to the IP$_3$ molecule allowed for increased spatial specificity with higher concentrations, ultimately uncovering IP$_3$ effects in dendrites (74). Thus, the methods described above for glutamate uncaging can be used to uncage other molecules now being produced, given they have appropriate 2P cross sections (both sensitive to 2P photo-cleaving and far enough removed from the maximum 2P excitation cross section used to image any other dyes used concurrently) and the caged compound is inert and stable.

2.5. Organelle Imaging in Neurons

Mitochondria have been implicated as key players in SNc dopaminergic neuron vulnerability and the pathogenesis of Parkinson's disease (75, 76). Recent work has underscored the importance of being able to monitor mitochondrial function of SNc dopaminergic neurons in situ (60). Besides being critical for cellular adenosine triphosphate (ATP) production, mitochondria are crucial players in other physiological tasks, such as regulation of cytosolic calcium, intracellular pH, apoptosis, and reactive oxygen species (ROS) production (77). Given that photo-stimulation can lead to unwanted ROS production (78, 79), 2PLSM has become an ideal method for studying mitochondrial function in brain slices because it limits photon dose and has the resolution needed to image subcellular organelles. Here, we outline the use of 2P microscopy in studying two important parameters of mitochondria physiology: the mitochondrial inner membrane potential ($\Delta\Psi m$) and ROS production.

Although the absolute $\Delta\Psi m$ is not measurable, it can be estimated on a relative scale using membrane permeable cationic dyes (80, 81). A variety of dyes are available for this purpose, the most commonly used being Rh123 and its derivatives tetramethyl rhodamine methylester (TMRM), tetramethyl rhodamine ethylester (TMRE), and the cyanine dye JC-1. TMRM has become the dye of choice because it displays better tissue penetration, low toxicity, and resistance to bleaching. Here, we will describe the use of TMRM to monitor inner mitochondrial membrane potential of SNc dopaminergic neurons in brain slices.

The first consideration is that of dye loading. A major difficulty in using brain slices (as opposed to cultured cells in a monolayer) is getting mitochondria of cells deep in the slice (~20–80 μm below the surface) adequately loaded with dye. This necessitates the use of high concentrations of TMRM to get mitochondria loaded quickly because of the limited lifetime of brain slices (approximately 4 h). We have found that incubating mouse brain slices (300-μm thick) with 2–4 μM TMRM for 30 min at 34–35°C is sufficient for loading.

Good labeling is of fundamental importance, as it allows one to use lower laser power (reducing photodamage) to obtain high-quality measurements. After loading, TMRM is washed out with TMRM-free ACSF for 20–60 min before imaging. This washout is a critical step as it keeps TMRM concentration low in neurons below the surface of the slice, reducing concerns about dye quenching. Measurements are then made in TMRM-free ACSF at 34–35°C. Fluorescent signals are acquired using an 830 nm excitation beam in a fixed focal plane with a pixel size between 0.18 and 0.21 μm. Time-series (t-series) measurements have been made at a rate of 2.5 frames per second and a dwell time of 4 μs. As a typical data set may include 1,000–2,000 frames, data analysis can becomes a limiting factor. To reduce this burden, fluorescent measurements are typically made of regions of interest (ROIs), which reduces the number of points per frame. ROIs must be monitored throughout the experiment to ensure the signal is stationary (i.e., no spatial drifting) and the baseline is constant (i.e. photobleaching and washout are minimal). Pharmacological manipulations can be performed using the same superfusion method described above.

Analysis of data collected during TMRM experiments can be performed on commercially available software (i.e., MATLAB) or shareware [i.e., Image J, PicViwer (University of Strathclyde)]. ROIs within the cell body can be chosen and monitored, and changes in TMRM fluorescence plotted over time. In our experiments, TMRM fluorescence intensity unexpectedly "flickered." Flickering frequency was measured in epochs of 100–200 s, depending on the design of the experiment. $\Delta\Psi$m can be estimated from the ROI fluorescence using the Nernst equation (60):

$$V = \frac{RT}{zF}\ln\left(\frac{\beta F_{\mathrm{m}}}{F_{\mathrm{n}}}\right),$$

where R is the gas constant, T is the temperature, F is the Faraday's constant, $z=1$, F_{m} is the fluorescence in the mitochondrial ROI, F_{n} is the fluorescence in the nucleus ROI at the same optical plane, and β is a scaling factor. The scaling factor can be calculated by assuming that the mitochondrial membrane potential is –100 mV at F_{maximal}. These techniques can be used with a variety of neuronal types in other brain regions.

As mentioned above, SNc dopaminergic neurons are autonomous pacemakers that flux calcium throughout the pacemaking cycle. Calcium entering during pacemaking must be extruded or sequestered through ATP-dependent processes, creating a metabolic burden on dopaminergic neurons. The demand for ATP is met primarily by mitochondria, through oxidative phosphorylation (OXPHOS). Increasing OXPHOS typically involves increased generation of superoxide and other ROS (82). Given the metabolic defenses against superoxide, whether or not this burden creates a measureable oxidant stress in SNc neurons in situ has been an open question. To address this question, our group generated a line of transgenic mice expressing a redox-sensitive variant of GFP (roGFP) with a mitochondrial matrix targeting sequence (mito-roGFP) (83, 84). To limit expression to monoaminergic neurons, mito-roGFP was expressed under the control of the tyrosine hydroxylase promoter. Dopaminergic neurons in the SNc and the adjacent ventral tegmental area (VTA) of these mice robustly expressed mito-roGFP that co-localized with mitochondrial markers, providing a reversible, quantitative means of monitoring oxidation of mitochondrial matrix proteins. Because the expression of mito-roGFP is restricted to a small set of neurons, it is possible to monitor the mitochondrial state of physiologically mature neurons in brain slices from young adult mice using 2PLSM. This GFP-based probe of redox potential has many advantages over previous methods of assessing redox status. For example, genetic encoding allows the probe to be introduced into any cell or organism that can express recombinant cDNA, and the proteins can be targeted to specific tissue and even subcellular locations.

Although roGFP is a ratiometric probe (83, 84), we have not used that capacity because of the difficulty in illuminating the sample at two wavelengths rapidly; this demands either two lasers or a rapidly tunable laser. Alternatively, we have excited roGFP with photons at a wavelength where the change in emission was most steeply related to redox status of the probe. In tissue sections, 2P excitation with 920 nm photons was found to be optimal with 2–3 µs pixel dwell times. ROIs were established over somatic or proximal dendritic regions. To establish redox status, sixty frames at 3–4 frames per second were acquired at one focal plane. This sampling strategy minimized photooxidation and tissue damage while maximizing signal reliability. Experiments can be performed either at room or physiological temperatures. At the end of each experiment, the maximum and minimum fluorescence of mito-roGFP was determined by fully reducing the mitochondria by bath application of 2 mM dithiothreitol (DTT) and then fully oxidizing the mitochondria with 100 µM aldrithiol (ald) (Fig. 5). This allows the relative oxidation (O_{rel}) to be calculated for each cell:

$$O_{rel} = 1 - \frac{F - F_{ald}}{F_{DTT} - F_{ald}}.$$

Although this limited us to one cell per slice, this calibration was critical for comparisons to be made between cells. The other important point is that the relationship between the dynamic range of the mito-roGFP probe and the dynamic range of the mitochondrial redox signal is unknown. All we know is that the mitochondrial redox signal changed within the range of our probe during the manipulations we performed.

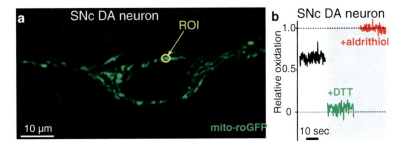

Fig. 5. Measuring mitochondrial redox state. (**a**) SNc Dopaminergic neuron in a brain slice from a mito-roGFP mouse. Mitochondria are labeled, and a region of interest (ROI) from which fluorescence measurements were taken is shown (*yellow circle*). (**b**) A representative fluorescence time series, showing fluorescence before (control, *black trace*) and after application of dithiothreitol (DTT; *green trace*) and Aldrithiol (*red trace*).

3. Optogenetic Approaches to Studying the Basal Ganglia

As mentioned above, the fine functional dissection of striatal circuitry has traditionally been difficult using traditional electrical stimulation. Optogenetic approaches to circuit analysis have burst upon the neuroscience scene and promise to revolutionize our approach to studying circuits. We briefly describe our experience with these approaches in the basal ganglia but the reader is referred to other recent reviews (85, 86). Optical manipulation of neuronal activity has been made possible with the advent of genetically encoded light-gated ion channels or ion pumps. The most widely used protein for this purpose is a microbial opsin, channelrhodopsin-2 (ChR2), a light-driven inwardly rectifying nonselective cation channel (87–89). ChR2 can be introduced into neurons with electroporation, viral vectors, particle-mediated gene transfer, or transgenic approaches. Exposure of ChR2 to a brief blue (~470 nm) light pulse leads to channel opening and a transient depolarization of the membrane potential; sufficiently strong stimulation allows the neuron to reach spike threshold (90–93). A train of action potentials can be triggered at frequencies up to 20 Hz by repetitive illumination. Hyperpolarization of neurons can be achieved with a bacteria-derived light-sensitive chloride pump, halorhodopsin (NpHR). This pump can be activated by yellow light (~580 nm) exposure, inhibiting neuronal activity (94, 95). Thus, ChR2 and NpHR can bidirectionally control neural activity with two different wavelengths of light (94). The optogenetics toolkit is rapidly expanding with newly engineered light-gated channel and pump variants. The properties of these variants differ in experimentally useful ways such as wavelength sensitivity, absolute light sensitivity, expression level, subcellular localization, and channel kinetics (92, 96, 97).

We are using optogenetic approaches to study the properties of corticostriatal and thalamostriatal microcircuits. Both populations of SPNs receive convergent glutamatergic inputs from the cortex and thalamus (69, 98–101), which terminate on both dendritic spines and shafts. Since these terminations cannot currently be distinguished, we have no means of using 2PLU of glutamate to contrast these synapses. The dendritic topography of projections arising from functionally distinct systems (e.g., frontal and motor corticostriatal microcircuits) is of major interest to us, given the recent discovery by our group using 2PLU that proximal and distal dendrites of SPNs have different properties. Recently, Karel Svoboda's group has described a novel strategy for using ChR2 to map functional synaptic inputs to dendrites (102). They referred to this approach as subcellular ChR2-assisted circuit mapping (sCRACM). The idea is simple and elegant. Using tetrodotoxin (TTX) to block generation of action potentials and 4-aminopyridine (4-AP) to boost terminal excitability, they used a focused beam of light to activate discrete ChR2 expressing presynaptic terminals along the dendrite while using a somatic patch electrode to monitor the cellular response. This allowed them to map the distribution of functional inputs within the entire dendritic arbor of a particular cell type. Not only could this strategy be used to map functional synaptic inputs onto SPNs, but also to build a full functional synaptic wiring diagram of the basal ganglia. Building this type of diagram can provide fundamental insight into the subcellular specificity of connectivity in the basal ganglia. This has obvious implications for our understanding of alterations in basal ganglia circuits in disease states, such as Parkinson's disease and Huntington's disease.

Despite the great advantages ChR2 affords in activating a defined set of axons and synaptic terminals, some discrepancies between electrically and optically induced neurotransmission have been reported (103–105). One explanation for some of the discrepancies is the desensitization of ChR2 with illumination, leading to diminished axonal sensitivity to illumination with repetitive stimulation. Another possible reason for divergence is the calcium permeability of ChR2, which could elevate the probability of transmitter release. One way of decreasing the chances of this happening is to stimulate axons at some distance away from the terminal of interest, allowing a normal action potential to propagate to the terminal and induce release. This requires a circumscribed illumination to activate ChR2 in a region several hundred microns from the cell being examined. Using 1P excitation with blue light, this is made difficult by the scattering of photons within the tissue (Rayleigh approximation for light scatter is proportional to $1/$ wavelength). Using a noncoherent light source, like an LED, worsens scattering. Using a coherent (laser) light source and a series of lenses, scattering and the size of the resulting illumination area can be as small as a few microns in diameter. The power delivered to the sample is another factor. As the intensity of the illumination is increased, the effective stimulation area will expand because of scattering, particularly with blue 1P excitation (106). An alternative is to use 2P illumination and longer wavelength light to diminish scattering. Several investigators have tried this approach with limited success. Using 2PLSM, ChR2 activation was insufficient to trigger APs; this was attributable to the low conductance and density of ChR2 achieved in the neuron and the small effective excitation volume (88, 107–109). To overcome this limitation, Rickgauer and Tank (109) used infrared 2P excitation (TPE) to increase the ChR2 activation area and elicit APs. However, repeated use of a small diameter, high-intensity light might be accompanied by accumulated ChR2 desensitization, especially in regions smaller than a soma. More recently two new light directing strategies have been developed to increase the area of ChR2 activation while still minimizing light scatter with 2P microscopy. Andrasfalvy et al. (110) have used temporally focused laser pulses (TEFO) and Papagiakoumou et al. (111) have combined generalized phase contrast with temporal focusing and digital holography to generate large illumination patterns that can activate ChR2 and produce APs. Each activation strategy possesses advantages and disadvantages for the end user to consider (108, 112).

4. On the Horizon: Optical Voltage Measurements?

Although still far from maturity, techniques allowing the optical measurement of voltage have been making strides. Many different approaches have been taken, each with strengths and limitations for particular applications (113). Though successful voltage imaging has yet to be demonstrated in SPN dendrites in intact slices, several recent studies in other systems offer a preliminary roadmap of how this might be tackled. One limiting factor of many common voltage-sensitive dyes, such as those derived from styryl and hemicyanine compounds, is that they can be toxic in mammalian neurons (114, 115). This effect can be diminished by heroism – patching a neuron, allowing the dye to load and equilibrate through the recording electrode, removing the electrode and letting the cell "rest" for several hours and then re-patching it with an electrode not containing the dye – but alternatives need to be developed for mortals. Palmer and Stuart (115) used the lipophilic voltage-sensitive dye JPW3028 to measure voltage changes in cortical pyramidal neuron dendritic spines using this approach. A more convenient method has been described by Bradley et al. (116) in Purkinje neurons using a hybrid voltage sensor pair. In this method, a lipophilic fluorophore (DiO) is applied to the membrane of the neuron of interest with the recording electrode and allowed to equilibrate. Dipicrylamine (DPA), a synthetic lipophilic photon acceptor ion, was then bath applied and acted as a membrane-partitioned nonfluorescent voltage sensor. The pair reported voltage-induced changes in fluorescence through a FRET mechanism. Thus far, this principle has only been successfully demonstrated in soma and proximal dendrites. One caveat to be noted is that DPA must be used at low concentrations, as higher concentrations (>5 µM) can alter membrane capacitance (116).

References

1. Larkman AU (1991) Dendritic morphology of pyramidal neurones of the visual cortex of the rat: III. Spine distributions. J Comp Neurol 306: 332–343

2. Surmeier DJ, Ding J, Day M, Wang Z, Shen W (2007) D1 and D2 dopamine-receptor modulation of striatal glutamatergic signaling in striatal medium spiny neurons. Trends Neurosci 30: 228–235

3. Wilson CJ (1992) Dendritic morphology, inward rectification, and the functional properties of neostriatal neurons. In: McKenna T, Davis, J, Zornetzer SF (ed) Single Neuron Computation. Academic Press, New York, 141–171

4. Magee JC, Johnston D (1995) Synaptic activation of voltage-gated channels in the dendrites of hippocampal pyramidal neurons. Science 268: 301–304

5. Stuart G, Spruston N, Sakmann B, Hausser M (1997) Action potential initiation and backpropagation in neurons of the mammalian CNS. Trends Neurosci 20: 125–131

6. Stuart GJ, Sakmann B (1994) Active propagation of somatic action potentials into neocortical pyramidal cell dendrites. Nature 367: 69–72

7. Denk W, Strickler JH, Webb WW (1990) Two-photon laser scanning fluorescence microscopy. Science 248: 73–76

8. Wokosin D, Squirrell JM, Eliceiri KW, White JG (2003) Optical workstation with concurrent, independent multiphoton imaging and experimental laser microbeam capabilities. Review of Scientific Instruments 74: 193–201

9. Svoboda K, Denk W, Kleinfeld D, Tank DW (1997) In vivo dendritic calcium dynamics in neocortical pyramidal neurons. Nature 385: 161–165

10. Svoboda K, Tank DW, Denk W (1996) Direct measurement of coupling between dendritic spines and shafts. Science 272: 716–719

11. Yuste R, Denk W (1995) Dendritic spines as basic functional units of neuronal integration. Nature 375: 682–684

12. Zipfel WR, Williams RM, Webb WW (2003) Nonlinear magic: multiphoton microscopy in the biosciences. Nat Biotechnol 21: 1369–1377

13. Svoboda K, Yasuda R (2006) Principles of two-photon excitation microscopy and its applications to neuroscience. Neuron 50: 823–839

14. Oheim M, Beaurepaire E, Chaigneau E, Mertz J, Charpak S (2001) Two-photon microscopy in brain tissue: parameters influencing the imaging depth. J Neurosci Meth 111: 29–37

15. Wokosin DL (2008) Nonlinear excitation fluorescence microscopy: source considerations for biological applications In: Clarkson WA, Hodgson N, Shori RH (eds), Solid State Lasers XVII: Technology and Devices (Proceedings of SPIE), pp 1–11.

16. Wokosin DL, Centonze V, White JG, Armstrong D, Robertson G, Ferguson AI (1996) All-solid-state ultra-fast lasers facilitate multi-photon excitation fluorescence imaging. IEEE J Sel Top Quant Electron 2: 1051–1065

17. Gerfen CR (1992) The neostriatal mosaic: multiple levels of compartmental organization. Trends Neurosci 15: 133–139

18. Kawaguchi Y, Wilson CJ, Augood SJ, Emson PC (1995) Striatal interneurons: chemical, physiological and morphological characterization. Trends Neurosci 18: 527–535

19. Tepper JM, Wilson CJ, Koos T (2008) Feedforward and feedback inhibition in neostriatal GABAergic spiny neurons. Brain Res Rev 58: 272–281

20. Smith Y, Bevan MD, Shink E, Bolam JP (1998) Microcircuitry of the direct and indirect pathways of the basal ganglia. Neuroscience 86: 353–387

21. Fujiyama F, Sohn J, Nakano T, Furuta T, Nakamura KC, Matsuda W, Kaneko T (2011) Exclusive and common targets of neostriatofugal projections of rat striosome neurons: a single neuron-tracing study using a viral vector. Eur J Neurosci 33: 668–677

22. Gertler TS, Chan CS, Surmeier DJ (2008) Dichotomous anatomical properties of adult striatal medium spiny neurons. J Neurosci 28: 10814–10824

23. Kreitzer AC, Malenka RC (2007) Endocannabinoid-mediated rescue of striatal LTD and motor deficits in Parkinson's disease models. Nature 445: 643–647

24. Lobo MK, Covington HE, Chaudhury D, Friedman AK, Sun H, Damez-Werno D, et al (2010) Cell type-specific loss of BDNF signaling mimics optogenetic control of cocaine reward. Science 330: 385–390

25. Koos T, Tepper JM (1999) Inhibitory control of neostriatal projection neurons by GABAergic interneurons. Nat Neurosci 2: 467–472

26. Plotkin JL, Wu N, Chesselet MF, Levine MS (2005) Functional and molecular development of striatal fast-spiking GABAergic interneurons and their cortical inputs. Eur J Neurosci 22: 1097–1108

27. Day M, Wokosin D, Plotkin JL, Tian X, Surmeier DJ (2008) Differential excitability and modulation of striatal medium spiny neuron dendrites. J Neurosci 28: 11603–11614

28. Heintz N (2004) Gene expression nervous system atlas (GENSAT) . Nat Neurosci 7: 483

29. Gerfen CR (1992) The neostriatal mosaic: multiple levels of compartmental organization. J Neural Transm Suppl 36: 43–59

30. Gittis AH, Nelson AB, Thwin MT, Palop JJ, Kreitzer AC (2010) Distinct roles of GABAergic interneurons in the regulation of striatal output pathways. J Neurosci 30: 2223–2234

31. Freiman I, Anton A, Monyer H, Urbanski MJ, Szabo B (2006) Analysis of the effects of cannabinoids on identified synaptic connections in the caudate-putamen by paired recordings in transgenic mice. J Physiol 575: 789–806

32. Bloodgood BL, Sabatini BL (2007) Nonlinear regulation of unitary synaptic signals by CaV(2.3) voltage-sensitive calcium channels located in dendritic spines. Neuron 53: 249–260

33. Carter AG, Sabatini BL (2004) State-dependent calcium signaling in dendritic spines of striatal medium spiny neurons. Neuron 44: 483–493

34. Conchello JA, Lichtman JW (2005) Optical sectioning microscopy. Nat Methods 2: 920–931

35. Shaw J (2006) Comparison of widefield/Deconvolution and confocal microscopy for three dimensional imaging. In Pawley JB (ed) Handbook of Biological Confocal Miicroscopy, 3rd edn. Springer, New York, 453–467

36. Holmes TJ, O'Connor N.J (2000) Blind Deconvolution of 3D Transmitted Light Brightfield Micrographs. J. Microsc 200: 114–127

37. Levoy M (1988) Display of surfaces from volume data. IEEE Comp Graph Applic 8: 29–37

38. Meißner M, Huang J, Bartz D, Mueller K, Crawfis R (2000) A practical evaluation of four popular volume rendering algorithms. In: Proc. of Symposium on Volume Visualization and Graphics. ACM Press 81–90

39. Lorensen, WE, Cline HE (1987) Marching Cubes: A High Resolution 3D Surface Construction Algorithm. Computer Graphics 21: 163–169

40. Cabral B, Cam N, Roran J (1994) in Proceedings of the Symposium on Volume Visualization. 91–98 (ACM Press)

41. Kasai H, Fukuda M, Watanabe S, Hayashi-Takagi A, Noguchi J (2010) Structural dynamics of dendritic spines in memory and cognition. Trends Neurosci 33: 121–129

42. Lee MC, Yasuda R, Ehlers MD (2010) Metaplasticity at single glutamatergic synapses. Neuron 66: 859–870

43. Matsuzaki M, Honkura N, Ellis-Davies GC, Kasai H (2004) Structural basis of long-term potentiation in single dendritic spines. Nature 429: 761–766

44. Tanaka J. Horiike Y, Matsuzaki M, Miyazaki T, Ellis-Davies GC, Kasai H (2008) Protein synthesis and neurotrophin-dependent structural plasticity of single dendritic spines. Science 319: 1683–1687

45. Tepper JM, Sharpe NA, Koos TZ, Trent F (1998) Postnatal development of the rat neostriatum: electrophysiological, light- and electron-microscopic studies. Dev Neurosci 20: 125–145

46. Wilson CJ, Groves PM (1980) Fine structure and synaptic connections of the common spiny neuron of the rat neostriatum: a study employing intracellular inject of horseradish peroxidase. J Comp Neurol 194: 599–615

47. Magee JC, Johnston D (1997) A synaptically controlled, associative signal for Hebbian plasticity in hippocampal neurons. Science 275: 209–213

48. Nakamura T, Barbara JG, Nakamura K, Ross WN (1999) Synergistic release of Ca2+ from IP3-sensitive stores evoked by synaptic activation of mGluRs paired with backpropagating action potentials. Neuron 24: 727–737

49. Nakamura T, Nakamura K, Lasser-Ross N, Barbara JG, Sandler VM, Ross WN (2000) Inositol 1,4,5-trisphosphate (IP3) -mediated Ca2+ release evoked by metabotropic

agonists and backpropagating action potentials in hippocampal CA1 pyramidal neurons. J Neurosci 20: 8365–8376

50. Nimchinsky EA, Sabatini BL, Svoboda K (2002) Structure and function of dendritic spines. Annu Rev Physiol 64: 313–353

51. Gee KR, Brown KA, Chen WN, Bishop-Stewart J, Gray D, Johnson I (2000) Chemical and physiological characterization of fluo-4 Ca(2+) -indicator dyes. Cell Calcium 27: 97–106

52. Minta A, Kao JP, Tsien RY (1989) Fluorescent indicators for cytosolic calcium based on rhodamine and fluorescein chromophores. J Biol Chem 264: 8171–8178

53. Murphy SN, Thayer SA, Miller RJ (1987) The effects of excitatory amino acids on intracellular calcium in single mouse striatal neurons in vitro. J Neurosci 7: 4145–4158

54. Murphy SN, Miller RJ (1989) Regulation of Ca++ influx into striatal neurons by kainic acid. J Pharmacol Exp Ther 249: 184–193

55. Garaschuk O, Milos RI, Konnerth A (2006) Targeted bulk-loading of fluorescent indicators for two-photon brain imaging in vivo. Nat Prot 1: 380–386

56. Carter AG, Soler-Llavina GJ, Sabatini BL (2007) Timing and location of synaptic inputs determine modes of subthreshold integration in striatal medium spiny neurons. J Neurosci 27: 8967–8977

57. Soler-Llavina GJ, Sabatini BL (2006) Synapse-specific plasticity and compartmentalized signaling in cerebellar stellate cells. Nat Neurosci 9: 798–806

58. Chan CS, Guzman JN, Ilijic E, Mercer JN, Rick C, Tkatch T, Meredith GE, Surmeier DJ (2007) 'Rejuvenation' protects neurons in mouse models of Parkinson's disease. Nature 447: 1081–1086

59. Guzman JN, Sanchez-Padilla J, Chan CS, Surmeier DJ (2009) Robust pacemaking in substantia nigra dopaminergic neurons. J Neurosci 29: 11011–11019

60. Guzman JN, Sanchez-Padilla J, Wokosin D, Kondapalli J, Ilijic E, Schumacker PT, Surmeier DJ (2010) Oxidant stress evoked by pacemaking in dopaminergic neurons is attenuated by DJ-1. Nature 468: 696–700

61. Kawaguchi Y (1997) Neostriatal cell subtypes and their functional roles. Neurosci Res 27: 1–8

62. Tepper JM, Sawyer SF, Groves PM (1987) Electrophysiologically identified nigral dopaminergic neurons intracellularly labeled with HRP: light-microscopic analysis. J Neurosci 7: 2794–2806

63. Alexander GE, Crutcher MD, DeLong MR (1990) Basal ganglia-thalamocortical circuits: parallel substrates for motor, oculomotor, "prefrontal" and "limbic" functions. Prog Brain Res 85: 119–146

64. Parent M, Parent A (2006) Single-axon tracing study of corticostriatal projections arising from primary motor cortex in primates. J Comp Neurol 496: 202–213

65. Higley MJ, Sabatini BL (2010) Competitive regulation of synaptic Ca2+ influx by D2 dopamine and A2A adenosine receptors. Nat Neurosci 13: 958–966

66. Canepari M, Nelson L, Papageorgiou G, Corrie JE, Ogden D (2001) Photochemical and pharmacological evaluation of 7-nitroindolinyl-and4-methoxy-7-nitroindolinyl-amino acids as novel, fast caged neurotransmitters. J Neurosci Methods 112: 29–42

67. Huang YH, Sinha SR, Fedoryak OD, Ellis-Davies GC, Bergles DE (2005) Synthesis and characterization of 4-methoxy-7-nitroindolinyl-D-aspartate, a caged compound for selective activation of glutamate transporters and N-methyl-D-aspartate receptors in brain tissue. Biochemistry 44: 3316–3326

68. Matsuzaki M, Ellis-Davies GC, Nemoto T, Miyashita Y, Iino M, Kasai H (2001) Dendritic spine geometry is critical for AMPA receptor expression in hippocampal CA1 pyramidal neurons. Nat Neurosci 4: 1086–1092

69. Doig NM, Moss J, Bolam JP (2010) Cortical and thalamic innervation of direct and indirect pathway medium-sized spiny neurons in mouse striatum. The Journal of neuroscience : the official journal of the Society for Neuroscience 30: 14610–14618

70. Smith AD, Bolam JP (1990) The neural network of the basal ganglia as revealed by the study of synaptic connections of identified neurones. Trends Neurosci 13: 259–265

71. Ellis-Davies GC (2008) Neurobiology with caged calcium. Chem Rev 108: 1603–1613

72. Kaplan JH, Somlyo AP (1989) Flash photolysis of caged compounds: new tools for cellular physiology. Trends Neurosci 12: 54–59

73. Shigeri Y, Tatsu Y, Yumoto N (2001) Synthesis and application of caged peptides and proteins. Pharmacol Ther 91: 85–92

74. Sarkisov DV, Gelber SE, Walker JW, Wang SS (2007) Synapse specificity of calcium release probed by chemical two-photon uncaging of inositol 1,4,5-trisphosphate. J Biol Chem 282: 25517–25526

75. Zhu J, Chu CT (2010) Mitochondrial dysfunction in Parkinson's disease. J Alz Dis 20,Suppl 2: S325–334

76. Schapira AH (2010) Future strategies for neuroprotection in Parkinson's disease. Neurodegener Dis 7: 210–212

77. Nicholls DG, Budd SL (2000) Mitochondria and neuronal survival. Physiol Rev 80: 315–360

78. Foster KA, Galeffi F, Gerich FJ, Turner DA, Muller M (2006) Optical and pharmacological tools to investigate the role of mitochondria during oxidative stress and neurodegeneration. Prog Neurobiol 79: 136–171

79. Zorov DB, Kobrinsky E, Juhaszova M, Sollott SJ (2004) Examining intracellular organelle function using fluorescent probes: from animalcules to quantum dots. Circ Res 95: 239–252

80. Nicholls DG (2002) Mitochondrial function and dysfunction in the cell: its relevance to aging and aging-related disease. Int J Biochem Cell Biol 34: 1372–1381

81. Nicholls DG (2002) Bioenergetics. 3 edn, Academic Press, New York

82. Lambert AJ, Brand M. D (2009) Reactive oxygen species production by mitochondria. Meth Mol Biol 554: 165–181

83. Dooley CT Dore TM, Hanson GT Jackson WC, Remington SJ Tsien RY (2004) Imaging dynamic redox changes in mammalian cells with green fluorescent protein indicators. J Biol Chem 279: 22284–22293

84. Hanson GT, Aggeler R Oglesbee D, Cannon M Capaldi RA, Tsien RY, Remington SJ (2004) Investigating mitochondrial redox potential with redox-sensitive green fluorescent protein indicators. J Biol Chem 279: 13044–13053

85. Deisseroth K (2011) Optogenetics. Nat Meth 8: 26–29

86. Diester I, Kaufman MT, Mogri M, Pashaie R, Goo W, Yizhar O, Ramakrishnan C, Deisseroth K, Shenoy KV (2011) An optogenetic toolbox designed for primates. Nat Neurosci 14: 387–397

87. Hegemann P, Moglich A (2011) Channelrhodopsin engineering and exploration of new optogenetic tools. Nat Meth 8: 39–42

88. Nagel G, Szellas T, Huhn W, Kateriya S, Adeishvili N, Berthold P, Ollig D, Hegemann P, Bamberg E (2003) Channelrhodopsin-2, a directly light-gated cation-selective membrane channel. Proc Natl Acad Sci USA 100: 13940–13945

89. Sineshchekov OA, Jung KH, Spudich JL (2002) Two rhodopsins mediate phototaxis to low- and high-intensity light in Chlamydomonas reinhardtii. Proc Natl Acad Sci USA 99: 8689–8694

90. Boyden ES, Zhang F, Bamberg E, Nagel G, Deisseroth K (2005) Millisecond-timescale, genetically targeted optical control of neural activity. Nature Neuroscie 8: 1263–1268

91. Lanyi JK. Halorhodopsin: a light-driven chloride ion pump (1986) Annu Rev Biophys Biophys Chem 15: 11–28

92. Zhang F, Gradinaru V, Adamantidis AR, Durand R, Airan RD, de Lecea L, Deisseroth K (2010) Optogenetic interrogation of neural circuits: technology for probing mammalian brain structures. Nat Prot 5: 439–456

93. Zhang F, Wang LP, Boyden ES, Deisseroth K (2006) Channelrhodopsin-2 and optical control of excitable cells. Nat Meth 3: 785–792

94. Han X, Boyden ES (2007) Multiple-color optical activation, silencing, and desynchronization of neural activity, with single-spike temporal resolution. PLoS One 2, e299

95. Zhang F, Wang LP, Brauner M, Liewald JF, Kay K, Watzke N, Wood PG, Bamberg E, Nagel G, Gottschalk A, Deisseroth K (2007) Multimodal fast optical interrogation of neural circuitry. Nature 446: 633–639

96. Gradinaru V, Zhang F, Ramakrishnan C, Mattis J, Prakash R, Diester I, Goshen I, Thompson KR, Deisseroth K (2010) Molecular and cellular approaches for diversifying and extending optogenetics. Cell 141: 154–165

97. Knopfel T, Lin MZ, Levskaya A, Tian L, Lin JY, Boyden ES (2010) Toward the second generation of optogenetic tools. J Neurosci 30: 14998–15004

98. Kemp JM, Powell TP (1971) The termination of fibres from the cerebral cortex and thalamus upon dendritic spines in the caudate nucleus: a study with the Golgi method. Philos Trans R Soc Lond B 262: 429–439

99. Kitai ST, Kocsis JD, Wood J (1976) Origin and characteristics of the cortico-caudate afferents: an anatomical and electrophysiological study. Brain Res 118: 137–141

100. McGeorge AJ, Faull RL (1989) The organization of the projection from the cerebral cortex to the striatum in the rat. Neuroscience 29: 503–537

101. Smith Y, Bennett BD, Bolam JP, Parent A, Sadikot AF (1994) Synaptic relationships between dopaminergic afferents and cortical or thalamic input in the sensorimotor territory of the striatum in monkey. J Comp Neurol 344: 1–19

102. Petreanu L, Mao T, Sternson SM, Svoboda K (2009) The subcellular organization of neocortical excitatory connections. Nature 457: 1142–1145

103. Cruikshank SJ, Urabe H, Nurmikko AV, Connors BW (2010) Pathway-specific feedforward circuits between thalamus and neocortex revealed by selective optical stimulation of axons. Neuron 65: 230–245

104. Ren J, Qin C, Hu F, Tan J, Qiu L, Zhao S, Feng G, Luo M (2011) Habenula "cholinergic" neurons corelease glutamate and acetylcholine and activate postsynaptic neurons via distinct transmission modes. Neuron 69: 445–452

105. Schoenenberger P, Scharer YP, Oertner TG (2011) Channelrhodopsin as a tool to investigate synaptic transmission and plasticity. Exp Physiol 96: 34–39

106. Schoenenberger P, Grunditz A, Rose T, Oertner TG (2008) Optimizing the spatial resolution of Channelrhodopsin-2 activation. Brain Cell Biol 36: 119–127

107. Helmchen F, Denk W (2005) Deep tissue two-photon microscopy. Nat Meth 2: 932–940

108. Peron S, Svoboda K (2011) From cudgel to scalpel: toward precise neural control with optogenetics. Nat Meth 8: 30–34

109. Rickgauer JP, Tank DW (2009) Two-photon excitation of channelrhodopsin-2 at saturation. Proc Natl Acad Sci USA 106: 15025–15030

110. Andrasfalvy BK, Zemelman BV, Tang J, Vaziri A (2010) Two-photon single-cell optogenetic control of neuronal activity by sculpted light. Proc Natl Acad Sci USA 107: 11981–11986

111. Papagiakoumou E, Anselmi F, Bègue A, de Sars V, Glückstad J, Isacoff EY, Emiliani V (2010) Scanless two-photon excitation of channelrhodopsin-2. Nat Meth 7: 848–854

112. Shoham S (2010) Optogenetics meets optical wavefront shaping. Nat Meth 7: 798–799

113. Peterka DS, Takahashi H, Yuste R (2011) Imaging voltage in neurons. Neuron 69: 9–21

114. Antic S, Cohen LB, Lam YW, Wachowiak M, Zecevic D, Zochowski M (1999) Fast multisite optical measurement of membrane potential: three examples. FASEB J 13, Suppl 2: S271-276

115. Stuart GJ, Palmer LM (2006) Imaging membrane potential in dendrites and axons of single neurons. Pflugers Arch 453: 403–410

116. Bradley J, Luo R, Otis TS, DiGregorio DA (2009) Submillisecond optical reporting of membrane potential in situ using a neuronal tracer dye. J Neurosci 29: 9197–9209

Chapter 11

Electrophysiological Analysis of Movement Disorders in Mice

Shilpa P. Rao, Véronique M. André, Carlos Cepeda, and Michael S. Levine

Abstract

Electrophysiological approaches have emerged as powerful tools to investigate neuronal activity in diseased states. The subtle changes that precede overt neurological dysfunction and development of behavioral abnormalities in movement disorders can be investigated in animal models using a wide range of electrophysiological recording techniques. These can provide critical information from the level of single neurons to coordinated activity in microcircuits and large networks in the brain. Since cellular dysfunction can be identified at a very nascent stage, potential prodromal markers and therapeutic targets can be identified. In this chapter, we review the application of electrophysiological approaches to the study of movement disorders with emphasis on what we have learned from rodent models of Huntington's and Parkinson's diseases.

Key words: Huntington's disease, Parkinson's disease, Mouse models, Electrophysiology

1. Introduction

While loss of neurons is the hallmark of movement disorders like Huntington's (HD) and Parkinson's diseases (PD), it has become increasingly clear that neuronal dysfunction precedes neuronal degeneration (1–3). Subtle pathophysiological changes in neurons and a clinical prodrome precede recognizable symptoms in human disease and these often cause many of the initial disease symptoms. Neuronal dysfunction at the earliest stage is characterized by subtle functional changes in ion channels, neurotransmitter release, and synaptic receptors rather than overt structural changes.

Electrophysiological approaches represent some of the most powerful tools to investigate cellular function and dysfunction in the nascent stage of the disease. Investigations at this stage also provide an excellent opportunity to uncover new targets for thera-

Emma L. Lane and Stephen B. Dunnett (eds.), *Animal Models of Movement Disorders: Volume I*, Neuromethods, vol. 61,
DOI 10.1007/978-1-61779-298-4_11, © Springer Science+Business Media, LLC 2011

peutic intervention which may differ markedly from such targets after overt neuronal degeneration sets in, further underlining the importance of performing electrophysiological studies in cellular and animal models of movement disorders, probably the best way of identifying prodromal targets.

In this chapter, we will review a number of the electrophysiological techniques applied to the study of movement disorders, with the primary focus on HD, but also PD, in rodent models. From single cell recordings in vitro, to recordings in whole animals in vivo, each technique provides unique but complementary glimpses into the types of neuronal alterations that can be correlated with behavioral deficits and with mechanisms that ultimately lead to cell death.

2. Preparations Used to Examine Electrophysiological Function

The fundamental function of neuronal networks is the integration of synaptic inputs that sculpt cell membrane excitability and allow faithful transmission of this information. When investigating neuronal dysfunction, it is important to understand whether the changes are reflective of altered inputs to the neuron, caused by altered neurotransmitter release, or due to intrinsic alterations in neuronal voltage-gated ion channels and/or neurotransmitter receptors. It is therefore essential to investigate neuronal function both with intact synaptic inputs and in isolation or semi-isolation. Various techniques and preparations have been designed based on which aspect of neuronal function is the focus of the study (Fig. 1).

2.1. Acute Brain Slices

The use of acute brain slices for electrophysiology has become widespread since it was first introduced in the 1960s (4). Brain slices can be obtained from a wide range of animal species and neural regions. Following removal of the brain and isolation of the region of interest, a vibrating microtome is used to obtain a thin slice (100–400 μm thickness) of brain tissue that can be used for electrophysiological experiments after 1–2 h of incubation to allow for neuronal recovery. The extracellular ionic environment can be precisely controlled and various drugs can be applied onto neurons. The local anatomy of the tissue is relatively intact, leaving microcircuits and therefore identification of subsets of local neuronal connections is possible. With the use of brain slices, it is also possible to record from two or more neurons simultaneously or from different parts of the same neuron such as the soma and dendrites. With the development of sophisticated imaging techniques, brain slices have become a versatile preparation for a variety of electrophysiological studies. The main shortcoming using brain slices is that, though the local cytoarchitecture is relatively preserved, there is loss of long-distance synaptic connections.

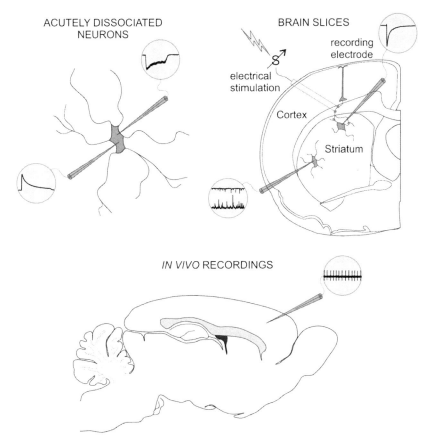

Fig. 1. Preparations used for electrophysiological recordings. Schematic representation of some of the most common electrophysiological techniques. *Top left*: Acutely dissociated neuron with two recording electrodes and sample traces of excitatory (*top inset*) and inhibitory (*bottom inset*) currents evoked by the application of NMDA and GABA, respectively. *Top right*: Brain slice showing a patch electrode and recordings of spontaneous currents (*lower inset*) from a MSSN and another electrode recording evoked currents (*upper inset*) in the MSSN in response to stimulation of inputs from the cortex. *Lower* schematic: Single- or multiunit in vivo recordings can be performed in anesthetized or awake behaving animals. The *inset* shows a sample trace of an in vivo, single-unit extracellular recording.

2.2. Acutely Dissociated Neurons

Neurons can be isolated from most regions of the brain using a combination of mild enzymatic digestion and gentle mechanical trituration (5). This method provides perhaps the most isolated preparation for relatively intact neurons. Also, because these acutely isolated neurons are free from cellular debris and ensheathing glia, the formation of gigaohm seals is facilitated in patch-clamp studies. Neurons obtained from this preparation are free of synaptic inputs and usually shorn of their dendritic processes except for the initial stumps of primary dendrites.

The advantage of using dissociated neurons is the absence of synaptic inputs, so intrinsic and postsynaptic changes in the neuron can be studied in relative isolation. The loss of dendrites reduces space clamp distortion of currents in voltage-clamp recordings but also leads to loss of information from ion channels and receptors located more peripherally on the dendrites.

2.3. Neuronal Cultures   Dissociated neurons and brain slices from many regions of the brain can be grown in culture for electrophysiological analysis (6–8). Since neurons in culture form a monolayer, they can be easily visualized and the extracellular solution in this preparation, as with that in acutely dissociated neurons, can be precisely maintained and quickly changed to record effects of various pharmacological manipulations. As these preparations have synaptic contacts, in addition to ionic currents, synaptic connectivity can be studied. Also, similar to acutely isolated neurons, the bare neuronal membranes facilitate gigaohm seal formation. Neurons from two regions of the brain can be co-cultured and they will form synaptic connections (9). Neuronal cultures are simple yet powerful tools to examine neuronal function; however, there are disadvantages. Though synaptic connections form in these preparations, they do not represent the total of all inputs the neurons receive. Neurons are maintained in an artificial environment in the absence of a majority of their influences from other neurons, hormones, and neurotransmitters and may have compensatory or abnormal properties that are not present in vivo. Finally, cultured preparations typically do not survive for long time periods minimizing the ability to evaluate long-lasting changes.

2.4. In Vivo Recordings   The field of electrophysiology has evolved to the stage where chronic in vivo recordings can be performed in awake animals in a variety of species. Using microwire electrodes, single- or multiunit extracellular recordings can be performed by surgically placing the electrode at desired coordinates inside the brain using a stereotaxic instrument. Action potentials in neurons are detected as small voltage fluctuations (0.1–1 mV). This technique can be coupled with iontophoresis and local drug administration, stimulation (e.g., electrical, chemical, light) of afferents, or antidromic activation of target areas. With the advent of advanced computer techniques, it has become possible to simultaneously record from large samples of single neurons distributed across multiple brain sites in fully awake and behaving animals over long periods of time to integrate changes in neuronal activity with behavioral alterations.

The obvious advantage of in vivo electrophysiology is that neurons can be studied in their natural milieu with their entire complement of inputs, targets, circulating hormones, and modulators. In vivo experiments either in awake or anesthetized preparations can be used to determine the effect of systemically administered drugs or physiological manipulations that might have several targets of action. On the other hand, it is much more difficult to precisely manipulate the local environment surrounding the cell groups under study in in vivo experiments. In awake, behaving animals there are several other variables such as behavioral state, stress, differences in training, and task ability that could affect the recorded activity.

3. An Overview of Electrophysiological Methods

3.1. Sharp Electrode Recordings

Classical electrophysiological techniques involve transmembrane recordings made using a fine-tipped "sharp" electrode inserted carefully inside the neuron. With this technique, a fine-tipped glass capillary filled with KCl or K-acetate is inserted into the cell to obtain transmembrane intracellular recordings. The potential difference between the tip of the electrode and a reference electrode placed in the bath provides the transmembrane potential of the neuron. When a constant or time-varying current is applied to a neuron and the change in membrane potential recorded, it is termed current-clamp recordings. This experimental protocol is more physiological than voltage-clamp recordings as it mimics the changes in membrane potential that a neuron undergoes in vivo in response to applied stimuli. Extracellular recordings can be obtained when the electrode is close to, but outside the neuron. In this configuration, the electrode records relatively large, fast electrical events (e.g., action potentials) from single neurons (single-unit) or groups of neurons (multiunit recordings) as well as slower time course events (field potentials).

Extracellular and intracellular recordings with sharp electrodes are relatively easy to perform. They are therefore the technique of choice for long-term recordings of synaptic plasticity like long-term potentiation (LTP) and long-term depression (LTD). However, due to the high electrical resistance of the pipettes, the recording conditions are not optimal because the cell membrane is typically damaged. Sharp electrode penetration leads to depolarization and sometimes injury discharge coupled with damage to the neuronal membrane which is recorded as a significant leakage resistance across the membrane of the recorded cell (10).

3.2. Patch-Clamp Recordings

In contrast to sharp electrode recordings in which an electrode is inserted through the cell membrane, the patch-clamp technique involves forming a high-resistance seal (gigaseal) by pushing a micropipette with a fire-polished tip against the cell membrane (11). Cells can be studied in current or voltage-clamp modes. In voltage-clamp mode, the membrane potential is held constant (holding potential) or a series of voltage commands, steps, or ramps are applied and the ionic currents through the membrane are recorded. Patch-clamp recording is a versatile method to study voltage- and time-dependent ionic conductances and their modulation by various pharmacological agents.

One of the main difficulties in performing patch-clamp recordings from brain slices is clogging of the patch pipette tip as it passes through the tissue overlying the neuron to be recorded within the slice. This problem can be surmounted by using imaging techniques like infrared (IR) videomicroscopy combined with differential interference contrast (DIC) optics and using a non fire-polished patch pipette to penetrate the slice and "clean" away the debris above the target neuron with a puff from the pipette solution (12, 13). Once the surface of the neuron of interest is clean, a gigaohm seal can be formed.

After formation of the gigaohm seal, currents from single ion channels can be recorded from a patch of the cell membrane in *cell attached* configuration or from single channels in a membrane patch isolated from the neuron in either *inside-out* or *outside-out* configurations. Following gigaohm seal formation, rupturing the cell membrane spanning the micropipette tip enables current recordings from all ion channels in the entire cell membrane and this configuration is termed *whole-cell* recording. The patch-clamp technique enables recordings from a wide variety of cell types and allows measurements of currents in the picoampere (pA) range. One of the drawbacks of this technique is that the internal milieu of the recorded cell is altered depending on the pipette solution used. Addition of gramicidin or nystatin in the pipette (*perforated patch*) can circumvent this problem by allowing movement of some small monovalent ions and preserving most of the ionic and second messenger constituents of the cell.

3.3. Electrotonic Potentials and Dye-Coupling

Besides chemical synaptic transmission between neurons, there exists a form of communication between neurons mediated by gap junctions termed electrical coupling. This can be investigated at the same time that intracellular or patch-clamp recordings are performed by including a low molecular weight dye like Lucifer Yellow (LY) in the pipette. LY crosses gap junctions and neurons that are connected by gap junctions appear "dye-coupled" in that more than one neuron is labeled even though the activity of only one neuron was recorded.

3.4. Combining Patch-Clamp and Optical Imaging Techniques

The versatility of the patch-clamp technique enables it to be combined with a variety of imaging techniques. These techniques include imaging of ions, especially Ca^{2+}, in living cells. The cytosolic Ca^{2+} level and its changes can be visualized using fluorescent, Ca^{2+} binding dyes, such as Fura-2 delivered to the cell in the patch pipette. Photobleaching and uncaging of compounds in localized regions are also associated techniques for visualization of ions. Fluorescent proteins like green, yellow, and red fluorescent proteins (GFP, YFP, and RFP, respectively) are powerful tools to label neurons. The fluorescent protein can be targeted to specific neurons to identify them in a preparation that permits visualization. The temporal resolution of electrophysiology can be combined with the spatial resolution of imaging techniques to define spatiotemporal characteristics of individual neurons and networks (14). The two techniques can be combined in a wide array of configurations. Electrophysiological recordings of synaptic inputs from the soma of a neuron can be correlated with the location of the input in the dendritic tree by measuring synaptically activated Ca^{2+} signals in the dendritic arbor. By simultaneous imaging activity in the local network using a fluorescent Ca^{2+} probe, reverse (presynaptic) or forward (postsynaptic) connectivity of an electrophysiologically recorded neuron can be obtained (14).

3.5. Selective Neuronal Ablation

The ability to selectively inactivate specific neurons or populations of neurons in vitro and in vivo provides a valuable tool to understand specific neuronal functions and the neuronal circuit as a whole. There are various methods for rapidly inducing and reversing modulation of neuronal excitability and neurotransmitter release (15). The allatostatin receptor (AlstR) system is based on the expression of the Drosophila AlstR that is selectively activated by the insect peptide hormone allatostatin. Coupling of allatostatin to AlstR leads to opening of a G protein-coupled inwardly rectifying K^+ channel that causes hyperpolarization and suppression of action potential firing (16). A second system to selectively silence neurons is the ivermectin-sensitive chloride channel. This involves co-expression of two subunits (α and β) of a *Caenorhabditis elegans* glutamate- and ivermectin-gated chloride channel (GluCl$\alpha\beta$) that is activated by ivermectin and results in a chloride current that suppresses neuronal activity. For example, in vivo, expression of GluCl$\alpha\beta$ unilaterally in the striatum and administration of ivermectin leads to perturbation of amphetamine-induced rotational behavior (17). The MIST (Molecular Systems for the Inactivation of Synaptic Transmission) system blocks neurotransmission by interfering with the function of presynaptic proteins like synaptobrevin, synaptophysin, and syntaxin (18). Diphtheria toxin can also be used to remove neurons. For example, when the diphtheria toxin receptor and diphtheria toxin injection ablated selectively dopamine (DA) D_2 receptor-containing neurons in the entire striatum hyperlocomotion occurred, whereas more selective ablation in the nucleus accumbens increased amphetamine conditioned place preference (19).

3.6. Optogenetics

Although electrical simulation has been the standard method for activation of neurons and fiber tracts, new techniques combining genetic expression of light-sensitive channels in selected neuronal populations and stimulation with light have been developed. These optogenetic approaches permit rapid stimulation and inhibition of genetically targeted neuronal populations on a millisecond timescale. This is achieved by genetically expressing the microbial opsins *Chlamydomonas reinhardtii* Channelrhodopsin-2 (ChR2) and *Natronomonas pharaonis* halorhodopsin (NpHR) in neurons (20, 21). ChR2 is a cation channel that is activated by ~470 nm blue light and NpHR is a chloride pump activated by ~580 nm yellow light. By varying the wavelength of light used, ChR2 and

NpHR can be activated independent of each other to evoke action potential firing or inhibit neuronal activity, respectively (22). The activity of specific neuronal subtypes can be modulated by targeting these opsin proteins to the neurons of interest. While brain slices with ChR2- and NpHR-expressing neurons can be studied, an important application of optogenetics is the ability to express and study these proteins in intact animals. Fiber optic-based systems have been developed to deliver light to precise ChR2- or NpHR-expressing neuronal targets in superficial and deep brain structures (23, 24).

4. Alterations in Cellular and Network Electrophysiological Properties in Models of Movement Disorders

A significant number of electrophysiological studies have been performed on brain slices obtained from rodent models of HD and PD (25–28) (Table 1). These studies have revealed important changes in passive and active membrane properties, as well as synaptic transmission, as the disease progresses. Of major interest in these studies, specific changes in DA D_1 and D_2 receptor-expressing medium-sized spiny neurons (MSSNs) of the direct and indirect striatal output pathways, respectively, can now be identified by using mice that express the EGFP gene under the control of the promoter for either DA D_1 or D_2 receptors. In these mice, D_1 and D_2 DA receptor neurons can be individually identified and the mice can be crossed to genetic models of HD or PD as well as determining the effects of DA depletion (29, 30).

4.1. Passive Membrane Properties

Intracellular and patch-clamp recordings in slices have provided useful information regarding changes in passive membrane properties such as cell capacitance and input resistance in mouse models of HD. These properties correlate with cell surface characteristics exemplified by total surface area as well as the number of conducting channels in the membrane. In striatal MSSNs, there is an increase in input resistance, which probably reflects the progressive loss of conductive channels (31, 32). Another observation is reduced cell capacitance in these neurons, an electrophysiological observation that has been supported by morphological evidence (32). In models of PD in which DA has been depleted chronically, differential alterations occur in passive membrane properties of MSSNs originating the direct or indirect pathways (those expressing D_1 or D_2 DA receptors, respectively). Neurons of the indirect pathway show no change in cell capacitance or input resistance. However, D_1 receptor-expressing MSSNs have a decreased capacitance and a higher input resistance compared to controls indicative of changes in the direct pathway output (33).

Table 1
Summary of electrophysiological outcomes in models of HD and PD using different approaches

Preparation	HD	Reference
Dissociated neurons and cultures	Decreased NMDAR currents and increased NMDAR Mg^{2+} sensitivity in cortical pyramidal neurons	(37)
	Increased NMDAR currents and decreased NMDAR Mg^{2+} sensitivity in striatal MSSNs	(65)
	Increased voltage-activated Ca^{2+} currents in cortex	(37)
	Decreased voltage-activated Ca^{2+} currents in MSSNs	(31)
	Decreased K^+ channel inward rectification in MSSNs	(36)
	Imbalance of synaptic and extrasynaptic NMDAR signaling leading to toxic extrasynaptic NMDAR stimulation and cell death	(67)
Brain slices	Alterations in synaptic activity in the cortex – increased spontaneous EPSC frequency and evoked EPSC amplitude; decreased spontaneous IPSC frequency	(41)
	Alterations in synaptic activity in the striatum – decreased spontaneous EPSC frequency and increased IPSC frequency	(38–40)
	Impaired LTP at CA1 and dentate granule cell synapses in the hippocampus	(58)
	Impaired LTD in hippocampus	(59)
	Aberrant LTD in cortex	(41)
	Abnormal striatal synaptic plasticity-loss of depotentiation in MSSNs and lack of LTP in cholinergic interneurons	(61)
	Elevated extrasynaptic NMDAR activity	(69)
	Differential changes in D1 and D2 MSSN glutamate neurotransmission and dopamine modulation	(70)
In vivo	Increased firing rate and decreased correlated firing between neurons both in the cortex and in striatum	(71, 72)

Preparation	PD	Reference
Brain slices	Increased spontaneous glutamatergic activity	(42, 43)
	Increased GABAergic IPSCs in cortex and decrease in striatum	(46)
	Impaired hippocampal synaptic responses to prolonged trains of repetitive stimulation	(55)
	Decreased glutamate release in CA1 region of hippocampus	(57)
	Deficits in excitatory transmission in the striatum	(2)
	Short-term and long-term synaptic plasticity affected in corticostriatal pathway	(56)
	Increased activity of indirect pathway MSSNs and decreased activity of direct pathway MSSNs	(33)
	Attenuation of collateral connectivity between MSSNs	(51)

4.2. Ionic Currents

Whole-cell recordings in voltage-clamp mode have been used to examine ionic currents mediated by the movement of Na^+, K^+, Ca^{2+}, and Cl^- through neuronal channels. Na^+ currents are mediated by movement of the ion through voltage-gated Na^+ channels and are critical for action potential generation. Na^+ also enters the cell through nonspecific cation channels that are ligand-gated (e.g., α-amino-3-hydroxyl-5-methyl-4-isoxazole-propionate (AMPA) and N-methyl-D-aspartate (NMDA) receptors). In the latter case, though these channels are permeable to both Na^+ and K^+ (34), binding of the ligand to the receptor channel leads to predominant Na^+ entry because the resting membrane potential (RMP) of most neurons is close to the equilibrium potential for K^+. Possible sources of Ca^{2+} currents in neurons are Ca^{2+} entry into the neuron through voltage-activated Ca^{2+} channels and NMDA receptor channels. Similarly, there are voltage-activated and ligand-activated K^+ channels and also Ca^{2+}-activated K^+ channels. Ca^{2+} also can be mobilized by changing the binding properties of intracellular Ca^{2+} stores. The anionic Cl^- current is predominantly caused by synaptic activation of γ-aminobutyric acid-A ($GABA_A$) receptors leading to membrane hyperpolarization. In very young animals, however, due to a Na^+-K^+-Cl^- cotransporter, the reversal potential for Cl^- is relatively depolarized (–35 mV) and GABA release is depolarizing and capable of producing excitation instead of inhibition (35).

Numerous alterations in ionic conductances have been observed in current and voltage-clamp recordings from MSSNs in slices from the R6/2 mouse model of HD leading to a depolarized RMP, a lower rheobase, and an alteration in action potential firing (32). In addition, decreases in voltage-activated Ca^{2+} currents and K^+ channel inward rectification occur (31, 32). In parallel to the decrease in inward and outwardly rectifying K^+ currents, MSSNs in R6/2 mice show decreased expression of K^+ channel subunit proteins Kir2.1, Kir2.3, and Kv2.1 (36). Attenuation of K^+ conductances may contribute to altered electrophysiological properties of MSSNs and their selective vulnerability in the striatum. In contrast to the striatum, voltage-activated Ca^{2+} currents are increased in cortical pyramidal neurons in symptomatic R6/2 mice, indicative of complex changes that could effectively increase cortical excitability and alter corticostriatal function (37).

4.3. Synaptic Activity

In the absence of external electrical stimuli, neurons display spontaneous membrane currents, which can be either excitatory or inhibitory. These spontaneous excitatory and inhibitory postsynaptic currents (EPSCs or IPSCs, respectively) can be recorded in voltage or current-clamp mode and provide information related to both presynaptic and postsynaptic elements. Even in the absence of presynaptic action potentials, neurotransmitter vesicles are occasionally released into the synapse and they generate miniature (m) EPSCs or mIPSCs. Miniature currents can be recorded by isolating the neuron using tetrodotoxin, a Na^+ channel antagonist, which blocks action potentials in neurons. A condition or compound that affects presynaptic neuron firing will affect spontaneous currents but not miniature currents. Changes in presynaptic transmitter release or the function of postsynaptic receptors are reflected by changes in frequency and amplitude of miniature currents. Another method to record evoked EPSCs and IPSCs is by electrically stimulating the inputs to the neuron.

Using mouse models of HD, time- and region-dependent changes in synaptic transmission in the corticostriatal pathway have been demonstrated. In the R6/2 mouse model of HD, dysregulation of glutamatergic inputs occurs early (5–7 weeks) and is manifested by the presence of large amplitude and complex synaptic events in the striatum (38). In parallel, there is a progressive decrease in spontaneous EPSCs and increase in the stimulus intensity required to evoke EPSPs in MSSNs (32, 38). These findings suggested a progressive disconnection of the cortex from the striatum (25). In addition, there is an increase in spontaneous GABAergic IPSCs in the striatum of these mice possibly leading to reduced output of MSSNs to their targets (39, 40). Large amplitude synaptic events in striatum suggested cortical hyperexcitability in R6/2 mice. In agreement, in the cerebral cortex, a progressive increase in frequency of spontaneous EPSCs in pyramidal neurons is observed (41). In contrast, the frequency of cortical spontaneous IPSCs initially increased and subsequently decreased in fully phenotypic mice. This imbalance leads to increased cortical excitability and is manifested by enhanced seizure susceptibility in R6/2 animals (41).

Studies have also demonstrated alterations in corticostriatal synaptic transmission in rats with neurotoxic or electrolytic lesions in the substantia nigra, which are the classical, and most studied, models of PD. DA-depleting lesions lead to increases in spontaneous synaptic activity mediated by glutamate (42–44). In contrast, in a genetic mouse model of PD, overexpression of human WT α-synuclein (α-Syn), a presynaptic protein whose expression is abnormal in familial PD, results in decreases in spontaneous and mEPSCs recorded in MSSNs. α-Syn overexpressing mice may be a good model of preclinical PD at early ages, but after 7 months of age these mice begin to display evidence of DA depletion, which is the more classical hallmark of PD (45). In addition when these mice are 3 months of age, the threshold for evoked EPSCs is higher and they are smaller. GABAergic events are reduced in MSSNs but increased in the cortex suggesting circuit disturbances in corticostriatal synaptic function as well (46). Finally, DA modulation is altered (45). These studies highlight differences in synaptic alterations depending on the use of neurotoxic versus genetic models of PD. Of more importance, they show that in genetic models, early targets for therapeutics can be identified that may differ significantly from targets that become evident later when neurodegeneration is more evident.

A recent study in mice chronically lesioned with 6-hydroxydopamine suggested that there are differential changes in the MSSNs originating the direct and indirect striatal output pathways. Following striatal DA depletion, there is no change in spontaneous synaptic activity in MSSNs of the indirect pathway, but these MSSNs show enhanced sensitivity to presynaptic stimulation and decreased paired-pulse facilitation indicating increased probability of release. MSSNs of the direct pathway showed decreased frequency of spontaneous activity and postsynaptic desensitization to excitatory stimuli. The overactivity of the indirect pathway and under-activity of the direct pathway might be instrumental in producing the parkinsonian state in these mice (33).

A recent study where DA D_1 and D_2 MSSNs of the direct and indirect pathway were stimulated using optogenetic techniques showed that bilateral excitation of D_2 MSSNs leads to a parkinsonian state with freezing, bradykinesia, and decreased locomotor initiations, whereas D_1 MSSN excitation reduced freezing and increased locomotion (47). When D_1 MSSNs were activated in a mouse model of PD, the parkinsonian state was abolished.

An interesting application of the patch-clamp technique is dual recordings from pairs of neurons, which provides information on synaptic function and plasticity, neuronal integration, and neuronal circuits (48). Action potentials are evoked in one neuron and the postsynaptic potential (or current, if the second neuron is held in voltage-clamp mode) is recorded in a connected neuron thereby observing presynaptic and postsynaptic function at the level of the two cells directly. Paired recordings have demonstrated recurrent collateral connections between MSSNs in the striatum (49, 50) and similar studies in animal models of PD have shown severe attenuation of collateral connectivity between MSSNs (51). In the R6/2 mouse model of HD, paired recordings from MSSNs reveal impaired feedback connectivity mediated by MSSN axon collaterals and the appearance of abnormal reciprocal connectivity between pairs of MSSNs in mutant mice (52). In contrast, interneurons connected to MSSNs evoke larger responses in R6/2 than in control mice (52). These results suggest differential changes in the feedback and feedforward striatal GABAergic microcircuits with overactivity in the feedforward circuit that could ultimately affect striatal output and motor function in HD.

Other studies have shown that both forms of corticostriatal plasticity, LTP and LTD, are affected in animal models of PD (28). An interesting aspect of striatal plasticity is its requirement of DA transmission (53). High-frequency stimulation (HFS) of corticostriatal glutamatergic fibers fails to induce LTP in MSSNs in PD-affected rats and the ability to depotentiate after LTP is lost in rats with l-DOPA-induced dyskinesias (54). Field recordings from the hippocampus in mice that lack α-Syn show impaired synaptic responses to a prolonged train of repetitive stimulation suggesting a role for the protein in synaptic vesicle release and possibly recycling (55). Overexpression of α-Syn in mice alters both short-term and long-term presynaptic plasticity in the corticostriatal pathway probably reflecting a reduction of glutamate at corticostriatal synapses (56). Parkin-deficient mice show smaller amplitude extracellular field excitatory postsynaptic potentials but normal LTP (57). Intracellular recordings from MSSNs in parkin-deficient mice showed that greater currents are required to induce synaptic responses (2), again suggesting a change in corticostriatal input but in a different genetic model of PD.

Studies also have helped to shed light on alterations in synaptic plasticity in HD. Hippocampal LTP and LTD are both altered in mouse models of HD with synaptic alterations occurring both at the CA1 and dentate granule cell synapses (58, 59). Synaptic alterations are also seen in the cortex of HD transgenic mice in the form of an age-dependent, but biphasic change in the ability of perirhinal synapses to support LTD and similar to changes in membrane properties in MSSNs, cortical neurons display depolarized RMPs and reduced cell capacitance (60). In the striatum, HFS-induced LTP can be reversed by low-frequency stimulation and this depotentiation is lost in HD models, possibly due to decreases in acetylcholine from striatal cholinergic interneurons (61).

Findings in the striatum from examination of dye-coupling mediated by gap junctions both in vitro (43, 62) and in vivo (63, 64) indicate a close relationship between gap junction function and DA. Dye-coupling is increased within the rat striatum after DA-depleting lesions (43). The increase in dye-coupling also has been demonstrated in vivo in a rat model of PD and following pharmacological manipulations of the DA system (63, 64). These studies indicate that one of the functions of DA in the striatum is to uncouple gap junctions and DA depletion leads to an increase in striatal electrotonic coupling (43).

Studies in acutely dissociated neurons from mouse models of HD have demonstrated changes in glutamate receptor function supportive of the excitotoxicity hypothesis to explain neurodegeneration seen in HD (31, 65). In the striatum, a subpopulation of MSSNs show increased responsiveness to NMDA. This is accompanied by a decrease in NMDA receptor (NMDAR) Mg^{2+} sensitivity (31, 65). Conversely, acutely dissociated pyramidal neurons from the cortex in a mouse model of HD display smaller NMDA currents and an increase in NMDAR Mg^{2+} sensitivity (37). Studies in dissociated neurons have also demonstrated alterations in AMPA currents, K^+ channels, and voltage-gated Ca^{2+} channels in mouse models of HD (36, 37). Support for the excitotoxicity hypothesis also comes from cell swelling experiments that reveal that NMDA-induced cell swelling, visualized by IR videomicroscopy and DIC optics, is enhanced in cortical and striatal neurons of transgenic HD mice (66).

Experiments in cultured neurons and brain slices have provided much information on the role of extrasynaptic NMDARs in the pathogenesis of HD (67). These studies have shown a functional dichotomy of NMDAR signaling resulting from activation of distinct genetic programs and opposing actions on intracellular signaling pathways (27, 68). Stimulation of synaptic NMDARs generates pro-survival signaling while stimulation of extrasynaptic NMDARs appears to be neurotoxic. In the context of HD, imbalance between synaptic and extrasynaptic NMDARs could have a role in excitotoxicity-induced neuronal degeneration (67). An important outcome of these studies is the emergence of memantine, a weak NMDAR antagonist, as a potential therapeutic agent for the treatment of HD symptoms (67, 69).

Electrophysiological recordings from identified direct and indirect pathway MSSNs (D_1 and D_2 receptor-expressing MSSNs, respectively) have revealed lower threshold for action potential firing, higher frequency of spontaneous EPSCs, and large amplitude inward currents (>100 pA) in D_2 MSSNs suggesting that D_2 MSSNs are more excitable and reflect ongoing cortical activity more faithfully than D_1 MSSNs (29, 30). Further, in transgenic HD mice major, age-dependent differential changes in glutamate and DA transmission to direct and indirect pathway MSSNs have been demonstrated suggesting a possible mechanism for differential vulnerabilities of subpopulations of neurons in HD. Recent evidence indicated that DA neurotransmission and its modulation of glutamate receptor-mediated activity in the striatum is affected in HD (70). In the early stages of the phenotype, there was selective dysfunction of both direct and indirect pathway MSSNs but in different ways. Direct pathway MSSNs displayed increased frequencies of spontaneous EPSCs and decreased paired-pulse ratios suggesting increased glutamate release onto those neurons. While these two parameters were not affected in MSSNs of the indirect pathway, these neurons displayed larger evoked glutamate currents that could account for the increased vulnerability of enkephalin-expressing MSSNs in HD. Direct pathway MSSNs also show a lack of D_1 receptor modulation. This alteration was rescued with the DA blocker tetrabenazine which is also a drug that has recently been approved for the treatment of chorea in HD. Indirect pathway MSSNs were unresponsive to D_2 receptor stimulation. In late-stage HD, both direct and indirect pathway MSSNs display decreased spontaneous EPSCs compared to age-matched controls. D_1 receptor agonist modulation was present in the late stage of HD and D_2 receptor modulation remained absent. This indicates differential imbalance in glutamate transmission, as well as DA modulation, in D_1 and D_2 receptor-expressing MSSNs in HD. Hyperkinesia at an early stage could be explained by increased glutamate activity and DA tone in direct pathway neurons, whereas hypokinesia at a late stage could result from reduced input onto these neurons (70).

In vivo extracellular recordings in freely behaving transgenic mouse and rat models of HD have provided information suggesting severe dysregulation of information processing in the cortex and striatum (71, 72). Multiunit recordings indicate that there is a decrease in correlated firing and coincident bursts between pairs of MSSNs in the striatum in models of HD (71, 73). Similarly, in the cortex, there is decreased synchrony between neuronal pairs in HD mice indicating a population-level deficit in cortical information processing (72). In both the cortex and the striatum, there is elevated neuronal firing at the single-unit level (71, 72). Recently, this technique has also been used to record from intranigral grafted cells in a mouse model of PD (74). Extracellular recordings within the transplanted graft using antidromic and orthodromic striatal stimulation suggest functional recovery of the nigrostriatal loop (74).

5. Summary and Future Directions

Electrophysiological techniques represent powerful and versatile tools to study nervous system function in animal models of disease. Alteration in function can be examined at the level of a single neuron in isolation to entire networks in intact animals. Integration of the information obtained from a variety of experimental configurations can provide a complete picture of the neuronal pathology in movement disorders like HD and PD. Introduction of new techniques and advancement and refinement of existing methods of electrophysiology, for example, by combining optogenetics and recordings from awake and behaving animals, will further advance our knowledge of mechanisms underlying disease progression. Pharmacological and/or genetic manipulations of affected neuronal systems and their targets could pave the way toward better therapeutic options in these diseases.

Acknowledgments

Supported by USPHS Grants NS41574, NS33538, NS38367, ES16732, and Contracts from the Hereditary Disease Foundation and the Cure Huntington's Disease Initiative, Inc.

References

1. Tobin AJ, Signer ER. (2000) Huntington's disease: the challenge for cell biologists. Trends Cell Biol 10:531–536
2. Goldberg MS, Fleming SM, Palacino JJ, Cepeda C, Lam HA, Bhatnagar A, Meloni EG, Wu N, Ackerson LC, Klapstein GJ, Gajendiran M, Roth BL, Chesselet MF, Maidment NT, Levine MS, Shen J (2003) Parkin-deficient mice exhibit nigrostriatal deficits but not loss of dopaminergic neurons. J Biol Chem 278:43628–43635
3. Levine MS, Cepeda C, Hickey MA, Fleming SM, Chesselet MF (2004) Genetic mouse models of Huntington's and Parkinson's diseases: illuminating but imperfect. Trends Neurosci 27:691–697
4. Yamamoto C. and McIlwain H. (1966) Electrical activities in thin sections from the mammalian brain maintained in chemically defined media in vitro. J Neurochem 13:1333–1343
5. Kay AR, Wong RK (1986) Isolation of neurons suitable for patch-clamping from adult mammalian central nervous systems. J Neurosci Methods 16:227–238
6. Gähwiler BH (1981) Organotypic monolayer cultures of nervous tissue J Neurosci Methods 4:329–342
7. Shahar A, de Vellis J, Vernadakis A, Haber B (1989) A dissection and tissue culture manual of the nervous system. Wiley, New York
8. Kettenmann H, Grantyn R (1992) Practical electrophysiological methods. A guide for in vitro studies in vertebrate neurobiology. Wiley, New York
9. Gähwiler BH, Brown DA (1985) Functional innervation of cultured hippocampal neurones by cholinergic afferents from co-cultured septal explants. Nature. 313:577–579
10. Li WC, Soffe SR, Roberts A (2004) A direct comparison of whole cell patch and sharp electrodes by simultaneous recording from single spinal neurons in frog tadpoles. J Neurophysiol 92:380–386
11. Hamill OP, Marty A, Neher E, Sakmann B, Sigworth FJ (1981) Improved patch-clamp techniques for high-resolution current recording from cells and cell-free membrane patches. Pflugers Arch 391:85–100
12. Blanton MG, Lo Turco JJ, Kriegstein AR (1989) Whole cell recording from neurons in slices of reptilian and mammalian cerebral cortex. J Neurosci Methods 30:203–210
13. Edwards FA, Konnerth A, Sakmann B, Takahashi T (1989) A thin slice preparation for patch clamp recordings from neurones of the mammalian central nervous system. Pflugers Arch 414:600–612
14. Scanziani M, Häusser M (2009) Electrophysiology in the age of light. Nature 461:930–939
15. Tervo D, Karpova AY (2007) Rapidly inducible, genetically targeted inactivation of neural and synaptic activity in vivo. Curr Opin Neurobiol 17:581–586
16. Birgül N, Weise C, Kreienkamp HJ, Richter D (1999) Reverse physiology in drosophila: identification of a novel allatostatin-like neuropeptide and its cognate receptor structurally related to the mammalian somatostatin/galanin/opioid receptor family. EMBO J 18:5892–5900
17. Lerchner W, Xiao C, Nashmi R, Slimko EM, van Trigt L, Lester HA, Anderson DJ (2007) Reversible silencing of neuronal excitability in behaving mice by a genetically targeted, ivermectin-gated cl(–) channel. Neuron 54:35–49
18. Karpova AY, Tervo DG, Gray NW, Svoboda K (2005) Rapid and reversible chemical inactivation of synaptic transmission in genetically targeted neurons. Neuron 48:727–735
19. Durieux PF, Bearzatto B, Guiducci S, Buch T, Waisman A, Zoli M, Schiffmann SN, de Kerchove d'Exaerde A. (2009) D2R striatopallidal neurons inhibit both locomotor and drug reward processes. Nat Neurosci 12:393–395
20. Boyden ES, Zhang F, Bamberg E, Nagel G, Deisseroth K (2005) Millisecond-timescale, genetically targeted optical control of neural activity. Nat Neurosci 8:1263–1268
21. Li X, Gutierrez DV, Hanson MG, Han J, Mark MD, Chiel H, Hegemann P, Landmesser LT, Herlitze S (2005) Fast noninvasive activation and inhibition of neural and network activity by vertebrate rhodopsin and green algae channelrhodopsin. Proc Natl Acad Sci USA 102:17816–17821
22. Zhang F, Wang LP, Brauner M, Liewald JF, Kay K, Watzke N, Wood PG, Bamberg E, Nagel G, Gottschalk A, Deisseroth K (2007) Multimodal fast optical interrogation of neural circuitry. Nature 446:633–639
23. Aravanis AM, Wang LP, Zhang F, Meltzer LA, Mogri MZ, Schneider MB, Deisseroth K (2007) An optical neural interface: in vivo control of rodent motor cortex with integrated fiberoptic and optogenetic technology. J Neural Eng 4:S143–156
24. Lobo MK, Covington HE 3rd, Chaudhury D, Friedman AK, Sun H, Damez-Werno D, Dietz DM, Zaman S, Koo JW, Kennedy PJ, Mouzon E, Mogri M, Neve RL, Deisseroth K, Han MH,

Nestler EJ (2010) Cell type-specific loss of BDNF signaling mimics optogenetic control of cocaine reward. Science 330:385–390

25. Cepeda C, Wu N, André VM, Cummings DM, Levine MS (2007) The corticostriatal pathway in Huntington's disease. Prog Neurobiol 81:253–271

26. Cepeda C, Cummings DM, André VM, Holley SM, Levine MS (2010) Genetic mouse models of Huntington's disease: focus on electrophysiological mechanisms. ASN Neuro 2(2):art:e00033. doi:10.1042/AN20090058

27. Milnerwood AJ, Raymond LA (2010) Early synaptic pathophysiology in neurodegeneration: insights from Huntington's disease. Trends Neurosci 33:513–523

28. Calabresi P, Galletti F, Saggese E, Ghiglieri V, Picconi B (2007) Neuronal networks and synaptic plasticity in Parkinson's disease: beyond motor deficits. Parkinsonism Relat Disord 13:S259–262

29. Cepeda C, André VM, Yamazaki I, Wu N, Kleiman-Weiner M, Levine MS (2008) Differential electrophysiological properties of dopamine D1 and D2 receptor-containing striatal medium-sized spiny neurons. Eur J Neurosci 27:671–682

30. Gertler TS, Chan CS, Surmeier DJ (2008) Dichotomous anatomical properties of adult striatal medium spiny neurons. J Neurosci 28:10814–10824

31. Cepeda C, Ariano MA, Calvert CR, Flores-Hernández J, Chandler SH, Leavitt BR, Hayden MR, Levine MS (2001) NMDA receptor function in mouse models of Huntington disease. J Neurosci Res 66:525–539

32. Klapstein GJ, Fisher RS, Zanjani H, Cepeda C, Jokel ES, Chesselet MF, Levine MS (2001) Electrophysiological and morphological changes in striatal spiny neurons in R6/2 Huntington's disease transgenic mice. J Neurophysiol 86:2667–2677

33. Warre R, Thiele S, Talwar S, Kamal M, Johnston TH, Wang S, Lam D, Lo C, Khademullah CS, Perera G, Reyes G, Sun XS, Brotchie JM, Nash JE (2011) Altered function of glutamatergic cortico-striatal synapses causes output pathway abnormalities in a chronic model of parkinsonism. Neurobiol Dis 41:591–604

34. Eccles JC (1964) Ionic mechanism of postsynaptic inhibition. Science 145:1140–1147

35. Leinekugel X, Khalilov I, McLean H, Caillard O, Gaiarsa JL, Ben-Ari Y, Khazipov R (1999) GABA is the principal fast-acting excitatory transmitter in the neonatal brain. Adv Neurol 79:189–201

36. Ariano MA, Cepeda C, Calvert CR, Flores-Hernandez J, Hernandez-Echeagaray E, Klapstein GJ, Chandler SH, Aronin N, DiFiglia M, Levine MS (2005) Striatal potassium channel dysfunction in Huntington's disease transgenic mice. J Neurophysiol 93:2565–2574

37. André VM, Cepeda C, Venegas A, Gomez Y, Levine MS (2006) Altered cortical glutamate receptor function in the R6/2 model of Huntington's disease. J Neurophysiol 95:2108–2119

38. Cepeda C, Hurst RS, Calvert CR, Hernández-Echeagaray E, Nguyen OK, Jocoy E, Christian LJ, Ariano MA, Levine MS (2003) Transient and progressive electrophysiological alterations in the corticostriatal pathway in a mouse model of Huntington's disease. J Neurosci 23:961–969

39. Cepeda C, Starling AJ, Wu N, Nguyen OK, Uzgil B, Soda T, André VM, Ariano MA, Levine MS (2004) Increased GABAergic function in mouse models of Huntington's disease: reversal by BDNF. J Neurosci Res 78:855–867

40. Centonze D, Rossi S, Prosperetti C, Tscherter A, Bernardi G, Maccarrone M, Calabresi P (2005) Abnormal sensitivity to cannabinoid receptor stimulation might contribute to altered gamma-aminobutyric acid transmission in the striatum of R6/2 Huntington's disease mice. Biol Psychiatry 57:1583–1589

41. Cummings DM, André VM, Uzgil BO, Gee SM, Fisher YE, Cepeda C, Levine MS (2009) Alterations in cortical excitation and inhibition in genetic mouse models of Huntington's disease. J Neurosci 29:10371–10386

42. Galarraga E, Bargas J, Martínez-Fong D, Aceves J (1987) Spontaneous synaptic potentials in dopamine-denervated neostriatal neurons. Neurosci Lett 81:351–355

43. Cepeda C, Walsh JP, Hull CD, Howard SG, Buchwald NA, Levine MS (1989) Dye-coupling in the neostriatum of the rat: I. Modulation by dopamine-depleting lesions. Synapse 4:229–237

44. Calabresi P, Centonze D, Bernardi G (2000) Electrophysiology of dopamine in normal and denervated striatal neurons. Trends Neurosci 23:S57–63

45. Lam HT, Wu N, Cely I, Hean S, Richter F, Magen I, Cepeda C, Ackerson LC, Walwyn W, Masliah E, Chesselet M-F, Levine MS, Maidment NT (2011) Elevated tonic extracellular dopamine concentration and altered dopamine modulation of synaptic activity precede dopamine loss in the striatum of mice overexpressing human α-synuclein. J Neurosci Res 89:1091–1102

46. Wu N, Joshi PR, Cepeda C, Masliah E, Levine MS (2010) Alpha-synuclein overexpression in

mice alters synaptic communication in the corticostriatal pathway. J Neurosci Res 88:1764–1776

47. Kravitz AV, Freeze BS, Parker PR, Kay K, Thwin MT, Deisseroth K, Kreitzer AC (2010) Regulation of parkinsonian motor behaviours by optogenetic control of basal ganglia circuitry. Nature 466:622–626

48. Miles R, Poncer JC (1996) Paired recordings from neurones. Curr Opin Neurobiol 6:387–394

49. Czubayko U, Plenz D (2002) Fast synaptic transmission between striatal spiny projection neurons. Proc Natl Acad Sci USA 99: 15764–15769

50. Tunstall MJ, Oorschot DE, Kean A, Wickens JR (2002) Inhibitory interactions between spiny projection neurons in the rat striatum. J Neurophysiol 88:1263–1269

51. Taverna S, Ilijic E, Surmeier D (2008) Recurrent collateral connections of striatal medium spiny neurons are disrupted in models of Parkinson's disease. J Neurosci 28:5504–5512

52. Rao SP, André VM, Cepeda C, Levine MS (2010) Altered inhibitory inputs onto striatal medium-sized spiny neurons in the R6/2 mouse model of Huntington's disease. Soc Neurosci Abstract 861.16

53. Bagetta V, Ghiglieri V, Sgobio C, Calabresi P, Picconi B (2010) Synaptic dysfunction in Parkinson's disease. Biochem Soc Trans 38: 493–497

54. Picconi B, Centonze D, Håkansson K, Bernardi G, Greengard P, Fisone G, Cenci MA, Calabresi P (2003) Loss of bidirectional striatal synaptic plasticity in L-DOPA-induced dyskinesia. Nat Neurosci 6:501–506

55. Cabin DE, Shimazu K, Murphy D, Cole NB, Gottschalk W, McIlwain, KL, Orrison B, Chen A, Ellis CE, Paylor R, Lu B, Nussbaum RL (2002) Synaptic vesicle depletion correlates with attenuated synaptic responses to prolonged repetitive stimulation in mice lacking alpha-synuclein. J. Neurosci 22:8797–8807

56. Watson JB, Hatami A, David H, Masliah E, Roberts K, Evans CE, Levine MS (2009) Alterations in corticostriatal synaptic plasticity in mice overexpressing human alpha-synuclein. Neuroscience 159:501–513

57. Itier JM, Ibanez P, Mena MA, Abbas N, Cohen-Salmon C, Bohme GA, Laville M, Pratt J, Corti O, Pradier L, Ret G, Joubert C, Periquet M, Araujo F, Negroni J, Casarejos MJ, Canals S, Solano R, Serrano A, Gallego E, Sanchez M, Denefle P, Benavides J, Tremp G, Rooney TA, Brice A, Garcia de Yebenes J (2003) Parkin gene inactivation alters behaviour and dopamine neurotransmission in the mouse. Hum Mol Genet 12:2277–2291

58. Murphy KP, Carter RJ, Lione LA, Mangiarini L, Mahal A, Bates GP, Dunnett SB, Morton AJ (2000) Abnormal synaptic plasticity and impaired spatial cognition in mice transgenic for exon 1 of the human Huntington's disease mutation. J Neurosci 20:5115–5123

59. Milnerwood AJ, Cummings DM, Dallérac GM, Brown JY, Vatsavayai SC, Hirst MC, Rezaie P, Murphy KP (2006) Early development of aberrant synaptic plasticity in a mouse model of Huntington's disease. Hum Mol Genet 15:1690–1703

60. Cummings DM, Milnerwood AJ, Dallérac GM, Waights V, Brown JY, Vatsavayai SC, Hirst MC, Murphy KP (2006) Aberrant cortical synaptic plasticity and dopaminergic dysfunction in a mouse model of Huntington's disease. Hum Mol Genet 15:2856–2868

61. Di Filippo M, Tozzi A, Picconi B, Ghiglieri V, Calabresi P (2007) Plastic abnormalities in experimental Huntington's disease. Curr Opin Pharmacol 7:106–111

62. Walsh JP, Cepeda C, Hull CD, Fisher RS, Levine MS, Buchwald NA (1989) Dye-coupling in the neostriatum of the rat: II. Decreased coupling between neurons during development. Synapse 4:238–247

63. Onn SP, Grace AA (1994) Dye coupling between rat striatal neurons recorded in vivo: compartmental organization and modulation by dopamine. J Neurophysiol 71:1917–1934

64. Onn SP, Grace AA (1999) Alterations in electrophysiological activity and dye coupling of striatal spiny and aspiny neurons in dopamine-denervated rat striatum recorded in vivo. Synapse 33:1–15

65. Starling AJ, André VM, Cepeda C, de Lima M, Chandler SH, Levine MS (2005) Alterations in N-methyl-D-aspartate receptor sensitivity and magnesium blockade occur early in development in the R6/2 mouse model of Huntington's disease. J Neurosci Res 82:377–386

66. Levine MS, Klapstein GJ, Koppel A, Gruen E, Cepeda C, Vargas ME, Jokel ES, Carpenter EM, Zanjani H, Hurst RS, Efstratiadis A, Zeitlin S, Chesselet MF (1999) Enhanced sensitivity to N-methyl-D-aspartate receptor activation in transgenic and knockin mouse models of Huntington's disease. J Neurosci Res 58:515–532

67. Okamoto S, Pouladi MA, Talantova M, Yao D, Xia P, Ehrnhoefer DE, Zaidi R, Clemente A, Kaul M, Graham RK, Zhang D, Vincent Chen HS, Tong G, Hayden MR, Lipton SA (2009) Balance between synaptic versus extrasynaptic NMDA receptor activity influences inclusions and neurotoxicity of mutant huntingtin. Nat Med 15:1407–1413

68. Hardingham GE, Bading H (2010) Synaptic versus extrasynaptic NMDA receptor signalling:

implications for neurodegenerative disorders. Nat Rev Neurosci 10:682–696

69. Milnerwood AJ, Gladding CM, Pouladi MA, Kaufman AM, Hines RM, Boyd JD, Ko RW, Vasuta OC, Graham RK, Hayden MR, Murphy TH, Raymond LA (2010) Early increase in extrasynaptic NMDA receptor signaling and expression contributes to phenotype onset in Huntington's disease mice. Neuron 65:178–190

70. André VM, Cepeda C, Fisher YE, Huynh M, Bardakjian N, Singh S, Yang XW, Levine MS (2011) Differential electrophysiological changes in striatal output neurons in Huntington's disease J Neurosci 31:1170–1182

71. Miller BR, Walker AG, Shah AS, Barton SJ, Rebec GV (2008) Dysregulated information processing by medium spiny neurons in striatum of freely behaving mouse models of Huntington's disease. J Neurophysiol 100:2205–2216

72. Walker AG, Miller BR, Fritsch JN, Barton SJ, Rebec GV (2008) Altered information processing in the prefrontal cortex of Huntington's disease mouse models. J Neurosci 28:8973–8982

73. Miller BR, Walker AG, Fowler SC, von Hörsten S, Riess O, Johnson MA, Rebec GV (2010) Dysregulation of coordinated neuronal firing patterns in striatum of freely behaving transgenic rats that model Huntington's disease. Neurobiol Dis 37:106–113

74. Besnard S, Decressac M, Denise P (2010) In-vivo deep brain recordings of intranigral grafted cells in a mouse model of Parkinson's disease. Neuroreport 21:485–48

Part II

Dopamine Systems

Chapter 12

Genetic Models of Parkinson's Disease

Ralf Kühn, Daniela Vogt-Weisenhorn, and Wolfgang Wurst

Abstract

Parkinson's disease (PD) is a chronic, progressive neurodegenerative movement disorder. To understand the pathomechanisms and to develop new drugs and therapies for PD, it is important to have animal models that recapitulate the slow progression and symptoms of the disease. The generation of genetic animal models of genes responsible for autosomal dominant but also autosomal recessive forms of PD has indeed accelerated our understanding of these pathomechanisms. To model the effect of dominant mutant alleles, transgenic mice were produced that express mutation-bearing proteins in neurons. To model loss-of-function alleles, knockout mice were generated and studied. However, none of these models recapitulate PD disease as it occurs in PD patients. The latest mouse genetic technology may offer, at least in part, a relief for these challenges through the timed control of protein expression or gene knockout. Both can be achieved by the use of the Tamoxifen inducible CreER^{T2} gene switch that enables the inducible activation of transgene expression or inducible gene knockout. Thereby the CreER^{T2} system can be used to generate genetic models of gain-of-function (dominant) disease-associated alleles by the regulated expression of mutant coding regions as well as to model loss-of-function (recessive) disease-associated alleles by inducible gene knockout. In this chapter, we cover the design of such Tamoxifen inducible transgene constructs and the use of premade conditional knockout alleles generated by large-scale mutagenesis projects. Furthermore, we provide an overview of the available brain-specific CreER^{T2} mouse lines, notes on the control groups for inducible knockout experiments and a protocol for Tamoxifen administration.

Key words: Parkinson's disease, Mouse model, Knockout, Transgenic, Tamoxifen

1. Introduction

Parkinson's disease (PD) is a chronic, progressive neurodegenerative movement disorder. Tremor, rigidity, slow movement (bradykinesia), poor balance and gait disturbances are characteristic primary symptoms of Parkinson's disease. In 1914, it was shown that PD is closely associated with the degeneration of neurons in the substantia nigra pars compacta and the formation of proteinaceous inclusions – Lewy bodies. The awareness that this brain region is harbouring

Emma L. Lane and Stephen B. Dunnett (eds.), *Animal Models of Movement Disorders: Volume I*, Neuromethods, vol. 61,
DOI 10.1007/978-1-61779-298-4_12, © Springer Science+Business Media, LLC 2011

dopaminergic neurons and that PD symptoms arise from low dopamine levels in the brain arose about 50–60 years ago (1, 2). Still, the underlying molecular mechanisms of this loss of dopaminergic neurons in the substantia nigra pars compacta are not yet understood. In order to understand these mechanisms and to develop new drugs and therapies for PD, it is important to have animal models available that recapitulate the slow progression and symptoms of the disease. Thus, animal models of PD, including rodent models, have aimed to reproduce the loss of dopaminergic neurons, initially by using neurotoxins that are known to kill more or less selectively the dopaminergic neurons, such as 6-OHDA, MPTP, paraquat and rotenone (3–7, see also chapters 13, 19–21 this volume). These models have been irreplaceable both for our understanding of the pathology of the disease and for the consequences of dopaminergic neuronal loss and are therefore of uttermost importance for developing symptomatic treatments (8). Unfortunately, they mainly show acute neurodegeneration and therefore do not mimic the long progressive nature of the disease and hence are rather limited in their usefulness to mimic molecular pathomechanisms acting long before the clinical manifestation of the disease. Indeed, over the last decade, it became apparent that clinically PD patients present with non-motor symptoms which often occur prior to the onset of motor symptoms and are not necessarily associated with a loss-of-function of the dopaminergic system (9). This is supported by the fact that neurodegeneration in PD affects brain areas beyond and prior the substantia nigra pars compacta (10). This indicates that other non-toxin-based animal models are needed in order to develop novel innovative therapies and to invest in preventive measures.

Such a novel approach became available by the identification of mutations in genes causing familial forms of PD, starting with the identification of a missense mutation in the gene encoding α-synuclein in 1997 (11, 12). Interestingly, some of the genes linked to monogenic forms of PD have now also been identified as risk factors for sporadic forms of the diseases (13). The generation of genetic animal models of genes responsible for autosomal dominant but also autosomal recessive forms of PD has indeed accelerated our understanding of these pathomechanisms (11), mainly by the analysis of cellular models derived from these animals. However, these cellular systems do not provide the advantage of studying the pathomechanisms in the context of living complex organisms. Therefore, it is indispensible to analyse these animal models also systemically at the behavioural and molecular level. Indeed for several neurodegenerative diseases, the genetic animal models – especially mice – have been quite instructive in understanding the pathogenesis (e.g. ALS (14), AD (15)). However, it also became evident that models have to be genetically optimised in order to generate an animal model that faithfully recapitulates all phases of the disease. This is specifically evident in the field of Parkinson's disease.

Various genetic mouse models for PD-associated alleles have been described over the last years. To model the effect of dominant mutant alleles, transgenic mice were produced that express mutation-bearing proteins in neurons. To model loss-of-function alleles, knockout mice were generated and studied (for review, see ref. 16). However, it is a hallmark of these studies that none of these models recapitulate PD disease as it occurs in PD patients. It is therefore discussed whether the genetic makeup and short lifespan of mice as compared to humans prevents the modelling of PD in rodents – and/or whether the presence of hitherto unknown compensatory mechanisms can keep the disease in check as long as no dramatic environmental stressors are acting on the organism. Indeed such potential compensatory mechanisms are coming more and more into the spotlight, even though the concept itself has been acknowledged about two decades ago (17, 18). Indeed such compensatory measures may in fact occur during development since the genetic alterations are expressed throughout embryonic life.

To answer this question, new genetic models need to be generated that enable the timed control of protein expression or gene knockout by external induction in adult or ageing mice in neuronal but also non-neuronal cell types of the brain (Fig. 1a). Furthermore, new techniques have to be developed in order to extend the genetic animal modelling to model Parkinson's disease in organisms closer to the human condition than the mouse, such as non-human primates. This will be enabled in future by the zinc-finger nuclease technology that is further discussed at the end of this section.

Mouse genetic technology offers a solution for the timed control of protein expression or gene knockout by the use of a Tamoxifen inducible gene switch that enables the inducible activation of transgene expression and inducible gene knockout. This switch for inducible and cell type-specific gene manipulation is provided by a fusion protein of Cre recombinase with the ER^{T2} mutant ligand binding domain of human oestrogen receptor (amino acids 282–595, Fig. 1b) (19–21). The $CreER^{T2}$ fusion protein can be constitutively expressed in mice from a transgene or knock-in allele using a cell type-specific promoter region, but does not exhibit recombinase activity unless a specific activating ligand is supplied. In the absence of this ligand, the ER^{T2} domain is bound by heat shock proteins that block recombinase activity (Fig. 1c). The ER^{T2} domain is unresponsive to the natural ligand estradiol due to the change of three amino acid residues. However, recombinase activity of the fusion protein can be induced by the synthetic ligand 4-hydroxy-tamoxifen, which releases the heat shock proteins from the ER^{T2} domain (Fig. 1c).

Thereby, the $CreER^{T2}$ protein can be used as an inducible gene switch by the administration of Tamoxifen to mice at a selected time window. The transient activation of the $CreER^{T2}$ recombinase enables excision of a gene segment from the genome that has been

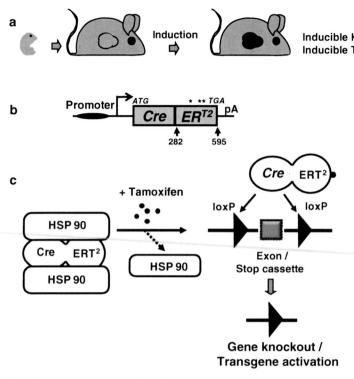

Fig. 1. The Tamoxifen inducible gene switch. (**a**) Gene inactivation in knockout mice or gene expression in standard transgenic mice is restricted to selected cell types but the timing of gene inactivation or transgene activation cannot be further manipulated. In contrast, in Tamoxifen inducible knockout or transgenic mice, the onset of gene inactivation or transgene expression can be regulated in a selected cell type, e.g. induced in adult mice. (**b**) The Tamoxifen inducible Cre-ER^{T2} system. Structure of a transgene construct for cell type-specific expression of the Cre-ER^{T2} fusion protein. In the Cre-ER^{T2} fusion gene, the coding region of Cre is fused to the ligand binding domain of the human oestrogen receptor (ER^{T2}, residues 282–595). The ER^{T2} ligand binding domain harbours three point mutations (***) that render ER^{T2} unresponsive to the natural ligand estradiol. (**c**) Principle of the Tamoxifen inducible Cre-ER^{T2} gene switch. Cre-ER^{T2} protein is constitutively expressed in a selected cell type but remains inactive by binding to heat shock proteins (HSP90) that interfere with recombinase activity. Cre-ER^{T2} becomes activated by the ER^{T2} ligand 4-OH-tamoxifen that leads to the dissociation of HSP90 proteins. Upon HSP90 dissociation, the fusion protein is released from the inactive complex and mediates the excision of a loxP-flanked DNA segment via the active Cre domain. The deletion of a loxP-flanked exon leads to the knockout of the target gene, whereas the deletion of a loxP-flanked stop cassette results in transgene expression.

marked for deletion by insertion of two Cre recombinase recognition (loxP) sites (Fig. 1c). The removal of a loxP-flanked exon from an endogenous gene allows the use of the CreER^{T2} switch for inducible gene inactivation in vivo. Furthermore, the same switch can be used to excise an inhibitory loxP-flanked DNA segment (stop element) from a transgene. The removal of the stop element leads to the expression of the transgene (Fig. 1c).

Thereby the CreER^{T2} gene switch can be used to generate genetic models of gain-of-function (dominant) disease-associated alleles by the regulated expression of mutant coding regions as well as to model loss-of-function (recessive) disease-associated alleles by inducible gene knockout (Fig. 2a, b).

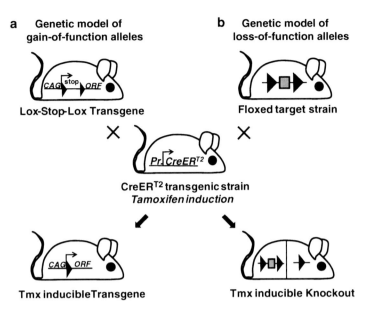

a   Genetic model of gain-of-function alleles

Lox-Stop-Lox Transgene

b   Genetic model of loss-of-function alleles

Floxed target strain

CreER^T2 transgenic strain
Tamoxifen induction

Tmx inducibleTransgene

Tmx inducible Knockout

Fig. 2. Tamoxifen inducible protein expression and gene knockout. (**a**) To obtain Tamoxifen inducible transgene expression, a mouse strain harbouring a loxP-flanked stop cassette between the CAG promoter and the open reading frame (ORF) of interest (lox-stop-lox-transgene) is crossed to mice expressing CreER^T2 fusion recombinase from a specific cell type promoter (Pr). In double transgenic offspring, transgene activation occurs in the selected cell type through Tamoxifen-induced Cre activation that deletes the stop cassette. (**b**) To generate Tamoxifen (Tmx) inducible knockout mice, a strain harbouring a loxP-flanked gene segment (floxed target strain) is crossed to mice expressing CreER^T2 fusion recombinase from a specific cell type promoter (Pr). In double transgenic offspring, gene inactivation occurs in the selected cell type through Tamoxifen-induced Cre activation that deletes the floxed gene segment. LoxP sites are shown as *filled triangles*.

Using CreER^T2 as a Tamoxifen-regulated switch, inducible transgenic mice can be conveniently generated by the insertion of a loxP-stop-loxP transgene into the *Rosa26* locus of ES cells (Fig. 2a). The targeted insertion of the conditional expression unit into the *Rosa26* locus ensures reproducible transgene expression that can be activated at any time in a cell type of interest. The expression vector for the protein of interest contains the respective open reading frame (ORF) located in between a loxP-flanked transcriptional stop element and a polyadenylation (pA) signal. The presence of the stop element initially prevents the production of mRNA containing the ORF. By the cross of lox-stop-lox transgenic mice with a CreER^T2 strain (Fig. 2a), the ORF-coded protein becomes expressed from the vector-based CAG promoter upon the Tamoxifen-induced, Cre-mediated excision of the loxP-flanked stop cassette. Thereby the production of the protein of interest is controlled in vivo in a cell type-specific manner, depending on the chosen CreER^T2 mouse line, and in a temporal manner by the administration of Tamoxifen at a chosen point in time. The inducible activation of *Rosa26*-based conditional transgenes has been used for the activation of reporter genes to characterise the recombination profile of CreER^T2 transgenic lines (22–28) and for the forced expression of endogenous proteins (29, 30). By replacing reporter gene expression with coding regions for disease-associated proteins, this approach represents a very convenient and straightforward way for the generation of inducible transgenic mice. In this chapter, Sect. 3.1 covers the design of such Tamoxifen inducible transgene constructs. We provide a single step cloning approach for the generation of conditional vectors and a recombinase-based protocol for the simple vector insertion into the *Rosa26* locus of ES cells.

Besides the activation of conditional transgenes, the Tamoxifen inducible CreER[T2] gene switch is often used for gene inactivation in conditional knockout mice (Fig. 2b) (19–21). For this purpose, two loxP sites are placed into intron regions of the target gene such that they flank an essential coding exon but do not interfere with gene function. To enable inducible gene knockout, the floxed target strain is crossed to a chosen CreER[T2] mouse line. Upon Tamoxifen administration, Cre-mediated excision of the loxP-flanked region results in the in vivo deletion of the coding segment from the genome of CreER[T2] expressing cells. Upon deletion of the target exon, the mutant gene codes for a non-functional protein either by loss of an essential domain or as result of a reading frame shift. In Sect. 3.2, we cover the design of premade conditional knockout alleles generated by the EUCOMM/KOMP large-scale mutagenesis projects. Since the number of conditional alleles available from these projects is continuously expanding, self-made vector constructions are no longer mandatory to obtain a conditional mouse mutant. However, we further provide a simple PCR-based approach for the generation of conditional targeting vectors applicable in any standard laboratory. Finally, Sect. 3.3 summarises the use of CreER[T2] transgenic mice. We include an overview of the available brain-specific CreER[T2] mouse lines, notes on the control groups for inducible knockout experiments and a protocol for Tamoxifen administration.

Gene targeting in other species than mice
Over the past two decades, the mouse has developed into the prime mammalian genetic model to study human biology and disease because methods are available that allow the production of targeted, predesigned mouse mutants. This reverse genetics approach that enables the production of germ line and conditional knockout mice by gene targeting relies on the use of murine embryonic stem (ES) cell lines. ES cell lines exhibit unique properties such that they are able, once established from the inner cell mass of a mouse blastocyst, to renew indefinitely in cell culture while retaining their early pluripotent differentiation state. Gene targeting in ES cells has revolutionised the in vivo analysis of mammalian gene function using the mouse as genetic model system (31). However, since germ line competent ES cell lines that can be genetically modified could be established only from mice, this reverse genetics approach is presently restricted to this rodent species. The exception from

this rule is achieved by homologous recombination in primary cells from pig and sheep followed by the transplantation of nuclei from recombined somatic cells into enucleated oocytes (cloning) (32, 33). Since this methodology is inefficient and time-consuming, it does not have the potential to develop into a simple routine procedure.

Experiments in model systems have demonstrated that the frequency of homologous recombination of a gene targeting vector is strongly increased if a double-strand break is induced within its chromosomal target sequence. Zinc-finger nucleases (ZFNs) were developed as a method to apply the stimulatory power of double-strand breaks to sequences of endogenous genes, without the need to introduce an artificial nuclease recognition site. Using zinc-finger nucleases in the absence of a gene targeting vector for homology directed repair, knockout alleles were generated in mammalian cell lines and knockout zebra fish and rats were obtained upon the expression of ZFN mRNA in one cell embryos (30, 34, 35).

Furthermore, ZFNs were used in the presence of exogeneous gene targeting vectors that contain homology regions to the target gene for homology-driven repair of the double-strand break through gene conversion. This methodology has been applied to gene engineering in mammalian cell lines and gene correction in primary human cells (36–38). More recently, zinc-finger nucleases together with gene targeting vectors were further used to generate targeted mutants in fertilised oocytes of mice and rats (39, 40). Therefore, ZFN-assisted gene targeting in zygotes provides a new, ES cell independent paradigm to manipulate the genome of mammals and other vertebrates with unprecedented freedom. In the near future, it will be therefore possible to create gain- and loss-of-function alleles as genetic PD models in non-murine species, e.g. the rat.

2. Materials

2.1. Tamoxifen Inducible Gene Expression

2.1.1. Plasmids

1. pEx-CAG-stop-bpA
2. pCAG-C31Int
3. pRosa26-5′probe

2.1.2. Cells

1. ES cell line IDG26.10-3
2. Feeder cells: G418 resistant murine embryonic fibroblasts

2.1.3. Primers

1. Neo: (5′-GTT GTG CCC AGT CAT AGC CGA ATA G-3′)
2. Pgk (5′-CAC GCT TCA AAA GCG CAC GTC TG-3′)
3. Hyg-1 (5′-GAA GAA TCT CGT GCT TTC AGC TTC GAT G-3′)
4. Hyg-2 (5′-AAT GAC CGC TGT TAT GCG GCC ATT G-3′)

5. SEQfor (5'-CATTATACGAAGTTATACC-3')

6. SEQrev (5'-ATC ATT TAC GCA ATT CCG C-3')

Plasmids and ES cells are available from the authors

2.2. Tamoxifen Inducible Gene Knockout

1. Plasmid pEasyfloxII-DTA (submitted to Addgene by R. Kühn)

2. Genomic BAC clones (e.g. from ImaGenes, http://www.imagenes-bio.de or Source BioScience, http://www.lifesciences.sourcebioscience.com)

3. Bioinformatic software for sequence analysis and vector design (e.g. VectorNTI, CLC Sequence Viewer, or Geneious)

4. Software for PCR primer design (e.g. VectorNTI or Primer3, http://frodo.wi.mit.edu/primer3/)

5. Proofreading DNA polymerase, e.g. Phusion (New England Biolabs) or Herculase-II (Stratagene)

6. DNA oligonucleotides as PCR primers

2.3. Inducible Cre Activity

Tamoxifen stock solution: for i.p. injection, a 10 mg/ml stock of tamoxifen free base is prepared. Suspend 1 g tamoxifen free base (Sigma #T5648) in 10 ml 100% ethanol. Add 90 ml of sunflower seed oil (Sigma #S5007) and stir the suspension for several hours until the tamoxifen crystals are completely dissolved. Store the stock solution in aliquots at –20°C for up to 4 weeks.

3. Description of Methods

3.1. Tamoxifen Inducible Gene Expression

This section describes a simple and straightforward protocol for the production of transgenic mice that harbour a Tamoxifen inducible expression cassette within the *Rosa26* locus on chromosome 6. To generate an expression vector for the protein of interest, a DNA fragment containing the respective ORF is ligated in between the loxP-flanked stop cassette and the polyadenylation (pA) signal of the generic vector backbone of pEx-CAG-stop-bpA (Fig. 3a). The presence of the stop cassette (a puromycin resistance coding region followed by triple pA signals) prevents the production of mRNA containing the ORF. The ORF-coded protein becomes expressed from the vector-based CAG promoter upon the Cre-mediated excision of the loxP-flanked stop cassette. By this means, the production of the protein of interest can be controlled in vivo in a cell type-specific and temporal manner, depending on the characteristics of the chosen CreER^{T2} transgenic mouse line. The pEx-CAG-stop-ORF-bpA conditional expression vector (Fig. 3b) is inserted by recombinase-mediated cassette exchange (RMCE) into the modified *Rosa26* allele of IDG26.10-3 ES cells (Fig. 3c). The targeted insertion of the vector into the *Rosa26* locus ensures reproducible transgene expression that can be activated in any cell type of interest. For the stable integration into the ES cell genome, we use a RMCE protocol based on the attB and attP recognition sites of phiC31 integrase that we initially developed for the targeted insertion of shRNA expression cassettes into *Rosa26* (41, 42). Figure 3b illustrates the C31Int/attB/attP-mediated cassette exchange at a pre-engineered *Rosa26* docking site. Since the pEx-CAG-stop-ORF-bpA plasmid includes a promoterless neomycin resistance gene (neo), which is activated only after correct cassette exchange at the *Rosa26* acceptor locus (Fig. 3d), RMCE positive ES cell clones can be conveniently selected in G418. Finally, RMCE ES cell clones are used for the generation of chimaeric mice and germline transmission of the conditional transgene.

3.1.1. Cloning of Inducible Expression Vectors

1. Digest DNA of plasmid pEx-CAG-stop-bpA with the single cutters *Nru*I, *Asi*SI, *Bgl*II or a combination of these enzymes and dephosphorylate the open ends. Generate a PCR-based DNA fragment containing the ORF of interest. Include into the PCR primers appropriate restriction sites for ligation into the pEX-CAG-stop-bpA plasmid and add, immediately upstream of the ATG start codon, a Kozak consensus sequence (GCCACC) to ensure efficient translation. Ligate 50 ng of the digested and purified PCR fragment with 100 ng opened pEx-CAG-stop-bpA in a 15 μl standard ligation reaction.

2. Transform the ligation reaction into *Escherichia coli* cells (we use DH5α), and digest miniprep DNA with a combination of ORF internal and external (*Nru*I, *Asi*SI, *Bgl*II) restriction enzymes to determine the presence and orientation of the insert.

3. Prepare Maxiprep (100 ml LB culture, Quiagen Plasmid Maxi kit, Quiagen) with one confirmed clone of pEx-CAG-stop-ORF-pA and use primers SEQfor/rev for the sequence verification of the ligated insert.

3.1.2. Stable Vector Integration into the Rosa26 Locus of ES Cells by RMCE

1. Co-electroporate 25 μg (circular) DNA of pEx-CAG-stop-ORF-pA together with 25 μg (circular) DNA of the C31 Integrase expression vector pCAG-C3Int into IDG26.10-3 ES cells grown on feeder layers. For electroporation of 50 μg plasmid DNA, use 4×10^6 ES cells in 0.8 ml PBS with a 0.4 cm BioRad cuvette (we use the BioRad Genepulser Xcell at 300 V and 2 ms time constant). Plate transfected cells into two 10-cm culture dishes containing embryonic fibroblasts as feeder layer.

2. On day 2 after transfection, add G418 (140 μg/ml) to the culture medium. Keep in selection for 8 days by changing medium daily.

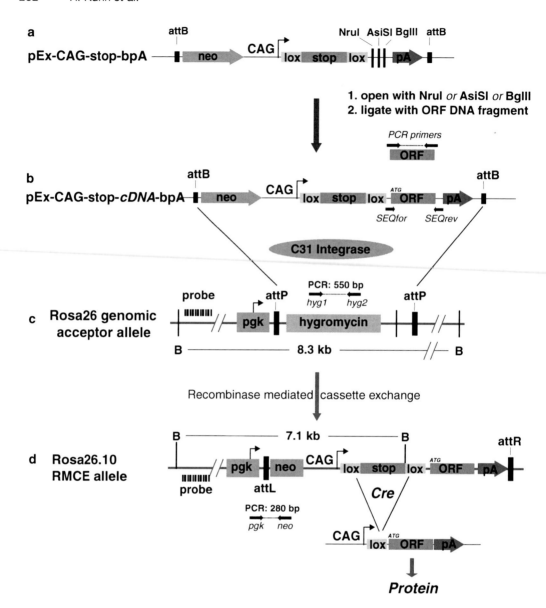

Fig. 3. Construction of Tamoxifen inducible transgenes and integration into the *Rosa26* locus of ES cells. (**a**) Schema of the conditional expression vector pEx-CAG-stop-bpA containing a promoterless neomycin resistance (neo) gene, the CAG promoter followed by a loxP-flanked transcriptional stop cassette, *Nru* I, *Asi* SI and *Bgl* II restriction sites for the insertion of a DNA fragment containing an open reading frame (ORF) for protein expression and a polyadenylation (bpA) signal. These elements are flanked by attB recognition sites for phiC31 integrase. (**b**) Completed conditional expression vector pEx-CAG-stop-ORF-bpA obtained by ligation of an ORF into the *Nru* I, *Asi* SI or *Bgl* II site of plasmid pEx-CAG-stop-bpA. Primers SEQfor/rev can be used for the sequence confirmation of the insert. For genomic integration, this vector is transfected together with a C31 integrase expression plasmid into Rosa26.10 ES cells harbouring a modified *Rosa26* allele containing two C31 integrase attP recognition sites. (**c**) The *Rosa26* acceptor locus in IDG3.2 Rosa26.10 ES cells contains a pgk promoter and a hygromycin resistance coding region flanked by two attP recognition sites. PCR genotyping can be performed with the primer pair hyg1/hyg2 resulting in a 550 bp PCR product. Using the Rosa-5′ genomic hybridization probe and *Bam*HI (B) digestion, the acceptor allele is recognised as a 8.3 kb band. (**d**) Upon C31 Integrase-mediated recombination of both pairs of attB and attP sites, the hygromycin resistance gene becomes exchanged by recombinase-mediated cassette exchange (RMCE) against the ORF expression unit. ES cell clones habouring RMCE alleles can be selectedby

3. On day 9: pick 12 (or more) colonies per RMCE experiment. Expand cells on feeders for freezing and on gelatine for DNA isolation.

4. Extract genomic DNA by standard procedures (43) and identify positive clones by PCR or Southern blotting as described below.

3.1.3. Identification of RMCE Positive ES Cell Clones

1. For PCR identification of RMCE positive ES cell clones use the neo and pgk primers listed in Sect. 2.1 (annealing temperature: 65°C; correct product size 280 bp). Note that feeder cells containing a pgk-neo resistance gene will yield an additional amplification product of 160 bp. Incorrectly recombined clones will still contain the hygro gene which can be checked by using the Hyg-1 and Hyg-2 primers listed in Sect. 2.1 (annealing temperature 65°C, product size 550 bp). Note that using feeders harbouring an additional hygro resistance gene precludes this screening procedure.

2. Southern blot verification of correct clones is mandatory for exclusion of partially recombined alleles or chromosomal rearrangements. To distinguish the wildtype and RMCE *Rosa26* alleles, cleave the genomic DNA with *Bam*HI and use the 450 bp *Eco*RI fragment from plasmid pRosa-5′probe as hybridization probe. The wildtype *Rosa26* locus is identified by a band of 5.8 kb. The parental IDG26.10-3 ES cells show an additional band of 8.3 kb derived from the modified *Rosa26* allele before RMCE (Fig. 3c). Positive RMCE clones show a new band of 7.1 kb due to the integration of the neo and CAG-stop-ORF gene cassettes (Fig. 3d). Usually, ~40% of the neomycin resistant clones undergo complete RMCE.

3. Use confirmed clones for generating transgenic mouse lines by blastocyst injection or tetraploid embryo aggregation as described in (44).

4. For the Tamoxifen inducible activation of the CAG expression vector, see Sect. 3.3.

3.2. Tamoxifen Inducible Gene Knockout

Conditional mutagenesis enables inactivation of a target gene via Cre/loxP recombination in somatic cells. In the conditional allele, a functionally essential (critical) exon is flanked by two loxP sites that do not interfere with gene function (Fig. 4c, 5e). Conditional

---

Fig. 3. (continued) the neo resistance gene that is expressed from the pgk promoter within *Rosa26*. Recombined ES cell clones can be identified by PCR genotyping using the primer pair pgk/neo that amplifies a 280 bp fragment. Using the Rosa-5′ genomic hybridization probe and *Bam*HI (B) digestion, the RMCE allele is recognised as a 7.1 kb band, in contrast to the second *Rosa26* wildtype allele that generates a 5.8 kb *Bam*HI fragment (data not shown). Upon the Cre-mediated excision of the loxP-flanked stop cassette in vivo, the ORF is transcribed and translated into the desired protein.

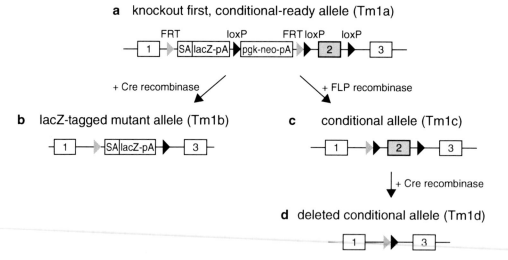

a   knockout first, conditional-ready allele (Tm1a)

b   lacZ-tagged mutant allele (Tm1b)

c   conditional allele (Tm1c)

d   deleted conditional allele (Tm1d)

Fig. 4. Gene targeting using "knockout first, conditional-ready" alleles. The "knockout first, conditional-ready" allele (Tm1a, **a**) consists of a β-galactosidase reporter cassette ("lacZ-pA") and a neomycin expression cassette ("pgk-neo-pA") flanked by two FRT sites. The critical exon 2 is flanked by two loxP sites, a third loxP site is located between the lacZ and neo genes. Upon Cre recombination of the Tm1a allele, exon 2 and the neomycin cassette are excised, leading to the lacZ-tagged knockout allele (Tm1b, **b**). If the knockout first allele is initially recombined using FLP recombinase, the β-galactosidase cassette becomes deleted, creating a conditional knockout allele (Tm1c, **c**). Subsequent recombination with Cre recombinase in specific cell types leads to the conditional inactivated allele (Tm1d, **d**). Exons of the target gene are shown as *numbered rectangles*.

mutant mice are obtained by crossing the conditional strain with a transgenic line that constitutively expresses Cre recombinase or the Tamoxifen inducible CreER^{T2} fusion protein under control of a cell type-specific promoter. Thereby, the inactivation of the target gene is either restricted to a selected cell type without temporal control or can be induced at a chosen time.

Due to the substantial existing and growing sources of information on targeting vectors, ES cell clones, and mutant mouse strains, a mutagenesis project should be started by searching the relevant databases for preexisting materials to avoid potential duplication. This can be conveniently achieved by searching for the name of the gene of interest at the MGI (http://www.informatics.jax.org), IGTC (http://www.genetrap.org) and IKMC (http://www.knock-outmouse.org) Web pages. If these databases do not yield materials relevant to the project requirements, it may be necessary to consider generating a novel conditional gene targeting vector, as described in Sect. 3.2.2.

3.2.1. Conditional Alleles Generated By Large-Scale Mutagenesis Programmes

The large-scale EUCOMM/KOMP mutagenesis programmes, which are coordinated by the International Knockout Mouse Consortium (IKMC), generate a genome-wide resource of gene-specific targeting vectors, targeted ES cell clones and mutant mouse strains. The EUCOMM/KOMP alleles are mostly of the "KO first, conditional-ready" type (Fig. 4) (45). This design initially disrupts the target gene by the intronic insertion of a gene disruption cassette that includes a splice acceptor element, a β-galactosidase reporter gene and a neo resistance gene. In these "knockout first" alleles, gene function is inactivated by splicing of upstream exons to the splice acceptor site of the targeting cassette (Targeted mutation 1a, Tm1a, Fig. 4a). Most EUCOMM/KOMP targeting vectors contain a targeting cassette with a neo resistance that is driven by its own β-actin promoter to allow the targeting of all genes, irrespective of their expression status in mouse ES cells. Promoterless vectors that utilise the promoter of the targeted gene to drive the expression of the neo resistance are also used for the targeting of genes that are expressed in ES cells. In addition to the gene disruption cassette, the targeted EUCOMM/KOMP alleles contain two loxP sites that flank a critical exon of the target gene (Fig. 4a). The loxP-flanked critical exon, as the target for Cre-mediated excision, is determined by these criteria: (1) its deletion causes a translational reading frame shift in the remaining mRNA that leads to the production of a shortened, mutant protein, (2) the exon is present in all transcript splice variants of the targeted gene, (3) the size of the exon is less than 1 kb, to ensure that the distance of the loxP sites is minimal for efficient Cre recombination, (4) the size of the flanking introns is at least 0.5 kb, so that loxP sites can be placed in non-conserved regions that are not required for endogenous splicing. In many cases, the second exon of a target gene fulfils these critical criteria, but in some instances several small exons are combined into a group of critical exons.

The "KO first," Tm1a allele can be converted into a classical KO allele by breeding the Tm1a mice with a Cre transgenic germline deleter strain. By this means, the neomycin selection cassette and the critical exon are deleted in the germline, resulting in a β-galactosidase-tagged, classical KO allele (Tm1b, "beta-Gal reporter," Fig. 4b). Furthermore, the Tm1a-type allele can be converted into a Tm1c allele for conditional mutagenesis by the use of the two FLP recombinase recognition (FRT) sites that are flanking the gene disruption cassette. The Flp-mediated deletion is typically performed by breeding the Tm1a mice with an Flp transgenic germline deleter strain (for a line in the C57BL/6N background, see ref. 46). Upon the removal of the gene disruption cassette, the remaining loxP sites that are flanking the critical exon delineate the configuration of a conditional allele (Tm1c, "conditional," Fig. 4c). This functional conditional allele can subsequently be inactivated in somatic cells by breeding Tm1c mice to a mouse line that expresses Cre recombinase in a time- and tissue-specific manner, resulting in the non-functional Tm1d allele ("deletion," Fig. 4d). At present, the EUCOMM/KOMP resource offers >10,000 targeting vectors, >8,000 genes targeted in the C57BL/6N-derived ES cell line JM8 (47) and >700 established mouse strains.

These resources are accessible through the programmes in the Web pages (http://www.eucomm.org; http://www.komp.org) and the IKMC (http://www.knockoutmouse.org).

3.2.2. Protocol for the Construction of Conditional Gene Targeting Vectors

The design of conditional knockout alleles follows two main objectives: (1) a pair of loxP sites must be inserted into intron regions of the target gene such that the encoded protein is inactivated through Cre-mediated excision of the loxP-flanked exon(s), (2) the target gene should not be disrupted by the presence of the loxP sites prior to Cre-mediated recombination. This section presents a protocol for vector construction based on PCR amplification of the gene targeting vector homology arms, as a method applicable in any laboratory equipped for molecular biology methods.

A conditional targeting construct can be assembled from three DNA segments of the target gene that can be amplified by PCR from a genomic template (Fig. 5a). The construct's segments A and

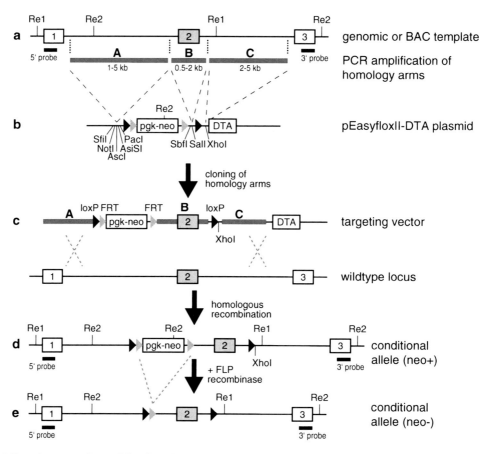

Fig. 5. Targeting vectors for conditional knockout alleles. (**a**) Genomic DNA or a BAC clone is used for the PCR amplification of the targeting vector's homology arms A, B and C, which are subsequently ligated into the backbone of the pEasyfloxII-DTA plasmid (**b**). (**c**) The resulting targeting vector is used for the mutagenesis of the target gene by homologous recombination, leading to a neomycin-positive mutant allele (**d**). Finally, the neomycin cassette is excised using FLP recombinase to obtain the conditional allele (**e**). Exons of the target gene are shown as *numbered rectangles*.

C serve as homology regions that mediate recombination with the chromosomal target gene. Segment B comprises the loxP-flanked region of a conditional allele that is selected for Cre-mediated excision. We recommend placing loxP sites at least ~250 bp apart from the exon boundaries to reduce the risk of interference with splicing signals. Within the size limit of ~2 kb often only a single exon can be flanked by loxP sites. Therefore, the target gene must be carefully analysed for its genomic structure using, e.g. the Ensembl genome database (http://www.ensembl.org) to define a critical exon, the deletion of which will disrupt gene function. In general, the critical exon is the first exon downstream of the gene's exon 1 that exhibits different reading frames upon splicing to its preceeding and its following exon. Thereby, the genomic deletion of the critical exon leads to reading frame shift within the mRNA of the recombined allele and to a premature stop codon. This will result either in the production of a truncated, non-functional protein or in the degradation of the mutant mRNA by the nonsense-mediated decay mechanism. Alternatively, if a functional domain of the target protein is known, a specific exon, e.g. coding for a protein kinase domain, can be chosen as critical exon.

For the construction of a conditional targeting vector, we use the generic plasmid backbone pEasyfloxII-DTA, which contains two loxP sites, a FRT-flanked neo expression cassette for positive selection and a DTA gene for negative selection (Fig. 5b). This vector harbours *Sfi*I, *Not*I, *Asc*I, *Asi*SI and *Pac*I sites for the insertion of the 5′ homology arm (segment A), an SbfI and SalI site for the insertion of the segment B (which contains the critical exon), and an *Xho*I site for the cloning of the 3′ homology arm (segment C). The *Sfi*I or *Not*I sites enable convenient linearization of the targeting vector prior to the electroporation of ES cells. The 5′ homology arm A should have a size of 1–5 kb and the loxP-flanked segment B should comprise 0.5–2 kb and include one or more exons of the target gene to inactivate the target gene by Cre-mediated deletion. The third genomic segment C represents the 3′ homology region with a size of 2–5 kb (Fig. 5a). The size of segment C should be at least equal to the length of segment B to provide sufficient sequence space for HR downstream of the second loxP site. The primers for the PCR amplification of the three segments must include the restriction sites chosen for cloning of the respective genomic fragment into pEasyfloxII-DTA. As part of the vector design, the absence of these sites within the amplified segments must be confirmed. After cloning of the homology arms into the three restriction sites of the plasmid, the targeting vector contains the 5′ homology region, the loxP-flanked exchange cassette, and the 3′ homology arm (Fig. 5c).

Recombination of the targeting vector with the wildtype allele can be easily identified by Southern blot analysis as shown in the example in Fig. 5e: as compared to the wildtype, digestion with restriction enzyme 1 ("Re1," detected by the 5′ probe) leads to an increased band size of ~2 kb, and digestion with restriction enzyme 2 ("Re2," detected by the 3′-probe) leads to a band of reduced size. The presence of the loxP site can be confirmed by double digestion with Re2 and *Xho*I, using the 3′-probe for detection (Fig. 5e). Finally, the pgk-neo cassette must be excised using FLP recombinase to obtain the functional (neo-free) allele for conditional knockout (Fig. 5e). This recombination event should be confirmed by Southern blot analysis using the 5′- and 3′-probes for hybridisation.

The frequency of homologous recombination between a targeting vector and its genomic target sequence is strongly reduced if the homologous sequences are not identical (48). To ensure proper recombination, the segments A–C for vector construction should be amplified from the genome of the same mouse strain as used to establish the ES cell line chosen for the targeting experiment. As template for PCR, we use BAC clones of the RPCI-23 genomic library, derived from the mouse strain C57BL/6J. BAC clones that cover the genomic region of interest can be found via the Ensembl mouse genome database (http://www.ensembl.org) and ordered from, e.g. Imagenes (http://www.imagenes-bio.de, Berlin, Germany). Such C57BL/6-based gene targeting vectors can be transfected into the C57BL/6-derived ES cell line JM8 (47) or into F_1 ES cell lines like IDG3.2 (49) or V6.5 (50), derived from (C57BL/6 × 129) hybrid blastocysts. Upon completion of vector cloning, the homology regions of the targeting construct should be sequenced to avoid the transfer of undesired mutations into the target gene. Before the transfection of ES cells, the gene targeting vector must be linearised at its 5′ end. For this purpose, pEasyflox-DTA contains a *Sfi*I and a *Not*I site (Fig. 5b).

The electroporation of ES cells and the selection and screening of neomycin resistant colonies is a multi-step procedure that exceeds the scope of this chapter and is described elsewhere (51). Identified recombinant ES cell clones are injected into blastocysts for the production of germline chimaeric mice to transmit the conditional allele to their offspring. In its initial configuration, a targeted allele includes the neomycin resistance expression cassette and that must be excised later from the genome to ensure the normal function of the conditional allele. The neo gene cassette can be conveniently removed from the targeted locus by crossing conditional (neo positive) offspring with FLP recombinase transgenic (deleter) mice (46, 52). Double transgenic offspring derived from this cross have lost the neo gene by recombination between the FRT sites flanking the selection marker. These offspring can then be further crossed to a Cre transgenic strain of choice resulting in the first mice that harbour one copy of the conditional allele (neo negative) and the Cre transgene. Half of these offspring will also contain the FLP transgene and should be excluded from breeding. Next, the intercross of conditional males and females will deliver the first homozygous conditional mice.

3.2.3. Steps for the Generation of a Targeting Vector

1. Download the genomic sequence of the target gene from a genome database (e.g. Ensembl).

2. Analyse and annotate the target sequence for exon positions, alternatively spliced exons, reading frames and functional protein domains to identify a critical exon using a sequence analysis software (follow guidelines described above in Sect. 3 Design rules for gene targeting vectors).

3. Develop a Southern blot screening strategy using 5'- and 3'-probes and identify suitable restriction enzymes following the guidelines described in Sect. 3.

4. Order an BAC clone containing the target gene (see Sect. 3) from a genomic library (e.g. RPCI-23 for C57BL/6-derived ES cells).

5. Design suitable primer pairs for the PCR amplification of the homology arms (see detailed information in following Sect. 3.2.2.

6. Amplify homology arms by high-fidelity PCR using a proof-reading polymerase.

7. Clone the homology arms into an appropriate vector backbone to obtain the targeting vector (see detailed information in following Sect. 3.2.2.

8. Confirm the integrity of targeting vector homology regions by sequence analysis to avoid the transfer of undesired mutations into the target gene.

9. Upon transfection of ES cells, select for recombinant clones using the positive selection marker provided by the vector backbone (most commonly neomycin).

10. Screen for correctly targeted clones by Southern blotting.

3.3. CreER^{T2} Transgenic Mice

Many hundred Cre-expressing mouse strains have been described that enable conditional mutagenesis in a variety of cell types and tissues. An overview of Cre mouse lines and their recombinase activity profiles is given by the Cre-X database (http://nagy.mshri.on.ca/cre) (53) and the Cre portal of the Jackson laboratory (http://www.creportal.org). In addition, a growing number (>30) of transgenic and knock-in mouse strains expressing CreER^{T2} in specific cell types were described in the last years, demonstrating the feasibility of this technique in all major organs including the brain (19, 20, 54). With regard to brain-specific CreER^{T2} mice, the available strain collection is presently limited to the strains that are compiled in Table 1. However, the recently launched EUCOMM-Tools programme plans to generate 250 CreER^{T2} driver lines within the next years. Therefore, in near future, inducible gene activation and knockout will be applicable also to many cell types and regions of the brain. For a protocol for the generation of BAC transgenic CreER^{T2} mice, see ref. (21).

Table 1
CreER^{T2} mouse lines

Specificity	Promoter	Short name (-CreERT2)	Insertion type	References
Astrocytes	Glial fibrillary acidic protein	GFAP	Transgene	(59, 60)
	Astrocyte-specific glutamate transporter	GLAST	Knock-in	(61)
Neuronal stem cells	Nestin	Nes	Transgene	(62)
Schwann cells	Proteolipid protein	PLP	Transgene	(63, 64)
	P0 fused to connexin 32	P0Cx	Transgene	(63)
	Calcium/calmodulin-dependent protein kinase 2α	CamKIIα	Transgene (BAC)	(57)
Neurons, forebrain	Frizzled10	Frizzled10	Transgene	(65)
Neurons, forebrain	CamKII	CamKII	Transgene (BAC)	(57)
Cortical neurons	$Na_V1.8$	$Na_V1.8$	Transgene	(66)
Nociceptive neurons in DRG	Villin	Vil	Transgene	(67)
Dopaminergic neurons	Dopamine transporter	DAT	Transgene (BAC)	(68)
Dopaminoceptive neurons	D1-receptor	Drd1a	Transgene (YAC)	(69)
Serotonergic neurons	Tph2	Tph2	Transgene (PAC)	(68)
Serotonergic neurons	Pet-1 ETS gene	Pet-1	Transgene	(70)
Noradrenergic Neurons	DBH	DBH-i	Transgene (PAC)	(71)

To characterise the recombinase expression profile of new and established Cre mouse strains, multiple Cre reporter alleles are available. Such indicator strains contain a loxP-flanked DNA segment that initially prevents the expression of a β-galactosidase, GFP, or other reporter gene from a coupled promoter region. Reporter strains based on randomly integrated transgenes may not exhibit reporter activity in all tissues but can be useful to map Cre activity in tissues of confirmed reporter activity. The most widely used indicator strains were constructed as knock-in alleles into the widely expressed *Rosa26* locus and express β-galactosidase or fluorescent proteins (Table 2). The expression of the β-galactosidase reporter can be monitored in tissue sections by the histochemical X-Gal stain or by immunohistochemistry using specific antibodies. The latter method allows the detailed documentation of reporter

Table 2
Cre reporter mice

Reporter name	Reporter prior recombination	Reporter upon recombination	Promoter region	Insertion type	References
Rosa-lacZ	None	β-Galactosidase	Rosa26	Rosa26 knock-in	(24, 26)
Rosa-GFP	None	eGFP	Rosa26	Rosa26 knock-in	(22)
Rosa-YFP	None	eYFP/eCFP	Rosa26	Rosa26 knock-in	(27)
Rosa-RFP	None	tdRFP	Rosa26	Rosa26 knock-in	(23)
mT/mG	RFP (tdTomato)	GFP	pCAGGS	Rosa26 knock-in	(25)
R26NZG	None	β-Galactosidase/GFP	pCAGGS	Rosa26 knock-in	(28)
Z/AP	β-Galactosidase	Alkaline phosphatase	pCAGGS	Transgene	(72)
Z/EG	β-Galactosidase	eGFP	pCAGGS	Transgene	(73)

expression by co-staining with cell type-specific markers. However, it has been noted that loxP-modified alleles at various genomic locations may be recombined by a given CreER^{T2} transgene at varying efficacy. Therefore, using conditional knockout alleles, it is important to retest the extent of recombination of a newly used loxP allele even if the specificity of recombinase activity in a CreER^{T2} line has been characterised with a reporter strain.

For the design of inducible gene switch experiments, it must be noted that the treatment of mice with tamoxifen, which acts as an antagonist of the wildtype oestrogen receptor, can cause side effects and experimental artefacts. It has been described that the treatment of CreER^{T2} transgenic mice with tamoxifen, but not the treatment of Cre negative controls, induces the death of lymphoma cells (55). Acute tamoxifen treatment may cause behavioural alterations but 4 weeks after the treatment most, if not all, of these effects are normalised (56). It is therefore essential to include control groups to test for the potential side effects of tamoxifen and the activation of Cre in a specific experimental setting. The treatment of littermates containing a CreER^{T2} transgene and the conditional allele of interest with the tamoxifen vehicle alone does not represent a sufficient control. In addition, Cre-ER^{T2} transgenic littermates and groups of mice that harbour only the conditional allele should be treated with tamoxifen and compared to the experimental group containing both genetic modifications.

3.3.1. Protocol: Tamoxifen Induction in Mice

The current standard method to induce CreER^{T2} activity is the intraperitoneal (i.p.) injection of the free base of tamoxifen solved in oil. The free base of tamoxifen is inactive on CreER^{T2} but is metabolised in the liver to the active ligand 4-OH-tamoxifen. Alternatively, 4-OH-tamoxifen can be directly injected, but this compound is more costly than the free base of tamoxifen. Both compounds were shown to induce recombination in all mouse tissues including the brain. For complete induction of CreER^{T2} in adult mice, the 10 mg/ml tamoxifen stock solution (see Sect. 2.3) is injected i.p. daily for 5 consecutive days at a dosage of 40 mg/kg body weight (this equals to 4 µl stock solution per 1 g body weight). However, for complete induction in the brain, it was shown that two injections per day for 5 consecutive days are required (57).

A more convenient and less stressful alternative to the intraperitoneal injection of tamoxifen is the oral administration of tamoxifen citrate with the food. For this purpose, tamoxifen citrate is mixed with the normal chow to a concentration of 360 mg/kg chow. The duration of the treatment depends on the desired tissue, but previous studies have shown that for most tissues 1–2 weeks are sufficient (58).

References

1. Carlsson A, Lindqvist M, Magnusson TOR (1957) 3,4-Dihydroxyphenylalanine and 5-Hydroxytryptophan as Reserpine Antagonists. Nature 180(4596):1200–1200.

2. Dahlstrom A, Fuxe K (1964) Localization of monoamines in the lower brain stem. Experientia 20(7):398–399.

3. Ungerstedt U, Arbuthnott GW (1970) Quantitative recording of rotational behavior in rats after 6-hydroxy-dopamine lesions of the nigrostriatal dopamine system. Brain Res 24(3):485–493.

4. Ungerstedt U (1968) 6-Hydroxy-dopamine induced degeneration of central monoamine neurons. Eur J Pharmacol 5(1):107–110.

5. Sonsalla PK, Heikkila RE (1988) Neurotoxic effects of 1-methyl-4-phenyl-1,2,3,6-tetrahydropyridine (MPTP) and methamphetamine in several strains of mice. Prog Neuropsychopharmacol Biol Psychiatry 12(2–3):345–354.

6. Betarbet R, Sherer TB, MacKenzie G, Garcia-Osuna M, Panov AV, Greenamyre JT (2000) Chronic systemic pesticide exposure reproduces features of Parkinson's disease. Nat Neurosci 3(12):1301–1306.

7. McCormack AL, Thiruchelvam M, Manning-Bog AB, Thiffault C, Langston JW, Cory-Slechta DA, Di Monte DA (2002) Environmental risk factors and Parkinson's disease: selective degeneration of nigral dopaminergic neurons caused by the herbicide paraquat. Neurobiol Dis 10(2):119–127.

8. Jenner P (2008) Functional models of Parkinson's disease: a valuable tool in the development of novel therapies. Ann Neurol 64 Suppl 2:S16–29.

9. Schapira AH, Tolosa E (2010) Molecular and clinical prodrome of Parkinson disease: implications for treatment. Nat Rev Neurol 6(6):309–317.

10. Braak H, Del Tredici K, Rub U, de Vos RA, Jansen Steur EN, Braak E (2003) Staging of brain pathology related to sporadic Parkinson's disease. Neurobiol Aging 24(2):197–211.

11. Cookson MR, Bandmann O (2010) Parkinson's disease: insights from pathways. Hum Mol Genet 19(R1):R21–27.

12. Polymeropoulos MH, Lavedan C, Leroy E, Ide SE, Dehejia A, Dutra A, Pike B, Root H, Rubenstein J, Boyer R, Stenroos ES, Chandrasekharappa S, Athanassiadou A, Papapetropoulos T, Johnson WG, Lazzarini AM, Duvoisin RC, Di Iorio G, Golbe LI, Nussbaum RL (1997) Mutation in the alpha-synuclein gene identified in families with Parkinson's disease. Science 276(5321):2045–2047.

13. Simon-Sanchez J, Schulte C, Bras JM, Sharma M, Gibbs JR, Berg D, et al. (2009) Genome-wide association study reveals genetic risk underlying Parkinson's disease. Nat Genet 41(12):1308–1312.

14. Gurney ME, Pu H, Chiu AY, Dal Canto MC, Polchow CY, Alexander DD, Caliendo J, Hentati A, Kwon YW, Deng HX, et al. (1994) Motor neuron degeneration in mice that express a human Cu,Zn superoxide dismutase mutation. Science 264(5166):1772–1775.

15. Kokjohn TA, Roher AE (2009) Amyloid precursor protein transgenic mouse models and Alzheimer's disease: understanding the paradigms, limitations, and contributions. Alzheimers Dement 5(4):340–347.

16. Dawson TM, Ko HS, Dawson VL (2010) Genetic Animal Models of Parkinson's Disease. Neuron 66(5):646–661.

17. Zigmond MJ, Berger TW, Grace AA, Stricker EM (1989) Compensatory responses to nigrostriatal bundle injury. Studies with 6-hydroxydopamine in an animal model of parkinsonism. Mol Chem Neuropathol 10(3):185–200.

18. Calne DB, Zigmond MJ (1991) Compensatory mechanisms in degenerative neurologic diseases. Insights from parkinsonism. Arch Neurol 48(4):361–363.

19. Anastassiadis K, Glaser S, Kranz A, Berhardt K, Stewart AF (A practical summary of site-specific recombination, conditional mutagenesis, and tamoxifen induction of CreERT2. Methods Enzymol 477:109–123.

20. Feil S, Valtcheva N, Feil R (2009) Inducible Cre mice. Methods Mol Biol 530:343–363.

21. Parkitna JR, Engblom D, Schutz G (2009) Generation of Cre recombinase-expressing transgenic mice using bacterial artificial chromosomes. Methods Mol Biol 530:325–342.

22. Kawamoto S, Niwa H, Tashiro F, Sano S, Kondoh G, Takeda J, Tabayashi K, Miyazaki J (2000) A novel reporter mouse strain that expresses enhanced green fluorescent protein upon Cre-mediated recombination. FEBS Lett 470(3):263–268.

23. Luche H, Weber O, Nageswara Rao T, Blum C, Fehling HJ (2007) Faithful activation of an extra-bright red fluorescent protein in "knock-in" Cre-reporter mice ideally suited for lineage tracing studies. Eur J Immunol 37(1):43–53.

24. Mao X, Fujiwara Y, Orkin SH (1999) Improved reporter strain for monitoring Cre recombinase-mediated DNA excisions in mice. Proc Natl Acad Sci USA 96(9):5037–5042.

25. Muzumdar MD, Tasic B, Miyamichi K, Li L, Luo L (2007) A global double-fluorescent Cre reporter mouse. Genesis 45(9):593–605.

26. Soriano P (1999) Generalized lacZ expression with the ROSA26 Cre reporter strain. Nat Genet 21(1):70–71.

27. Srinivas S, Watanabe T, Lin CS, William CM, Tanabe Y, Jessell TM, Costantini F (2001) Cre reporter strains produced by targeted insertion of EYFP and ECFP into the ROSA26 locus. BMC Dev Biol 1:4.

28. Yamamoto M, Shook NA, Kanisicak O, Yamamoto S, Wosczyna MN, Camp JR, Goldhamer DJ (2009) A multifunctional reporter mouse line for Cre- and FLP-dependent lineage analysis. Genesis 47(2):107–114.

29. Rodriguez P, Da Silva S, Oxburgh L, Wang F, Hogan BL, Que J (BMP signaling in the development of the mouse esophagus and forestomach. Development 137(24): 4171–4176.

30. Santiago Y, Chan E, Liu PQ, Orlando S, Zhang L, Urnov FD,et al. (2008) Targeted gene knockout in mammalian cells by using engineered zinc-finger nucleases. Proc Natl Acad Sci USA 105(15):5809–5814.

31. Capecchi MR (2005) Gene targeting in mice: functional analysis of the mammalian genome for the twenty-first century. Nat Rev Genet 6(6):507–512.

32. Gong M, Rong YS (2003) Targeting multicellular organisms. Curr Opin Genet Dev 13(2):215–220.

33. Lai L, Prather RS (2003) Creating genetically modified pigs by using nuclear transfer. Reprod Biol Endocrinol 1:82.

34. Doyon Y, McCammon JM, Miller JC, Faraji F, Ngo C, Katibah GE, et al. (2008) Heritable targeted gene disruption in zebrafish using designed zinc-finger nucleases. Nat Biotechnol 26(6):702–708.

35. Geurts AM, Cost GJ, Freyvert Y, Zeitler B, Miller JC, Choi VM, et al. (2009) Knockout rats via embryo microinjection of zinc-finger nucleases. Science 325(5939):433.

36. Hockemeyer D, Soldner F, Beard C, Gao Q, Mitalipova M, DeKelver RC, et al. (2009) Efficient targeting of expressed and silent genes in human ESCs and iPSCs using zinc-finger nucleases. Nat Biotechnol 27(9):851–857.

37. Porteus MH, Carroll D (2005) Gene targeting using zinc finger nucleases. Nat Biotechnol 23(8):967–973.

38. Urnov FD, Miller JC, Lee YL, Beausejour CM, Rock JM, Augustus S, Jamieson AC, Porteus MH, Gregory PD, Holmes MC (2005) Highly efficient endogenous human gene correction using designed zinc-finger nucleases. Nature 435(7042):646–651.

39. Cui X, Ji D, Fisher DA, Wu Y, Briner DM, Weinstein EJ (Targeted integration in rat and mouse embryos with zinc-finger nucleases. Nat Biotechnol 29(1):64–67.

40. Meyer M, de Angelis MH, Wurst W, Kuhn R (Gene targeting by homologous recombination in mouse zygotes mediated by zinc-finger nucleases. Proc Natl Acad Sci USA 107(34): 15022–15026.

41. Delic S, Streif S, Deussing JM, Weber P, Ueffing M, Holter SM, Wurst W, Kuhn R (2008) Genetic Mouse Models for Behavioral Analysis through Transgenic RNAi Technology. Genes Brain Behav.

42. Hitz C, Wurst W, Kuhn R (2007) Conditional brain-specific knockdown of MAPK using Cre/loxP regulated RNA interference. Nucleic Acids Res 35(12):e90.

43. Sambrook J, Macallum P, Russell D (2001) Molecular Cloning: A Laboratory Manual (Cold Spring Harbour Press, Cold Spring Harbour) 3. Ed.

44. Reid SW, Tessarollo L (2009) Isolation, microinjection and transfer of mouse blastocysts. Methods Mol Biol 530:269–285.

45. Friedel RH, Seisenberger C, Kaloff C, Wurst W (2007) EUCOMM--the European conditional mouse mutagenesis program. Brief Funct Genomic Proteomic 6(3):180–185.

46. Kranz A, Fu J, Duerschke K, Weidlich S, Naumann R, Stewart AF, Anastassiadis K (An improved Flp deleter mouse in C57Bl/6 based on Flpo recombinase. Genesis 48(8):512–520.

47. Pettitt SJ, Liang Q, Rairdan XY, Moran JL, Prosser HM, Beier DR, Lloyd KC, Bradley A, Skarnes WC (2009) Agouti C57BL/6N embryonic stem cells for mouse genetic resources. Nat Methods 6(7):493–495.

48. te Riele H, Maandag ER, Berns A (1992) Highly efficient gene targeting in embryonic stem cells through homologous recombination with isogenic DNA constructs. Proc Natl Acad Sci USA 89(11):5128–5132.

49. Hitz C, Steuber-Buchberger P, Delic S, Wurst W, Kuhn R (2009) Generation of shRNA transgenic mice. Methods Mol Biol 530:101–129.

50. Eggan K, Akutsu H, Loring J, Jackson-Grusby L, Klemm M, Rideout WM, Yanagimachi R, Jaenisch R (2001) Hybrid vigor, fetal overgrowth, and viability of mice derived by nuclear cloning and tetraploid embryo complementation. Proc Natl Acad Sci USA 98(11):6209–6214.

51. Southon E, Tessarollo L (2009) Manipulating mouse embryonic stem cells. Methods Mol Biol 530:165–185.

52. Rodriguez CI, Buchholz F, Galloway J, Sequerra R, Kasper J, Ayala R, Stewart AF, Dymecki SM (2000) High-efficiency deleter mice show that FLPe is an alternative to Cre-loxP. Nat Genet 25(2):139–140.

53. Nagy A, Mar L, Watts G (2009) Creation and use of a cre recombinase transgenic database. Methods Mol Biol 530:365–378.

54. Friedel RH, Wurst W, Wefers B, Kuhn R (2010) Generating conditional knockout mice. Methods Mol Biol 693:205–231.

55. Schmidt-Supprian M, Rajewsky K (2007) Vagaries of conditional gene targeting. Nat Immunol 8(7):665–668.

56. Vogt MA, Chourbaji S, Brandwein C, Dormann C, Sprengel R, Gass P (2008) Suitability of tamoxifen-induced mutagenesis for behavioral phenotyping. Exp Neurol 211(1):25–33.

57. Erdmann G, Schutz G, Berger S (2007) Inducible gene inactivation in neurons of the adult mouse forebrain. BMC Neurosci 8:63.

58. Kiermayer C, Conrad M, Schneider M, Schmidt J, Brielmeier M (2007) Optimization of spatiotemporal gene inactivation in mouse heart by oral application of tamoxifen citrate. Genesis 45(1):11–16.

59. Hirrlinger PG, Scheller A, Braun C, Hirrlinger J, Kirchhoff F (2006) Temporal control of gene recombination in astrocytes by transgenic expression of the tamoxifen-inducible DNA recombinase variant CreERT2. Glia 54(1):11–20.

60. Casper KB, Jones K, McCarthy KD (2007) Characterization of astrocyte-specific conditional knockouts. Genesis 45(5):292–299.

61. Mori T, Tanaka K, Buffo A, Wurst W, Kuhn R, Gotz M (2006) Inducible gene deletion in astroglia and radial glia--a valuable tool for functional and lineage analysis. Glia 54(1):21–34.

62. Imayoshi I, Ohtsuka T, Metzger D, Chambon P, Kageyama R (2006) Temporal regulation of Cre recombinase activity in neural stem cells. Genesis 44(5):233–238.

63. Leone DP, Genoud S, Atanasoski S, Grausenburger R, Berger P, Metzger D, Macklin WB, Chambon P, Suter U (2003) Tamoxifen-inducible glia-specific Cre mice for somatic mutagenesis in oligodendrocytes and Schwann cells. Mol Cell Neurosci 22(4):430–440.

64. Doerflinger NH, Macklin WB, Popko B (2003) Inducible site-specific recombination in myelinating cells. Genesis 35(1):63–72.

65. Gu X, Yan Y, Li H, He D, Pleasure SJ, Zhao C (2009) Characterization of the Frizzled10-CreER transgenic mouse: an inducible Cre line for the study of Cajal-Retzius cell development. Genesis 47(3):210–216.

66. Zhao J, Nassar MA, Gavazzi I, Wood JN (2006) Tamoxifen-inducible NaV1.8-CreERT2 recombinase activity in nociceptive neurons of dorsal root ganglia. Genesis 44(8):364–371.

67. el Marjou F, Janssen KP, Chang BH, Li M, Hindie V, Chan L, Louvard D, Chambon P, Metzger D, Robine S (2004) Tissue-specific and inducible Cre-mediated recombination in the gut epithelium. Genesis 39(3):186–193.

68. Weber T, Bohm G, Hermann E, Schutz G, Schonig K, Bartsch D (2009) Inducible gene manipulations in serotonergic neurons. Front Mol Neurosci 2:24.

69. Lemberger T, Parlato R, Dassesse D, Westphal M, Casanova E, Turiault M, Tronche F, Schiffmann SN, Schutz G (2007) Expression of Cre recombinase in dopaminoceptive neurons. BMC Neurosci 8:4.

70. Liu C, Maejima T, Wyler SC, Casadesus G, Herlitze S, Deneris ES (Pet-1 is required across different stages of life to regulate serotonergic function. Nat Neurosci 13(10):1190–1198.

71. Stanke M, Duong CV, Pape M, Geissen M, Burbach G, Deller T, Gascan H, Otto C, Parlato R, Schutz G, Rohrer H (2006) Target-dependent specification of the neurotransmitter phenotype: cholinergic differentiation of sympathetic neurons is mediated in vivo by gp 130 signaling. Development 133(1):141–150.

72. Lobe CG, Koop KE, Kreppner W, Lomeli H, Gertsenstein M, Nagy A (1999) Z/AP, a double reporter for cre-mediated recombination. Dev Biol 208(2):281–292.

73. Novak A, Guo C, Yang W, Nagy A, Lobe CG (2000) Z/EG, a double reporter mouse line that expresses enhanced green fluorescent protein upon Cre-mediated excision. Genesis 28(3–4):147–155.

Chapter 13

6-OHDA Lesion Models of Parkinson's Disease in the Rat

Eduardo M. Torres and Stephen B. Dunnett

Abstract

The 6-hydroxydopamine (6-OHDA) lesion of the rat nigrostriatal pathway is the most widely used animal model of Parkinson's disease. 6-OHDA is a highly specific neurotoxin which targets catecholamine neurones via the dopamine active transporter (DAT). When injected stereotaxically into the brain, either into the median forebrain bundle (MFB) or into the neostriatum, it causes extensive, irreversible loss of dopamine neurones in the ventral midbrain. The corresponding loss of dopamine innervation in target areas is associated with a range of long-term, behavioural deficits that form the target of experimental therapies, aimed at protecting or restoring dopaminergic deficits. In this chapter, the two most widely used 6-OHDA lesion protocols are described: (1) The MFB lesion that results in >97% unilateral depletion of dopamine neurones, principally in the ipsilateral striatum and nucleus accumbens. (2) The unilateral striatal lesion resulting in partial dopamine denervation of the striatum only. In vivo assessment of both lesion types by drug-induced rotation is also covered.

Key words: 6-Hydroxydopamine, Dopamine lesion, Parkinson's, Amphetamine rotation, Apomorphine rotation

1. Introduction

The specific neurotoxic effects of 6-OHDA rely on its structural similarity to dopamine, whereby dopamine reuptake mechanisms, principally the dopamine active transporter (DAT), selectively take up the neurotoxin, which becomes concentrated within dopamine neurones. Subsequent metabolism of the neurotoxin results in the production of toxic free-radicals within the cell, causing rapid dysfunction and eventual cell death. As 6-OHDA does not cross the blood–brain barrier, it must be injected stereotaxically into the brain. Intracerebral injection of 6-OHDA to induce lesions of the dopamine system in the brain dates back more than 40 years to work done by Urban Ungerstedt, who injected the neurotoxin

Emma L. Lane and Stephen B. Dunnett (eds.), *Animal Models of Movement Disorders: Volume I*, Neuromethods, vol. 61, DOI 10.1007/978-1-61779-298-4_13, © Springer Science+Business Media, LLC 2011

directly into the rat substantia nigra compacta (SNc) to induce "anterograde degeneration of the whole nigrostriatal dopamine system" (1). The modern version of the lesion targets the median forebrain bundle (MFB), the anterior efferent from the dopaminergic substantia nigra compacta, and ventral tegmental area (2). Injection of 6-OHDA into the MFB results in rapid uptake of the toxin into dopamine neurones, and a near complete loss of the dopamine neurones in these nuclei. There is extensive dopamine depletion of target areas on the side of injection, including the striatum, nucleus accumbens, septum, prefrontal cortex and olfactory bulb. Unfortunately, bilateral MFB lesions produce severe akinesia, aphagia and adipsia (3) and are unsuitable for most experiments. By contrast, unilaterally lesioned animals are outwardly normal, and capable of eating and drinking normally, and for this reason unilateral lesions are generally favoured. A large number of similar stereotaxic coordinates for injection into the MFB are available in the scientific literature. The recent literature also contains coordinates for double injection strategies aimed at the MFB, but it is the authors' experience that a well targeted single injection is considerably more straightforward, and, provided that the injection is targeted accurately, more than capable of causing near complete lesions of the MFB.

Partial lesions of the dopamine system may also be achieved by injection of 6-OHDA directly into the striatum. Using this approach results in a loss of dopamine neurones in the substantia nigra and corresponding dopamine denervation in the striatum. But dopamine innervation to other brain areas, notably the nucleus accumbens, is spared. Additionally, using the striatal route of toxin administration, the extent of the lesion may be varied, according to the dose of 6-OHDA injected, and so-called "partial lesions" may be achieved (4, 5).

The behavioural deficits produced by 6-OHDA lesions have been described previously and are not discussed in detail here. (see Fleming and Schallert in this volume). However, as no lesion protocol is 100% efficient, lesioned animals require behavioural testing in order to determine whether or not the lesion has been successful. The standard method of assessment is to observe and record turning behaviour under the influence of dopaminergic stimulant drugs. In rats with unilateral lesions, injections of pharmacological doses of amphetamine or apomorphine induce a striking turning behaviour known as "rotation" and which correlates closely with the extent of dopamine denervation (6). Amphetamine stimulates dopamine release and blocks reuptake from dopaminergic terminals in the intact striatum to a greater extent than on the lesioned side of the brain, resulting in a profound motor asymmetry. Together with the increase in activity caused by the drug, this causes the animal to adopt a classic nose-to-tail posture and to rotate vigorously towards the side of the lesion (ipsilateral). Apomorphine, on the other hand, stimulates supersensitive receptors in the lesioned striatum in preference to the normal receptor complement on the intact side, and results in rotation in the opposite direction, towards the intact (contralateral) side. (For a detailed description of rotation, see Dunnett and Torres, this volume).

In the present chapter, we describe the methods to carry out unilateral dopamine-depleting lesions in the rat by injection of 6-OHDA into the forebrain, either into the MFB or into the corpus striatum, and for subsequent rotation testing using both amphetamine and apomorphine.

2. Materials

2.1. Preparation of 6-OHDA

6-OHDA is highly toxic, and should be handled using the appropriate protective clothing (see the material safety data sheet that comes with the product for full details). The active neurotoxin is the 6-OHDA hydrobromide salt (for control experiments, the less active hydrochloride salt may be used). This can be obtained from a number of suppliers, each of which supplies a slightly different formulation. For the purposes of determining dosage, the final concentration of the neurotoxin must therefore be calculated on the weight of the free base. In solid form, 6-OHDA is sensitive to both light and temperature, and should be stored in a –20°C freezer. In solution, it oxidizes rapidly and all solvents should contain 0.01–0.1% ascorbic acid as an antioxidant. The 6-OHDA formulation supplied by Sigma-Aldrich (UK) has a molecular weight of 250.09 and contains ascorbic acid as a stabilizer. This can therefore be dissolved, without further addition of an antioxidant, using 300 mM sodium chloride (0.9% saline) solution. Sterile, nonpyrogenic saline in infusion bags is ideal.

A working concentration of Sigma-Aldrich 6-OHDA is achieved by dissolving 5 mg of the formulation in 0.8 ml of 300 mM sodium chloride (sterile saline infusion solution). This gives a solution containing 30 mM 6-OHDA in 0.03% ascorbic acid and free-base weight of 6-OHDA of 5.14 mg/ml. After dissolving the toxin, quickly divide the solution into 50 μl aliquots in 1.5 ml Eppendorf tubes and store at –20°C until needed. Aliquots are stable for at least 1 month. On the day of surgery, a single aliquot is thawed quickly, and then stored on ice during surgery. After 2 h, the original aliquot should be discarded and a fresh aliquot thawed. Repeat as necessary.

2.2. Preparation of Dopamine Active Drugs

2.2.1. Amphetamine

Amphetamine is a presynaptic dopamine agonist which activates the intact (contralateral) side of the brain to a much greater extent than the lesioned (ipsilateral) side. Crossover of striatal outputs leads to a principally attentional neglect on the contralateral side and causes the animal to rotate ipsilaterally (towards the lesion) (7).

Several different forms of amphetamine have been used in different studies, most commonly either d-amphetamine or the more potent methamphetamine. In our laboratory, we use a standard dose of 2.5 mg/kg methamphetamine hydrochloride for rotation screening of 6-OHDA lesions. Make up the drug at 2.5 mg/ml in 0.9% sterile saline and inject intraperitoneally at a dose of 1 ml/kg immediately prior to rotation testing (see below).

2.2.2. Apomorphine

Apomorphine is a dopamine receptor agonist which acts on both D1 and D2 receptors located primarily on the post-synaptic medium spiny projection neurons in the striatum. Following a dopamine lesion, the receptors in the lesioned striatum upregulate to compensate for the loss of dopamine, a phenomenon known as supersensitivity (8). Consequently, in a unilaterally lesioned animal, it is the lesioned side (ipsilateral) which is preferentially activated by the agonist, in particular at very low doses that are subthreshold for activating normo-sensitive receptors on the intact contralateral side of the brain. Because of the crossover of striatal output, activation in the lesioned striatum drives the animal to rotate toward, the opposite (contralateral) side. Whilst doses of 0.25–1.0 mg/kg induce locomotor activation in intact rats, a dose an order of magnitude lower is sufficient to activate supersensitive receptors in the 6-OHDA lesioned striatum. In our laboratory, we use a standard dose of 0.05 mg/kg to test for agonist-induced rotation. Make up the drug at a concentration of 0.1 mg/ml in 0.1% ascorbic acid in 0.9% sterile saline, and inject subcutaneously in the neck scruff at a dose of 0.5 ml/kg. Note that this dose of apomorphine is designed to produce maximal striatal asymmetry by preferential activation of receptors on the lesioned (supersensitive) side, with minimal receptor activation on the intact side. Higher doses may be used, some workers preferring 0.25 mg/kg, but this is less effective because it involves a competition between levels of activation on the two sides, and higher doses also induce focal stereotypy responses which compete with the biased locomotor activation which underlies rotation.

3. Methods

3.1. General Surgical Procedure

The preferred method for anaesthesia is by inhalation, depending on the surgical setup available. In our laboratory, we use the gaseous anaesthetic isoflurane vaporized into a 2:1 mixture of O_2:NO carrier gas delivered via an inhalation mask fixed around the incisor bar of the stereotaxic frame. Long-lasting injectables may also be used (e.g. equithesin, ketamine), but dopamine interacting drugs such as barbiturates should be avoided. The 6-OHDA neurotoxin is delivered using stereotaxic methods. We use a Kopf stereotaxic frame with adjustable nose bar height, blunt (45°) ear bars, cannula holder and drill holder. Note that although many experimenters prefer 17° ear bars, as they are easer to apply, these are more likely to cause inner ear damage, resulting in vestibular balance impairments and torsional body twisting, which compromises the reliability of subsequent rotation tests.

6-OHDA is delivered via a 30 guage, stainless steel cannula connected by 0.5 m of polythene tubing (Portex internal diameter 0.28 mm, outer diameter 0.61 mm) to a 10 µl microsyringe (SGE Europe Ltd, Milton Keynes, UK or Hamilton Co., USA) (see Fig. 1). As 30 guage cannula is easily bent, this can be reinforced with an outer reinforcing sleeve of 23 guage steel tubing, either crimped or glued to the injection cannula, and allowing at least 15 mm of the 30 guage tubing free at both ends. Prior to use, the syringe, cannula and tubing must be primed with sterile, isotonic saline solution. Any air remaining in the system is liable to compression during injection and may prevent expulsion of the toxin from the cannula. The syringe is driven by a motorized syringe pump calibrated to deliver 1 µl/min. Note that direct injection using a suitably sized microsyringe is possible, by repeated small depressions of the plunger over 3 min to deliver 3 µl. Holes in the skull are made using a motorized, surgical drill attached to the stereotaxic frame, with a size 1/2 drill bit. Because of the small size of the hole in the skull, the cannula can be accurately positioned for injection, without the need for repositioning from bregma. If a larger drill bit is used, or if hand drilling is necessary, the needle should be zeroed using bregma and positioned using the x–y coordinates.

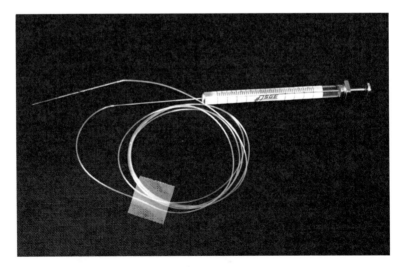

Fig. 1. Ten-microlitre microsyringe setup for MFB lesion injections. The 30-gauge cannula is connected to the needle of the syringe using polythene tubing. The syringe, tubing and cannula are filled with sterile infusion saline prior to use (priming).

3.2. Detailed Surgical Methodology

3.2.1. Unilateral Lesion of the MFB

Table 1 gives two sets of coordinates; the first is based on Dunnett et al. (9) and is designed for use on rats over 250 g in weight. The second is based on recent modifications made by Torres et al. [38] and is recommended for rats in the 150–250 g weight range.

1. Anaesthetize the rat, and place the rat in the stereotaxic frame with the nose bar set at the appropriate height (see Table 1).

2. Shave the scalp using hair clippers, and make an incision using a scalpel, through the skin along the midline (maximum 2.5 cm long) starting from just behind the inter-ocular line, and cutting caudally.

3. Open the incision and use the scalpel blade to pare back the underlying connective tissue and scrape the exposed surface of the skull clean.

4. Position the tip of the drill bit directly over bregma; the zero point for x–y coordinates.

5. Raise the drill bit from the skull and move it to the x–y coordinates (see Table 1). Drill through the skull, taking care not to damage the underlying dura matter.

6. Load the cannula with 4–5 μl of 6-OHDA, first drawing a small "spacing" air bubble (0.5–1 μl) into the cannula to prevent mixing of the neurotoxin and the priming solution.

7. Lower the cannula into the drilled hole until it is just touching the dura mater. If the dura matter is undamaged, the correct depth is determined when the meniscus of fluid in the drill hole descends slightly, as the dura membrane is touched by the cannula.

8. Lower the cannula into the brain, allowing the cannula to pierce the dura mater to the desired depth from the dura mater.

9. Start the syringe pump and run for 3 min to deliver 3 μl of neurotoxin. Smooth flow of the toxin without blockage of the cannula is monitored by movement of the spacing bubble in the transparent polyethylene tubing.

Table 1
Stereotaxic coordinates for MFB injection of 6-OHDA

Coordinates from bregma (mm)

	Nose bar	A=	L=	V=[a]
MFB lesion I	−2.3	−4.4	−1.0	−7.8
MFB lesion II	−4.5	−4.0	−1.2	−7.0

[a]Negative lateral coordinates denote right-sided lesions

10. Stop the pump and leave the cannula in place for a further 2 min to allow fluid diffusion, before slowly retracting from the brain.

11. Purge any unused neurotoxin from the cannula and flush several times by repeated drawing and expulsion of clean saline solution.

12. Suture the skin and administer analgesia according to the local SOP.

13. Place animal in a heated recovery cage until fully conscious. With gaseous anaesthesia this usually takes 5–10 min, after which the rat can be returned to its home cage.

The 6-OHDA has an immediate effect; recovering animals will have a postural bias towards the lesion side and rotate ipsilaterally. Checking for rotation in recovering animals is the first index of lesion success. Post surgery complications with this type of lesion are very rare but animals should be given a thorough check 24 and 48 h later.

3.2.2. Partial Lesions by Striatal Injection of 6-OHDA

Table 2 shows coordinates for multiple injection of the toxin into the striatum. The coordinates are based on those described by Kirik et al. (4) modified so that the four injection sites are in a straight line, 1 mm apart, enabling four cannulae (each attached to a different syringe) to be used simultaneously. This requires a cannula holder like the one shown in Fig. 2 that has vertical grooves, exactly 1 mm apart, in which the cannulae are held tightly at the correct spacing. The cannulae are adjusted to the same length in the cannula holder by loosening the clamp and lowering the cannula tips onto the top surface of the ear bar before re-clamping. If a single syringe and cannula are used, drill all four holes before making the first injection.

Table 2
Stereotaxic coordinates for striatal injection of 6-OHDA

Coordinates from bregma (mm)[a]

	A=	L=	V=
Rat Striatal lesion[b]	+1.0	−2.0	−5.5, −5.0, −4.5
	+0.4	−2.8	−5.5, −5.0, −4.5
	−0.2	−3.6	−5.5, −5.0, −4.5
	−0.8	−4.4	−5.5, −5.0, −4.5

Note that, because the drill holes are evenly spaced, after the positioning and drilling of the first hole, movement between subsequent drill holes is the same: $A = -0.6$, $L = -0.8$
[a] Negative lateral coordinates denote right-sided lesions
[b] Modified from Kirik et al. (4)

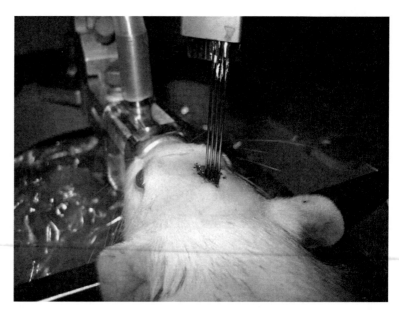

Fig. 2. Four-cannula setup for simultaneous injection of 6-OHDA at four sites in the striatum.

Note that, because the drill holes are evenly spaced, after positioning and drilling the first hole, movement between subsequent drill holes is always the same: $A = -0.6$, $L = -0.8$.

1. Place the rat in the frame with the nose bar set at the appropriate height (see Table 2).

2. Shave the scalp and make an incision in the skin along the midline (maximum 2.5 cm).

3. Open the incision and clean the surface of the skull with a scalpel blade.

4. Position the tip of the drill bit over bregma.

5. Move the drill bit to the $x-y$ coordinates for the first hole (see Table 2) and drill through the skull, taking care not to damage the underlying dura matter.

6. Repeat the process for holes 2–4.

7. Load each cannula with 3–4 µl of 6-OHDA, leaving a small air bubble in each syringe setup, between the neurotoxin and the priming solution.

8. Lower the cannulae until just touching the dura matter (see above).

9. Lower the cannula into the brain to the maximum depth from the dura matter.

10. Start the syringe pump and run for 2 min, raising the cannula(e) 0.5 mm after 40 s and again after 1 m 20 s to deliver 2 µl of neurotoxin at three different levels per injection.

11. Stop the pump and leave the cannula(e) in place for a further 2 min, before slowly retracting from the brain.

12. Flush the cannulae as above.

13. Suture the skin and administer analgesia according to your local SOP.

14. Place animals in a heated recovery cage until fully conscious.

3.3. Rotation Testing

In our laboratory, operated rats are tested for lesion efficacy using amphetamine (methamphetamine hydrochloride), at 2 weeks and 4 weeks post-lesion, and using apomorphine (apomorphine hydrochloride dehydrate) at 5 weeks post-lesion. Rotation is usually assessed using an automated rotometer system, over a 90 min test session for amphetamine and a 60 min session for apomorphine. Immediately following injection, rats are placed in 30 cm diameter circular, flat bottomed bowls, enclosed in 30–50 cm high Perspex or aluminium cylinders. Rats are attached via a harness to a rotometer head which sends information to a computer-controlled, automated rotometer system (e.g. Rotomax System, AccuScan Instruments Inc.). The system records both ipsilateral and contralateral turns, from which the net rotation (ipsilateral minus contralateral) can be calculated. Rotation scores are reported as either net rotations over the entire session, or as mean net rotations per minute. When an automated system is not available, alternative methods of measuring rotation may be used. Method one is to make a video recording of each animal's rotation session and then count rotations from the recording using high speed playback. A second method is to sample the rotation of each animal by recording rotation at regular intervals over the session. Animals are observed for 1 min at a time at either 10 min or 15 min intervals, recording both ipsilateral and contralateral rotations. Scores may then be extrapolated for the entire session.

In our early studies, we used a minimum criterion for lesion success as seven turns per minute under 5 mg/kg methamphetamine. However, this dose was associated with high levels of stereotypic behaviour and occasional mortality, As a result, we now use a somewhat lower dose (2.5 mg/kg), and six net rotations per minute (540 net rotations in a 90 min session) as the criterion for a successful 6-OHDA lesion corresponding to approximately 95% depletion of striatal dopamine in studies based on post-mortem neurochemical assay.

Re-lesioning of rats which do not meet this criterion is seldom successful, and as a result they are usually excluded from the experiment. For subsequent treatments, lesioned rats are allocated into counterbalance groups matched either on the 4-week amphetamine, or the 5-week apomorphine scores, such that prior to treatment, all groups have approximately the same mean rotation scores.

Following experimental treatments to ameliorate or reverse the effects of the 6-OHDA lesion, rotation scores are also used for in vivo assessment of treatment efficacy. Pre-lesion, protective treatments, such as growth factors (e.g. GDNF) or anti-apoptotic agents (e.g. caspase inhibitors), may reduce lesion-induced dopamine cell death in the ventral midbrain (10, 11). This is reflected in reduced levels of amphetamine rotation in the treated groups. Post-lesion treatments such as replacement of dopamine function by implantation of embryonic dopamine cells, or dopaminergic cells derived from stem cells, restore dopamine innervation to the lesioned striatum, and reduce the levels of net ipsilateral rotation. Interestingly, large dopamine grafts in the striatum can cause a net contralateral rotation (so-called over-compensatory rotation) due to their action on residual supersensitive receptors in the ipsilateral striatum (12, 13).

3.4. Behavioural Assessment

Unilateral lesions of the dopamine system induce well characterized deficits on the contralateral side of the body. The principal methods used to assess these deficits include the stepping, cylinder, staircase, and corridor tests, all of which detect either preferred use of the ipsilateral limbs or neglect of contralateral space. For full details of behavioural testing of unilateral lesioned animals, see the chapters by Fleming and Schallert, Smith, and Heuer, this volume and Farr and Trueman Volume II in this series.

3.5. Post-Mortem Assessment

The efficacy of the 6-OHDA lesion may assess *post-mortem* using immunohistochemical detection of the dopamine synthesizing enzyme, tyrosine hydroxylase (TH). Successful MFB lesions induce massive loss of TH immunoreactive neurons in the ipsilateral substantia nigra compacta and ventral tegmental area. Following striatal injection of the toxin, the majority of SNc neurons are lost. Fixation, sectioning, and staining of brain tissues are beyond the scope of the current chapter. Briefly, animals are sacrificed and perfused transcardially using formaldehyde fixative (usually 4% in phosphate-buffered saline). Frozen sections may then be stained immunohistochemically using antibodies against tyrosine hydroxylase. For a detailed methodology, see Torres et al. (14).

4. Conclusions

The rat, 6-OHDA lesion of the dopamine system is a widely used and extremely useful model for Parkinson's disease research. The unilateral depletion of the dopamine system has little effect on the rat's health and, after recovering from surgery, lesioned animals require no more extra care than a normal rat. Behaviourally, the loss of dopamine on one side of the brain leads to well characterized deficits, chiefly affecting use of the contralateral limbs and an attentional preference for ipsilateral space. To a greater or lesser extent, these deficits may be affected, pre- or post-lesion, by a range of therapies, either by amelioration of the effect of the lesion or by replacement of dopamine in the target area.

This chapter has detailed two, well tested and effective methods for unilateral lesioning of the rat dopamine system. The coordinates used have been developed over many years and are used by a number of laboratories in the field. However, there is a considerable variation in the coordinates used, and the results of a brief scan of papers published in 2009/2010 are shown in Table 3 for reference.

Table 3
MFB lesion coordinates in the literature 2009/2010

Coordinates (mm): reference	A=	L=	V=	Nose bar
Single injection coordinates				
Iczkiewicz et al. (15)	−4.8	+2.0	−8.0	−3.3
Suzuki et al. (16)	−4.8	+1.5	−7.8	N/A
Li et al. (17)	−4.4	−1.4	−7.8	N/A
Shim et al. (18)	−4.0	−1.5	−8.5[a]	N/A
Lehmkuhle et al. (19)	−4.4	−1.2	−7.5	N/A
Goren et al. (20)	−5.0	−2.0	−7.4	N/A
Mertens et al. (21)	−3.2	−1.5	−8.7[a]	N/A
Wang et al. (22)	−5.1	+2.0	−7.2	N/A
Gu et al. (23)	−4.4	−1.2	−7.8	N/A
Jadavji et al. (24)	−4.0	−1.5	−8.5[a]	N/A
Chung et al. (25)	−4.5	+0.9	−7.5	N/A
Warraich et al. (26)	−4.4	±1.0	−8.0	N/A
Pierucci et al. (27)	−5.2	+2.0	−8.1	N/A
Parr-Brownlie et al. (28)	−4.4	−1.8	−8.3[a]	N/A
Song et al. (29)	−4.5	+0.9	−7.5	N/A
Jimenez et al. (30)	−4.0	−1.2	−8.4[a]	−4.5
Avila et al. (31)	−4.4	+1.2	−8.3[a]	N/A
Casteels et al. (32)	−4.8	+2.1	−7.2	N/A
Double injection coordinates				
Nikkhah et al. (33)	−4.0	−0.8	−8.0	+3.4
	−4.4	−1.2	−7.2	−2.3
Rauch et al. (34)	−4.0	−0.8	−7.5	+3.4
	−4.4	−1.2	−7.8	−2.4
Jungnickel et al. (35)	−4.0	−0.8	−8.0	+3.4
	−4.4	−1.2	−7.8	−2.5
Silvestrin et al. (36)	−4.4	−1.8	−8.8[a]	−3.3
	−4.0	−1.6	−9.0[a]	−3.3
Bordia et al. (37)	−4.0	+0.75	−8.0	+3.4
	−4.4	+1.2	−7.8	−2.4

[a]Marked depths are reported, or assumed, to be from skull surface

Acknowledgements

Our experiments in this field are supported by grants from the UK Medical Research Council, Parkinson's UK, and the European Union Seventh Framework TransEUro, Replaces and NeuroStemCell programmes.

References

1. Ungerstedt U (1968) 6-Hydroxy-dopamine induced degeneration of central monoamine neurons. Eur. J. Pharmacol. 5: 107–110

2. Bjorklund A, Wiklund L, Descarries L (1981) Regeneration and plasticity of central serotoninergic neurons: a review. J. Physiol (Paris) 77: 247–255

3. Zigmond MJ, Stricker EM (1973) Recovery of feeding and drinking by rats after intraventricular 6-hydroxydopamine or lateral hypothalamic lesions. Science 182: 717–720

4. Kirik D, Rosenblad C, Bjorklund A (1998) Characterization of behavioral and neurodegenerative changes following partial lesions of the nigrostriatal dopamine system induced by intrastriatal 6-hydroxydopamine in the rat. Exp.Neurol. 152: 259–277

5. Sauer H, Oertel WH (1994) Progressive degeneration of nigrostriatal dopamine neurons following intrastriatal terminal lesions with 6-hydroxydopamine: a combined retrograde tracing and immunocytochemical study in the rat. Neuroscience 59: 401–415

6. Kelly PH, Roberts DC (1983) Effects of amphetamine and apomorphine on locomotor activity after 6-OHDA and electrolytic lesions of the nucleus accumbens septi. Pharmacol. Biochem. Behav. 19: 137–143

7. Ungerstedt U (1971) Striatal dopamine release after amphetamine or nerve degeneration revealed by rotational behaviour. Acta Physiol Scand. Suppl 367: 49–68

8. Ungerstedt U (1971) Postsynaptic supersensitivity after 6-hydroxy-dopamine induced degeneration of the nigro-striatal dopamine system. Acta Physiol Scand. Suppl 367: 69–93

9. Dunnett SB, Bjorklund A, Stenevi U, Iversen SD (1981) Grafts of embryonic substantia nigra reinnervating the ventrolateral striatum ameliorate sensorimotor impairments and akinesia in rats with 6-OHDA lesions of the nigrostriatal pathway. Brain Res. 229: 209–217

10. Dowd E, Monville C, Torres EM, Wong LF, Azzouz M, Mazarakis ND, Dunnett SB (2005) Lentivector-mediated delivery of GDNF protects complex motor functions relevant to human Parkinsonism in a rat lesion model. Eur.J.Neurosci. 22: 2587–2595

11. Kirik D, Georgievska B, Rosenblad C, Bjorklund A (2001) Delayed infusion of GDNF promotes recovery of motor function in the partial lesion model of Parkinson's disease. Eur. J. Neurosci. 13: 1589–1599

12. Torres EM, Dunnett SB (2007) Amphetamine induced rotation in the assessment of lesio ns and grafts in the unilateral rat model of Parkinson's disease. Eur. Neuropsycho-pharmacol. 17: 206–214

13. Bjorklund A, Dunnett SB, Stenevi U, Lewis ME, Iversen SD (1980) Reinnervation of the denervated striatum by substantia nigra transplants: functional consequences as revealed by pharmacological and sensorimotor testing. Brain Res. 199: 307–333

14. Torres EM, Monville C, Lowenstein PR, Castro MG, Dunnett SB (2005) Delivery of sonic hedgehog or glial derived neurotrophic factor to dopamine-rich grafts in a rat model of Parkinson's disease using adenoviral vectors. Increased yield of dopamine cells is dependent on embryonic donor age. Brain Res. Bull. 68: 31–41

15. Iczkiewicz J, Broom L, Cooper JD, Wong AM, Rose S, Jenner P (2010) The RGD-containing peptide fragment of osteopontin protects tyrosine hydroxylase positive cells against toxic insult in primary ventral mesencephalic cultures and in the rat substantia nigra. J. Neurochem.

16. Suzuki K, Okada K, Wakuda T, Shinmura C, Kameno Y, Iwata K, Takahashi T, Suda S, Matsuzaki H, Iwata Y, Hashimoto K, Mori N (2010) Destruction of dopaminergic neurons in the midbrain by 6-hydroxydopamine decreases hippocampal cell proliferation in rats: reversal by fluoxetine. PLoS. One. 5: e9260

17. Li Y, Huang XF, Deng C, Meyer B, Wu A, Yu Y, Ying W, Yang GY, Yenari MA, Wang Q (2010) Alterations in 5-HT2A receptor binding in

various brain regions among 6-hydroxydopamine-induced Parkinsonian rats. Synapse 64: 224–230

18. Shim JS, Kim HG, Ju MS, Choi JG, Jeong SY, Oh MS (2009) Effects of the hook of Uncaria rhynchophylla on neurotoxicity in the 6-hydroxydopamine model of Parkinson's disease. J. Ethnopharmacol. 126: 361–365

19. Lehmkuhle MJ, Bhangoo SS, Kipke DR (2009) The electrocorticogram signal can be modulated with deep brain stimulation of the subthalamic nucleus in the hemiparkinsonian rat. J. Neurophysiol. 102: 1811–1820

20. Goren B, Mimbay Z, Bilici N, Zarifoglu M, Ogul E, Korfali E (2009) Investigation of neuroprotective effects of cyclooxygenase inhibitors in the 6-hydroxydopamine induced rat Parkinson model. Turk. Neurosurg. 19: 230–236

21. Mertens B, Massie A, Michotte Y, Sarre S (2009) Effect of nigrostriatal damage induced by 6-hydroxydopamine on the expression of glial cell line-derived neurotrophic factor in the striatum of the rat. Neuroscience 162: 148–154

22. Wang S, Zhang QJ, Liu J, Wu ZH, Wang T, Gui ZH, Chen L, Wang Y (2009) Unilateral lesion of the nigrostriatal pathway induces an increase of neuronal firing of the midbrain raphe nuclei 5-HT neurons and a decrease of their response to 5-HT(1A) receptor stimulation in the rat. Neuroscience 159: 850–861

23. Gu S, Huang H, Bi J, Yao Y, Wen T (2009) Combined treatment of neurotrophin-3 gene and neural stem cells is ameliorative to behavior recovery of Parkinson's disease rat model. Brain Res. 1257: 1–9

24. Jadavji NM, Metz GA (2009) Both pre- and post-lesion experiential therapy is beneficial in 6-hydroxydopamine dopamine-depleted female rats. Neuroscience 158: 373–386

25. Chung EK, Chen LW, Chan YS, Yung KK (2008) Downregulation of glial glutamate transporters after dopamine denervation in the striatum of 6-hydroxydopamine-lesioned rats. J. Comp Neurol. 511: 421–437

26. Warraich ST, Allbutt HN, Billing R, Radford J, Coster MJ, Kassiou M, Henderson JM (2009) Evaluation of behavioural effects of a selective NMDA NR1A/2B receptor antagonist in the unilateral 6-OHDA lesion rat model. Brain Res. Bull. 78: 85–90

27. Pierucci M, Di M, V, Benigno A, Crescimanno G, Esposito E, Di Giovanni G (2009) The unilateral nigral lesion induces dramatic bilateral modification on rat brain monoamine neurochemistry. Ann. N.Y. Acad. Sci. 1155: 316–323

28. Parr-Brownlie LC, Poloskey SL, Bergstrom DA, Walters JR (2009) Parafascicular thalamic nucleus activity in a rat model of Parkinson's disease. Exp. Neurol. 217: 269–281

29. Song L, Kong M, Ma Y, Ba M, Liu Z (2009) Inhibitory effect of 8-(3-chlorostryryl) caffeine on levodopa-induced motor fluctuation is associated with intracellular signaling pathway in 6-OHDA-lesioned rats. Brain Res. 1276: 171–179

30. Jimenez A, Bonastre M, Aguilar E, Marin C (2009) Effect of the metabotropic glutamate antagonist MPEP on striatal expression of the Homer family proteins in levodopa-treated hemiparkinsonian rats. Psychopharmacology (Berl) 206: 233–242

31. Avila I, Reilly MP, Sanabria F, Posadas-Sanchez D, Chavez CL, Banerjee N, Killeen P, Castaneda E (2009) Modeling operant behavior in the Parkinsonian rat. Behav. Brain Res. 198: 298–305

32. Casteels C, Lauwers E, Baitar A, Bormans G, Baekelandt V, Van Laere K (2010) In vivo type 1 cannabinoid receptor mapping in the 6-hydroxydopamine lesion rat model of Parkinson's disease. Brain Res. 1316: 153–162

33. Nikkhah G, Rosenthal C, Falkenstein G, Roedter A, Papazoglou A, Brandis A (2009) Microtransplantation of dopaminergic cell suspensions: further characterization and optimization of grafting parameters. Cell Transplant. 18: 119–133

34. Rauch F, Schwabe K, Krauss JK (2010) Effect of deep brain stimulation in the pedunculopontine nucleus on motor function in the rat 6-hydroxydopamine Parkinson model. Behav. Brain Res. 210: 46–53

35. Jungnickel J, Kalve I, Reimers L, Nobre A, Wesemann M, Ratzka A, Halfer N, Lindemann C, Schwabe K, Tollner K, Gernert M, Grothe C (2011) Topology of intrastriatal dopaminergic grafts determines functional and emotional outcome in neurotoxin-lesioned rats. Behav. Brain Res. 216: 129–135

36. Silvestrin RB, de Oliveira LF, Batassini C, Oliveira A, Souza TM (2009) The footfault test as a screening tool in the 6-hydroxydopamine rat model of Parkinson's disease. J. Neurosci. Methods 177: 317–321

37. Bordia T, Campos C, Huang L, Quik M (2008) Continuous and intermittent nicotine treatment reduces L-3,4-dihydroxyphenylalanine (L-DOPA)-induced dyskinesias in a rat model of Parkinson's disease. J. Pharmacol. Exp. Ther. 327: 239–247

38. Torres EM, et al. (2011) Increased efficacy of the 6-hydroxydopamine lesion of the median forebrain bundle in small rats, by modification of the stereotaxic coordinates. J. Neurosci. Methods 200: 29–35

Chapter 14

6-OHDA Toxin Model in Mouse

Gaynor A. Smith and Andreas Heuer

Abstract

The unilateral 6-hydroxydopamine mouse model has received considerable attention of late as a model complementary to the hemi-parkinsonian rat. Although both species are similar in nature, there are significant differences between the two when conducting stereotaxic surgery, such as anaesthesia maintenance, technical procedure and differences in lesion co-ordinates. In the present chapter, we therefore discuss detailed methods, problems and suitability of mouse lesion techniques. Mice are also more prone to high post-lesion mortality rates and weight loss, therefore requiring more vigilant care. We describe basic behavioural tests that determine the level of dopaminergic cell death in mice, namely: drug-induced and spontaneous rotations, elevated beam test, staircase test, cylinder test, corridor, and rotarod for assessing lesion-induced deficits in mice. A number of other tests used to assess the rat model cannot however be adapted for use in the mouse.

Key words: 6-OHDA, Mouse, Motor asymmetry, Behaviour

1. Introduction

Mice can be rendered parkinsonian with the use of several toxins including 6-hydroxydopamine (6-OHDA) and 1-methyl-4-phenyl-1,2,3,6-tetrahydropyridine (MPTP), which cause the partial or complete destruction of the dopamine pathway (1) (for details on the use of MPTP, the reader should refer to chapter by Petzinger et al., Jackson and Jenner and Redmond, this volume). The commonality of the unilateral 6-OHDA lesioned rat, from its development by Ungerstedt in 1968 to date, is owed to its diverse use in motor, cognitive and therapeutic tests and hence remains the "gold standard" model to examine Parkinson's disease (PD) experimentally (2). Although the efficacy of the MPTP lesion in rodents has been shown to be dependent on factors including species, gender, strain and age of the animal (3, 4), 6-OHDA can be used to produce relatively stable lesions. Although MPTP creates a bilateral

Emma L. Lane and Stephen B. Dunnett (eds.), *Animal Models of Movement Disorders: Volume I*, Neuromethods, vol. 61,
DOI 10.1007/978-1-61779-298-4_14, © Springer Science+Business Media, LLC 2011

depletion of the nigro-striatal pathway in mice, this lesion is rarely complete and animals display a significant degree of spontaneous recovery after the lesion. Further, the highly toxic nature of MPTP requires implementation of stringent health and safety procedures. The major advantage of 6-OHDA as a toxin in mouse models of PD is that it can be applied unilaterally and be delivered stereotaxically into each sub-component of the nigro-striatal pathway (substantial nigra pars compacta (SNpc), medial forebrain bundle (MFB) and striatum (Str)). A unilateral model has the major advantage that each animal can serve as its own within-subject control (side ipsilateral and contralateral to the lesion) in addition to between-subject effects (lesioned animals and untreated/sham-operated controls). The majority of tests that have been used in the unilateral 6-OHDA rat model of PD utilise the side bias that is induced by a unilateral lesion (see chapters by Torres and Dunnett, this volume and Lindgren and Lane in volume II of this series) therefore, the translation of the unilateral mouse model has received more interest in recent years. The effectiveness of some of these behavioural tests designed for the rat is now being translated to the 6-OHDA mouse; however, further optimisation is needed. The further development of the unilateral 6-OHDA mouse is much lauded as it will allow behavioural comparisons to other genetic mice with PD traits and will provide a host for the transplantation for mouse-derived cell lines.

There are a number of key issues to consider when translating rat surgery, behavioural testing and therapeutics to the mouse. Behavioural tests indicating the presence of complete or partial lesions have shown conflicting results in the mouse model (5–7). New sensory-motor and motor behavioural indicators of nigro-striatal damage and methodological optimising to enhance survival in the mouse model will enhance the efficacy of the model. In the present chapter, we describe the unilateral 6-OHDA mouse model with reference to the rat model. This chapter focuses on differences between the mouse and rat models regarding surgery, rotational assessment and behavioural testing.

2. Methods

2.1. Surgery

Unilateral lesions to the nigro-striatal pathway in the mouse are usually aimed at one of the three sites, the SNpc, from which dopaminergic cell bodies reside which send out their projections to innervate the striatum, the MFB or the striatum itself (see Note 1). Lesions to each of these areas will result in mildly different neurotoxic models of PD and these have different uses behaviourally and histochemically. While lesions to the MFB will result in the greatest degree of degeneration in the ventral mesencephalon (SNpc and parts of the ventral tegmental area), terminal lesions, where the toxin is injected into the striatum, will more likely result in partial lesions (or complete lesions of selected regions of the striatum). The former is generally considered a late stage model of the disease, whereas the latter will resemble the earlier stages of the disorder. The toxin is injected via stereotaxic co-ordinates into the mouse brain to ensure consistent placement of the lesion cannula (see Table 1 for variation in striatal co-ordinates between groups).

Table 1
Parameters of mouse 6-OHDA lesions in published studies of the unilateral striatal lesion model

Conc. (μg/μl)[a]	Volume (μl)	Stereotaxic co-ordinates (mm)[b]	Strain	Sex	Reference
16	4	n/r	n/r	M	(31)
4	4	AP = +5 (to occipital structure), ML = −2.2, DV = −3.5	Swiss	M	(32)
4	4	AP = +0.5, ML = −2.4, DV = −3.1	C57Bl/6	M/F	(33)
3.32	4	AP = +4.8 (Lambda), ML = −1.7, DV = −3.0 and −2.7 (dura)	MNRI	M	(34)
4	4	AP = +3 (to occipital structure), ML = −2.2, DV = −3.5	Swiss	M	(35)
4	4	AP = +0.5, ML = −2.4, DV = −3.1	C57Bl/6	M	(36)
2	2	AP = +0.4, ML = −1.8, DV = −3.5	C57Bl/6	M	(37)
3	2 × 2	(1) AP = +1.0, ML = −2.1, DV = −2.9 (2) AP = +0.3, ML = −2.3, DV = −2.9	C57Bl/6	M	(38)
3	2 × 2	(1) AP = +1.0, ML = −2.1, DV = −2.9 (2) AP = +0.3, ML = −2.3, DV = −2.9	C57Bl/6	M	(38)
3	2 × 2	(1) AP = +1.0, ML = −2.1, DV = −2.9 (2) AP = +0.3, ML = −2.3, DV = −2.9	C57Bl/6	M	(39)
6	n/r	AP = +0.8, ML = −1.0, DV = −2.5	A/J	M/F	(40)
3	3.9 μg or 5.4 μg	AP = −1.2, ML = ±1.1, DV = −5.0, with incisor bar set at ±0.0	CBA	F	(6)

(continued)

**Table 1
(continued)**

Conc. (µg/µl)[a]	Volume (µl)	Stereotaxic co-ordinates (mm)[b]	Strain	Sex	Reference
2.5	2×2	AP = +0.5, ML = +2.4, DV = −4.0 and −3.0 (dura)	C57BL/6J/ OlaHsd	M	(41)
3	2×2	(1) AP = +1.0, ML = − 2.1, DV = −3.2 (2) AP = +0.3, ML = −2.3, DV = −3.2	C57Bl/7	M	(42)
2	2	AP = +0.4, ML = −1.8, DV = −3.5	C57Bl/6	M	(43)
4	2	AP = +0.8, ML = −1.9, DV = −2.6	C57Bl/6	M	(44)
2	1	AP = +0.4, ML = − 1.8, DV = −3.5	C57Bl/6/ SWISS/ svl29	M	(45)

AP anterior-posterior, *ML* medial-lateral, *DV* dorsal-ventral, *M* male, *F* female, *n/r* not reported
[a] *Conc.*: Free-base concentration of 6-OHDA·HBr
[b] Co-ordinates are calculated relative to Bregma, unless otherwise stated

As the focus of this chapter is on the unilateral mouse model and its deviation from the standard rat model, the readers is advised to refer to the instruction on rat surgery (discussed in Chap. 13) with respect to making up the toxin.

1. The surgery suite needs to be equipped for mouse surgery, including anaesthesia, analgesia, heated recovery blanked/box, scalpel (size 10) and suture (thread 5.0), drill, micro-drive pump, syringe and lesion cannula.

2. A Hamilton syringe is connected to the lesion cannula via polyethylene tubing. The Hamilton syringe is driven by a micro-drive pump.

3. Make up 6-hydroxydopamine to the required concentration (calculate from free-base weight).

4. Place mouse in induction chamber to induce anaesthesia (using 1–2% Isoflurane in a mixture of 2:1 oxygen to nitrous oxide as carrier gas (see Note 1)).

5. Shave head and place animal into the stereotaxic apparatus.

6. Clean the head and preferably swap with antiseptic (e.g. iodine).

7. Make incision with scalpel along the midline of the animals' skull.

8. Using the scalpel, scrape away dura and connective tissue so that skull surface can be seen clearly.

9. Locate bregma on the animals head and place the tip of the drill on bregma and adjust co-ordinates on the stereotaxic apparatus and gently burr hole.

10. Load lesion cannula with 2–3 μl of fresh toxin and gently lower lesion cannula through the burr hole to the required depth and infuse the lesion site slowly with the toxin and leave needle in place to allow the toxin to diffuse for additional 3 min.

11. Slowly remove lesion cannula to prevent drawback of the toxin, clean the animals' head and carefully suture up to ensure swift recovery.

12. Provide analgesia (Metacam s.c. injection or Paracetamol (1 mg/ml) added to the drinking water) and where necessary gather advice from local veterinary.

2.2. Post-Operative Care

Post-operative mortality is generally much higher in mice compared to rats receiving a similar lesion, as the result of aphagia and adipsia from both the decline in general health and spontaneous turning behaviour in the home cage; however, this can be overcome by intensive animal husbandry (see Note 1). It is essential to provide wet food/mash and injections of glucose-saline if required and monitor the animals' weight on a daily basis for 14 days post-lesion surgery. Housing mice 2 per cage following surgery is also helpful to prevent bullying and facilitate feeding. Bullying in mice is common after surgery and if noticed mice should be separated permanently.

3. Assessment of the Lesion and Resulting Motor Impairments

3.1. Drug-Induced Rotation

As described in chapter by Dunnett and Torres, this volume animals that receive unilateral dopamine lesions to the nigro-striatal pathway will respond with a typical turning behaviour when challenged with various drugs that modulate the dopamine system (8–10). Dopamine losses of 90% or more in the rat relate to a peak rotation rate of 5/7 turns per minute following an amphetamine or apomorphine challenge (11–13), quantified with the use of automated rotometer bowls. In mice, a 31.5-fold change in directional bias is seen following an apomorphine challenge (1, 5). Generally, amphetamine- and apomorphine-induced rotations have been shown to correlate highly with the percent of TH positive cell loss from the SNpc and are useful in determining the severity of a SNpc 6-OHDA lesion in mice (7). Non-pharmacologically induced, spontaneous rotation in a novel environment in mice was first noted by observations in an open field environment (14), where dopamine released by mild anxiety causes turning toward the ipsilateral direction.

3.2. Recording Rotational Behaviour in Mice

1. The video recorder is set up on a tripod facing downward above the laboratory floor.

2. Each mouse is placed in standard laboratory beakers (H: 14 cm, D: 11.5 cm) and allowed to explore environment for 20 min. Mice are injected with 0.05 mg/kg apomorphine (s.c.) or 2.5 mg/kg methamphetamine (i.p.) and placed back into the beaker (see Fig. 1b). Drugs are dissolved into physiological saline and used immediately. Apomorphine is dissolved with the addition of 0.2% ascorbate and kept on ice in the dark before use.

3. The investigator will then record peak rotation over 20 min immediately following apomorphine injection and 20 min after methamphetamine injection (see Note 2 for variations in the dose and administration procedures that can be used for each drug).

4. These recordings then can be analysed post hoc and full rotations can be counted using fast-forward video playback reporting net rotations over the period (No. of clockwise rotations – No. of anticlockwise rotations).

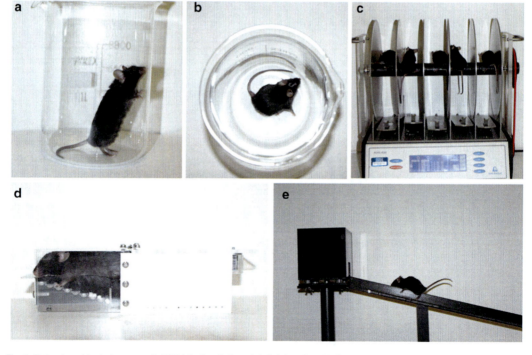

Fig. 1. Behavioural tests to assess 6-OHDA lesion-induced deficit in mice. (**a**) Cylinder test. (**b**) Apomorphine-induced rotation. (**c**) Rotarod. (**d**) Staircase. (**e**) Elevated beam.

3.3. Assessment of Motor and Sensory-Motor Deficits

Unilateral dopamine loss in the 6-OHDA lesioned rat model produces a number of deficits (sensory, motor and cognitive) on the contralateral side including spontaneous circling toward the (ipsilateral) lesioned side, reduced use of the contralateral forelimb and a neglect of the contralateral side. For general behavioural considerations, see Note 3. The unilateral lesioned rat is simplistic, flexible and allows for unilateral, partial and full denervation of the nigrostriatal tract, enabling its suitability for a diverse range of motor, cognitive and therapeutic strategy driven tests (9, 15–18). We describe below a selected range of behavioural hand tests that assess these deficits in mice. There also remain a number of rat tests used for the assessment of unilateral lesion motor deficits that are not easily translated to the mouse (see Note 4).

3.3.1. Elevated Beam Test

The elevated beam test analyses the motor coordination and dexterity of right and left hindlimbs and forelimbs (see Fig. 1e). In beam-based studies, deficits have been reported in hemi-parkinsonian rats, mice transgenic for Huntington's disease (19) and of late, unilateral lesioned mice (Heuer and Smith, unpublished observations). Here, foot slips are recorded on ipsilateral and contralateral sides on an inclined slope of decreasing width, until a designated point just before the animal reaches an enclosed platform at the top (for specific construction details, see Note 5). A foot slip is defined as paw misplacement where all digits slip off or entirely miss the ledge of the beam causing a slouch in the body posture of the mouse.

1. The elevated beam is placed on a bench top such that the investigator can walk around the apparatus and see both sides.

2. Training day 1: mice are placed pointing outward on the end of the elevated beam and have to balance and turn to face inward. Each mouse is required to do this three times.

3. Training day 2: each mouse is positioned to balance on the inner ledge to traverse the beam to at least to the half way point three times. Mice are also placed in the enclosed space at the top of the beam and allowed to explore it for 3 min.

4. Training day 3: mice must traverse the whole beam and reach the enclosed platform at the top twice.

5. On one side of the beam, a video recorder is set up on a tripod to record foot slips on this side to be scored post hoc.

6. Test day: mice are required to traverse the beam three times without turning in the opposite direction after the initial turn. The investigator must be on the opposite side of the elevated beam to the video recorder to count foot slips on that side, record the time to transverse the beam and note the direction of the initial turn.

7. For analysis, the two quickest times over a designated 80 cm distance of beam and the corresponding foot-slip counts to

these runs are averaged (foot-slip counts = No. of ipsilateral – No. of contralateral). Hindlimb and forelimb counts can be added together or used independently depending on the experimental question.

*3.3.2. Skilled Reaching/
Staircase Test*

The staircase test, see Fig. 1d, has been developed for objective assessment of independent forelimb use in skilled reaching behaviour (20). The rat staircase apparatus has been scaled down from the rat version and validated for use in the mice (21). The testing apparatus consists of a home box that is connected to a narrow corridor, just wide enough for the animal to crawl into, but not wide enough for the animal to turn around. In the corridor, a double staircase with eight steps per side is separated by a central platform. The animal can climb on the platform and reach down to the steps which are baited using 20 mg precision sucrose pellets. With the exception of the first two steps, pellets can only be retrieved via grasping by the respective arm. Outcome parameters that are assessed during a 15 min session are: total number of pellets eaten; total number of pellets displaced and side bias in the amount of pellets retrieved (for general considerations regarding the use of staircase test, see Note 6).

1. Food restrict animals to 85–90% of their free-feeding weight at least 1 week prior to starting behavioural testing maintaining them at this weight for the duration of testing (see Note 3).

2. On the first day of training only, bait both staircases with as many pellets as possible whilst on all other testing days the steps are baited with two 20 mg precision sucrose pellets per step.

3. Leave the animals for 15 min before putting them back into their home cages.

4. Count and record the number of pellets remaining per step and side.

5. Train the animals on a regular basis until the number of pellets eaten reaches asymptote.

6. Clean apparatus before commencing testing other animals in the same apparatus.

7. Present data as number of pellets eaten (total and side bias); number of pellets replaced (replace but not eaten); maximum distance reached (last step with two or more pellets). Side bias can be expressed as (Pellets retrieved left – Pellets retrieved right)/(Pellets retrieved left + Pellets retrieved right).

3.3.3. Cylinder Test in Mice

Mice can also be tested for paw preference in a novel cylindrical environment (6, 7). A beaker (H: 14 cm, D: 11.5 cm) is placed in front of two mirrors at 90° so that a 360° view can be seen by the observer so that the left/right forelimb bias is recorded (see Fig. 1a).

As with the rat protocol (see Fleming and Schallert, this volume) the session is videotaped and scored post hoc by detailed freeze frame analysis. Simultaneous ipsilateral and contralateral forelimb touches were excluded and mice failing to reach 20 touches were removed from the cylinder after 10 min. Mouse models with lesions to the MFB, SNpc and Striatum typically express a 70% usage of the ipsilateral limb correlating to greater than 85% dopamine depletion (6); comparable to levels recorded in the rat.

3.3.4. Corridor

The corridor test is used to assess deficits in sensory-motor function and proprioception. The mouse is placed in one of the two corridors, with sugar pellet baited pots placed at intervals along the length of the passageway. The apparatus is ergonomically modified for the mouse 2× (60 × 4 cm) from rat-based models (15). An ipsilateral bias of 75% is seen in the mouse when using the corridor correlating with near complete neuronal degeneration of the Striatum and SNpc (7) comparable with deficit levels recorded in the rat (15, 22). As with the rat test (see Fleming and Schallert, this volume), the forelimb bias is noted for the first 20 retrievals from sugar pellets placed in wells along the side of the corridor in either parallel pairs or staggered.

1. Prior to testing, mice are maintained at 85–90% of their free-feeding body weight (problems occurring with the food restriction of mice are discussed in Note 3).

2. Training day 1: sugar pellets (20 mg) are scattered along the floor of the corridor and the mouse is placed at one end to explore the novel environment for 5 min.

3. Training day 2: sugar pellets are placed in 20 pots of a 1 cm diameter (ten next to each wall in either a staggered or parallel fashion) and secured with double-sided sticky tape. The mouse is placed in the same starting location to explore the corridor over 5 min.

4. Test day: the mouse is placed in an empty corridor for 5 min and transferred to a parallel corridor baited with sugar pellet pots. The investigator must note what forelimb the mice use to retrieve each sugar pellet until a total of 20 retrievals has been made or until 5 min has elapsed. Mice may move up and down the corridor any number of times.

5. Total forelimb bias and therefore motor function and sensory neglect are calculated by No. of ipsilateral – No. of contralateral retrievals.

3.3.5. Rotarod

The rotarod test assesses learned motor coordination and balance, requiring the coordinated function of all limbs by the use of a motorised cylinder, turning at either a series of fixed or accelerating angular velocities (see Fig. 1c). Typically deficits manifest in all unilateral lesion models, and significant decreases in the latency to fall correspond to 60% loss of dopaminergic cells or greater within the SNpc (6). The rotarod test does not assess motor asymmetry but is still sensitive to unilateral lesions in rat and mouse (6, 19, 23). There are two protocols that can be used to assess the animals "behaviour, a fixed speed and an accelerating speed protocol." Where the fixed speed has been shown to be more sensitive in the rat (23), the accelerating version does allow for quicker assessment.

1. Animals are trained on rotarod for three daily sessions over three consecutive days. On the first day, the rotation speed is set at a fixed value of 8 rpm for 300 s. When an animal falls off during this period, it is placed back on the rod until the end of the trial. The speed for the second and third days is set to 12 and 24 rpm (see Note 7 for alternate protocols).

2. Baseline data is taken over 1 day allowing the animals three trials on the accelerating version of the rotarod. The rotation is set to increase gradually from 4 to 44 rpm over 300 s.

3. When the animal is placed on the rod, a timer is started and the timer is stopped when the animal falls off. Then the animal is placed back into its home cage and the latency on the rod recorded.

4. Behaviour deviating from the intention of the test as animals jumping off, making full rotations by clinging on to the rod or turning around is scored as invalid trial. In this case, the animal is removed from the rod and the trial is not counted.

5. After testing has finished, the rod is cleaned and dried to ensure a non-slippery surface for the following animal.

6. Average the latency to fall from the best two out of the three conducted trials.

4. Notes

1. *Considerations for surgery in mice.* Lesion surgery is carried out stereotaxically under gaseous anaesthesia in much the same way as the rat model (see Torres and Dunnett, this volume) where surgery is typically performed under a 1–3% Isoflurane in 2:1 oxygen/nitrous oxide gaseous anaesthesia mixture. It should be noted that the margin for error in modulating the anaesthetics for mice is small, as the result of small body mass, and experimenters should be extremely vigilant in monitoring their breathing and making small adjustments to the anaesthetic concentration where necessary. Animals are placed in a stereotaxic frame and cannula placements are determined using the co-ordinates of the 2001 Paxinos and Franklin mouse brain atlas (24). Stainless

steel tooth and ear bars used for this surgery are ergonomically designed for mice. The stereotaxic co-ordinates are adjusted for position along the nigro-striatal tract from landmarks on the skull surface. Standard co-ordinates in our laboratory used for 6-OHDA catecholamine neurotoxic injections are: striatal-terminal lesions: (1) AP = +1.0, ML = −2.1, DV = −2.9; (2) AP = +0.3, ML = −2.3, DV = −2.9, MFB: AP = −1.2, ML = −1.2, DV = −4.75 and SNpc: AP = −3.0, ML = −1.2, DV = −4.5 relative to bregma, with the nose bar set at 0 mm relative to the inter-aural line. Although these co-ordinates have been optimised for use in the C57/Bl6 strain, co-ordinates and background strain vary between groups depending upon supplier, age, gender and weight. Although co-ordinates can easily be calculated from the Paxinos and Watson Atlas, it is advised to pilot lesions before starting to work on a larger cohort. While MFB and SNpc 6-OHDA lesions have been used, the majority of published hemi-parkinsonian mouse studies use unilateral striatal lesions. Co-ordinates used for such lesions, however, are highly variable in the literature (Table 1), although it can be noted that consistency is maintained within groups. In accordance with the standard rat protocol (see Torres and Dunnett, this volume), injections are carried out over 3 min with 3 min allowed for diffusion of the neurotoxin into the surrounding area. For the mouse, unilateral lesions are typically carried out by single or dual injections of 6-OHDA HBr into the right MFB (1 μl), striatum (2 × 1.5 μl) or SNpc (1.5 μl) using a 30-gauge cannula, connected via fine polyethylene tubing to a 10 μl Hamilton syringe, delivered at a 0.5 μl/min flow rate using a micropump. The neurotoxin is typically used at a concentration of 3 μg/μl (calculated from free-base weight) dissolved in a solution of 0.2 mg/ml ascorbic acid in 0.9% sterile saline. Comparisons between studies must also be taken with caution since differences in 6-OHDA dose and volume with striatal lesions (Table 1) impact on the size, location, rate of neurotoxin uptake and hence dopaminergic cell survival. Intensive post-operative care is essential to maintain low mortality rates in mice. Good survival rates are necessary to insure meaningful experimental outcomes and in countries such as the UK death rates of 80% as previously reported would be deemed unacceptable by ethical standards and home office regulations. Typically, mortality rates of 25% or less are achieved through a combination of daily subcutaneous injections with a solution of 0.9% glucose in sterile saline and highly palatable wet mash for at least 2 weeks post-lesion, where daily weight measurements are essential. In general, mortality rates are higher in mice with MFB lesions compared to other lesion types. Health is dependent on the degree on the site of injection and hence the degree of dopaminergic depletion.

2. *Considerations for rotations in hemi-parkinsonian mice.* The dosage and volume of psycho-stimulants is adjusted for lesion type and severity from 0.015 to 0.1 mg/kg (s.c) for apomorphine and 0.63–2.5 mg/kg (i.p) for amphetamine, whilst the concentration is halved in favour of doubling the volume for ease to the investigator and accuracy, i.e. drugs should be given at the volume of 2 ml/kg. The rotational capacity of the mouse is highly variable between studies and the typical apomorphine doses used previously are often up to five times as high, enough to induce stereotypic behaviour and physiological adaptive changes. Up to nine mice can be tested simultaneously using a 3×3 beaker arrangement (with psycho-stimulant-induced rotation, a new mouse is injected every minute and the video recorder kept running throughout). Rotations in mice can also be automatically quantified (as described in Dunnett and Torres, this volume), however, this is more difficult in comparison to the rat as mice tend to escape more easily from their harnesses; therefore, the peak rotation technique is commonly used, although automation has successfully applied in some laboratories (7). The summation of three apomorphine challenges over 3 days is sometimes used to increase the rotational potential of the mouse (7, 25); however, this compromises the subsequent use of the models for pharmacological testing. Spontaneous net rotation can also be recorded within the first 5 min of the habituation phase in beakers used for psycho-stimulant rotation (see Sect. 2.2). Spontaneous rotation is useful where pharmacologically naïve animals are required for subsequent experiments. It should be noted that spontaneous rotation requires a novel environment and where numerous forms of rotation are needed spontaneous testing must be carried out first.

3. *General considerations with respect to behavioural testing.* Mice exhibit a strong freezing response when frightened; it is therefore advised to habituate them to the testing environment before commencing the behavioural test. The testing environment should be kept constant with respect to temperature, humidity, background noise level and smell. When possible, to avoid unintentional bias, all behavioural testing should be conducted in a manner in which the investigator is blind to the treatment group. For all tests where video recording is required ensure that taping is in slow-play mode to ensure adequate frame-by-frame playback when analysing the data (26). Food restriction is of increased difficulty in mice compared to the rat. Body weight must be measured three times a week, but preferably daily. It is advised to separate animals or house them in small groups for the duration of feeding as bullying is common in mice resulting in unequal food consumption. Mice should be food restricted to 85% of their free-feeding weight whilst allowing

for natural growth (growth curves for strains can be obtained through the breeder). In general, precision sugar pellets (20 mg) are used as a reward to motivate animals to perform the given task. A sample of the sugar pellets (e.g. 15/animal) can be provided in the home cage to prevent neophobia during the test.

4. Not all basic hand tests can be translated to the mouse. There are a number of tests designed for the rat which cannot directly be translated to the mouse. The grip strength test automatically assesses the lateralised grip strength between each forelimb, differences up to 83% are usually found following total obliteration of the nigro-striatal tract (27). The efficiency of this test requires the placement of the hind limbs on the bench and grip dexterity of higher order rodents and is therefore unsuitable for mice. The inverted grid test had been proposed as a viable alternative to the grip strength test, although mice with a high degree of neuronal loss have only negligible deficits on this test. Stride length in rats can also be calculated between each step and between forelimbs and hind limbs of the same side to uncover fine motor changes in gait and are often recorded simultaneously on the beam apparatus. Due to the modified nature of the beam used for mice, simultaneous scoring of paw placements is not possible, and when tested separately in a horizontal narrow corridor mice are prone to stop, groom and turn during testing. Motor responses can also be recorded immediately following the stimulation of the whisker, reduced movement capacity in this task reflect somatosensory deficits in the activation of the barrel cortex, however this is not usually tested in mice. As with the sideways adjusted stepping test, difficulties attest in the constraint of small rodents and in stimulating the whisker without visual awareness.

5. *Construction of the elevated beam.* The mouse beam walking protocol is a modified version of the rat beam test, previously used exclusively for rat where two key modifications are apparent, namely: the apparatus is inclined at 20° to encourage them to traverse and the central beam decreases in diameter from 1.5 to 0.5 cm toward the highest point to modulate task difficulty. To stop mice from jumping of the apparatus, the start of the beam must be constructed to be off the floor, we have found that 33 cm off the bench top appears to be sufficient. Soft towels are placed under the beam to dampen any falls that may occur.

6. *Notes on the staircase.* For mice to reach asymptotic performance of the staircase, test for baseline data can take 8–15 days. A high level of performance is advised for baseline data to be able to divide animals into well-matched groups before lesion surgery. Recently, it has been shown that by providing a higher number of pellets per staircase (e.g. 10 instead of 2),

the test becomes more discriminative (28). Furthermore, colouring the pellets using a small quantity of food colouring can provide information about pellets that are eaten and which are displaced by ineffective grasping by the mouse (29). Unfortunately, the available literature on this test in the mouse is sparse and/or has focused on other lesion models than 6-OHDA (21). Particular attention has to be paid to the dimensions of the staircase apparatus as differences in size between mouse strains affect performance in the apparatus more than differences in the rat. When mice can use their head to retrieve pellets from the first 2–3 steps out of the eight steps, it is therefore advised to either adapt the apparatus and/or not to bait the first two steps. Furthermore, mice do perform the task differently to the rat. A rat will move into the corridor and remaining there until all pellets are eaten, whereas a mouse will retrieve one pellet at a time and retreat to the home box to consume the reward.

7. Adaptations and problems with the rotarod test. Rather than using the increasing velocity protocol, the rotarod may alternatively be set at a fixed speed for the 5 min duration and retested at increasing speeds. However, this is generally only used in mice with severe motor phenotypes and in general the sensitivity of fixed speed protocol is insufficient to detect a deficit in unilaterally deficit mice. The acquisition of the task can be recorded to assess motor learning behaviour, in this protocol mice are tested 2–3 times per day over 1–4 training days, before a final test day. 6-OHDA-leisoned rodent models have deficits in learning to co-ordinate movements over the full 5 min, whereas in intact control animals improvement is seen daily. In rats, however, there is a discrimination between accelerating and fixed modes where in the former sensitivity to the extent of the lesion is apparent, but the latter protocol is able to deduce the presence of a lesion compared to intact controls (30). Moreover, in mice, this distinction between protocols has not been made in 6-OHDA unilateral lesioned models and typically the incrementing protocol is often used to increase the difficult of the task due to the higher agility levels in mice.

Acknowledgments

Our experiments in this field are supported by grants from the UK Medical Research Council, the UK Biotechnology and Biological Science Research Council, and the European Union Seventh Framework programme.

References

1. Jacobowitz, D.M., Burns, R.S., Chiueh, C.C., and Kopin, I.J. (1984). N-methyl-4-phenyl-1,2,3,6-tetra-hydropyridine (MPTP) causes destruction of the nigrostriatal but not the mesolimbic dopamine system in the monkey. Psychopharmacol Bull. 20(3): 416–22.

2. Ungerstedt, U. (1968). 6-Hydroxy-dopamine induced degeneration of central monoamine neurons. Eur J Pharmacol. 5(1): 107–10.

3. Sundstrom, E., Fredriksson, A., and Archer, T. (1990). Chronic neurochemical and behavioral changes in MPTP-lesioned C57BL/6 mice: a model for Parkinson's disease. Brain Res. 528(2): 181–8.

4. Sedelis, M., Hofele, K., Auburger, G.W., Morgan, S., Huston, J.P., and Schwarting, R.K. (2000). MPTP susceptibility in the mouse: behavioral, neurochemical, and histological analysis of gender and strain differences. Behavior Genetics. 30(3): 171–82.

5. Mandel, R.J. and Randall, P.K. (1985). Quantification of lesion-induced dopaminergic supersensitivity using the rotational model in the mouse. Brain Res. 330(2): 358–63.

6. Iancu, R., Mohapel, P., Brundin, P., and Paul, G. (2005). Behavioral characterization of a unilateral 6-OHDA-lesion model of Parkinson's disease in mice. Behav Brain Res. 162(1): 1–10.

7. Grealish, S., Mattsson, B., Draxler, P., and Bjorklund, A. (2010). Characterisation of behavioural and neurodegenerative changes induced by intranigral 6-hydroxydopamine lesions in a mouse model of Parkinson's disease. Eur J Neurosci. 31(12): 2266–78.

8. Ungerstedt, U. and Arbuthnott, G.W. (1970). Quantitative recording of rotational behavior in rats after 6-hydroxy-dopamine lesions of the nigrostriatal dopamine system. Brain Res. 24(3): 485–93.

9. Torres, E.M. and Dunnett, S.B. (2007). Amphetamine induced rotation in the assessment of lesions and grafts in the unilateral rat model of Parkinson's disease. Eur Neuropsychopharmacol. 17(3): 206–14.

10. Voigtlander, P.F. and Moore, K.E. (1971). Nigro-striatal pathway: stimulation-evoked release of (3 H)dopamine from caudate nucleus. Brain Res. 35(2): 580–3.

11. Costall, B., Fortune, D.H., and Naylor, R.J. (1976). Biphasic changes in motor behaviour following morphine injection into the nucleus accumbens (proceedings). Br J Pharmacol. 57(3): 423P.

12. Costall, B., Marsden, C.D., Naylor, R.J., and Pycock, C.J. (1976). The relationship between striatal and mesolimbic dopamine dysfunction and the nature of circling responses following 6-hydroxydopamine and electrolytic lesions of the ascending dopamine systems of rat brain. Brain Res. 118(1): 87–113.

13. Schmidt, J. and Westermann, K.H. (1980). Effects of preceding sensibilization by reserpine and haloperidol on toxicity of dopaminergic agonists. Arch Toxicol Suppl. 4: 479–81.

14. Cenci, M.A. and Lundblad, M. (2007). Ratings of L-DOPA-induced dyskinesia in the unilateral 6-OHDA lesion model of Parkinson's disease in rats and mice. Curr Protoc Neurosci. Chapter 9: Unit 9 25.

15. Dowd, E., Monville, C., Torres, E.M., and Dunnett, S.B. (2005). The Corridor Task: a simple test of lateralised response selection sensitive to unilateral dopamine deafferentation and graft-derived dopamine replacement in the striatum. Brain Res Bull. 68(1–2): 24–30.

16. Olsson, M., Nikkhah, G., Bentlage, C., and Bjorklund, A. (1995). Forelimb akinesia in the rat Parkinson model: differential effects of dopamine agonists and nigral transplants as assessed by a new stepping test. J Neurosci. 15(5 Pt 2): 3863–75.

17. Schallert, T., Fleming, S.M., Leasure, J.L., Tillerson, J.L., and Bland, S.T. (2000). CNS plasticity and assessment of forelimb sensorimotor outcome in unilateral rat models of stroke, cortical ablation, parkinsonism and spinal cord injury. Neuropharmacology. 39(5): 777–87.

18. Schallert, T., Whishaw, I.Q., Ramirez, V.D., and Teitelbaum, P. (1978). Compulsive, abnormal walking caused by anticholinergics in akinetic, 6-hydroxydopamine-treated rats. Science. 199(4336): 1461–3.

19. Brooks, S., Higgs, G., Janghra, N., Jones, L., and Dunnett, S.B. (2010). Longitudinal analysis of the behavioural phenotype in YAC128 (C57BL/6J) Huntington's disease transgenic mice. Brain Research Bulletin.

20. Montoya, C.P., Campbell-Hope, L.J., Pemberton, K.D., and Dunnett, S.B. (1991). The "staircase test": a measure of independent forelimb reaching and grasping abilities in rats. J Neurosci Methods. 36(2–3): 219–28.

21. Baird, A.L., Meldrum, A., and Dunnett, S.B. (2001). The staircase test of skilled reaching in mice. Brain Res Bull. 54(2): 243–50.

22. Dowd, E., Monville, C., Torres, E.M., Wong, L.F., Azzouz, M., Mazarakis, N.D., and Dunnett, S.B. (2005). Lentivector-mediated delivery of GDNF protects complex motor functions relevant to human Parkinsonism in a

rat lesion model. Eur J Neurosci. 22(10): 2587–95.

23. Monville, C., Torres, E.M., and Dunnett, S.B. (2006). Comparison of incremental and accelerating protocols of the rotarod test for the assessment of motor deficits in the 6-OHDA model. J Neurosci Methods. 158(2): 219–23.

24. Paxinos, G., and Franklin, K.B.J., ed. The Mouse Brain in Stereotaxic Coordinates. 2 ed. 2001, Academic Press: London.

25. Winkler, J.D. and Weiss, B. (1986). Reversal of supersensitive apomorphine-induced rotational behavior in mice by continuous exposure to apomorphine. The Journal of Pharmacology and Experimental Therapeutics. 238(1): 242–7.

26. Gaspar, P., Febvret, A., and Colombo, J. (1993). Serotonergic sprouting in primate MTP-induced hemiparkinsonism. Exp Brain Res. 96(1): 100–6.

27. Dunnett, S.B., Torres, E.M., and Annett, L.E. (1998). A lateralised grip strength test to evaluate unilateral nigrostriatal lesions in rats. Neurosci Lett. 246(1): 1–4.

28. Cordeiro, K.K., Jiang, W., Papazoglou, A., Tenorio, S.B., Dobrossy, M., and Nikkhah, G. (2010). Graft-mediated functional recovery on a skilled forelimb use paradigm in a rodent model of Parkinson's disease is dependent on reward contingency. Behav Brain Res. 212(2): 187–95.

29. Kloth, V., Klein, A., Loettrich, D., and Nikkhah, G. (2006). Colour-coded pellets increase the sensitivity of the staircase test to differentiate skilled forelimb performances of control and 6-hydroxydopamine lesioned rats. Brain Res Bull. 70(1): 68–80.

30. Monville, C., Torres, E.M., and Dunnett, S.B. (2006). Comparison of incremental and accelerating protocols of the rotarod test for the assessment of motor deficits in the 6-OHDA model. J Neurosci Methods.

31. Von Voigtlander, P.F. and Moore, K.E. (1973). Turning behavior of mice with unilateral 6-hydroxydopamine lesions in the striatum: effects of apomorphine, L-DOPA, amanthadine, amphetamine and other psychomotor stimulants. Neuropharmacology. 12(5): 451–62.

32. Torello, M.W., Czekajewski, J., Potter, E.A., Kober, K.J., and Fung, Y.K. (1983). An automated method for measurement of circling behavior in the mouse. Pharmacology, Biochemistry, and Behavior. 19(1): 13–7.

33. Randall, P.K. (1984). Lesion-induced DA supersensitivity in aging C57BL/6J mice. Brain Research. 308(2): 333–6.

34. Brundin, P., Isacson, O., Gage, F.H., Prochiantz, A., and Bjorklund, A. (1986). The rotating 6-hydroxydopamine-lesioned mouse as a model for assessing functional effects of neuronal grafting. Brain Research. 366(1–2): 346–9.

35. Thermos, K., Winkler, J.D., and Weiss, B. (1987). Comparison of the effects of fluphenazine-N-mustard on dopamine binding sites and on behavior induced by apomorphine in supersensitive mice. Neuropharmacology. 26(10): 1473–80.

36. Mandel, R.J. and Randall, P.K. (1990). Bromocriptine-induced rotation: characterization using a striatal efferent lesion in the mouse. Brain Research Bulletin. 24(2): 175–80.

37. Bensadoun, J.C., Deglon, N., Tseng, J.L., Ridet, J.L., Zurn, A.D., and Aebischer, P. (2000). Lentiviral vectors as a gene delivery system in the mouse midbrain: cellular and behavioral improvements in a 6-OHDA model of Parkinson's disease using GDNF. Experimental Neurology. 164(1): 15–24.

38. Lundblad, M., Picconi, B., Lindgren, H., and Cenci, M.A. (2004). A model of L-DOPA-induced dyskinesia in 6-hydroxydopamine lesioned mice: relation to motor and cellular parameters of nigrostriatal function. Neurobiology of Disease. 16(1): 110–23.

39. Lundblad, M., Usiello, A., Carta, M., Hakansson, K., Fisone, G., and Cenci, M.A. (2005). Pharmacological validation of a mouse model of l-DOPA-induced dyskinesia. Experimental Neurology. 194(1): 66–75.

40. Liang, Q., Smith, A.D., Pan, S., Tyurin, V.A., Kagan, V.E., Hastings, T.G., and Schor, N.F. (2005). Neuroprotective effects of TEMPOL in central and peripheral nervous system models of Parkinson's disease. Biochemical Pharmacology. 70(9): 1371–81.

41. Pavon, N., Martin, A.B., Mendialdua, A., and Moratalla, R. (2006). ERK phosphorylation and FosB expression are associated with L-DOPA-induced dyskinesia in hemiparkinsonian mice. Biological Psychiatry. 59(1): 64–74.

42. Santini, E., Valjent, E., Usiello, A., Carta, M., Borgkvist, A., Girault, J.A., Herve, D., Greengard, P., and Fisone, G. (2007). Critical involvement of cAMP/DARPP-32 and extracellular signal-regulated protein kinase signaling in L-DOPA-induced dyskinesia. J. Neurosci. 27(26): 6995–7005.

43. Alvarez-Fischer, D., Blessmann, G., Trosowski, C., Behe, M., Schurrat, T., Hartmann, A., Behr, T.M., Oertel, W.H., Hoglinger, G.U., and Hoffken, H. (2007). Quantitative ((123) I)FP-CIT pinhole SPECT imaging predicts striatal dopamine levels, but not number of nigral neurons in different mouse models of Parkinson's disease. NeuroImage. 38(1): 5–12.

44. Richter, F., Hamann, M., and Richter, A. (2008). Moderate degeneration of nigral neurons after repeated but not after single intrastriatal injections of low doses of 6-hydroxydopamine in mice. Brain Research. 1188: 148–56.

45. Alvarez-Fischer, D., Henze, C., Strenzke, C., Westrich, J., Ferger, B., Hoglinger, G.U., Oertel, W.H., and Hartmann, A. (2008). Characterization of the striatal 6-OHDA model of Parkinson's disease in wild type and alpha-synuclein-deleted mice. Experimental Neurology. 210(1): 182–93.

Chapter 15

Rotation in the 6-OHDA-Lesioned Rat

Stephen B. Dunnett and Eduardo M. Torres

Abstract

Rotation is one of the most widely used tests in behavioural neuroscience. It is designed to detect motor turning and side biases in animals with lesions of basal ganglia circuits of the brain, and most notably following unilateral dopamine-depleting 6-OHDA lesions of the nigrostriatal bundle. When activated by stimulant or dopamine agonist drugs, rats and mice turn vigorously in circles for the duration of drug activity, yielding a reliable and easily quantifiable measure of the extent of the lesion and mechanism of drug action on the unbalanced dopamine system in the host brain. The design of automated rotometer test apparatus is discussed in detail, with advice for selecting the features most appropriate for different experimental applications. The selection of drugs, doses, time frames and testing protocols are then described, again in terms of the general principles that are readily adapted to particular experimental applications. Finally, a series of notes consider frequently asked questions related to practical issues involving drug selection, drug sensitisation, multiple testing and test spacing, and theoretical issues related to why and how does an animal rotate, and conditioning effects on the rotation response.

Key words: Rotation, Rotometer, Skilled turning, Conditioned turning, Amphetamine, Apomorphine, 6-OHDA lesions, Unilateral lesions, Motor asymmetry

1. Introduction

Rotation was first introduced as a systematic behavioural outcome in experimental animals in the context of measuring the turning response of rats with unilateral lesions of ascending nigrostriatal dopamine pathways. Whereas, bilateral lesions of forebrain dopamine systems yield an akinetic animal that engages in no spontaneous or goal-directed behaviour, unilateral lesions yield a neglect of contralateral side of the body and a tendency to initiate all voluntary responses towards the ipsilateral side (1). If such animals are then treated with a stimulant drug such as amphetamine, that stimulates locomotor activity in an intact rat, a rat with unilateral lesions of the nigrostriatal bundle turns vigorously in tight head-to-tail

Emma L. Lane and Stephen B. Dunnett (eds.), *Animal Models of Movement Disorders: Volume I*, Neuromethods, vol. 61, DOI 10.1007/978-1-61779-298-4_15, © Springer Science+Business Media, LLC 2011

circles (1, 2), a response known as "rotation". Ungerstedt and Arbuthnott (3) introduced a mechanised apparatus to allow the automated recording of rotation, and this provided the basis for the first systematic characterisation of the rotation response to amphetamine. Thus, animals exhibit a dose-dependent turning for the duration of drug action that can reach as high as 15 or more turns per minute at the peak response of a high (5 mg/kg) dose of d- or meth-amphetamine (3, 4). With the introduction of the 6-OHDA-lesion method that allowed selective destruction of the nigrostriatal dopamine pathway (see Chap. 13 by Torres, this volume), along with sensitive histofluorescence and immunocytochemical tools for characterising, the simplicity and reliability of rotation has become one of the most widely used tools in behavioural neuroscience (1, 5, 6).

As a reliable animal model of the motor disorder resulting from dopamine deficiency in Parkinson's disease, rotation is widely used to characterise potential anti-parkinsonian drugs that act on or interact with dopamine receptors and postsynaptic signalling mechanisms, to understand the dyskinetic side effects of therapeutic drugs, to evaluate the efficacy of surgical treatments such as lesions or deep brain stimulation of striatal output pathways, and to provide a functional readout for novel therapeutic strategies involving neuroprotection, gene and cell replacement therapy. More generally, following its validation in the unilateral dopamine lesion model in rats, rotation can also provide a useful functional outcome in studies of striatal lesion models of Huntington's disease and stroke, and is increasingly being utilised in mice as well as rats.

2. Apparatus

2.1. Manual Observation

In early studies, animals were simply observed, described and rated for postural asymmetry, spontaneous turning, and turning following stimulation whether by a drug or a stressor (such as tail pinch) (2, 7). More quantitative measurement can be undertaken by direct observation of individual rats in standard circular bowls (we used waste paper bins in our first experiments) and manually counting turns per minute, usually with a sampling procedure (e.g. for 1 min in every 10 min) to allow several animals to be tested at the same time (8). In contrast to the financial cost of manufacturing or purchasing an automated rotation system, this low tech solution can be implemented with minimal equipment, training or delay at any time and is suitable, where only a single experiment or testing of only a few animals on a limited number of occasions are envisaged. However, there are significant non-financial costs in terms of experimenter patience and boredom, and if rotation is to be used as a recurrent experimental tool, an automated rotometer system should be considered.

2.2. Rotometer

The term "rotometer" was introduced by Ungerstedt and Arbuthnott (3) in their first description of an automated device to monitor and record rotation in unilateral 6-OHDA-lesioned rats tested under amphetamine. Their apparatus comprised testing rats in a hemispheric bowl (Fig. 1a). The rat wore a metal loop harness connected to a flexible wire connected to a cam and pivot transducer that turned within a microswitch connected to a counter for recording left and right 360° turns. Crucially, the cam and pivot assembly was located in the centre of curvature of the bowl, so that as the animal moved about the flex in the tether wire remained approximately constant avoiding the connector getting stretched or the animal getting tangled (see Fig. 1a).

In the early years, most laboratories fabricated similar rotometer apparatus in the university or medical workshop, following the same basic principles and several protocols for the mechanical and electronic components of rotometers have been published (9–13), and nowadays a range of commercial systems are available, at a price (Table 1).

Nevertheless, although all the rotometers are based on a similar design principle, the different rotometers differ in a number of details, each of which can prove important for their use and utility. The following points should be considered, whether purchasing a commercial system or designing a system to be custom-built.

2.2.1. Test Chamber

Rats are versatile animals, and are very adept at escape from confinement, in particular, when under the influence of stimulant drug. Lost data due to animals escaping from the apparatus were a significant problem when using the open hemispheric bowls used in early designs. One approach is to try and fit a lid, or to design a globe-shaped bowl with only a small opening for the wire at the

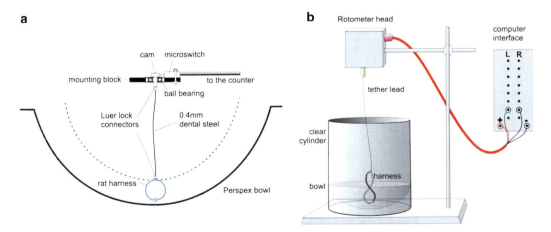

Fig. 1. Schematic illustration of the prototypical rotometer apparatus, **(a)** design based on the original Ungerstedt and Arbuthnott (3) description. **(b)** Design as currently used in the Brain Repair Group laboratory (redrawn from (1) with permission).

Table 1
Commercial suppliers of rotometer and locomotor tracking apparatus

Manufacturer	Model	Country	Website	Chamber	Connector	Harness	Species	Parallel[a]
Rotometer systems								
Accuscan Instruments	Rotometer	USA	www.accuscan-usa.com	Bowl, Cyl	Tethered	Metal loop	Rat, mouse	192
Bioseb	Rotameter	France	www.bioseb.com	Bowl, Cyl	Tethered	Velcro	Rat, mouse	15
Columbus Instruments	Rota-Count-8	USA	www.colinst.com	Bowl, Cyl	Tethered	Adjustable	Rat	32
Harvard Instruments	Rotameter	USA	www.harvardapparatus.com	Bowl, Cyl	Tethered	Velcro	Rat	15
Kinder Scientific	Motor-Rotometer	Germany	www.kinderscientific.com	Encl bowl	Tethered			
Panlab	LE 902 Rotometer	Spain	www.panlab-sl.com	Bowl, Cyl	Tethered	Velcro	Rat	15
San Diego Instruments	Rotometer System	USA	www.sandiegoinstruments.com	Encl bowl	Tethered	Adjustable	Rat	16
TSE	TSE Rotameter System	Germany	www.tse-systems.com	Bowl, Cyl	Tethered	Velcro jacket	Rat, mouse	32
Animal tracking systems								
Noldus	EthoVision XT	USA	www.noldus.com	Open field	Video tracker	Freely moving	Rat, mouse	96
Panlab	SMART system	Spain	www.panlab-sl.com	Open field	Video tracker	Freely moving	Rat, mouse	16
Med Associates	MED-OFA-MS	USA	www.med-associates.com	Open box	Photo beams	Freely moving	Rat, mouse	Individual
Ugo Basile	Rotometer	Italy	www.ugobasile.com	Cylinder	Magnet	Freely moving	Mouse	Individual

Cyl cylinder, *Encl bowl* enclosed globe-shaped bowl
[a] Maximum number of rotometers monitored in parallel by control apparatus

top of the sphere. In each case, ease of putting the animals in the bowls and removing them afterwards (especially, if retrieving a still hyperactive animal) becomes an issue. Third is to use a cylinder with high straight sides, which combines difficulty of escape and ease of changing animals. This can result in excessive flexing of the tether wire between times when the animal rears and when it extends into the outer corners of the floor, with an increased chance of the animal getting tangled in the tether. In our laboratory, we use polystyrene mixing bowls, containing a handful of sawdust, fitting snugly inside a tall transparent Perspex cylinder (Fig. 1b). This allows a good compromise for maintaining a centre of curvature, ease of change and ease of cleaning and changing sawdust (including the use of different materials for environmental conditioning and sensitisation of rotational responses). The reason for using a transparent cylinder is so that the animals' behaviour can be monitored by observation or video, in particular, to allow rating of L-dopa- and graft-induced dyskinesias simultaneous with recording rotation behaviour, under alternative pharmacological and cell-based treatment regimes (14) (see also Lindgren and Lane, this volume).

2.2.2. Animal Harness

The harness must be both comfortable for the animal, not restricting its movements, but also designed to avoid escape. Early systems involve a bent or curved wire belt, positioned tightly around the chest behind the elbows. However, these are difficult to fix reliably so that the animal cannot escape without over-tightening. The most secure systems are those involving a fabric jacket with holes to accommodate the forelimbs, and generally closed using Velcro. These are difficult for the animal to escape from, but can be quite fiddly or time consuming to fit, especially, if needing to change multiple animals in a batch between runs. A compromise between these two approaches is to use a broad fabric belt passed around the chest behind the elbows and fixed with velcro fastening for a firm, but not overtight adjustment. The harness or jacket may either be permanently fixed to the tether wire, in which case the animal must be placed in the harness while standing within reach of the tether wire very close to the rotometer, or the harness may be attached by a removable connector (such as a syringe type Luer connector) which then allows the rats to be harnessed more conveniently at a nearby bench top. In our laboratory, we have adopted a slightly different strategy, using a large rubber band (US: elastic band) – in the UK the standard Post Office band for bundling letters is ideal (Fig. 1b). The band is simply and efficiently applied by passing around the tips of the five fingers of one hand, fanning opening the hand to stretch the band, quickly passing over the head of the rat held in the other hand, and positioning snugly behind the elbows. The band is then connected to the tether by a small crocodile clip soldered to the end of the tether wire to allow a low pressure-elasticated grip, which can be adjusted to the size of the rat by pinching a fold of the band in the clip.

2.2.3. Cam and Pivot Assembly

Free turning of the pivot assembly is essential, and this joint should be a ball race to allow long-term free turning without resistance. Equally, the tether wire has to be reasonably flexible, but with no opportunity for twisting; piano wire is frequently used. Early assemblies used a cam and mechanical switch assembly, but this is better replaced by an optical switch that again applies no resistance. The most simple systems involve a single switch to detect 360° turns, but modern systems invariably involve the ability to detect smaller angles of turn, for example, by recording photo-beam interruptions by multiple holes or notches in a turning disc. There is a distinct advantage in a system with two closely adjacent switches or sensors that allows the direction of turning to be distinguished and recorded separately. In our laboratory, although our system has the capacity to record quarter turns, drug-induced rotation in 6-OHDA-lesioned animals is reliable, consistent and almost exclusively in one direction, so we routinely collect only data on 360° turns, separately for ipsilateral and contralateral direction, and then analyse the data in terms of net difference measures. However, there are occasions where it may be relevant to record quarter turns or to analyse data separately for the two directions, in particular, when looking at the effects of spontaneous turning, small lesions, intact animals, stereotyped responses, or other lesions with as yet uncharacterised turning morphology.

2.2.4. Data Recording

Early systems used separate electromechanical counters attached to each pivot assembly, with manual recording of data registers. Nowadays, rotometer control systems are almost exclusively micro-electronic, typically with a small laptop computer or dedicated processor receiving inputs from an interface panel connected to multiple rotometer pivot heads. The choice of hardware and software is endless; in established behavioural laboratories, the logical solution is to extend other standard control equipment to cover rotometer data recording; alternatively, the alternative commercial systems will provide the control unit and associated software, and this typically constitutes a significant proportion of the total system cost. Rotation experiments often involve some dozens of experimental animals, often with the individual drug tests running for several hours, and experimental designs often involving multiple tests on separate days. Consequently, it is considerably more efficient to test and run animals in a bank of rotometer bowls, rather than individually. So how many systems should be purchased/constructed and run in parallel? As a rule of thumb, a bank of eight is suitable for a small laboratory running rotation experiments on an occasional or regular basis. Just purchasing four bowls would do, but then most of the cost is taken up in the control equipment.

In our laboratory, we run a bank of 16 rotometers from one control computer, since we frequently have ten or more people engaged in long-term experiments with rotation as a primary readout. A significant factor in scaling up is the time it takes to change animals between test runs. With a relatively slow onset and long-acting drug such as amphetamine, it is no problem setting up the batch software, changing eight animals and then hitting the run key. However, with 16 changes which may take up to 5–10 min, or with a rapid onset and relatively short acting drug (such as the mixed dopamine receptor agonist apomorphine), then it is better to have hardware and software designed to start each bowl separately (perhaps with a remote trigger) within the same batch run. The capacity of some systems to provide for 32 or more parallel inputs is really only applicable to a pharmaceutical scale of high-throughput screening.

2.2.5. Species Issues

With the rise in a range of genetic models of disease, in particular in mice, there is considerable emerging interest in methods to evaluate motor behaviours, including rotation, in mice (15) (see also Brooks or Smith and Heuer, this volume). Following unilateral 6-OHDA lesions, mice rotate in response to the same stimulant drugs as do rats (16, 17). Indeed, most equipment suppliers offer scaled down bowls, jackets and tethers for mice. However, mice are not just little rats. It has proved more difficult to achieve stable and reliable unilateral lesions in mice, it is considerably more difficult physically to fit them with rotation harnesses, and they are much more efficient at escaping, getting tangled or reacting badly to the constraint. So just because multiple mouse rotometers are available on the market, it must not be assumed that they work reliably or well. Our advice would be to only purchase such a device if you have clear-cut independent evidence from an experienced mouse behaviour laboratory that the particular system does work without complication as advertised. In our laboratory, we have sought to adapt two consecutive systems to mice without great success. Consequently, for our current experiments in mice, we have reverted to an updated variant on the manual observation and sampling strategy (see Sect. 2.1), by video recording the test sessions with the treated mice placed without harnesses in our existing bowls and transparent cylinders, and then manually documenting individual animals' turning on ten times fast forward video play back (see Heuer and Smith, this volume).

2.3. Automated Freely Moving Systems

Although most rotation monitors have been based on the rotometer principle of mechanical recording of turning in tethered animals, a number of alternative systems developed for locomotor tracking and movement analysis have employed software to detect and quantify the turning behaviour and rotation of freely moving animals in open arenas. The techniques of animal tracking can involve a number of different technologies. Ethovision XT and the Panlab SMART systems (see Table 1) involve video analysis of locomotor paths using image analysis to detect the location of the rat by contrast against its background, to determine the point centre of gravity, and to record locomotor tracks within the test arena. The location of the rat in the arena can, alternatively, be monitored by movement of the animal through an electromagnetic field or by breaks in a Cartesian array of photocell detectors (Med Associate, Table 1). Although such open field systems are largely used to characterise parameters such as locomotor activity, speed of movement, stereotyped run paths, rearing, etc., "rotation" can usually be extracted by software algorithms analysing locomotor paths in small or wide angled circles. In most such systems explicit turning of the body axis rather than turning of the point centre of gravity in space is difficult, although the Ethovision TriWise software does permit inference of true rotation by extracting a three point head-back-rump longitudinal axis from the recorded images. Such systems have much broader application to a wide range of behavioural classes, but they are not as reliabile, precise or as easy to use as a mechanical rotometer, when the specific purpose of the research programme is to use the rotational asymmetry of unilateral lesioned animals as the primary outcome in designed experiments.

Another freely moving system is worth considering. In the light of the practical difficulties of testing mice in conventional rotometer apparatus, Ugo Basile have introduced a mouse rotometer system (see Table 1) based on detecting the magnetic field disturbance of a small magnet affixed to a mouse, either by subcutaneous implantation or taping to the base of the tail. The mice are then allowed to turn freely in an experimental test cylinder, accurately recording explicit body turns in either direction, and with similar readouts identical to that used in a conventional rotometer. The authors have no personal experience of this system, and so recommend gathering further evidence for direct validation, but it appears a priori to address the major difficulties of extending automated rotation testing to mice.

3. Methods

Once the apparatus is selected, the other features of testing rotation in unilateral lesioned rats follow the general principles well established in behavioural neuroscience.

3.1. Experimental Lesions

Unilateral lesions of the nigrostriatal dopamine pathway are typically made by stereotaxic placement, and slow injection of the toxin via a fine stainless steel glass cannula implanted in the appropriate target – the substantia nigra, the nigrostriatal bundle or the striatal terminals (see Torres and Dunnett, this volume). Lesioned animals will exhibit torsional and postural biases as soon as they rouse from anaesthesia, and frequently exhibit significant levels of spontaneous contralateral turning associated with dopamine release from degenerating terminals over several days following lesion (18). Thereafter, overt spontaneous turning in lesioned animals typically wanes, and the side of the lesion is not obvious to casual inspection. However, spontaneous turning can readily be elicited in animals with established lesions by activating the animals by means of a stressor (e.g. tail pinch, sudden loud noise, placing on a cold surface or in a shallow tray of crushed ice cold water) (7, 18). At this stage, turning is triggered by dopamine release from the intact side, and the turning is in the ipsilateral direction.

3.2. Drugs and Doses   The conventional stimulant drug used to activate stable rotation in the unilateral 6-OHDA rat is amphetamine. Amphetamines stimulate dopamine release and block reuptake from intact dopamine terminals in the forebrain. In the absence of a corresponding dopaminergic activation on the lesioned side, the resulting asymmetry induces rotation in the ipsilateral direction (3, 4) (Fig. 2a). Conversely, dopamine agonists (such as apomorphine or bromocriptine) induce rotation in the opposite, contralateral direction, which Ungerstedt hypothesised was due to the development of supersensitivity of the dopamine receptors located postsynaptically on striatal neurons of the denervated side (19) (Fig. 2b), an inspired prediction from the behavioural data that has subsequently been

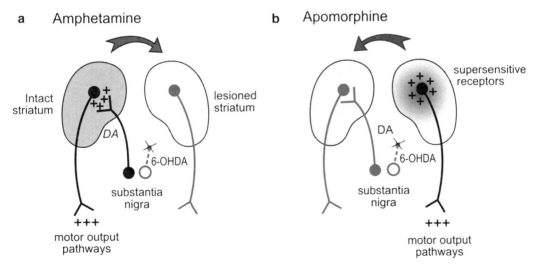

Fig. 2. (**a**) Effect of amphetamine in the bilateral nigrostriatal system viewed schematically from above, acting presynaptically to induce heightened functional output from the intact striatum on the side contralateral to the lesion and producing an ipsilateral motor bias. (**b**) Schematic effect of apomorphine to induce a heightened functional output on the ipsilateral side and contralateral turning bias (redrawn from (1) with permission).

well validated in pharmacokinetic, gene expression and receptor binding studies (1). The leading anti-parkinsonian drug L-dopa also produces contralateral rotation (20), believed to be due to its conversion to dopamine in both hemispheres, but inducing a greater activation on the side of receptor supersensitivity. Notably, receptor agonists and L-dopa are effective in inducing strong rotation even at very low doses that are sub-threshold for activating intact animals, further strengthening the behavioural evidence for the supersensitivity interpretation.

3.3. Rotation Testing

The experimental protocol for running rotation tests is straightforward, once the associated apparatus and software are installed and fully tested. Assuming a bank of eight rotometers run by a single control computer or dedicated processor.

Preparation

1. Prior to each days testing, thoroughly test the apparatus to ensure that:

   The bowls are clean and contain fresh sawdust.
   The tethers and harnesses are intact and without kinks.
   Each pivot turns freely and without resistance.

2. Start up the control system and load the rotometer programme. Enter parameters for a 5-min test run to record data in 1 min time bins. Run the programme, and manually check that turns of the wires (two or three times in each direction) are accurately registered both in real time on screen and in the data file saved at the end of the 5-min test run. Trouble-shoot any faults.

3. In the laboratory, dissolve samples of the drug in Eppendorf or small glass vials, at the appropriate experimental doses and in the appropriate buffers. Amphetamine can be dissolved in sterile saline and used for the whole day keeping the vial on the bench top at room temperature. Apomorphine rapidly oxidises and so must be weighed into several Eppendorf vials, kept on crushed ice along with a separate vial of the 0.1% ascorbate saline solvent, and mixed fresh prior to each set of injections and test run.

4. Collect the batch of animals for experimental testing. It is most convenient to place all animals in a single large cage for transfer to the test room, and for rapid coordinated injection as a batch, rather than trying to sort out the individual animals for testing from group cages at the time of injection. Animals for several test runs can be collected in several large cages and kept in a rack outside the test room if more convenient.

For Each Test Run

5. Run the start programme and enter parameters for the test run, which will typically include: experiment name; test details; drug

and dose; test duration (e.g. 90 or 120 min for an amphetamine test, 60 min for an apomorphine test); number of time bins into which the data is collected (e.g. 9 x 10 min bins, 12 x 5 min bins, etc.); animal code in each test bowl; and file name for storing the data output. Prime the programme as ready to start. Most computer programmes will have some parameters preset: whether to record data in 360° or 90° turns; separation of data into clockwise and anticlockwise directions; data file type (e.g. Excel format).

6. Inject the rats with the drug (i.p. for amphetamine, s.c. in the neck for apomorphine), fix the animal in the harness, and place in the bowl relevant to the individual animal code.

7. Run the data recording programme. In some systems, a switch at each rotometer station allows data recording to commence as soon as each individual animal is placed into its bowl, for accurate synchronisation with the time of drug delivery. In our system, bowls may be started singly or in multiples using on screen commands on the computer. In other systems, the only option is to hit the Enter button on the computer keyboard, starting all recordings simultaneously. In the latter situation, if time is of the essence (e.g. with a short acting drug), it can be helpful to have two people working together to inject and place in the bowls more efficiently. Alternatively, on such test runs, it may be appropriate to only test two or four animals at a time. The precise arrangement also depends on speed and efficiency of making the injections, which comes with experience, and the particular drug and test parameters used.

8. At the end of the test run, return to the test room, remove the animals from the bowls, ensure that data are properly collected and stored (electronically and/or by hard copy printout), note any unusual events, such as an animal's escape from a bowl, and prepare for the next test batch.

4. Notes

1. Which amphetamine?
   There are multiple isomers and salts of amphetamine available for experimental research. The most potent isoform is meth-amphetamine, which produces peak dose–response of turning rates in excess of 20 turns/min approx. 1 h after injection of 7.5 mg/kg. Thereafter, turning rates decline as the animals are increasingly locked in focal stereotyped responding, which competes for expression with open turning. Drug action lasts with high rates of turning up to 4–5 h, which can be stressful for the rats, and most labs experience occasional deaths during

testing with such high doses. As a consequence, most laboratories typically use the somewhat less potent *d*-amphetamine at doses of either 2.5 or 5.0 mg/kg i.p. for screening motor asymmetries after 6-OHDA lesion. Since the peak turning response occurs 30–60 min after injection, for efficiency of screening, most labs use 90 or 120 min test sessions to capture peak response, and then return the animals to home cages; however, in experiments in which the rotation behaviour is itself the subject of investigation rather than an empirical screen of lesion deficits, testing for at least 240 min is advised to capture the full period of the drug response (3).

2. Which agonist?

Apomorphine is the most widely used experimental mixed dopamine receptor agonist used in rotation tests, although there has been increasing attention to use of rotation for screening the pharmacokinetics and functional effects of a range of more selective agonists for clinical use. In particular, different drugs can act over very different time frames: in a study comparing the interactions of several agonists with L-dopa, piribedil showed a peak response at 10 min and had largely waned at 30 min, whereas the corresponding times for apomorphine were 10/45 min, ropinirole 60/120 min, L-dopa 60/180 min and bromocriptine 4 h continuing up to 12–18 h (14). Consequently, close attention must be paid to the time frame of experimental sessions in order to adequately capture peak responses and their interactions. Attention must also be paid to the fact that only very low doses of agonist are required to stimulate highly supersensitive receptors; 0.05 mg/kg s.c. apomorphine is more than sufficient. It is not uncommon to see in the literature studies using 0.5 or 1.0 mg/kg doses of apomorphine to test rotation in unilateral 6-OHDA rats, whereas this dose exerts a major activation of dopamine receptors of the intact as well as the lesion side, confounding the interpretation of any rotation asymmetries that are observed over the competing marked stereotypies that such doses induce.

3. Validity of rotation as a surrogate variable.

A major reason for the popularity of rotation as a measure is that the rate of turning in response to both amphetamine and apomorphine correlates very highly with the extent of dopamine depletion induced by the lesion. Correlation rates between the behavioural response and postmortem assay can often be as high as $r^2 = 0.8$ (21–24). Such a level of correlation between structural change and functional outcome is unusual in behavioural neuroscience – where large groups and statistical comparisons to detect effects are the norm – and makes rotation a powerful non-invasive tool to screen the effectiveness of lesions (and of any therapeutic intervention) in vivo.

4. Use of rotation for screening and group allocation.

   In the light of the closeness of this structural–functional association, rotation has been widely used to allocate animals into counterbalanced groups based on comparable or matched lesions, and to screen out animals with incomplete lesions. This is particularly important in long-term experimental studies of neuroprotective, trophic or cell therapies. The dopamine system is itself remarkably plastic such that if the lesions are incomplete spared neurons exhibit a considerable capacity for compensation and recovery of function (25). As a consequence, if partial lesioned animals are included it can be extremely difficult to separate treatment effects from long-term spontaneous recovery. This confound is efficiently and effectively removed by restricting experimental and control treatments to animals pre-determined to have a functionally complete lesion (e.g. turning in excess of 7 turns/min over 90 min following injection of 5 mg/kg amphetamine (8, 24)).

5. Why does an animal rotate?

   It remains unclear why an animal rotates away from the side of higher postsynaptic activation at dopamine targets in the two striata. Kelly and Moore (26) proposed a two-process model, judiciously supported by a series of elegant focal and combined lesions, in which dopamine activation in the nucleus accumbens (ventral striatum) provides a non-lateralised locomotor activation of the animal, and imbalance between the dorsal striata on the two sides determined the direction of turning; activation of both systems is then required to translate a postural, motor and attentional bias into active rotation. But why should they rotate? It might simplistically be thought that heightened activation of the left striatum, would produce increased vigour in stepping of the right limbs, which would induce turning to the left – the opposite to that which is actually observed.

   Alternatively, it has been suggested that animals neglect the side contralateral to the lesions, and so direct all voluntary responding due to a sensory or attentional bias to the ipsilateral side, but reaction time experiments that separate sensory and attentional from motor biases in these animals clearly indicate that the 6-OHDA lesion induces a failure to initiate responses to the contralateral side rather than a neglect of input from that side (27). Thus, although the phenomenon are clear, reliable and reproducible, the underlying systems foundation of rotation remains ambiguous.

6. How does an animal rotate?

   A number of authors have provide elegant descriptions of the behavioural movement sequences involved in rotating rats, and it is clear that the precise morphology of turning under

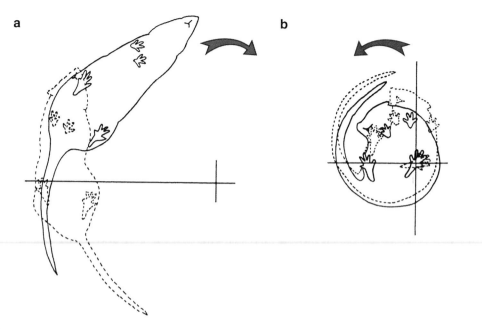

Fig. 3. Amphetamine (**a**) and apomorphine (**b**) not only induce turning asymmetries in opposite directions, but there are very different morphologies to the asymmetric postures during turning to the two drugs (drawings of ventral views of turning rats from Koshikawa (28) with permission).

amphetamine and apomorphine is quite distinct (28) (Fig. 3). However, such qualitative characterisation have not taken us any further in resolving why an activated animal actually rotates, beyond the general systems model of Kelly and Moore (26).

7. How often to test rotation.

And now to further practical matters. Although there has been some concern that intensive repeat rotation tests can cause further sensitisation of the receptors, confounding the long tracking of lesion and therapeutic changes, in our experience testing the animals once a week, for example, alternating between amphetamine and apomorphine on alternative weeks, provides stable long-term lesion baselines against which the longitudinal profile of experimental change can be monitored (29).

8. When to test rotation.

Unlesioned animals exhibit little rotation, and it is not useful to collect a baseline measure prior to making unilateral 6-OHDA lesions. The amphetamine rotation response stabilises over 2–4 weeks after lesion, and we would typically conduct two repeat tests at 2 and 4 weeks to establish baseline lesion data on which to match animals into groups for alternative therapeutic or experimental treatments. Similarly, following a treatment such as cell implantation or gene therapy, the effects of which may be expected to evolve over time, testing once every 3 weeks over 12 weeks or longer provides a good balance between the power of multiple measurements to establish reliability and the loss of power associated with making strict statistical corrections for multiple comparisons.

9. Associative conditioning.

Whereas, at a pharmacological level, significant receptor sensitisation is only apparent with massed testing (such as by daily treatments of agonist drugs), rotation can exhibit behavioural conditioning to its environmental context. Thus, if a unilateral 6-OHDA rat is repeatedly exposed to amphetamine in a rotometer bowl, in each case expressing marked ipsilateral rotation, then that same animal will exhibit a modest level of ipsilateral turning following a saline injection in that bowl, but not in another neutral environment (30, 31). Thus, the behavioural response to the drug is associated with the environment, which can enhance the response to the drug and induce a mild drug-like response in an untreated animal.

10. Operant conditioning.

Naïve animals can also be trained to turn in circles in one direction for reward, which is associated with a rise in dopamine release in the contralateral striatum. Moreover, unilateral 6-OHDA lesions abolish the animals' ability to learn or maintain conditioned contralateral turning, while enhancing their ability to perform the ipsilateral conditioned response. Thus, reinforced turning responses are also dependent upon the integrity of intact dopamine signalling pathways, although the voluntary emission of lateralised responding in such paradigms probably involves quite separate processes to the more stereotyped turning observed in conventional drug-induced rotation.

Acknowledgements

Our experiments in this field are supported by grants from the UK Medical Research Council, Parkinson's UK, and the European Union Seventh Framework programme.

References

1. Dunnett SB (2005) Motor functions of the nigrostriatal dopamine system: studies of lesions and behaviour. In: Dunnett SB, Bentivoglio M, Björklund A, Hökfelt T (eds) Handbook of Chemical Neuroanatomy. Vol. 21. Dopamine. Elsevier, pp 235–299

2. Andén N-E, Dahlström A, Fuxe K, Larsson K (1966) Functional role of the nigro-neostriatal dopamine neurons. Acta Pharmacol. Toxicol. 24: 263–274

3. Ungerstedt U, Arbuthnott GW (1970) Quantitative recording of rotational behaviour in rats after 6-hydroxydopamine lesions of the nigrostriatal dopamine system. Brain Res. 24: 485–493

4. Ungerstedt U (1971) Striatal dopamine release after amphetamine or nerve degeneration

revealed by rotational behaviour. Acta Physiol. Scand. suppl. 367: 49–68

5. Glick SD, Jerussi TP, Fleisher LN (1976) Turning in circles: the neuropharmacology of rotation. Life Sci. 18: 889–896

6. Pycock CJ (1980) Turning behaviour in animals. Neuroscience 5: 461–514

7. Ungerstedt U, Butcher LL, Butcher SG, Andén N-E, Fuxe K (1969) Direct chemical stimulation of dopaminergic mechanisms of the neostriatum of the rat. Brain Res. 14: 461–471

8. Björklund A, Dunnett SB, Stenevi U, Lewis ME, Iversen SD (1980) Reinnervation of the denervated striatum by substantia nigra transplants: functional consequences as revealed by pharmacological and sensorimotor testing. Brain Res. 199: 307–333

9. Schwarz RD, Stein JW, Bernard P (1978) Rotometer for recording rotation in chemically or electrically stimulated rats. Physiol. Behav. 20: 351–354

10. Jerussi TP (1982) A simple, inexpensive rotometer for automatically recording the dynamics of circling behavior. Pharmacol. Biochem. Behav. 16: 353–357

11. Kehinde LO, Buraimoh-Igbo LA, Ude OU, Makanjuola RO (1984) Electronic rotameter for quantitative evaluation of rotational behaviour in rats after unilateral lesions of the nigrostriatal dopamine system. Med. Biol. Eng. Comput. 22: 361–366

12. Hudson JL, Levin DR, Hoffer BJ (1993) A 16-channel automated rotometer system for reliable measurement of turning behavior in 6-hydroxydopamine lesioned and transplanted rats. Cell Transplant. 2: 507–514

13. Heredia-Lopez FJ, Bata-Garcia JL, Alvarez-Cervera FJ, Gongora-Alfaro JL (2002) A novel rotometer based on a RISC microcontroller. Behav. Res. Methods Instrum. Comput. 34: 399–407

14. Lane EL, Dunnett SB (2010) Pretreatment with dopamine agonists influence L-dopa mediated rotations without affecting abnormal involuntary movements in the 6-OHDA lesioned rat. Behav. Brain Res. 213: 66–72

15. Brooks SP, Dunnett SB (2009) Tests to assess motor phenotype in mice: a user's guide. Nat. Rev. Neurosci. 10: 519–529

16. Pycock C, Tarsy D, Marsden CD (1975) Inhibition of circling behavior by neuroleptic drugs in mice with unilateral 6-hydroxydopamine lesions of the striatum. Psychopharmacologia 45: 211–219

17. Grealish S, Mattsson B, Draxler P, Björklund A (2010) Characterisation of behavioural and neurodegenerative changes induced by intranigral 6-hydroxydopamine lesions in a mouse model of Parkinson's disease. Eur. J. Neurosci. 31: 2266–2278

18. Ungerstedt U (1976) 6-hydroxydopamine-induced degeneration of the nigrostriatal dopamine pathway: the turning syndrome. Pharmacol. Ther. B 2: 37–40

19. Ungerstedt U (1971) Postsynaptic supersensitivity after 6-hydroxydopamine-induced degeneration of the nigro-striatal dopamine system. Acta Physiol. Scand. suppl. 367: 69–93

20. Heikkila RE, Shapiro BS, Duvoisin RC (1981) The relationship between loss of dopamine nerve terminals, striatal (3H)spiroperidol binding and rotational behavior in unilaterally 6-hydroxydopamine-lesioned rats. Brain Res. 211: 285–292

21. Costall B, Marsden CD, Naylor RJ, Pycock CJ (1976) The relationship between striatal and mesolimbic dopamine dysfunction and the nature of circling responses following 6-hydroxydopamine and electrolytic lesions of the ascending dopamine systems of rat brain. Brain Res. 118: 87–113

22. Dunnett SB, Hernandez TD, Summerfield A, Jones GH, Arbuthnott GW (1988) Graft-derived recovery from 6-OHDA lesions: specificity of ventral mesencephalic graft tissues. Exp. Brain Res. 71: 411–424

23. Hefti F, Melamed E, Sahakian BJ, Wurtman RJ (1980) Circling behavior in rats with partial, unilateral nigro-striatal lesions: effects of amphetamine, apomorphine, and DOPA. Pharmacol. Biochem. Behav. 12: 185–188

24. Schmidt RH, Björklund A, Stenevi U, Dunnett SB, Gage FH (1983) Intracerebral grafting of neuronal cell suspensions. III. Activity of intrastriatal nigral suspension implants as assessed by measurements of dopamine synthesis and metabolism. Acta Physiol. Scand. suppl. 522: 19–28

25. Zigmond MJ, Abercrombie ED, Berger TW, Grace AA, Stricker EM (1990) Compensations after lesions of central dopaminergic neurons: some clinical and basic implications. Trends Neurosci. 13: 290–296

26. Kelly PH, Moore KE (1976) Mesolimbic dopaminergic neurones in the rotational model of nigrostriatal function. Nature 263: 695–696

27. Carli M, Evenden JL, Robbins TW (1985) Depletion of unilateral striatal dopamine impairs initiation of contralateral actions and not sensory attention. Nature 313: 679–682

28. Koshikawa N (1994) Role of the nucleus accumbens and the striatum in the production of turning behaviour in intact rats. Rev. Neurosci. 5: 331–346

29. Brown VJ, Dunnett SB (1989) Comparison of adrenal and fetal nigral grafts on drug-induced rotation in rats with 6-OHDA lesions. Exp. Brain Res. 78: 214–218

30. Annett LE, Reading PJ, Tharumaratnam D, Abrous DN, Torres EM, Dunnett SB (1993) Conditioning versus priming of dopaminergic grafts by amphetamine. Exp. Brain Res. 93: 46–54

31. Carey RJ (1986) Conditioned rotational behaviour in rats with unilateral 6-hydroxydopamine lesions of the substantia nigra. Brain Res. 365: 379–382

Chapter 16

Of Rats and Patients: Some Thoughts About Why Rats Turn in Circles and Parkinson's Disease Patients Cannot Move Normally

Gordon W. Arbuthnott

Abstract

Animal behaviours that are easy to measure make great test systems for drug development, but we sometimes neglect to try to understand how their four-legged world view translates to our own. In this brief essay, I try to relate the turning behaviour that has been so useful in the development of drugs that act on Parkinsonian symptoms to the actual symptoms themselves. The thoughts led to a couple of predictions about Parkinsonian behaviour that help to link the bradykinesia that both patients and animals show. In conclusion, I suggest the general idea that dopamine acts to facilitate the learning and expression of the predicted outcomes of simple motor acts: perhaps as a different expression of the reward prediction for which dopamine is already thought to be important.

Key words: Motor behaviour, Prediction of outcomes, Parkinsonian bradykinesia, Turning behaviour, Skill learning, Handedness

1. Why Rats Are Not Parkinson's Disease Models

When Bill Langston claimed that his 1-methyl-4-phenyl-1,2,3,6-tetrahydropyridine (MPTP)-treated monkeys were the first model of Parkinson's disease, he did not really mean that he was ignorant of all the work till then. The problem with earlier rodent models of the disease was their lack of "face validity". The animals did not have the classical clinical symptoms of the disease. The monkeys had profound bradykinesia, which could be unilateral if the toxin was limited to one side of the brain. They had altered gait and sleep and even developed tremor in some cases (1).

In the earlier models, rodents were dead in less than a week if the lesions were bilateral (2) and if not, then they tended to turn

Emma L. Lane and Stephen B. Dunnett (eds.), *Animal Models of Movement Disorders: Volume I*, Neuromethods, vol. 61, DOI 10.1007/978-1-61779-298-4_16, © Springer Science+Business Media, LLC 2011

one way if disturbed and to run in tight circles in response to drug treatments (3). Those circling behaviours were instrumental – are instrumental – in developing drugs capable of stimulating dopamine receptors that have had palliative effects in patients. However, the rats did not have the classical symptoms; they were a model of the end-stage neuropathology and could be useful in examining that end stage (4, 5), but the behaviour of a four-legged rodent was just not similar enough to two-legged primates, for the analogy to be obvious.

Even in the days before MPTP, there was an awareness of that problem. Turning rats were good test systems for drugs. They were excellent as models of replacement strategies because they could be transplanted with replacement cells. The behavioural "recovery" was easy to measure – the drugs resulted in the rats turning in the opposite direction to that before the transplant ((6) and many more before that). But did the rats have any bradykinesia?

One of the tests of motor dysfunction did indeed show one kind of bradykinesia. The animal is held in the experimenter's hand, including both hind legs. By watching carefully how the front legs are used in stepping (or in the "placement reaction" as some surface presents itself within the reach of the forepaws), it is easy to see, even without drug treatment, that the animals are abnormally bradykinetic on the side contralateral to the damage of the dopamine neurons (7).

2. Loss of Paw Use Is Not Only Bradykinesia

When I asked Daniel Ogura-Okorie to train rats to use one forepaw in an operant task, he said, "I don't need to train them, they always use the same paw!" We found that most of the animals were "left handed". What was obvious was that as soon as hungry animals learned to press a lever for food, they quickly preferred to do it with one forepaw. Following a 6-hydroxydopamine (6-OHDA) lesion of the brain on the side opposite to the preferred paw, they stopped responding, but slowly recovered their learned behaviour – now with the unpreferred paw (8). My first reaction was "Good – now we have a lesion that produces bradykinesia. Our rats are more like patients than we thought". We could show that the animals were only incapacitated if the lesion disrupted the use of their preferred paw (it did not matter which was the preferred) but that the behaviour reverted back – the animals used the previously unpreferred paw to press. The recovery was so complete that I leaned towards the idea that here was an effect of dopamine that was "purely motor".

Two things were a little puzzling. Firstly, if we made the lesion and then trained them to press, they did exactly what we would have predicted from our analysis – they learned to press normally, but always used the paw not affected by the lesion (i.e. ipsilateral to the lesion). Then, I had the idea of trying to train them to use the "affected bradykinetic" paw (i.e. contralateral to the lesion).

In stark contrast to the usual 10-min learning curve, if I insisted that the rats were only able to get food if they used the "bradyki-netic" paw, I could not get some of them to press at all. After several days of training, the "clever" ones would lean against the wall above the lever and slide down it so that the disabled paw hit the lever. I guess I was soft enough to count that as a success, though com-pared to the way animals usually press levers, in an operant box, this was not really a "press". All this was extremely time consuming; I needed to watch the animals throughout the training session.

The first problematic result was that the rats did not learn a new behaviour involving the "affected bradykinetic" extremity (i.e. contralateral to the lesion) even though they did use it to walk – and run in circles. I had spent many happy hours lying on the floor in Stockholm making sure that they used all four paws when running in circles under the influence of apomorphine, for instance. Therefore, learning seemed affected after all, even in the "purely motor" context.

The second problem was pointed out to me by Marianela Garcia-Munoz. She was puzzled that the animals took a week to make the switch between paws after the lesion of the dopamine system: the experiment that proved her point she did with Margot Hamilton in the same boxes that Daniel had used. They trained normal rats and just anaesthetised their preferred paw. Now the switch was instantaneous. The animals could not use the paw, they would walk three legged with the forepaw raised and they instantly switched to the other paw to press the lever. Seven weeks later – at a time when the 6-OHDA-treated animals are still using their pre-viously non-preferred paw – these rats returned to their preferred paw, the effect of the anaesthetic having been long gone (9).

Restricting rats to a single paw was not only our idea. Many others have studied the problems with paw use in rodents with 6-OHDA lesions. In general, they avoid using the paw and have systematic "neglect" of the world contralateral to the destroyed dopamine cells in many series of experiments (10) in most the "non-motor" aspects of the behaviour is also remarked upon.

Some experiments also tried to find out how much of the turning behaviour was "learned". The second dose of apomorphine caused a much more rapid onset of turning behaviour than it did on the first occasion (11, 12). That early burst of turning was reduced by the action of anxiolytics of the adenylate cyclase blocking type. At the simple direct level that was evidence to us that cyclase was not the dopamine receptor itself, (11) but what did it mean for that early response? Carey (12) suggested that the response was a learned reaction to the coincidence of drug and situation. Also in our experiments, the early response to a second dose of apomorphine was blunted if it was given in a very different environment. There is a competing explanation for these results that assigns the early response to D1 receptors and the later peak in the turning response to activation of D2 receptors (13). Such an explanation may actually be compatible with the "learning" one, if D1 receptors are mainly involved in the response to burst release to dopamine while the D2 response is linked to baseline release of dopamine. For the purposes of this discussion, the phenomenology of the responses highlights one way in which the turning too contains an element of experiential learning.

Aside from the obvious relationship to modern theories about the action of dopamine in reward-related learning (14), what can these experiments tell us about the symptoms of Parkinson's disease? Surely, the movements that are hard for patients to do (like walking, standing up, using a spoon) are well-learned. Their problems are usually in starting to move. Once walking is initiated, it progresses not quite as effortlessly as normal (15–17). We all learn to walk before we can run, but running is easier for patients.

What seems to be clear from the patients' viewpoint is that self-driven movements are difficult while movements in response to external signals are usually better. There were early stories about being able to walk across lines but not in plain floors – about climbing stairs being easier than walking on a flat surface (18).

3. Recent Studies on Normal Movement Might Help in Understanding What Goes Wrong

Recent studies of human movement may help the understanding of different kinds of movement and their requirement for basal ganglia activity. When people are perturbed in a planned movement by a sudden change in stiffness of the manipulandum, functional magnetic resonance imaging (fMRI) shows an activation in the basal ganglia (putamen and globus pallidus internal segment, GPi, and subthalamic nucleus, STN) that is related to the degree of the corrective movements made. If, however, the perturbation happens at the target, the changes in brain energy requirement (fMRI signal) that correlate with the changes of movement pattern are not related to the basal ganglia (19). It is important that in these studies similar rates of movement and similar numbers of "correcting movements" were made in both tasks; but the brain areas activated in parallel with the movement corrections were different.

Human skilled movement can be shown to use Bayesian statistics of conditional probabilities when estimating errors in target position (20). The motor system seems to be able to learn to predict errors and accommodate for them in planning movement. Unlike in the Grafton (19) experiments described above, the subjects in

Kording (20) have learned to make smooth movements without the corrections that typify the response to unpredictable changes.

To learn skilled movements, i.e. to develop the ability to predict the outcomes of actions, humans require a lot of practice (ask any sports coach!). In cricket, for example, it needs a minimum of 10,000 throws to activate the "muscle memory" needed for a fast bowler, according to the English Cricket coaching training. Parkinsonian patients seem unable to anticipate predictable patterns. Although the experiment about Bayesian prediction has not been done on patients, there is evidence from earlier studies that prediction is something that even treated patients are poor at doing (21). Given a predictable pattern, the latency of following the target drops to zero in normal people – though they cannot always report having "learned" the pattern. Parkinsonian patients could not reach zero latency even with sine waves – probably the simplest of patterns. Movement speed was a little slower and accuracy was slightly compromised, but the major difference was certainly the lack of the much-reduced latency of following, in response to predictable signals (21).

4. A Speculative Conclusion

In an idiosyncratic review like this, there is a great temptation to end with some speculation. Suppose that the basal ganglia are normally the holders of the "expected results" of movements. They store the Bayesian predictors of the consequences of movement. It is in their job description to tune the motor response to the situation – not according to what is available to the peripheral exteroreceptors, but according to past experience.

In a normal rat, the predicted response to walking forward is the parallel flow of the world past its whiskers, eyes and even its ears. But with no dopamine in one striatum, the prediction on one side is wrong; the animal has the impression that its actions on that side are ineffective – the predicted effect on the world does not happen. Therefore, it turns in a way to equalise the flow of the external cues. It turns in circles with the dopamine-deficient side on the inside of the circle (the effects of the basal ganglia lesions are contralateral and so it is the "normal" side of the brain that is on the side of the animal with disturbed predictions).

It could be proposed that animals turn because of a prediction error. Not a "reward prediction error" in this case. That is, a serious mismatch between expectations about the effects of actions and their perceived consequences. A far-fetched explanation for sure, but perhaps more realistic than thinking that rats are caterpillar tractors with the two caterpillar tracks going at different speeds. There are some interesting supporting experiments that involve reducing input to one side of normal animals and indeed the animals' turn (22). This idea might allow some insight into why the motor symptoms of the disease are so "mysterious". Parkinsonian patients, for example, complain of "not being sure that when leaving the interview room they can get through the door that they just passed to enter". This symptom only makes sense if the prediction of how much space is needed to move around is shot to pieces. In "Awakenings" too, it was reported how a patient explained his long inactivity as "being caused by an unpredictable space in which to move". In general, Sacks describes Parkinsonians as having lost their "inner sense of scale" (23).

If there are special external cues (like stairs), then the Parkinsonian patients can cope. The changes in the target size in the Grafton and Tunik (19) experiment can be handled without the basal ganglia and there are only external cues in that situation. In contrast, when the prediction of how the muscles interact with the manipulandum is disturbed (viscous load), it takes the involvement of the basal ganglia to fix it. The clear prediction is that Parkinsonian patients should be worse at coping with unpredictable changes in stiffness than in learning to change their movements to a changing target size.

There is also the evidence from studies of saccadic eye movements. Patients are a little slow, but they succeed in making saccades to visible targets. If they have to move to a "remembered" (no longer visible) target, then they are severely impaired (24, 25). These experiments represent another case where prediction must control the movement (the target is invisible) – and the behaviour is disrupted in the disease.

The errors that make rats turn in circles might indeed be instructive about human movement – only time and patient, thoughtful, experimentation tells.

Acknowledgements

Thanks for many readings of this text to Marianela Garcia-Munoz as well as for the encouragement to stay with handedness as a measure. Thanks too to Steve Dunnett – the only other person with whom I have really discussed this crazy idea.

References

1. Fox SH, Brotchie JM. The MPTP-lesioned non-human primate models of Parkinson's disease. Past, present, and future. In: Bjorklund A, Cenci MA (eds). *Progress in Brain Research.* Amsterdam: Elsevier; 2010. pp. 133–157.

2. Ungerstedt U. Adipsia and Aphagia after 6-Hydroxdopamine induced Degeneration of the Nigro-striatal Dopamine System. *Acta Physiol Scand Suppl* 1971; **367**: 95–122.

3. Ungerstedt U, Arbuthnott GW. Quantitative recording of rotational behaviour in rats after 6- hydroxy-dopamine lesions of the nigrostriatal dopamine system. *Brain Research* 1970; **24**: 485–493.

4. Ingham CA, Hood SH, Taggart P, Arbuthnott GW. Plasticity of synapses in the rat neostriatum after unilateral lesion of the nigrostriatal dopaminergic pathway. *Journal of Neuroscience* 1998; **18**: 4732–4743.

5. Stephens B, Mueller AJ, Shering AF, Hood SH, Taggart P, Arbuthnott GW *et al.* Evidence of a breakdown of corticostriatal connections in Parkinson's disease. *Neuroscience* 2005; **132**: 741–754.

6. Torres EM, Monville C, Gates MA, Bagga V, Dunnett SB. Improved survival of young donor age dopamine grafts in a rat model of Parkinson's disease. *Neuroscience* 2007; **146**: 1606–1617.

7. Tillerson JL, Cohen AD, Philhower J, Miller GW, Zigmond MJ, Schallert T. Forced limb-use effects on the behavioral and neurochemical effects of 6-hydroxydopamine. *Journal of Neuroscience* 2001; **21**: 4427–4435.

8. Ugura-Okorie DC, Arbuthnott GW. Altered paw preference after unilateral 60hydroxydopamine injections into lateral hypothalamus. *Neuropsychologia* 1981; **19**: 463–467.

9. Hamilton MH, Garcia-Munoz M, Arbuthnott GW. Separation of the motor consequences from other actions of unilateral 6-hydroxydopamine lesions in the nigrostriatal neurones of rat brain. *Brain Research* 1985; **348**: 220–228.

10. Dunnett SB, Lelos M. Behavioural analysis of motor and non-motor symptoms in rodent models of Parkinson's disease. In: Bjorklund A, Cenci MA (eds). *Progress in Brain Research.* Amsterdam: Elsevier; 2010. pp. 35–51.

11. Arbuthnott GW, Attree TJ, Eccleston D, Loose RW, Martin MJ. Is adenylate cyclase the dopamine receptor? *Medical Biology* 1974; **52**: 350–353.

12. Damianopoulos EN, Carey RJ. Apomorphine sensitization effects: Evidence for environmentally contingent behavioral reorganization processes. *Pharmacol Biochem Behav* 1993; **45**: 655–663.

13. Herrera-Marschitz M, Arbuthnott G, Ungerstedt U. The rotational model and microdialysis: Significance for dopamine signalling, clinical studies, and beyond. *Progress in Neurobiology* 2010; **90**: 176–189.

14. Schultz W, Dickinson A. Neuronal coding of prediction errors. *Annual Review of Neuroscience* 2000; **23**: 473–500.

15. Plotnik M, Giladi N, Hausdorff J. A new measure for quantifying the bilateral coordination of human gait: effects of aging and Parkinson's disease. *Experimental Brain Research* 2007; **181**: 561–570.

16. Plotnik M, Giladi N, Hausdorff JM. Bilateral coordination of gait and Parkinson's disease: the effects of dual tasking. *J Neurol Neurosurg Psychiatry* 2009; **80**: 347–350.

17. Plotnik M, Giladi N, Dagan Y, Hausdorff J. Postural instability and fall risk in Parkinson's disease: impaired dual tasking, pacing, and bilateral coordination of gait during the "ON" medication state. *Experimental Brain Research* 2011: 1–10.

18. Martin JP. The Basal Ganglia and Posture. *The Basal Ganglia and Posture.* Philadelphia: Lippincott; 1965.

19. Grafton ST, Tunik E. Human Basal Ganglia and the Dynamic Control of Force during On-Line Corrections. *Journal of Neuroscience* 2011; **31**: 1600–1605.

20. Kording KP, Wolpert DM. Bayesian integration in sensorimotor learning. *Nature* 2004; **427**: 244–247.

21. Wing AM, Miller E. Basal ganglia lesions and psychological analyses of the control of voluntary movement. *Ciba Foundation Symposium* 1984; **107**: 242–257.

22. Steiner H, Huston JP. Control of turning behavior under apomorphine by sensory input from the face. *Psychopharmacology (Berl)* 1992; **109**: 390–394.

23. Sacks, O. Epilogue pp 284, Awakenings, Vintage Books, 1999.

24. Briand KA, Strallow D, Hening W, Poizner H, Sereno AB. Control of voluntary and reflexive saccades in Parkinson's disease. *Experimental Brain Research* 1999; **129**: 38–48.

25. Hodgson TL, Dittrich WH, Henderson L, Kennard C. Eye movements and spatial working memory in Parkinson's disease. *Neuropsychologia* 1999; **37**: 927–938.

Chapter 17

Comparing Behavioral Assessment of Sensorimotor Function in Rat and Mouse Models of Parkinson's Disease and Stroke

Sheila M. Fleming and Timothy Schallert

Abstract

To maximize the success of any translational research endeavor, sensitive and reliable behavioral outcome measures in valid animal models are essential. A common goal of preclinical studies in both Parkinson's disease and stroke is to reduce or reverse sensorimotor impairments associated with damage to the striatum and sensorimotor cortex. Here, we describe several behavioral tests of sensorimotor function that have been shown to be sensitive to varying degrees of nigrostriatal and cortical damage in both rats and mice and are highly useful for preclinical studies of potential therapeutics.

Key words: Limb-use asymmetry, Motor coordination, Sensory neglect, Vibrissae-elicited limb placing, Movement disorders, Preclinical testing

1. Introduction

In movement disorders such as Parkinson's disease (PD) and stroke, animal models have been integral to the development of current clinical treatments, rehabilitative therapies, and symptomatic treatments commonly used in patients. For example, administration of 3,4-dihydroxyphenylalanine (L-DOPA) remains the gold standard treatment for PD and was originally shown by Carlsson et al. (1) to reverse reserpine-induced akinesia in animals. Several decades later, Bergman et al. (2) showed that lesions of the subthalamic nucleus in nonhuman primates could reverse motor impairments induced by the toxin 1-methyl-4-phenyl-1,2,3,6-tetrahydropyridine (MPTP) leading to the development of deep brain stimulation as a treatment for PD. In stroke, tissue plasminogen activator treatment, currently the only established clinical treatment for stroke, was shown to reduce neurological damage following

Emma L. Lane and Stephen B. Dunnett (eds.), *Animal Models of Movement Disorders: Volume I*, Neuromethods, vol. 61,
DOI 10.1007/978-1-61779-298-4_17, © Springer Science+Business Media, LLC 2011

cerebral embolism in rabbits (3). However, despite the progress made, improved treatments are still needed in both PD and stroke. Sensitive behavioral outcome measures in animal models continue to be essential in the discovery and development of potential novel therapies. In this chapter, we highlight several behavioral tests that are uniquely useful in rat or mouse models of PD and stroke. A key characteristic is that these tests are not only highly sensitive to cell loss, but also can detect small gradations in the extent of damage and treatment effects across a wide range of brain degeneration and dysfunction.

1.1. Overview of Rodent Models of Parkinsonism and Stroke

Toxins that relatively selectively disrupt or destroy nigrostriatal dopaminergic neurons in rodents have been used to model Parkinsonism for over 40 years (4). The drug reserpine and toxins, 6-hydroxydopamine (6-OHDA) and MPTP, continue to contribute to our understanding of basal ganglia circuitry and voluntary movement (1, 5, 6). More recently, toxins, such as paraquat and rotenone, highlight selective vulnerability of nigrostriatal dopaminergic neurons (7–9). Administration of these toxins results in largely selective loss of nigrostriatal dopaminergic neurons and sensorimotor impairments in rats and mice (1, 5, 6, 10–15). These models tend to have good face validity in that they display impairments in movement initiation, weight shifting, and postural stability that are reversible with L-DOPA similar to Parkinson patients (16–18).

Although most cases of PD are sporadic, the discovery of specific mutations in genes that cause familial forms of PD has led to new animal models of PD. Several genes, including α-synuclein, parkin, DJ-1, UCHL1, Pink1, LRRK2, and ATP13A2, have been implicated in familial PD (19–27). Over the last decade, with the discovery of each mutation, mice with similar gain or loss of function mutations have been developed (28–32). Behavioral tests that can be used in both rats and mouse models would be very helpful in evaluating potential therapeutics in multiple models.

The most commonly studied models of stroke include focal ischemia created by transient or permanent occlusion of the middle cerebral artery. Occlusion can be induced with electrocoagulation and introduction of an embolus, vasoconstricting agents, or photothrombosis (33–35). These models can range in their severity of cortical and striatal damage depending on the time of occlusion and reperfusion and whether the proximal or distal branch of the MCAO is affected. Similarly, the extent and permanency of sensorimotor impairment can vary, especially depending on the degree of striatal damage (36).

Following injury or disease, both humans and animals use compensatory strategies to perform tasks accurately, making it difficult to detect impairments (37–39). Therefore, in animal models, it is important to use tests that minimize deficit masking via motor strategies. Here, we highlight several tests that are effective in detecting varying degrees of sensorimotor injury, but also minimize the influence of compensatory movements (see also http://www. schallertlab.org for relevant movies).

2. Materials and Methods

2.1. Limb-Use Asymmetry During Wall Exploration

The limb-use asymmetry test is a measure of voluntary limb use for weight shifting and maintaining stability during vertical exploration of the walls of an enclosure (36, 40–44). This test is commonly used in the unilateral 6-OHDA rat and mouse models of Parkinsonism and focal ischemia (36, 45). Forelimb use during wall explorative activity can be assessed by videotaping rats in a transparent cylinder (20 cm diameter and 30 cm height) until 20 limb usages occur. A mirror placed behind and slightly to the side of the cylinder at an angle enables the rater to record all forelimb movements, including when the animal is turned away from the camera. The image of the entire cylinder and its unobscured mirror image should fill the video frame without overlapping. The camera should be placed several feet away so that its zoom feature can be employed to ensure nonblurred limb usage. The cylindrical shape encourages vertical exploration of the walls with the forelimbs. Several behaviors are scored to determine the extent of the forelimb-use asymmetry displayed by the animal. These behaviors include independent and simultaneous or rapidly alternating use of the left and/or right forelimb for contacting the wall during a full rear. This test has been shown to be highly sensitive to dopamine cell loss and focal ischemia (16, 36, 43, 46) and is widely used to assess the efficacy of potential treatment interventions (46–52). With unilateral dopamine deficiency or middle cerebral artery occlusion, the animal relies primarily on the ipsilateral forelimb for wall support and lateral movements. In contrast, the contralateral forelimb is rarely used (depending on the extent of cell loss) or is used simultaneously for weight support on the wall or alternating with the ipsilateral forelimb, and even with mild degeneration it is not used independently in successive steps (44, 53).

This test can also be adapted to measure spontaneous movement and limb use in bilateral models of Parkinsonism, including MPTP or rotenone-treated rats and genetic mouse models (9, 54–56). Here, the cylinder for mice (height = 15.5 cm, diameter = 12.7 cm) or rats is placed on a piece of glass with a mirror positioned at an angle beneath the cylinder to allow a clear view of movements along the floor and walls of the cylinder. Videotapes are viewed and rated in slow motion and the number of wall-dependent vs. wall-free rears, forelimb wall exploratory movements, both forelimb and hind limb steps, and time spent grooming can be measured within a designated amount of time. In this version of the cylinder test,

MPTP or rotenone-treated rats show decreased rearing (especially wall-free (57)) while several genetic mouse models of Parkinsonism show reduced stepping, especially with the hind limbs (16, 54–56).

2.2. Ledged Beam Test

The ledged beam test has been used in both rats with a unilateral 6-OHDA nigrostriatal microinfusion or middle cerebral artery occlusion. An adapted version has also been shown to be highly sensitive in the unilateral 6-OHDA mouse and several genetic mouse models of Parkinsonism (30, 52, 54–56). For most adult rats, the beam typically should be 130 cm long and tapered in width from 10 to 1.5 cm. The beam is divided and marked into three 43.3-cm bins of increasing difficulty. The animal's home cage (including bedding), placed on its side at the end of the beam, is used as a potent reinforcer to encourage repeated traverses. Tapered widths should be adjusted to fit the size of the animals tested (for example, for rats weighing 300–500 g, the beam should taper from a width of 6–1.5 cm keeping the length the same). Large, aged rats require a more generous starting width to accommodate the distance between their hind limbs. The surface of the beam is covered with a rubber mat for optimal purchase, and underhanging ledges placed 2.0 cm below the upper surface of the beam permit foot faults to occur without the animal slipping off the beam or needing to compensate with postural adjustments with the intact limb or tail deviations for balance. That is, the ledges (2 cm wide) provide a support for the animal to use if it cannot keep all limbs on the top of the beam. The extent of ledge use gives an indication of the degree of injury or true recovery of function.

The beam is placed at a 15° angle three feet above the ground, and a mirror is placed on one side of the beam to allow simultaneous viewing of all limbs. Animals are trained to walk from the widest, lowest part of the beam to the narrowest, highest part of the beam, which leads directly into the animal's home cage. To maximize motivation to traverse the beam, it is important that the animals remain in the home cage for a few moments after each trial rather than being picked up immediately for another trial. During pre-training, each animal is placed at the narrow end of the beam in close proximity to the home cage. The distance from the home cage should be progressively increased until the animal is at the widest part of the beam and can traverse the entire length without lateral movements or assistance from the experimenter. Following training trials (approximately 10–15), rats are videotaped as they walk across the beam. Only steps and faults by the animal during forward movement should be scored. Hind limb and forelimb errors are measured separately by recording (1) the number of times an animal places a limb (fore or hind) on the support ledge (foot faults), (2) the location of the fault on the beam, and (3) the type of fault made. In addition, the total number of steps is recorded, which is an important detail because larger rats cross the beam using fewer steps than smaller rats. Steps and faults are not counted if an animal's head is oriented left or right. This does not alter lesion vs. sham results, but is done because even sham animals use the ledge when oriented laterally during pauses. The number of times the animal steps onto the ledge instead of walking along the beam surface is measured. Foot faults are divided into two categories: full faults, in which the animal places the entire surface of the limb flat upon the support ledge, and half-faults, where the limb lands in between the beam and ledge surfaces (i.e., lands partially on the beam and partially on the ledge). A full fault is given a score of 1.0, and a half-fault is given a score of 0.5. A step is not scored as a foot fault if the animal's digits extend over the edge of the surface but the rest of the limb rests wholly on the beam surface. Ipsilateral and contralateral hind limb foot faults are recorded. The scores are averaged across five trials for each animal and the percent ipsilateral faults per step and percent contralateral faults per step are calculated. To derive one score, the percent ipsilateral faults per step is subtracted from the percent contralateral faults per step for each bin. Videotapes are viewed using a camcorder with slow motion and frame-by-frame capabilities and rated by an experimenter blind to all experimental conditions.

The often-called challenging beam in mice is an adaptation of the ledged beam described above (54–56). Briefly, the beam consists of four sections (25 cm each, 1 m total length), each section having a different width. The beam starts at a width of 3.5 cm and gradually narrows to 0.5 cm by 1-cm increment. Animals are trained to traverse the length of the beam starting at the widest section and ending at the narrowest, most difficult, section. On the day of the test, a mesh grid (1-cm squares) of corresponding width is placed over the beam surface leaving approximately a 1-cm space between the grid and the beam surface, which (like the ledge described above) serves as a crutch to prevent compensatory motor learning that can otherwise mask extant deficits. Animals are then videotaped while traversing the grid-surfaced beam. Videotapes are rated for errors, number of steps made by each animal, and time to traverse. This test has been shown to be highly sensitive in mice with mutations associated with familial PD (30, 54, 56) and in mice with a developmental loss of nigrostriatal dopamine neurons (55). In addition, impairments in mice with dopamine cell loss can be reversed with L-DOPA (55).

2.3. Somatosensory Asymmetries

Sensory impairments are common in Parkinson's disease and stroke. Somatosensory asymmetry can be assessed in rats using a bilateral tactile stimulation test that is sensitive to both nigrostriatal damage and focal ischemia (16, 58, 59). Here, animals are tested to indicate the presence of a somatosensory asymmetry. This is done by removing the animal from the home cage and attaching adhesive stimuli (e.g., Tough Spots, a common lab test tube adhesive label that varies commercially in size) to the distal-radial aspect of each forelimb in random (left–right) order. After being returned to the home cage, rats rapidly contact and remove the stimuli one at a time using the teeth. The order and latency of stimulus contact and removal are recorded for each of five trials. The order of contact is used to determine whether animals show a bias for the stimulus on the forelimb unaffected by the injury. Rats with a unilateral 6-OHDA lesion or middle cerebral artery occlusion repeatedly contact and remove the stimulus from the ipsilateral limb prior to the contralateral forelimb (16, 58, 59). However, the Parkinson's model animals are much slower than stroke model animals to remove the contralateral stimulus.

To increase sensitivity, the magnitude of asymmetry can also be measured by increasing the size of the adhesive stimulus placed on the affected forelimb while decreasing the size of the stimulus placed on the unaffected forelimb. That is, the size of the impaired limb stimulus is progressively increased over trials and the size of the nonimpaired limb stimulus is simultaneously decreased by an equal amount (e.g., 14.1 mm^2). A sufficient increase in the impaired/nonimpaired ratio causes a reversal of the original bias when the rat responds first to the stimulus placed on the impaired limb. The ratio necessary to reverse the initial bias is proportional to the degree of nigrostriatal dopamine degeneration or intrinsic striatal neuronal injury (36, 59, 60).

Similar to the test of somatosensory asymmetry in rats, response to sensory stimuli can be measured in mice with genetic mutations. In this test, small adhesive stimuli are placed on the snout of the mouse and the time to make contact and remove the stimulus is recorded. If the animal does not remove the stimulus within 60 s, the experimenter removes it and the trial for the next mouse is initiated. Stimulus contact and removal times are calculated for each animal. This test has been shown to be sensitive in many genetic mouse models of PD, including α-synuclein over-expressing mice (54), DJ-1 knockout mice (52), parkin knockout mice (30), and most recently mice with a parkin Q113X mutation (56).

2.4. Vibrissae-Elicited Limb Placing

Rats use their vibrissae to gain bilateral information about their immediate environment. When there is no stable surface for support, the rat makes a placing response to the first surface that one set of vibrissae contacts. The vibrissae-elicited forelimb placing test takes advantage of this automatic placing response. This test is sensitive to both unilateral 6-OHDA lesions and focal ischemia (12, 13, 61). In this test, the rat's torso is supported by the investigator and suspended in a way that allows all four legs to hang freely in the air. The experimenter then brings the rat toward the edge of a tabletop and the rat's vibrissae are brushed against the table edge on the same side of the body in which forelimb placing is being evaluated. The percentage of trials in which the rat successfully places its forepaw onto the tabletop is recorded for each side.

A cross-midline placing response can also be measured and can differentially distinguish between impairment and rate of recovery after 6-OHDA lesions vs. proximal middle cerebral artery occlusion in rats. In middle cerebral artery occlusion, vibrissae stimulation applied to the ipsilateral side of the body is able to trigger a placing response in the impaired forelimb many weeks before stimulation of the contralateral vibrissae can. After extensive unilateral ischemic damage to striatum, recovery of the capacity for contralateral vibrissae stimulation to induce placing of the contralateral forelimb may be absent chronically. Thus, sensory input to the intact hemisphere can, with time, activate neural substrates for limb placing linked to motor programs in the injured hemisphere, but sensory input to the injured hemisphere cannot.

In contrast, unilateral nigrostriatal lesions lead to a lack of placing response in the contralateral limb regardless of whether the ipsilateral or contralateral vibrissae are stimulated. These data are consistent with the view that the contralateral forelimb impairment caused by nigrostriatal dopamine depletion is not mediated by a vibrissae-related sensory deficit but is a motor deficit comparable to Parkinsonian akinesia. In addition, although the cross-midline ipsilateral vibrissae-induced contralateral forelimb placing deficit can recover following total striatal ischemic injury, in the severe unilateral 6-OHDA-induced Parkinson's model the incapacity to evoke contralateral forelimb placing persists indefinitely, regardless of which vibrissae are stimulated (61).

3. Notes

3.1. Limb-Use Asymmetry During Wall Exploration

Different strains of rats and mice can have different activity levels, which can impact the number of limb placements an animal makes along the wall of the cylinder. In addition, rodents can habituate to the cylinder environment with repeated testing also reducing the amount of limb use. However, rearing and limb use can be encouraged in rodents without affecting the asymmetry score. Here are a few strategies we have found to encourage movement within the cylinder. Test in the dark cycle under red light. If that is not possible, then briefly switch off and on the lights in the testing room, place bedding from the rat's home cage onto the cylinder floor, pick up the rat and place it back into the cylinder, perform multiple tests like the somatosensory asymmetry test prior to testing in the cylinder, or lightly touch a cotton-tipped applicator to the snout of the animal.

3.2. Ledged Beam Test   Similar to limb-use asymmetry, different motivation levels can reduce movement along the beam. If a rat or a mouse does not readily move down the beam, it can be encouraged to move by testing in the dark cycle under low light, making sure that cagemates are in the home cage placed at the end of the beam, or lightly tapping on the backside of the animal.

3.3. Somatosensory Asymmetries and Vibrissae-Elicited Limb Placing   Both of these tests require rats and mice to be well-handled. We advise practicing the procedures on nonlesioned or wild-type animals for several trials until the animal and experimenter perform the test consistently across trials.

4. Conclusion

The success of preclinical trials of potential therapeutics for both PD and stroke depends, in part, on the choice of behavioral outcome measures utilized. Tests that are sensitive in multiple models and species are particularly attractive because potential beneficial effects can be directly compared across different models. In addition, it is important to employ multiple tests for assessment as they all measure different aspects of sensorimotor function and can provide important information on the benefits and limits of a potential treatment.

References

1. Carlsson A, Lindqvist M, Magnusson T (1957) 3,4-Dihydroxyphenylalanine and 5-hydroxytryptophan as reserpine antagonists. Nature 180: 1200

2. Bergman H, Wichmann T, DeLong MR (1990) Reversal of experimental parkinsonism by lesions of the subthalamic nucleus. Science 249: 1436–1438

3. Zivin JA, Fisher M, DeGirolami U, Hemenway CC, Stashak JA (1985) Tissue plasminogen activator reduces neurological damage after cerebral embolism. Science 230: 1289–1292

4. Ungerstedt U (1968) 6-Hydroxy-dopamine induced degeneration of central monoamine neurons. Eur J Pharmacol 5: 107–110

5. Ungerstedt U (1971) Adipsia and aphagia after 6-hydroxydopamine induced degeneration of the nigro-striatal dopamine system. Act Physiol Scand Suppl 367: 95–122

6. Burns RS, Chiueh CC, Markey SP (1983) A primate model of parkinsonism: selective destruction of dopaminergic neurons in the pars compacta of the substantia nigra by N-methyl-4-phenyl-1,2,3,6-tetrahydropyridine. Proc Natl Acad Sci USA 80: 4546–4550

7. Betarbet R, Sherer TB, MacKenzie G Ebert MH, Jacobowitz DM, Kopin IJ. (2000) Chronic systemic pesticide exposure reproduces features of Parkinson's disease. Nat Neurosci 3: 1301–1306

8. Brooks AI, Chadwick CA, Gelbard HA, Cory-Slechta DA, Federoff HJ (1999) Paraquat elicited neurobehavioral syndrome caused by dopaminergic neuron loss. Brain Res 823: 1–10

9. Fleming SM, Zhu C, Fernagut PO, Mehta A, DiCarlo CD, Seaman RL, Chesselet MF (2004) Behavioral and immunohistochemical effects of chronic intravenous and subcutaneous infusions of varying doses of rotenone. Exp Neurol 187: 418–422

10. Marshall JF, Turner BH, Teitelbaum P (1971) Sensory neglect produced by lateral hypothalamic damage. Science 174: 523–525

11. Schallert T, Whishaw IQ, Ramirez VD, Teitelbaum P (1978) Compulsive, abnormal walking caused by anticholinergics in akinetic,

6-hydroxydopamine-treated rats. Science 199: 1461–1463

12. Tillerson JL, Cohen AD, Philhower J, Miller GW, Zigmond MJ, Schallert T (2001) Forced limb-use effects on the behavioral and neurochemical effects of 6-hydroxydopamine. J Neurosci 21: 4427–4435

13. Tillerson JL, Cohen AD, Caudle WM, Zigmond MJ, Schallert T, Miller GW (2002) Forced nonuse in unilateral parkinsonian rats exacerbates injury. J Neurosci 22: 6790–6799

14. Ogawa N, Hirose Y, Ohara S, Ono T, Watanabe Y (1985) A simple quantitative bradykinesia test in MPTP-treated mice. Res Commun Chem Pathol Pharmacol 50: 435–441

15. Heikkila RE, Hess A, Duvoisin RC (1984) Dopaminergic neurotoxicity of 1-methyl-4-phenyl-1,2,5,6-tetrahydropyridine in mice. Science 224: 1451–1453

16. Fleming SM, Delville Y, Schallert T (2005) An intermittent, controlled-rate, slow progressive degeneration model of Parkinson's disease: antiparkinson effects of Sinemet and protective effects of methylphenidate. Behav Brain Res 156: 201–213

17. Tillerson JL, Caudle WM, Reverón ME, Miller GW (2002). Detection of behavioral impairments correlated to neurochemical deficits in mice treated with moderate doses of 1 methyl-4-phenyl 1,2,3,6-tetrahydropyridine. Exp Neurol 178: 80–90

18. Alam M, Schmidt WJ (2004) L-DOPA reverses the hypokinetic behaviour and rigidity in rotenone-treated rats. Behav Brain Res 153: 439–446

19. Bonifati V, Rizzu P, van Baren MJ, Schaap O, Breedveld GJ, Krieger E, et al. (2003) Mutations in the DJ-1 gene associated with autosomal recessive early-onset parkinsonism. Science 2994: 256–259

20. Kitada T, Asakawa S, Hattori N, Matsumine H, Yamamura Y, Minoshima S, Yokochi M, Mizuno Y, Shimizu N. (1998) Mutations in the parkin gene cause autosomal recessive juvenile parkinsonism. Nature 392: 605–608

21. Krüger R, Kuhn W, Müller T, Woitalla D, Graeber M, Kösel S, Przuntek H, Epplen JT, Schöls L, Riess O (1998) Ala30Pro mutation in the gene encoding alpha-synuclein in Parkinson's disease. Nat Genet 18: 106–108

22. Polymeropoulos MH, Lavedan C, Leroy E, Ide SE, Dehejia A, Pike B, et al (1997) Mutation in the alpha-synuclein gene identified in families with Parkinson's disease. Science 276: 2045–2047

23. Singleton AB, Farrer M, Johnson J (2003) alpha-Synuclein locus triplication causes Parkinson's disease. Science 302: 841

24. Wintermeyer P, Krüger R, Kuhn W Müller T, Woitalla D, Berg D, et al. (2000) Mutation analysis and association studies of the UCHL1 gene in German Parkinson's disease patients. Neuroreport 11: 2079–2078

25. Valente EM, Abou-Sleiman PM, Caputo V, Muqit MM, Harvey K, Gispert S, et al. (2004) Hereditary early-onset Parkinson's disease caused by mutations in PINK1. Science 304: 1158–1160

26. Zimprich A, Biskup S, Leitner P, Lichtner P, Farrer M, Lincoln S, et al. (2004) Mutations in LRRK2 cause autosomal-dominant parkinsonism with pleomorphic pathology. Neuron 44: 601–7

27. Ramirez A, Heimbach A, Gründemann J, Stiller B, Hampshire D, Cid LP, et al (2006) Hereditary parkinsonism with dementia is caused by mutations in ATP13A2, encoding a lysosomal type 5 P-type ATPase. Nat Genet 38: 1184–1191

28. Masliah E, Rockenstein E, Veinbergs I, Mallory M, Hashimoto M, Takeda A, Sagara Y, Sisk A, Mucke L (2000) Dopaminergic loss and inclusion body formation in alpha-synuclein mice: implications for neurodegenerative disorders. Science 287: 1265–1269

29. Rockenstein E, Mallory M, Hashimoto M, Song D, Shults CW, Lang I, Masliah E (2002) Differential neuropathological alterations in transgenic mice expressing alpha-synuclein from the platelet-derived growth factor and Thy-1 promoters. J Neurosci Res 68: 568–578

30. Goldberg MS, Fleming SM, Palacino JJ, Capeda C, Lam HA, Bhatnagar A, et al. (2003) Parkin-deficient mice exhibit nigrostriatal deficits but not loss of dopaminergic neurons. J Biol Chem 278: 43628–43635

31. Chen L, Cagniard B, Mathews T, Jones S, Koh HC, Ding Y, Carvey PM, Ling Z, Kang UJ, Zhuang X (2005) Age-dependent motor deficits and dopaminergic dysfunction in DJ-1 null mice. J Biol Chem 280: 21418–21426

32. Li Y, Liu W, Oo TF, Wang L, Tang Y, Jackson-Lewis V, et al. (2009) Mutant LRRK2(R1441G) BAC transgenic mice recapitulate cardinal features of Parkinson's disease. Nat Neurosci 12: 826–828

33. Watson BD, Dietrich WD, Busto R , Wachtel MS, Ginsberg MD (1985) Induction of reproducible brain infarction by photochemically initiated thrombosis. Ann Neurol 17: 497–504

34. Zhang L, Zhang ZG, Zhang RL, Lu M, Krams M, Chopp M (2003) Effects of a selective CD11b/CD18 antagonist and recombinant human tissue plasminogen activator treatment alone and in combination in a rat embolic model of stroke. Stroke 34: 1790–1795

35. Tamura A, Graham DI, McCulloch J, Teasdale GM (1981) Focal cerebral ischaemia in the rat: 1. Description of technique and early neuropathological consequences following middle cerebral artery occlusion. J Cereb Blood Flow Metab 1: 53–56

36. Schallert T, Fleming SM, Leasure JL, Tillerson JL, Bland ST (2000) CNS plasticity and assessment of forelimb sensorimotor outcome in unilateral rat models of stroke, cortical ablation, parkinsonism and spinal cord injury. Neuropharmacology 39: 777–787

37. LeVere TE (1988) Neural system imbalances and the consequences of large brain injuries. In: Finger S, LeVere TE, Almi CR and Stein DG, eds. Brain Injury and Recovery, Theoretical and Controversial Issues. New York : Plennum Press, 15–28

38. Schallert T (1988) Aging-dependent emergence of sensorimotor dysfunction in rats recovered from dopamine depletion sustained early in life. Ann N Y Acad Sci 515: 108–120

39. Whishaw IQ (2000) Loss of the innate cortical engram for action patterns used in skilled reaching and the development of behavioral compensation following motor cortex lesions in the rat. Neuropharmacology 39: 788–805

40. Ariano MA, Grissell AE, Littlejohn FC Buchanan TM, Elsworth JD, Collier TJ, Steece-Collier K (2005) Partial dopamine loss enhances activated caspase-3 activity: differential outcomes in striatal projection systems. J Neurosci Res 82: 387–396

41. Johnson RE, Schallert T, Becker JB (1999) Akinesia and postural abnormality after unilateral dopamine depletion. Behav Brain Res 104: 189–196

42. Schallert T, Kozlowski DA, Humm JL, Cocke RR (1997) Use-dependent structural events in recovery of function. Adv Neurol 73: 229–238

43. Schallert T, Tillerson JL (2000) Intervention strategies for degeneration of dopamine neurons in parkinsonism: Optimizing behavioral assessment of outcome. In: Emerich DF, Dean RL III, Sandberg PR (eds), Central Nervous System Diseases. Totowa, NJ: Humana Press; 131–151

44. Schallert T, Woodlee MT (2005) Orienting and placing. In: Whishaw IQ, Kolb B (eds), The behavior of the laboratory rat. New York: Oxford University Press;129–140

45. Li X, Blizzard KK, Zeng Z, DeVries AC, Hurn PD, McCullough LD (2004) Chronic behavioral testing after focal ischemia in the mouse: functional recovery and the effects of gender. Exp Neurol 187: 94–104

46. Kawamata T, Dietrich WD, Schallert T, Gotts JE, Cocke RR, Benowitz LI, Finklestein SP (1997) Intracisternal basic fibroblast growth factor enhances functional recovery and up-regulates the expression of a molecular marker of neuronal sprouting following focal cerebral infarction. Proc Natl Acad Sci USA 94: 8179–8184

47. Yang M, Stull ND, Berk MA, Snyder EY, Iacovitti L (2002) Neural stem cells spontaneously express dopaminergic traits after transplantation into the intact or 6-hydroxydopamine-lesioned rat. Exp Neurol 177: 50–60

48. Kozlowski DA, Connor B, Tillerson JL, Schallert T, Bohn MC (2000) Delivery of a GDNF gene into the substantia nigra after a progressive 6-OHDA lesion maintains functional nigrostriatal connections. Exp Neurol 166: 1–15

49. Connor B, Kozlowski DA, Schallert T, Tillerson JL, Davidson BL, Bohn MC (1999) Differential effects of glial cell line-derived neurotrophic factor (GDNF) in the striatum and substantia nigra of the aged Parkinsonian rat. Gene Ther 6: 1936–1951

50. Luo J, Kaplitt MG, Fitzsimons HL, Zuzga DS, Liu Y, Oshinsky ML, During MJ (2002) Subthalamic GAD gene therapy in a Parkinson's disease rat model. Science 298: 425–429

51. Shi LH, Woodward DJ, Luo F, Anstrom K, Schallert T, Chang JY (2004) High-frequency stimulation of the subthalamic nucleus reverses limb-use asymmetry in rats with unilateral 6-hydroxydopamine lesions. Brain Res 1013: 98–106

52. Chan CS, Glajch KE, Gertler TS, Guzman JN, Mercer JN, Lewis AS, et al (2011) HCN channelopathy in external globus pallidus neurons in models of Parkinson's disease. Nat Neurosci14: 85–92

53. Schallert T (2006) Behavioral tests for preclinical intervention assessment. NeuroRx 3: 497–504

54. Fleming SM, Salcedo J, Fernagut PO, Rockenstein E, Masliah E, Levine MS, Chesselet MF (2004) Early and progressive sensorimotor anomalies in mice overexpressing wild-type human alpha-synuclein. J Neurosci 24: 9434–9340

55. Hwang DY, Fleming SM, Ardayfio P, Moran-Gates T, Kim H, Tarazi FI, Chesselet MF, Kim KS (2005) 3,4-dihydroxyphenylalanine reverses the motor deficits in Pitx3-deficient aphakia mice: behavioral characterization of a novel genetic model of Parkinson's disease. J Neurosci 25: 2132–2137

56. Lu XH, Fleming SM, Meurers B, Ackerson LC, Mortazavi F, Lo V, et al (2009) Bacterial artificial chromosome transgenic mice expressing a truncated mutant parkin exhibit age-dependent hypokinetic motor deficits, dopaminergic neuron degeneration, and accumulation of proteinase K-resistant alpha-synuclein. J Neurosci 29: 1962 1967

57. Goldberg NR, Haack AK, Lim NS, Janson OK, Meshul CK (2011) Dopaminergic and behavioral correlates of progressive lesioning of the nigrostriatal pathway with 1-methyl-4-phenyl-1,2,3,6-tetrahydroppyridine. Neuroscience180: 256–271

58. Schallert T, Upchurch M, Lobaugh N, Farrar SB, Spirduso WW, Gilliam P, Vaughn DM, Wilcox RE (1982) Tactile extinction: distinguishing between sensorimotor and motor asymmetries in rats with unilateral nigrostriatal damage. Pharmacol Biochem Behav 16: 455–462

59. Schallert T, Upchurch M, Wilcox RE, Vaughn DM (1983) Posture-independent sensorimotor analysis of inter-hemispheric receptor asymmetries in neostriatum. Pharmacol Biochem Behav. 18: 753–759

60. Barth TM, Jones TA, Schallert T (1990) Functional subdivisions of the rat somatic sensorimotor cortex. Behav Brain Res 39: 73–79

61. Woodlee MT, Asseo-Garcia AM, Zhao X, Liu SJ, Jones TA, Schallert T (2005) Testing forelimb placing "across the midline" reveals distinct, lesion-dependent patterns of recovery in rats. Exp Neurol 191: 310–317.

Chapter 18

Rodent Models of L-DOPA-Induced Dyskinesia

Hanna S. Lindgren and Emma L. Lane

Abstract

A common side effect of the pharmacotherapy for treatment of the movement disorder Parkinson's disease is the development of L-DOPA-induced dyskinesia (LID). These are abnormal, involuntary, choreic and dystonic movements that can be very debilitating, and therefore new treatment options are sought. LID can be modelled in rodents (rats and mice) with a unilateral 6-hydroxydopamine lesion by chronic administration of L-DOPA. This chapter provides a detailed description of the most commonly used rating scale for abnormal involuntary movements in rodents as well as experimental considerations important for setting up this model.

Key words: L-DOPA, Abnormal involuntary movements, Rating scale, Dyskinesia

1. Introduction

At present, the dopamine precursor L-DOPA remains the most effective pharmacotherapy for Parkinson's disease (PD), but its use in long-term therapy is hampered by the development of motor fluctuations and L-DOPA-induced dyskinesia (LID; choreic and dystonic abnormal involuntary movements, AIMs) in the majority of patients within 10 years of L-DOPA treatment (1, 2). Although the precise mechanisms underlying LID have remained elusive, two main risk factors have been conclusively identified in clinical trials, namely the disease severity (reflecting the extent of dopamine denervation) and the use of high doses of L-DOPA (3, 4). These risk factors can easily be reproduced in rodent and non-human primate models of PD by injection of neurotoxins targeting the dopaminergic (DA) nigrostriatal system followed by daily injections of L-DOPA at sufficient, but clinically relevant, doses (5–9). With large DA depletion either bilaterally or unilaterally (MPTP and 6-OHDA models, respectively, see chapters by Torres &

Dunnett, Smith & Heuer, Petzinger et al., and Jackson & Jenner, this volume), repeat dosing with L-DOPA will begin to induce AIMs, which can be quantified both on the basis of their persistence and amplitude (6, 10–14). These abnormal movements provide a well-validated equivalent of peak dose dyskinesia seen in patients, and can be attenuated by drugs with known anti-dyskinetic efficacy (12, 15). Although non-human primate models of LID have a notable advantage of sharing similar movement patterns as humans, the rodent model of LID is relatively simple and easy to set up, thus being both available and affordable to many laboratories over the world. In addition, they allow for modelling of LID using larger, more consistent groups of animals under strictly controlled experimental conditions, thereby providing an optimal starting point for reliable results. In this chapter, we will focus exclusively on the rodent model of LID (the MPTP-treated primate model of PD and LID is covered in detail in the chapter by Jackson and Jenner, this volume). Here, we will cover detailed protocols for LID assessment in rats and mice, a relatively straightforward protocol if certain principles are adhered to, and we describe the potential pit falls and caveats, and ways they can be avoided.

2. Experimental Design

Experimental design will depend on the rationale of the experiment, whether to explore mechanisms underlying LID development, to assess the potential of anti-parkinsonian drugs to cause LID, or to evaluate the putative anti-dyskinetic agents. The model is used with both rats and mice, and there is some experience of deviation away from the standard strains of Sprague Dawley rats and C57/BL6 mice (see Note 1). The standard lesion type is the medial forebrain bundle (mfb) 6-OHDA lesion, but other lesion sites have been used and characterised in both rat and mouse strains (see Note 2).

To induce AIMs in a population of lesioned rats, the standard study design is typically 10–28 days of daily L-DOPA injections at doses in a range considered therapeutically relevant (6–20 mg/kg co-administered with benserazide HCl, see Note 3), in either one or two injections per day. During that period, AIMs ratings should be performed every third day (or more frequently) in order to monitor the development of dyskinesia and to confirm its stabilisation, usually occurring after 10–14 days of treatment (see Fig. 1a).

Fig. 1. (a) During a chronic treatment period with daily injections of L-DOPA at low doses, the rodents will gradually develop abnormal involuntary movements, reaching a plateau after 10–14 days of treatment. (b) Both the duration and the peak severity of the AIMs will gradually change after chronic L-DOPA administration. The amount of time the rodents show AIMs will decrease, while the peak severity increases as the treatment period progresses. (c) Subcutaneous (s.c. in the scruff) and intraperitoneal (i.p.) injections of L-DOPA result in a similar onset, time profile and severity grade of AIMs. Note that there may be a slight shift in the temporal profile of AIMs depending on where the injection is made.

a

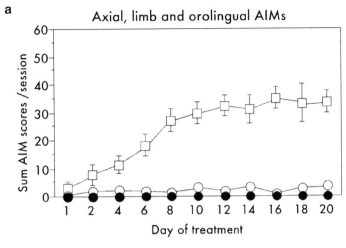

Axial, limb and orolingual AIMs

—□— Dyskinetic   —○— Non-dyskinetic   —●— Saline-treated controls

b

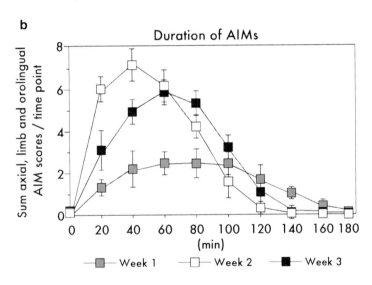

Duration of AIMs

—■— Week 1   —□— Week 2   —■— Week 3

c

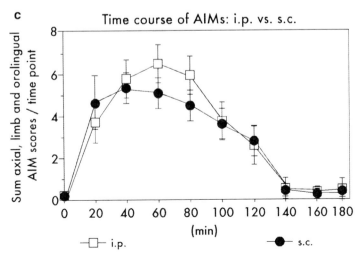

Time course of AIMs: i.p. vs. s.c.

—□— i.p.   —●— s.c.

The dose of L-DOPA used will also be influenced by the desired outcome. Using lower doses of L-DOPA will favour a more gradual development of dyskinetic movements, whereas higher doses will result in a rapid and robust appearance of AIMs with a more extensive distribution across the body (9). Two injections per day will result in a longer time "on" L-DOPA (~6 h day, compared to 3 h from a single injection). This may be considered more representative of the clinical situation, as patients will take L-DOPA several times a day. For experimental purposes, however L-DOPA once daily is sufficient for the induction and monitoring of AIMs. To maintain a stable expression of AIMs over time, it is sufficient to inject the animals with L-DOPA 2–4 times a week after the first induction period (16).

When testing potential anti-dyskinetic drugs, animals should be pre-treated with L-DOPA (primed) until they reach a plateau in their AIMs development (usually around 10–14 days). Animals are allocated into matched experimental groups based on their AIMs scores and one group given L-DOPA together with the anti-dyskinetic drug for one to several days. The other group will receive L-DOPA plus the drug vehicle for the same period of time. Thereafter, the group allocation is switched and the animals are tested again. In general, 3 days of washout between different drugs and doses is sufficient, but this is of course dependent on the pharmacokinetics of the drug. The crossover design allows for within-animal comparisons and counter-acts the fact that both the severity and expression pattern of AIMs may differ between animals.

When testing anti-dyskinetic properties of a drug, it is crucial to include tests of sensorimotor performance in the experimental design (see Schallert, this volume for well-validated tasks for motor performance in rats with a 6-OHDA lesion) since the anti-dyskinetic property of the drug must not come at the expense of L-DOPA efficacy. It is also of great importance that an observer blinded to the treatment of the animals makes the AIMs ratings. In addition, it is essential to keep all testing parameters consistent in order to get reliable results and to fully comprehend the efficacy of the anti-dyskinetic drug.

3. Methods

3.1. Methodology

Using either translucent rotometers or transparent Perspex cages (see Note 4), the animal enclosures are positioned in such a way that the animals can be observed from all directions, since they will move around during the testing session. Rodents are placed individually in the cages and allowed to habituate for at least 15 min. During the habituation period, the L-DOPA and benserazide HCl are dissolved in saline (see Note 3) and the syringes and needles are prepared. The timer is started upon injection of the first animal, and thereafter one animal is injected per minute (for a group of up to 20 animals) until all animals have received their ʟ-DOPA (or saline) injection (see Note 5 for route of administration). AIMs ratings are then started 20 min after the first animal was injected and each animal is rated for 1 min in turn (see Note 6). This continues for 180 min following drug administration (or until all dyskinetic movements are replaced by a normal motor pattern). For each animal and observation period, a score (and an amplitude if the complimentary amplitude rating scale is used, see below) is given for each AIM subtype, based on the duration and persistence of the abnormal movement during the 1-min observation period.

The description of the scales and classification of the different subtypes has been taken from the publication made by Cenci and Lundblad in *Current Protocols in Neuroscience* (17), in order to keep the conformity of the rating scale with the addition of hindlimb AIMs.

3.2. Basic Rating Scale

In the following description, five different classes of abnormal involuntary movement (axial, limb, orolingual, hindlimb and loco-motor) are each rated, within each 1-min time bin, according to a five point scale:

0 = no AIMs.
1 = occasional AIMs, which are present during less than half of the observation time.
2 = frequent signs of AIMs, which are present more than half of the observation time.
3 = AIMs are present during the entire observations time, but can be interrupted by external stimuli (e.g. sudden, load opening of the lid of the cage).
4 = continuous AIMs that are not suppressible by external stimuli.

If the scores are summed for each rat across all AIM classes and time points, the theoretical maximum score in one test (nine monitoring periods over 180 min) is 180 if locomotion is included, 144 if it is considered separately, see specific comment below.

3.2.1. Axial AIMs

At the beginning of the treatment period when the AIMs are mild, the axial component can be described as a lateral flexion of the trunk or neck towards the side contralateral to the lesion. When the severity increases, the flexion will be of a more dystonic and choreiform-twisted character (dystonic refers to a slow forcing movement of a body part into an abnormal position).

3.2.2. Limb AIMs

In mild cases, limb AIMs are expressed as hyperkinetic, purpose-less, jerky stepping movements of the forelimb contralateral to the lesion. Another common feature is small circular movements around the snout. As the severity increases, AIMs may be of a more hyperkinetic and/or dystonic character with prolonged circular movements involving the full arm and shoulder.

3.2.3. Orolingual AIMs

The facial, tongue and masticatory muscles are involved in this type of AIMs and are recognized as empty jaw movement, twisting of facial muscles and occasionally tongue protrusion on the side contralateral to the lesion. When this subtype reaches substantial severity, it may involve self-mutilating biting of the contralateral forepaw, easily detectable as a round bald patch on the forelimb. In extraordinary circumstances with chronic testing, this can lead to sores, since these animals are typically severely dyskinetic, refraining from injecting on non-test days will allow the skin to heal between sessions without adversely affecting behaviour.

3.2.4. Hindlimb AIMs

The contralateral hindlimb may also be taken into consideration. Scoring of the hindlimb was not included in the original Cenci scale, but some researchers have begun to use this as an additional measure (18). In some animals following L-DOPA administration, the hindlimb is abnormally postured, twisted but still supporting weight or elevated and non-weight bearing.

3.2.5. Locomotive AIMs

The locomotive AIM subtype is described as increased locomotion with contralateral side bias, i.e. the animal will turn towards the side contralateral to the lesion. Locomotive activity requires tactile contact of at least three paws on the floor, which rules out scoring this subtype of AIMs when rodents are sitting or standing on their hindlimbs, thus keeping their forelimbs off the floor.

It has been repeatedly shown that the amount of contralateral rotation induced by L-DOPA is directly related to the locomotive AIM score and is a less representative AIM than the other three subtypes. Locomotive AIMs are included in the ratings in conformity with the first description of the rodent AIMs scale, but it generally accepted that this is excluded when analysing the combined AIMs score, and considered as a separate measure. It is useful to maintain this measure so that a score can be obtained of locomotor activity if rotometers are not used to assess locomotion independently.

3.3. Ratings of AIMs Using the Amplitude Scale

As the original rating scale only takes into account the amount of time that the rodents are displaying abnormal movements and not that the quality of these movements may change with chronic L-DOPA administration, Winkler and co-authors implemented a complementary amplitude scale (14) to further expand the dynamic range of the ratings. This scale is very sensitive to variations in the amplitude of the dyskinetic movements between different phases of the L-DOPA treatment and/or between individual animals.

Amplitude scores are given to axial, limb and orolingual AIMs, and the score should reflect the maximum amplitude of the AIM observed during the 1-min observation period.

3.3.1. Amplitude of Limb AIMs

Score L1: Tiny movements of the paw around a fixed position, either as lateral or circular movements around the snout or as a repetitive tapping of the forepaw on the floor (as if the rodent was about to start walking, but could not).

Score L2: Movements resulting in visible displacement of the whole limb, e.g. the paw looses contact with the snout, and reaches halfway to the floor.

Score L3: Large displacement of the whole limb with visible contraction of shoulder muscles.

Score L4: Vigorous limb displacement of maximal proximal amplitude, with conspicuous contraction of both shoulder groups and extensor muscles.

- – *Extensor muscles*: those on the backside of the paw.
- – *Vigorous*: the movement is very energetic.
- – *Maximal proximal amplitude*: if the movement is circular, the limb is displaced around approximately half of the circumference around the body, if the movement is sagittal and/or frontal, the limb is lifted up to reach an angle of >90° with respect to the body.

3.3.2. Amplitude of Axial AIMs

Score A1: Sustained deviation of the head and neck at ~30° angle.

Score A2: Sustained deviation of the head and neck at angle ≤60°.

Score A3: Sustained of the head, neck and upper trunk at an angle >60° but ≤90°.

Score A4: Sustained twisting of the head, neck and upper trunk at an angle >90° (maximal amplitude), causing the animal to loose balance (from a bipedal position).

Angle is calculated as a deviation from the longitudinal axis of the body.

- • In A1 and A2, the rodent can still maintain a quadrupedal position, whereas A3 and A4 refer to torsion of the upper body forcing the rat to assume a bipedal position.

3.3.3. Amplitude of Orolingual AIMs

Score O1: Twitching of facial muscles accompanied by small masticatory movements without jaw opening.

Score O2: Twitching of facial muscles accompanied by noticeable masticatory movements, occasionally leading to jaw opening.

Score O3: Movement with broad involvement of facial and masticatory muscles. Jaw opening is frequent, but tongue protrusion is occasional.

Score O4: All of the above muscle categories are involved to a maximal possible degree. Tongue protrusion is more frequent.

3.3.4. Amplitude
of Hindlimb AIMs

Score H1: Twisting of the hindlimb but maintained under the body in proximity to the ipsilateral hindlimb.

Score H2: Extension of the hindlimb out from under the body, generally only partially weight bearing and may be twisted.

Score H3: (a) Hindlimb is elevated and non-weight bearing, but not fully dystonic (i.e. only partially extended) or (b) maximally dystonic but remaining in contact with the floor.

Score H4: The hindlimb is maximally dystonic and elevated, i.e. non-weight bearing.

• In H3 and H4, the animal may be supporting itself against the wall of the observation cage or on the floor.

By using the amplitude scale, the dynamics of the AIMs ratings is expanded. For example, the maximum possible sum of axial, orolingual, limb and hindlimb AIMs increases from 16 to 64 and the theoretical maximum score in one test from 144 to 576 (see Note 7).

4. Analysis of the Data

Bearing in mind that the locomotive AIM subtype is more related to contralateral turning than the expression of AIMs, it is strongly advised to analyse this subtype separately from the other three subtypes.

The AIMs scores recorded from one testing session can be expressed in three different ways.

1. As the sum of axial, orolingual and limb AIMs on all observation periods.

2. As the integral of AIM scores over time (area under the curve, AUC, referred to as the "integrated AIMs score").

3. As the sum of the products (basic score × amplitude score) on each observation period (referred to as "global AIMs score").

The most commonly used way of expressing data is as the sum of axial, orolingual and limb AIMs. More recently, global AIMs have become more popular in particular when comparing different lesion types and treatment groups as it expands the dynamic range of the AIMs ratings. In many studies, AIMs are observed during a chronic L-DOPA treatment period and in those studies, the development of AIMs is easily presented as a line graph over time (see Fig. 1a). Other interesting parameters may be the duration of the dyskinetic response (see Fig. 1b), and the peak severity of AIMs (see Fig. 1b).

The development of AIMs during the treatment period is often analysed using repeated-measures analysis of variance (ANOVA) even though non-parametric statistics are often more appropriate to use when handling rating scales which are considered ordinal data. However, the basic AIM scores reflect the time a specific movement is present unlike many other scales where a score often represent if a behaviour is present or not. It has also been shown that the AIM scores are strongly linearly correlated to parametric quantities such as the number of FosB-positive cells and levels of prodynorphin mRNA in the striatum (5, 6). Consequently, a basic score of 2 can be regarded as a quantity representing half as much as a basic score of 4. So using parametric statistics is not necessarily incorrect and it does have a descriptive power superior to non-parametric tests. We therefore suggest using parametric statistics (as long as basic statistical rules are followed, i.e. normal distribution of data) and in particular using repeated-measures ANOVA to analyse the development of AIMs over time. Repeated-measures ANOVA provides valuable information on the interaction between a time factor (testing session) and a group category (treatment), which is not readily available in non-parametric tests. However, it is recommended to corroborate key results with non-parametric statistics.

5. Experimental Considerations

Despite having full lesions, in every cohort of rats a proportion will not develop AIMs in response to therapeutic doses of ʟ-DOPA. Typically, around 20% of the rats will remain "non-dyskinetic", showing behavioural activation (contralateral rotations) and functional improvement in response to ʟ-DOPA, but showing no or few AIMs with low amplitude. Non-dyskinetic animals have been defined as having a score of 0–1 on each AIM subtype on all time points for the duration of the ʟ-DOPA challenge (19, 20). It has also been shown that animals with AIM severity grade ≤1 do not exhibit molecular markers of LID, such as the activation of extracellular signal-regulated kinases 1/2 and up-regulation of ΔFosB in striatal neurons (20, 21), and can thus be considered comparable to the animals showing absolutely no AIMs in response to ʟ-DOPA.

6. Notes

1. *Strain and species.* In contrast to the widely used rat model of LID, only a few groups have studied AIMs in the mouse although the use of mice is increasing rapidly with the need to use transgenic and knock out models (22–25). The lesion type

is an important consideration as highlighted by Smith and Heuer (Chap. 16, this volume) since both survival rates and post-surgical care are variable with the degree of DA depletion and location of injection site. In addition, non-DA damage has also to be considered when choosing the site of lesion. The strain of rodent used may also influence the development of AIMs and the response to L-DOPA. In studies comparing rat strains, we have found no difference between Sprague Dawley and Lister Hooded rats with lesions of the mfb, but Wu and colleagues reported a significantly lower propensity to develop AIMs in Lewis rats (26), which may be related to the expression of dopamine transporter in this strain. When considering mouse strains, there are vast differences in levels of spontaneous activity, free dopamine, dopamine transporters and dopamine receptors among different mouse strains. AIMs experiments with C57/BL6 and CD1 strains have been reported, but since these have not been compared systematically, it is quite likely that the expression of AIMs will vary, so caution would be advised when working with new strains.

2. *Dopamine depletion and AIMs:* Modelling of LID in rodents was first characterised in unilaterally dopamine-depleted rats with 6-OHDA lesions of the mfb (described by Torres and Dunnett, this volume) by Cenci and co-workers in the late 1990s (5, 6). When given to rats, L-DOPA can only induce AIMs when the DA depletion in striatum exceeds >80%, and AIMs of the highest severity appear only when more than 90% of the DA innervation is lost (14). Rats with striatal DA lesions have been used in studies of LID (14), as have lesions of the substantia nigra (27), but typically these produce a less severe lesion, and subsequently less severe abnormal movements. Similarly, the terminal and cell body lesions can be more selective to nigrostriatal dopamine with less damage to mesolimbic DA projections, thereby altering the locomotor responses. In addition to DA consequences, it is important to consider that these different types of lesions have distinct effects on non-DA systems, for example lesions of the mfb are more likely to cause non-specific damage to the serotonergic (19) and the noradrenergic neurons of the midbrain (28). The implications of such non-specific degeneration are uncertain, some suggesting that this is a closer representation of the true PD, so it is important to know in each case what has been lesioned and to what extent for accurate interpretation of results. Generating rodents with extensive bilateral 6-OHDA lesions can be ethically and practically difficult as the survival of the animals is compromised by their reduced ability to eat and drink (29, 30). Few studies have been studying AIMs in bilaterally lesioned rodents, but as expected the animals showed AIMs on both sides of the body (31, 32).

Herrmann et al left a gap of 2 weeks between the lesion of each hemisphere and interestingly, they report that AIMs actually worsened on the original side following the second lesion, suggesting a degree of interhemispheric coupling in nigrostriatal function (31). Selecting lesioned animals for AIMs studies can be done using the same techniques as testing for loss of motor function (e.g. using cylinder, stepping or paw reaching tests, see Schallert, this volume), but drug-induced rotation using apomorphine should be avoided. As a non-selective D_1/D_2 receptor agonist with a short half-life, this drug vigorously activates striatal dopamine D_1 receptors, the major pathway through which LID is thought to be established. Therefore, apomorphine administration can accelerate the onset of L-DOPA-induced AIMs, and it is therefore preferable to test rotation using agents that preferentially stimulate the intact hemisphere, such as d-amphetamine and methamphetamine. As they are dopamine releasing and reuptake inhibiting agents, they will act predominantly through the intact hemisphere with significantly less propensity to sensitise the dopamine receptors on the lesioned side of the striatum.

3. *L-DOPA/Benserazide dosage*: L-DOPA is available both as the L-DOPA base and as L-DOPA methyl ester hydrochloride. The methyl ester is most commonly used experimentally due to its high water solubility, and therefore ease of dissolution. The L-DOPA solution should be freshly made every day and used within an hour of dissolving, since it rapidly oxidises, evidenced by a yellow/brown colouration of the solution. The usage of benserazide in the rodent studies is in line with the usage of peripheral amino acid dopa decarboxylase (AADC) inhibitors in the clinic, usually either benserazide or carbidopa at a fixed ratio of 1:5. In rodents, however doses of the inhibitors higher than 8 mg/kg are required to block most of the peripheral AADC activity (33). Lower doses of benserazide result in shorter L-DOPA duration so optimal doses range between 12 and 15 mg/kg/day, which is sufficient to block most peripheral AADC activity without affecting the central activity of the enzyme (34).

4. *Testing environment*: It is well established that the environment in which behavioural observations are made strongly affects the outcome of the test. The traditional testing environment for the rodent model of LID has been in transparent rectangular cages, but recent studies have examined the influence of different testing environments on the expression of AIMs (35). When rats are tested in hemispherical bowls, originally used for assessing drug-induced rotations, they are more prone to rotate than to display AIMs affecting the forelimbs and trunk. In contrast, those latter motor features are more promoted in rectangular

cages where the rats can support themselves against the walls and corners of the cage (35). As long as the rats are tested consistently in the same environment during the chronic treatment period, the decision of which type of testing cage will not influence the overall result. Several recent studies have shown that if rodents are only exposed to the test environment in the presence of L-DOPA, they may be vulnerable to two forms of conditioning, over and above the pharmacological sensitisation associated with repeated dosing (18). They may show context dependent conditioning, i.e. mild AIMs in the absence of L-DOPA when placed in the environment again, or, more importantly, the sensitisation of the behavioural response may become linked to the environmental context, i.e. context-specific sensitisation (18). To reduce the impact of these conditioning events on the behavioural scores care should be taken to acclimatise animals to the chambers for at least 15 min prior to L-DOPA administration.

5. *Route of administration*: Intraperitoneal injections of L-DOPA produce a robust behavioural response and sensitisation, but the responses can be variable, with the occurrence of dose failure episodes, i.e. an unpredictable failure to respond to single L-DOPA doses. Dose failure episodes have repeatedly been reported in the literature (36–38), but can be avoided by using the subcutaneous route of administration (9). This may produce a slight shift in the temporal course of the response, which may vary depending on where the injection is given, but will otherwise produce the same profile (see Fig. 1c).

6. *Training on the rating scales*: Each experiment should include at least one control group of drug naive 6-OHDA-lesioned rodents injected with saline (and possibly a second control group of intact rats) that should be rated alongside the L-DOPA-treated animals. This provides a baseline control group and during the learning processes enables the discrimination between normal and abnormal motor patterns. During the learning phase, it is also advisable not to include too many animals in one testing session, thus allowing for a longer observation time per animal instead of the standard 1-min that can be managed by an experienced rater. Inter-rater reliability is important to control for videos can be rated and kept to validate different raters, and ensuring that this is maintained regularly, especially if new parameters are assessed.

7. *Theoretical maxima*: As mentioned, it is common to score the locomotor AIM subtype, but to exclude this from any summation of the AIMs scores. Importantly, it should be remembered that especially if locomotion is included, the theoretical maximum will never be reached because if locomotion is present it is not possible to score a 4 on some of the non-locomotor elements, for example the definition of axial movements 4 is loss of

balance. Similarly, it is unlikely that animals will be maximally dyskinetic for the duration of the experiment, as shown in Fig. 1c, animals will tend to start lower, peak after 40–60 min for an hour or more, then decline over time. Examining the time course is therefore as important as the total AIM scores to determine where or how the behaviours have altered, whether it is a reduction in the peak response, or a shift in the overall time course.

7. Concluding Remarks

Ratings of L-DOPA-induced AIMs can sometimes be troublesome in all species, not only in rodents. Basic knowledge of vital experimental parameters such as lesion type, L-DOPA-dose, treatment duration and testing environment, and how changes in these will affect the outcome of the study, is crucial when predicting results for new therapeutic approaches. In conclusion, consistency is the key to studies of AIMs and maintaining strict experimental parameters, in particular on the testing day, will result in reliable and reproducible results.

Acknowledgements

HSL is currently supported on a fellowship provided by the Swedish Research Council. Work carried out in these labs is supported by EU FP7 framework grants REPLACES (EC contract no. 222918) and TRANSEURO (EC contract no. 222918).

References

1. Ahlskog JE, Muenter MD (2001) Frequency of levodopa-related dyskinesias and motor fluctuations as estimated from the cumulative literature. Mov Disord 16:448–458
2. Marsden CD (1990) Parkinson's disease. Lancet 335:948–952
3. Grandas F, Galiano ML, Tabernero C (1999) Risk factors for levodopa-induced dyskinesias in Parkinson's disease. J Neurol 246:1127–1133
4. Schrag A, Quinn N (2000) Dyskinesias and motor fluctuations in Parkinson's disease. A community-based study. Brain 123 (Pt 11): 2297–2305
5. Andersson M, Hilbertson A, Cenci MA (1999) Striatal fosB expression is causally linked with L-DOPA-induced abnormal involuntary movements and the associated upregulation of striatal prodynorphin mRNA in a rat model of Parkinson's disease. Neurobiol Dis 6:461–474
6. Cenci MA, Lee CS, Bjorklund A (1998) L-DOPA-induced dyskinesia in the rat is associated with striatal overexpression of prodynorphin- and glutamic acid decarboxylase mRNA. Eur J Neurosci 10:2694–2706
7. Clarke CE, Sambrook MA, Mitchell IJ, Crossman AR (1987) Levodopa-induced dyskinesia and response fluctuations in primates rendered parkinsonian with 1-methyl-4-phenyl-1,2,3,6-tetrahydropyridine (MPTP). J Neurol Sci 78:273–280
8. Kuoppamaki M, Al-Barghouthy G, Jackson MJ, Smith LA, Quinn N, Jenner P (2007) L-dopa

dose and the duration and severity of dyskinesia in primed MPTP-treated primates. J Neural Transm 114:1147–1153

9. Lindgren HS, Rylander D, Ohlin KE, Lundblad M, Cenci MA (2007) The "motor complication syndrome" in rats with 6-OHDA lesions treated chronically with L-DOPA: relation to dose and route of administration. Behav Brain Res 177:150–159

10. Bezard E, Ferry S, Mach U, et al. (2003) Attenuation of levodopa-induced dyskinesia by normalizing dopamine D3 receptor function. Nat Med 9:762–767

11. Di Monte DA, McCormack A, Petzinger G, Janson AM, Quik M, Langston WJ (2000) Relationship among nigrostriatal denervation, parkinsonism, and dyskinesias in the MPTP primate model. Mov Disord 15:459–466

12. Lundblad M, Andersson M, Winkler C, Kirik D, Wierup N, Cenci MA (2002) Pharmacological validation of behavioural measures of akinesia and dyskinesia in a rat model of Parkinson´s disease. Eur. J. Neurosci. 15:120–132

13. Pearce RK, Banerji T, Jenner P, Marsden CD (1998) De novo administration of ropinirole and bromocriptine induces less dyskinesia than L-dopa in the MPTP-treated marmoset. Mov Disord 13:234–241

14. Winkler C, Kirik D, Bjorklund A, Cenci MA (2002) L-DOPA-induced dyskinesia in the intrastriatal 6-hydroxydopamine model of parkinson's disease: relation to motor and cellular parameters of nigrostriatal function. Neurobiol Dis 10:165–186

15. Dekundy A, Lundblad M, Danysz W, Cenci MA (2007) Modulation of L-DOPA-induced abnormal involuntary movements by clinically tested compounds: further validation of the rat dyskinesia model. Behav Brain Res 179:76–89

16. Westin JE, Andersson M, Lundblad M, Cenci MA (2001) Persistent changes in striatal gene expression induced by long-term L-DOPA treatment in a rat model of Parkinson's disease. Eur J Neurosci 14:1171–1176

17. Cenci MA, Lundblad M (2007) Ratings of L-DOPA-induced dyskinesia in the unilateral 6-OHDA lesion model of Parkinson's disease in rats and mice. Curr Protoc Neurosci Chapter 9:Unit 9

18. Lane EL, Daly CS, Smith GA, Dunnett SB (2011) Context-driven changes in L-DOPA-induced behaviours in the 6-OHDA lesioned rat. Neurobiol Dis, 42:99–107

19. Lindgren HS, Andersson DR, Lagerkvist S, Nissbrandt H, Cenci MA (2009) l-DOPA-induced dopamine efflux in the striatum and the substantia nigra in a rat model of Parkinson's

disease: temporal and quantitative relationship to the expression of dyskinesia. J Neurochem 112: 1465–76

20. Westin JE, Vercammen L, Strome EM, Konradi C, Cenci MA (2007) Spatiotemporal pattern of striatal ERK1/2 phosphorylation in a rat model of L-DOPA-induced dyskinesia and the role of dopamine D1 receptors. Biol Psychiatry 62:800–810

21. Westin JE, Lindgren HS, Gardi J, Nyengaard JR, Brundin P, Mohapel P, Cenci MA (2006) Endothelial proliferation and increased blood-brain barrier permeability in the basal ganglia in a rat model of 3,4-dihydroxyphenyl-L-alanine-induced dyskinesia. J Neurosci 26:9448–9461

22. Lundblad M, Picconi B, Lindgren H, Cenci MA (2004) A model of L-DOPA-induced dyskinesia in 6-hydroxydopamine lesioned mice: relation to motor and cellular parameters of nigrostriatal function. Neurobiol Dis 16:110–123

23. Pavon N, Martin AB, Mendialdua A, Moratalla R (2006) ERK phosphorylation and FosB expression are associated with L-DOPA-induced dyskinesia in hemiparkinsonian mice. Biol Psychiatry 59:64–74

24. Santini E, Valjent E, Usiello A, et al. (2007) Critical involvement of cAMP/DARPP-32 and extracellular signal-regulated protein kinase signaling in L-DOPA-induced dyskinesia. J Neurosci 27:6995–7005

25. Xiao D, Bastia E, Xu YH, et al. (2006) Forebrain adenosine A2A receptors contribute to L-3,4-dihydroxyphenylalanine-induced dyskinesia in hemiparkinsonian mice. J Neurosci 26: 13548–13555

26. Wu Q, Levine N, Maries N, Sortwell C, Steece-Collier K (2005) Strain differences in development of levodopa induced dyskinesia between parkinsonian Lewis and Fischer 344 rats. Experimental Neurology 193:263–264

27. Paille V, Brachet P, Damier P (2004) Role of nigral lesion in the genesis of dyskinesias in a rat model of Parkinson's disease. Neuroreport 15:561–564

28. Fulceri F, Biagioni F, Lenzi P, Falleni A, Gesi M, Ruggieri S, Fornai F (2006) Nigrostriatal damage with 6-OHDA: validation of routinely applied procedures. Ann N Y Acad Sci 1074:344–348

29. Ungerstedt U (1971) Adipsia and aphagia after 6-hydroxydopamine induced degeneration of the nigro-striatal dopamine system. Acta Physiol Scand Suppl 367:95–122

30. Zigmond MJ, Stricker EM (1972) Deficits in feeding behavior after intraventricular injection of 6-hydroxydopamine in rats. Science 177:1211–1214

31. Herrmann J, Timmer M, Lundblad M, Cenci MA, Nikkah G, Winkler C (2008) L-DOPA-induced dyskinesia after uni- and bilateral dopamine-denervating lesions in the rat Parkinsonian model. Abstract at the 10th International Conference on Neural Transplantation and Repair Sep 10–13, Freiburg, Germany

32. Paille V, Henry V, Lescaudron L, Brachet P, Damier P (2007) Rat model of Parkinson's disease with bilateral motor abnormalities, reversible with levodopa, and dyskinesias. Mov Disord 22:533–539

33. Da Prada M, Kettler R, Zurcher G, Schaffner R, Haefely WE (1987) Inhibition of decarboxylase and levels of dopa and 3-O-methyldopa: a comparative study of benserazide versus carbidopa in rodents and of Madopar standard versus Madopar HBS in volunteers. Eur Neurol 27 Suppl 1:9–20

34. Shen H, Kannari K, Yamato H, Arai A, Matsunaga M (2003) Effects of benserazide on L-DOPA-derived extracellular dopamine levels and aromatic L-amino acid decarboxylase activity in the striatum of 6-hydroxydopamine-lesioned rats. Tohoku J Exp Med 199: 149–159

35. Pinna A, Pontis S, Morelli M (2006) Expression of dyskinetic movements and turning behaviour in subchronic L-DOPA 6-hydroxydopamine-treated rats is influenced by the testing environment. Behav Brain Res 171:175–178

36. Marin C, Aguilar E, Bonastre M, Tolosa E, Obeso JA (2005) Early administration of entacapone prevents levodopa-induced motor fluctuations in hemiparkinsonian rats. Exp Neurol 192:184–193

37. Mura A, Mintz M, Feldon J (2002) Behavioral and anatomical effects of long-term L-dihydroxyphenylalanine (L-DOPA) administration in rats with unilateral lesions of the nigrostriatal system. Exp Neurol 177:252–264

38. Papa SM, Engber TM, Kask AM, Chase TN (1994) Motor fluctuations in levodopa treated parkinsonian rats: relation to lesion extent and treatment duration. Brain Res 662:69–74

Chapter 19

Using the MPTP Mouse Model to Understand Neuroplasticity: A New Therapeutic Target for Parkinson's Disease?

Giselle M. Petzinger, Beth E. Fisher, Garnik Akopian, Ruth Wood, John P. Walsh, and Michael W. Jakowec

Abstract

Since the first identification of 1-methyl-4-phenyl-1,2,3,6-tetrahydropyridine (MPTP) as a selective neurotoxin for nigrostriatal dopaminergic neurons in 1983, there have been over 2,000 manuscripts published utilizing this compound in mice, attesting to the value of this model. Most of this work is focused on neuroprotection and mechanisms of cell death. While MPTP may not replicate all the features of PD, it provides a key means by which depletion of striatal dopamine can be achieved and by which brain repair mechanism(s) involved in compensation may be studied, including the role of exercise. In this chapter, we have outlined some of the methods utilized in our laboratories to investigate the physiological and molecular correlates of exercise-induced neuroplasticity (brain repair processes) in the MPTP mouse model of PD. Specifically, we have shown that both treadmill and voluntary running wheel can be used to modulate motor and nonmotor-related behaviors. We also demonstrate methods used in our laboratory to examine the effects of experience-dependent neuroplasticity (exercise) on both pre- (dopaminergic/glutamatergic) and post- [striatal medium spiny neurons (MSNs)] pathways. Our studies demonstrate that intensive treadmill exercise can improve motor and nonmotor behaviors through modulation of both glutamatergic and dopaminergic neurotransmission. We have also shown that voluntary running wheel may be used to examine the nonmotor benefits of exercise.

Key words: Exercise, Basal ganglia, Neurophysiology, Dopamine, Glutamate, Dendritic spines, Motor behavior, Mood disorders

1. Introduction

The inadvertent self-administration of 1-methyl-4-phenyl-1,2,3,6-tetrahydropyridine (MPTP) by heroin addicts in the late 1970s and early 1980s induced an acute form of parkinsonism whose clinical features were indistinguishable from idiopathic Parkinson's disease (PD) (1, 2). Immediately following its identification, MPTP

Emma L. Lane and Stephen B. Dunnett (eds.), *Animal Models of Movement Disorders: Volume I*, Neuromethods, vol. 61, DOI 10.1007/978-1-61779-298-4_19, © Springer Science+Business Media, LLC 2011

was administered to both rodents and nonhuman primates and some of the most valuable animal models of PD were developed (3). Currently, the MPTP mouse remains one of the most commonly used animal models of PD. Since its inception, the majority of PD studies have utilized this model as a means to examine mechanisms of cell death of nigrostriatal dopaminergic neurons and neuroprotection studies (4). More recently, however, there has been a general interest in understanding how the brain following injury or neurodegenerative disease processes (such as PD) may benefit from repair mechanisms induced by experience, including enriched environment, physical activity, and exercise. As such, our laboratories have been particularly interested in understanding the effects of intensive, challenging exercise and skill acquisition on repair processes of the striatum. For these studies, we have focused on the "acute" MPTP mouse model and provide rationale below for its particular utility in exercise studies. In this chapter, we highlight methods in our laboratories that are used to study exercise and neuroplasticity in the MPTP mouse that will include (1) the administration, and safety issues regarding MPTP, and assessment of lesioning; (2) the forced treadmill and voluntary running wheel exercise paradigms; (3) motor and nonmotor assessments to evaluate behavioral benefits of exercise; (4) electrophysiological studies to assess synaptic function of the striatum; and (5) biocytin labeling to evaluate exercise effects on spine morphology and dendritic spine density.

2. Methods

2.1. Safety Issues with MPTP

MPTP is a neurotoxic substance that must be handled with a high degree of caution to protect research personnel. We have published a thorough review on issues of safety when handling MPTP (5). Since Sigma has begun packaging MPTP-HCl preweighed in 100 mg per vial enclosed with a rubber septum, there is little risk of exposure to toxin. Water or saline can be introduced through the rubber septum and syringes easily loaded without potential risk of exposure due to aerosol or spillage. Injections can be performed in a fume hood either in the animal facility or laboratory. Following injections of MPTP, mice are housed in the same cages and then transferred to clean cages after 48 h. The half-life of MPTP due to clearance is 3 h and therefore no detectable MPTP remains in the mice after the 2-day period (6). Intra-nasal passage of MPTP through aerosolization is unlikely to occur using the delivery method outlined above. In addition, studies examining normal cage mates of animals lesioned with MPTP in the vicinity of a potential aerosolized exposure due to urine or saliva failed to display any detectable levels of dopamine depletion (7). Thus, the hazard risk of using MPTP with the technique outlined is low. While nonhuman primates show relatively high sensitivity to MPTP (typically 0.2 mg/kg via intracarotid injection routes) such exposures cannot normally be achieved by researchers through passive exposure and provide little risk if proper precautions are taken.

2.2. MPTP Preparation and Injections

MPTP is a meperidine derivative available in a number of different conjugates. The free-base form of MPTP-HCl is soluble in aqueous solutions. MPTP-HCl is purchased from Sigma (St. Louis, MO) in vials containing 100 mg (equal to 84.6 mg free-base). Water or saline (18.4 ml) can be safely introduced into the vial by needle and syringe through the rubber septum to generate a solution of 5 mg/ml (free-base). Preweighing mice to screen for those weighing 25 g (the typical weight for 10-week-old C57BL/6 male mice) allows injection of 100 μl of MPTP solution per mouse. Mice receive four injections of 20 mg/kg free-base with 2–3 h between injections. Injections target the intraperitoneal space using a tuberculin syringe with a 26-gauge needle. We have used subcutaneous injections between the shoulders but have found the intraperitoneal delivery easier to perform and not requiring a restrainer. Recently, a number of laboratories, including ours, have noted increased animal death with the "acute" regimen beyond the typical 10–20% level. In an attempt to increase animal survival, some researchers have elected to decrease the concentration of MPTP from 20 to 10–15 mg/kg per injection. However, one must be cautious when making such an adjustment since the degree of lesioning will likely be diminished. Animal deaths are not strictly due to central effects, but rather are more likely caused by the cumulative or acute systemic effects on the heart, blood pressure, kidney, and liver (8–10). In our laboratory, we have identified several parameters that may facilitate animal survival including (1) handling mice routinely during the week before injections to reduce stress, (2) maintaining an ambient temperature around 72°F with warming pads to avoid hypothermia (take caution with heating lamps that may be too harsh), (3) increasing the time between injections from 2 to 2.5 or 3 h (the half-life of MPTP in the mouse is 3 h and will not alter brain concentrations dramatically), thus allowing the animal to recover from systemic effects, and finally (4) ensuring easy access to food and water (placement at the bottom of the cage), thus avoiding dehydration or starvation.

2.3. Selection of Species, Age, and Sex

Mice (*Mus musculus*) are the most common rodent species for which MPTP is used. Rat species are typically resistant to systemic lesioning with MPTP and require stereotactic delivery of MPP+ targeting the striatum or ventricles. While the C57BL/6 strain is often used, many different mouse strains can be lesioned by MPTP to varying degrees (11–13). When using different strains it is important to assess the degree of MPTP-lesioning by examining the loss of striatal dopamine, depletion of striatal TH immunoreactivity, and the number of SNpc neurons lost (see below). This is an important issue, especially when using transgenic strains that are often derived from a number of different strains including FVB/N, and others, often in a mixed background. To avoid possible variability in lesioning or mild lesions due to strain backgrounds that show partial resistance to MPTP, transgenic strains can be backcrossed into the C57BL/6 background. Typically ten generations are required to achieve congenic background. MPTP is typically administered to mice that are 8–10 weeks of age, since younger mice are resistant to MPTP due to low levels of monoamine oxidase B (MOA-B) and older mice (>1 year of age) show increased lethality to MPTP (14, 15). When using older mice such as retired breeders up to 1 year of age, no adjustments in the lesioning regimen are necessary. However, mice greater than 18 months of age show increased lethality toward MPTP with increased susceptibility to lesioning. For example, in 18-month-old mice, a regimen of four injections of 10 mg/kg free-base i.p. will result in a degree of lesioning similar to four injections of 20 mg/kg free-base i.p. in 10-week-old mice (14, 15). Finally, most studies utilize male mice since changes in estrogen levels within the estrous cycle can influence both the degree of MPTP-lesioning and the levels of expression of indicators of basal ganglia integrity including TH protein (16). The important issue is that within an experiment using different strains of mice, one must assess the degree of lesioning with respect to dopamine depletion and cell loss.

2.4. Metabolism of MPTP

Following systemic administration, MPTP crosses the blood–brain barrier. The meperidine analog MPTP is converted to 1-methyl-4-phenyl-2,3-dihydropyridinium (MPDP+) by monoamine oxidase B (primarily in astrocytes) and then spontaneously or enzymatically oxidizes to 1-methyl-4-pyridinium (MPP^+), its toxic form. MPP^+ acts as a substrate of the dopamine transporter (DAT) and is taken up by SNpc neurons, leading to the inhibition of mitochondrial complex I, the depletion of ATP, and cell-specific death of nigrostriatal dopaminergic neurons. DAT is the primary uptake mechanism and accounts for the specificity of MPP^+ to dopaminergic neurons. We and others have found that DAT expression or activity can be altered by pharmacological or behavioral manipulations and therefore influence the degree of lesioning (dopamine depletion). As a cautionary note, these factors should be considered in interpreting any experimental study using MPTP. For example, our studies have shown that exercise itself can down-regulated DAT expression on nigrostriatal dopaminergic terminals and therefore reduce cell-specific uptake of MPP^+ (17). Hence, such an alteration in MPP^+ bioavailability will reduce the degree of MPTP-lesioning despite no change in the lesioning regimen itself. Another factor influencing the bioavailability of MPP^+ is the vesicular monoamine transporter 2 (VMAT-2) that is responsible for sequestering MPP+ within dopaminergic striatal terminals. Increased expression of VMAT-2 leads to protection from MPP+ toxicity while decreased expression can increase susceptibility (18, 19).

2.5. Assessing the Degree of Lesioning

While, in general, consistency in the degree of nigrostriatal dopaminergic neuron cell death and striatal dopamine depletion can be achieved in mice when parameters of the lesioning regimen including concentration, route of administration, age, and strain of mice are all maintained constant, there may still be some variability in the degree of lesioning during MPTP administration. Thus, we routinely assess the degree of MPTP-lesioning in our mouse model during any set of experiments. Six to eight mice are typically sufficient to examine several parameters including (1) striatal dopamine depletion, (2) loss of striatal TH- or DAT-immunoreactivity, and (3) loss of midbrain dopaminergic neurons. We routinely assess all three of these parameters at 7 or 10 days postlesioning when cell death is complete and maximum dopamine depletion is reached.

Neurotransmitter concentrations are determined by HPLC according to an adaptation of the method of Kilpatrick et al. (20, 21). Tissues for analysis are homogenized in 0.4 N perchloric acid and centrifuged at $12,000 \times g$ to separate precipitated protein. The protein pellet is resuspended in 0.5 N NaOH and the total protein concentration determined using the Coomassie Plus protein assay system (Pierce, Rockford, IL) in a Biotek Model Elx800 microplate reader (Biotek Instruments, Wincoski, VT) and KCjunior software. The concentrations of dopamine, 3,4-dihydroxyphenylacetic (DOPAC) and homovanillic acid (HVA) are assayed by HPLC with electrochemical detection. Samples are injected with an ESA (Chelmsford, MA) autosampler. Dopamine and its metabolites are separated by a 150×3.2 mm reverse phase 3-µm-diameter C-18 column (ESA) regulated at 28°C. The mobile phase MD-TM (ESA) consists of acetyl nitrile in phosphate buffer and an ion-pairing agent delivered at a rate of 0.6 ml/min. The electrochemical detector is an ESA model Coularray 5600A with a four-channel analytical cell with three set potentials at −100, 50, and 220 mV. The HPLC is integrated with a Dell GX-280 computer with analytical programs including ESA Coularray for Windows software and the statistics package InStat (GraphPad Software, San Diego, CA). Concentration of dopamine is reported as ng/mg protein and typically ranges from 150 to 200 ng/mg protein in saline animals and 10–15 ng/mg protein for MPTP-lesioned mice.

The number of nigrostriatal dopaminergic neurons in the SNpc is determined using unbiased stereology with the computer-imaging program BioQuant Nova Prime (BioQuant Imaging, Nashville, TN) and an Olympus BX-50 microscope (Olympus Optical, Tokyo, Japan) equipped with a motorized stage and digital Retiga-cooled CCD camera (Q-Imaging, Burnaby, British Columbia, Canada).

Brain tissue is prepared from at least three mice in each group, tissue sliced at 30 μm thickness and every sixth section is collected and stained for TH-ir using a rabbit polyclonal antibody (Millipore) and counterstained for Nissl substance (20, 22). Section collection is started rostral to the SNpc at bregma 2.50 mm before the closure of the third ventricle through the prominence of the pontine nuclei at bregma 4.24 mm according to the stereotaxic atlas of the mouse brain (23). Each stained ventral mesencephalon section is viewed at low magnification (10× objective) and the SNpc outlined and delineated from the ventral tegmental TH-ir neurons using the third nerve and cerebral peduncle as landmarks. Neurons are viewed at high magnification (80× objective) and counted if they display TH-ir and have a clearly defined nucleus, cytoplasm, and nucleolus. The total number of SNpc dopaminergic neurons is determined based on the method of Gundersen and Jensen (24).

Western immune-blotting of protein samples prepared from the contralateral side of brains harvested for HPLC analysis is separated by gel electrophoresis, transferred in Towbin buffer to nitrocellulose filters, and probed with antibodies against TH or DAT with an antibody against alpha-tubulin to control for protein loading. Analysis on the Licor Odyssey (Lincoln, NB) permits fast quantitative analysis of protein levels. Some investigators preform immunohistochemical staining against TH in tissue sections. This method is not as reliably quantitative as western blotting but it can indicate a relative degree of lesioning. We have found that immunohistochemical staining for TH within a few days of MPTP will still generate some immune-reactivity, despite a severe loss of striatal dopamine. We suspect that immune-reactive epitopes may persist for a few days before their clearance and therefore do not directly correlate with dopamine depletion.

2.6. Selection of MPTP-Lesioning Regimen

The acute MPTP regimen of four times 20 mg/kg free-base injections given i.p. every 2 h over an 8-h period, cell death starts immediately, peaks at 12–24 h, and is complete by 72 h (25). Since studies in our laboratories have been particularly interested in understanding the effects of intensive and challenging exercise on repair processes of the striatum and motor learning, we delay the initiation of exercise to 5 days after MPTP-lesioning to a point when toxin-induced cell death is complete (17). Such a study design is selected to determine the effects of experience-dependent plasticity in the form of intensive treadmill exercise on restoration of motor behavior, and reparative mechanisms of both remaining dopaminergic neurons and postsynaptic striatal neuronal function and connectivity (17, 26). Other reports in the literature have examined the effects of exercise on providing neuroprotection from toxin-induced cell death. In such studies, exercise is initiated either before or during the lesioning period and continues during the phase of toxin-induced cell death (3–5 days for the acute regimen of MPTP and up to 1 month for 6-hydroxydopamine) (25, 27). For interpreting the effects of an early exercise intervention, one must be cognoscente of the potential impact of behavior on the bioavailability of toxin including its uptake, metabolism, or storage.

3. Functional Assessment

3.1. Exercise Paradigms

In our laboratories, we are interested in examining how either voluntary or forced exercise facilitates brain repair processes. For these studies and behavioral testing, mice are group-housed four per cage on a 12:12 reversed light–dark schedule (lights off at 11 a.m.). All exercise and behavioral testing are conducted in the animal room. Daily exercise begins 2 h after lights are turned off and under red light during the mouse active phase. Voluntary exercise is performed with running wheels in separate cages each containing a Fast-Trac running wheel (18 cm diameter, Bio-Serv Inc., Frenchtown, NJ) equipped with a bicycle odometer (CatEye, Boulder, CO) to measure distance and speed. Controls typically include mice housed in cages with either a locked wheel or no wheel. For interpretation of the degree of exercise, it is necessary that mice be housed singly and that an odometer record distance traveled. Typically, the most active phase for running is during the night. While few studies continuously monitor the velocity of the running wheel with more sophisticated programs, some laboratories are beginning to more carefully collect running data, since intensity, speed, and duration parameters have been established in human trials as promoting training-induced benefits.

For forced exercise studies, our laboratories have utilized a motorized 6-lane treadmill apparatus, inclined at 5° (Model EXER 6 M, Columbus Instruments, Columbus, OH). Each daily exercise session typically begins with an initial warm-up at slower speed, followed by an incremental increase in treadmill speed. Five days after MPTP-lesioning or saline administration, mice begin running on the treadmill at a speed of 8 m per minute (m/min) for 10 min with an incremental increase of 1 or 2 m/min every 10 min for a total of 30 min, a 5 min rest, and repeat. Mice are monitored for their ability to maintain a forward position on the treadmill and speed is increased gradually. Mice run for 6 weeks, 5 days per week, and achieve a typical running speed of 22–24 m/min. On the first two days, running incentive is maintained by a shock grid and beaded metal curtain as a tactile stimulus. Within a few days of running, we have found that mice do not require the shock plate and maintain their forward position for the entire hour.

A mouse that is unable to maintain a forward position in the treadmill and cannot run at the requisite speed is excluded from the study. A log of running speed, maximum velocity, and duration is recorded. Since we are interested specifically on the effects of intensive treadmill exercise on basal ganglia plasticity, mice are housed in traditional home cages and not in housing containing components of an enriched environment (28). Of note, we and others have observed that mice subjected to exercise may manifest transient hyperactivity within the first week after MPTP and immediately following the exercise intervention (29). Sedentary mice are exposed to either a nonmoving treadmill or a fixed running wheel on the same daily schedule as exercising mice.

3.2. Assessing Motor Behavior

A number of tests are available to assess motor behavior in mice including motorized treadmill, rotarod, and open field (17, 29, 30, see also Brooks, Holter and Glasl, Smith and Heuer, this volume). While the motorized treadmill is used as an exercise intervention in our laboratories, specific outcomes can be documented to monitor motor behavior on the treadmill including maximum running speed and running duration (endurance). Duration is captured on a daily basis and during the exercise intervention. The maximum running speed of animals can be captured weekly or at the end of the exercise paradigm. Maximum velocity is defined as the highest velocity at which the animal can maintain a forward position on the treadmill for 75% of a 10-min running trial, and is usually recorded prior to the commencement of the exercise intervention.

An accelerating rotarod (five-lane accelerating rotarod; Ugo Basile, Comerio, Italy) can be used to assess motor balance and coordination as well as a test of learning a novel motor task. Mice are placed on the rotarod (3 cm diameter) (described in Chap. 1, this volume) and they are challenged to stay on the apparatus when both speed and direction are altered. We have found that setting the velocity to 30 rpm and changing direction (forward and reverse) every 24 s is sufficiently difficult for both saline and MPTP mice as a starting point (17). The test consists of five consecutive trials each separated by 1 min, with a maximum cutoff latency of 200 s. The latency to fall in each trial is recorded. The rotarod can also be used to test for motor endurance. This protocol uses a spindle designed for rats (7.3 cm diameter) to minimize dependence on motor coordination. For these studies, mice are subjected to seven trials at increasing speed from 12 to 24 rpm at 2 rpm intervals with each trial lasting 150 s with 150 s rests between trials. For these studies, animals are acclimated to the rotarod at a low speed (5 rpm) for 10 min, and prior to MPTP administration.

Several laboratories have used the open field test as a means to assess motor behavior (described in Chap. 7, this volume). Typically, total exploratory distance traveled is recorded as well as other parameters including mid-field crossings and time spent exploring along the wall. Since this behavioral test is also sensitive to nonmotor features such as anxiety, its reliability as a pure motor test is interpreted with caution and we do not routinely use this test for the assessment of motor behavior.

3.3. Assessing Nonmotor Behavior

While the majority of studies with exercise in patients with PD and in animal models have focused on motor behavior benefits, there is growing interest in the potential benefits of exercise on nonmotor behavior, including cognitive function and mood disorders (anxiety and depression). In our laboratories, we have focused our studies on examining the effects of exercise on both anxiety and depression and have used tests that have sensitivity to detect changes in mood (31, 32). Sucrose preference and tail suspension are sensitive instruments for depression and are typically measured at the beginning of the dark phase (active). Elevated plus maze and marble burying are used to assess features of anxiety and examined at the end of the light phase (less active). Our studies in the MPTP mouse have shown that both forced and voluntary exercises reduce mood disorders (31).

The Sucrose Preference test can be used to examine anhedonia, a key component of depression, by evaluating any reduction in the preference of a mouse to drink their preferred dilute sucrose solution. Typically, animals are tested weekly during the exercise intervention. Mice are first exposed to the sucrose solution overnight and 2 days before testing, by placing two 50-ml bottles fitted with ball-point drinking spouts, each containing 2% sucrose solution. Mice are then water-deprived over night prior to testing to increase their desire for fluid consumption. On testing day, and during the dark phase, animals are then individually housed with a 1-h access to two bottles, one containing 2% sucrose and the other containing water. For animals tested multiple times over a course of exercise, sucrose bottles are alternated to minimize a preference for the animal to only move toward one side of the testing cage. Fluid consumption is determined by weight. Tail suspension measures behavioral despair, another symptom of depression. For testing, mice are individually suspended from their tails at a height of 20 cm using a piece of adhesive tape wrapped around the tail 2 cm from the tip and behavior is videotaped for 6 min. Duration of immobility (completely motionless), which is thought to represent despair, is measured.

Elevated Plus maze is a test of anxiety and takes advantage of a mouse preference for enclosed space versus its drive to explore a novel environment. The maze consists of two open arms perpendicular to two closed arms and located 50 cm above the floor. For testing, a mouse is placed on a central platform facing an open arm, and the behavior is videotaped for 6 min. The number of open and closed arm entries, and duration of time on the open arm are scored.

Marble burying is a test of defensive behavior and an act of coping strategy in response to a discrete threat and is considered a measure of anxiety. An individual mouse is placed in a novel cage filled with a 10-cm thick pine bedding and containing a cluster of twenty-five, 10-mm diameter blue glass marbles. After 30 min, the mouse is removed and the number of buried marbles scored.

3.4. Electrophysiology Studies and Biocytin Labeling

A focus in our laboratory has been to investigate pathological changes and exercise-induced adaptions occurring at glutamatergic corticostriatal synapses following MPTP-lesioning of the nigrostriatal pathway. We routinely examine the physiological properties of striatal medium spiny neurons (MSNs) in acute brain slices in our MPTP and exercise studies. We rely on the physiological readouts of spontaneous excitatory postsynaptic currents (sEPSCs) and stimulus-evoked EPSCs (eEPSCs) to look for experimentally induced changes at corticostriatal synapses. Each experiment begins with the establishment of an input–output relationship to estimate the relative corticostriatal innervations of MSNs. This is accomplished by gradually increasing intensity of extracellular stimulation and recording respective evoked EPSCs. We next acquire an estimate of the presynaptic probability of transmitter release from excitatory inputs using a traditional paired-pulse paradigm with interstimulus intervals (ISI) of 50 ms. To measure changes in subunit composition of postsynaptic AMPA receptors, we capitalize on known properties of GluR2 lacking versus GluR2 containing AMPA receptors by examining current–voltage relationship for synaptically evoked EPSCs and measuring the rectification index (RI) for evoked synaptic responses.

Brain slices (hemi- or bilateral hemispheres) are prepared after decapitation following halothane or isoflurane anesthesia. Brains are quickly removed and placed in ice-cold modified artificial cerebral spinal fluid (aCSF) containing (in mM) 124 sucrose, 62 NaCl, 3 KCl, 2.6 $MgCl_2$, 1.2 $CaCl_2$, 26 $NaHCO_3$, 1.25 NaH_2PO_4, 10 glucose. The reduced sodium concentration and reduced Ca^{2+}/Mg^{2+} ratio, with respect to the normal aCSF (see below), lowers the excitability of neurons during the cutting procedure, which protects them from excitotoxicity. Coronal corticostriatal slices (350 μm thick) are cut in modified aCSF with a Vibratome-1000 (Vibratome Co., St Louis, MO) and transferred to an individual chamber with a nylon net submerged in normal aCSF containing (in mM) 124 NaCl, 3 KCl, 1.3 $MgCl_2$, 2.4 $CaCl_2$, 26 $NaHCO_3$, 1.25 NaH_2PO_4, 10 glucose. Normal and modified aCSFs are continuously bubbled with 95% O_2/5% CO_2 to maintain pH 7.4. Slices are allowed to recover from cutting in an incubation chamber at least 45 min at room temperature and transferred to a submerged brain slice recording chamber kept at 30–32°C, as outlined in Akopian and Walsh (33). Picrotoxin (50 μM) is used to block $GABA_A$ receptor-mediated inhibition to isolate excitatory synaptic events.

We routinely use whole-cell voltage clamp and electrical stimulation methods to examine corticostriatal synaptic input. Voltage clamp is chosen as the recording method to reduce possible activation of postsynaptic conductance, which can contribute to changes in synaptic strength under current clamp conditions (34). Whole-cell recordings are obtained from striatal MSNs identified visually using infrared differential interference contrast microscopy (IR-DIC). Recording electrodes containing 0.5% biocytin (Sigma-Aldrich) are routinely used to verify cell type based on morphology and for analysis of dendritic spine density (see below). Patch electrodes are pulled with a Flaming-Brown P-87 pipette puller (Sutter Instrument, Novato, CA) from thin-wall borosilicate capillary glass of 1.5-mm outer diameter (WPI, Sarasota, FL). The electrodes have resistances ranging between 4 and 6 MΩ. The pipette internal solution is composed of (in mM): 130 Cs gluconate, 10 CsCl, 5 EGTA, 2 MgCl$_2$, 10 HEPES, 5 QX-314, 2 ATP-Mg, 0.25 GTP-Na, pH 7.25, 285 mOsm. Cesium (Cs) is used to block K$^+$ conductance, which dramatically increases the space clamp of synaptic events occurring in dendrites. Spermine (100 µM) (Sigma-Aldrich) is included to provide polyamine modulation of GluR2 lacking AMPA receptors. Liquid junction potential with this internal solution is estimated to be around 14.7–15 mV.

Most recently, we have modified our pipette solution by increasing the Cl$^-$ concentration to 40 mM and decreasing gluconate concentration to 100 mM to help reduce both the liquid junction potential to ~10 mV and the averaged series resistance (Rs). The concentration of spermine has been increased to 200 µM to ensure that free concentration of spermine is not reduced by increasing concentration of negative Cl$^-$ ions. We have found that this modified internal solution reduces voltage and space clamp errors, improves seal formation and the time for establishing lower Rs after getting into whole-cell configuration in brains of adult mice. Rs are gradually reduced from initial values of 40–35 MΩ to typically 10–20 MΩ over a 5–15 min period. Rs are monitored throughout the experiment by measuring the instantaneous current response to 5-ms hyperpolarizing (–5 mV) pulses delivered before synaptic stimulation and (1) in experiments performed with older generation patch-clamp amplifiers (Axopatch-D) is not compensated in cases when there is no need to change the holding potential and (2) is automatically compensated when using new amplifiers (Multiclamp 700B) equipped with auto compensation of pipette and cell capacitances. We routinely compensate Rs up to 80%. The ground electrode consists of a salt bridge constructed from a glass electrode filled with agar. Passive membrane properties of the cells in slices are determined in voltage clamp mode with the Membrane Test option of Clampex 10 software by using a 10 mV depolarizing step voltage command from a holding potential of –70 mV.

Changes in the amplitude and frequency of sEPSCs are sampled for 3 min. Neurons are voltage clamped at –70 mV during sEPSC recording. The membrane currents are filtered at 2 kHz and digitized at 5 kHz. sEPSCs are analyzed off-line using the threshold detection option found in Clampfit 10 (Molecular Devices, Sunnyvale, CA). The threshold amplitude for the detection of an event is set at above 5 pA. Cumulative frequency histograms are generated for sEPSC amplitude and inter-event intervals. The bin size for sEPSC amplitude is 2 pA, and the bin size for sEPSC inter-event interval is 100 ms. Stimulus-evoked EPSCs in the lateral portion of the dorsal striatum are generated using stimulation electrodes filled with aCSF and positioned 100–200 μm from the recording electrode at the border between the striatum and the overlying corpus callosum. Rectangular current pulses of 0.1 ms duration are applied to the stimulation electrodes relative to a reference electrode placed in the recording chamber using the Stimulus Isolation Unit A365 (WPI, Sarasota, FL) triggered by digital output of Digidata 1320 (Molecular Devices, Sunnyvale, CA). Input (stimulation intensity)–output (synaptic response) relationships are determined for corticostriatal synapses by applying a standard ascending sequence of stimulus intensities and recording evoked EPSCs. Neurons are voltage clamped at –70 mV during periods of stimulation. The slope of the input–output relationship is determined for each cell and compared by one-way repeated ANOVA.

To estimate the paired-pulse ratio (PPR), five paired-pulse synaptic stimulations are delivered through the stimulating electrode with an ISI of 50 ms, at holding potential of –70 mV, at 20-s intervals. The intensity of synaptic stimulation is set at about 50% of maximum responses obtained from the I/O curve. The five traces are averaged, and the PPR is expressed as a percentage of the ratio of the second pulse to the first one of the pair.

The rectification index (RI) for evoked AMPA receptor-mediated synaptic responses is determined as the slope of the synaptic current–voltage relationship (I–V) curve at positive potentials (0 to +60 mV) divided by the slope of synaptic I–V curve at negative potentials (–80 to 0 mV) (35, 36). Synaptic current–voltage relationships (I–V) are obtained by generating synaptic currents with electrical stimulation of cortical afferents every 20 s at different holding potentials ranging from –80 to +60 mV, with increments of 20 mV to the previous step holding level. The stimuli is set at an intensity of about 50% of maximum responses obtained from the I–O curve and delivered 5 s after the stepped change in holding potential. All RI experiments are performed in slices bathed in picrotoxin (50 μM) to block $GABA_A$ receptor-mediated responses and AP-5 (50 μM) to block NMDA receptor-mediated responses. Estimating rectification properties of AMPA receptors can be contaminated by unavoidable errors introduced by incomplete voltage and space clamp control, especially when recording MSNs in slices from adult animals. However, when using the spermine-free pipette solution on MSNs, we constantly obtain linear voltage–current relationships within the rage from −80 to +60 mV, with a reversal potential close to 0 mV after liquid junction potential correction. Under our experimental conditions, we are able to repeat I–V paradigms several times in the same cell and obtain average I–V relationships for that cell. Also, we believe that expression of RI as ratios of I–V slopes at positive to negative voltages is more accurate and representative than expressing RI as ratios of responses at individual positive to negative holding levels.

3.5. Injection of Biocytin to MSNs

In some experiments, the density of dendritic spines and dendritic branch arborization is determined by inclusion of biocytin-HCl (Sigma-Aldrich) in the pipette solution at a concentration of 0.5%. This serves as a morphological verification of MSN targets in the physiology studies as well as for the determination of structural correlates of striatal neuroplasticity. Caution is taken to prevent excess spillover of biocytin from the pipette into the slice to avoid background staining. This is achieved by first filling the pipette tip with biocytin-free internal solution and then back-filling the remaining pipette volume with the biocytin containing solution and by minimizing the time for obtaining cell attached configuration. In our experience, rapid cell attached configuration using reduced positive pressure inside the patch pipette is sufficient for high-quality biocytin labeling. In most cases, cells are kept in whole-cell configuration for 10–15 min after obtaining the desired physiological information and labeled with biocytin by delivering 10 ms −10 mV hypo-polarizing pulses at 0.5 Hz applied to the pipette. Following biocytin delivery, tissue slices are immersed in 4% paraformalde-hyde/phosphate-buffered saline (PFA/PBS), pH 7.2 for 24 h, into 20% sucrose until sunk, and then they are frozen flat onto a microscope slide in isopentane on dry ice. Frozen slices are dislodged and mounted on chucks of OCT for sectioning at 60-μm thickness in a cryostat (Leica 1950). Slices are processed for visualization of biocytin filled neurons using the ABC Elite Kit from Vector Bioscience (Burlingame, CA) with diaminobenzoic acid as the precipitate. Labeled cells are visualized at low magnification with an Olympus BX51 microscope and the number of dendritic spines and/or dendritic arborization quantified using the computer-assisted program Bioquant (Bioquant Imaging, Nashville, TN).

3.6. Fast-Scan Cyclic Voltammetry

Fast-scan cyclic voltammetry is used for the analysis of dopamine release from coronal in vitro brain slices of the striatum (37). Coronal corticostriatal slices are obtained with the same procedure as outlined for electrophysiological experiments. Single slices are transferred to the recording chamber and bathed continuously with the oxygenated aCSF solution maintained at a temperature of 32°C. Disc carbon fiber microelectrodes (CFMEs) are made from

7-mm unsized carbon fiber (Goodfellow Corporation, Devon, PA) by electrophoretic anodic deposition of paint (38). Extracellular dopamine is monitored at the CFME every 100 ms by applying a triangular waveform (0.4–1.0 V vs. Ag/AgCl, 300 V/s). Currents are recorded with a modified VA-10 Voltammetric and Amperometric Amplifier (NPI Electronic, Tamm, Germany). Data acquisition is controlled by Clampex 7.0 software (Molecular Devices, Menlo Park, CA). Electrical stimulation is used to elicit dopamine efflux with a twisted, bipolar, nichrome electrode placed on the surface of the slice. Single pulses (0.1 ms, 200 µA) are generated with a Master-8 pulse generator (AMPI, Jerusalem, Israel) connected to an A360R Constant Current Stimulus Isolator Unit (World Precision Instruments, Sarasota, FL). Stimulus intervals between pulses are not less than 5 min. The CFMEs are inserted 75–100 µm into the slice at a position 100–200 µm from the stimulating electrode pair (39). Each slice is sampled for dopamine at five sites, which represents medial to lateral and dorsal to ventral dimensions. Three rostral slices are examined in each mouse and the values were averaged for each animal. Changes in extracellular dopamine are determined by monitoring the current over a 200 mV window at the peak oxidation potential for dopamine. Background-subtracted cyclic voltammograms are created by subtracting the current obtained before stimulation from the current obtained in the presence of dopamine. To convert oxidation current to dopamine concentration, electrodes are calibrated with dopamine standard solutions after experimental use. The kinetics of the dopamine signal evoked by intrastriatal stimulation are studied by monitoring the cyclic voltammetry signal for 1 s before and 5 s after intrastriatal stimulation at a sampling rate of once every 100 ms (10 Hz). The decay of the dopamine signal is determined by normalizing post-peak dopamine measurements to the peak dopamine measured. The decay constant is then determined from a single exponential fit of the decay in dopamine signal according to the following equation: $y = Ae^{-kt}$, where A is the peak dopamine signal at time 0 and the constant $(-k)$ is the decay rate for exponential decay of the dopamine signal. ANOVA is performed between all groups for the decay rate constant $(-k)$ (40).

4. Conclusion

Since the first identification of MPTP as a selective neurotoxin for nigrostriatal dopaminergic neurons in 1983 (2), there have been over 2,000 manuscripts published utilizing this compound in mice, attesting to the value of this model. Most of this work is focused on neuroprotection and mechanisms of cell death (41, 42). While MPTP may not replicate all the features of PD, it provides a key means by which depletion of striatal dopamine can be achieved, and by which brain repair mechanism(s) involved in compensation may be studied, including the role of exercise. In this chapter, we have outlined some of the methods utilized in our laboratories to investigate the physiological and molecular correlates of exercise-induced neuroplasticity (brain repair processes) in the MPTP mouse model of PD. Specifically, we have shown that both treadmill and voluntary running wheel can be used to modulate motor- and nonmotor-related behaviors. We also demonstrate methods used in our laboratory to examine the effects of experience-dependent neuroplasticity (exercise) on both pre- (dopaminergic/glutamatergic) and post- (striatal MSNs) pathways. Our studies have shown that intensive treadmill exercise can improve motor and nonmotor behaviors through modulation of both glutamatergic and dopaminergic neurotransmission.

References

1. Davis GC, Williams AC, Markey SP, Ebert MH, Caine ED, et al. (1979) Chronic parkinsonism secondary to intravenous injection of meperidine analogues. Psychiatry Research 1:249–54

2. Langston JW, Ballard P, Tetrud JW, Irwin I. (1983) Chronic parkinsonism in humans due to a product of meperidine-analog synthesis. Science 219:979–80

3. Petzinger GM, Jakowec MW. (2004) The MPTP-lesioned model of Parkinson's disease with emphasis on mice and nonhuman primates. Comp.Med 54:497–513

4. Dauer W, Przedborski S. (2003) Parkinson's disease: mechanisms and models. Neuron 39:889–909

5. Przedborski S, Jackson-Lewis V, Naini AB, Jakowec M, Petzinger G, et al. (2001) The parkinsonian toxin 1-methyl-4-phenyl-1,2,3,6-tetrahydropyridine (MPTP): a technical review of its utility and safety. Journal of Neurochemistry 76:1265–74

6. Irwin I, Ricaurte GA, DeLanney LE, Langston JW. (1988) The sensitivity of nigrostriatal dopamine neurons to MPP+ does not increase with age. Neurosci Lett 87:51–6

7. Lau YS, Novikova L, Roels C. (2005) MPTP treatment in mice does not transmit and cause Parkinsonian neurotoxicity in non-treated cagemates through close contact. Neurosci Res 52:371–8

8. Davey GP, Tipton KF, Murphy MP. (1992) Uptake and accumulation of 1-methyl-4-phenylpyridinium by rat liver mitochondria measured using an ion-selective electrode. Biochem J 288:439–43

9. Fuller R, Hemrick-Luecke S. (1987) Depletion of heart norepinephrine in mice by some analogs of MPTP (1-methyl-4-phenyl-1,2,3,6-tetrahydropyridine). Res Commun Chem Pathol Pharmacol 56:147–56

10. Fuller RW, Hemrick-Luecke SK. (1986) Depletion of norepinephrine in mouse heart by 1-methyl-4-phenyl-1,2,3,6-tetrahydropyridine (MPTP) mimicked by 1-methyl-4-phenylpyridinium (MPP+) and not blocked by deprenyl. Life Sci 39:1645–50

11. Heikkila RE. (1985) Differential neurotoxicity of 1-methyl-4-phenyl-1,2,3,6-tetrahydropyridine (MPTP) in Swiss-Webster mice from different sources. Eur. J. Pharmacol. 117:131–3

12. Muthane U, Ramsay KA, Jiang H, Jackson-Lewis V, Donaldson D, et al. (1994) Differences in nigral neuron number and sensitivity to 1-methyl-4-phenyl-1,2,3,6-tetrahydropyridinium in in C57/bl and CD-1 mice. Experimental Neurology 126:195–204

13. Sedelis M, Hofele K, Auburger GW, Morgan S, Huston JP, Schwarting RK. (2000) MPTP susceptibility in the mouse: behavioral, neurochemical, and histological analysis of gender and strain differences. Behav Genet 30:171–82

14. Irwin I, Finnegan KT, DeLanney LE, Di Monte D, Langston JW. (1992) The relationship between aging, monoamine oxidase, striatal dopamine and the effects of MPTP in C57BL/6 mice: a critical reassessment. Brain Research 572:224–31

15. Ricaurte GA, DeLanney LE, Irwin I, Langston JW. (1987) Older dopaminergic neurons do not recover from the effects of MPTP. Neuropharmacology 26:97–9

16. Ookubo M, Yokoyama H, Kato H, Araki T. (2009) Gender differences on MPTP (1-methyl-4-phenyl-1,2,3,6-tetrahydropyridine) neurotoxicity in C57BL/6 mice. Mol Cell Endocrinol 311:62–8

17. Fisher BE, Petzinger GM, Nixon K, Hogg E, Bremmer S, et al. (2004) Exercise-induced behavioral recovery and neuroplasticity in the 1-methyl-4-phenyl-1,2,3,6-tetrahydropyridine-lesioned mouse basal ganglia. J Neurosci Res 77:378–90

18. Guillot TS, Miller GW. (2009) Protective Actions of the Vesicular Monoamine Transporter 2 (VMAT2) in Monoaminergic Neurons. Mol Neurobiol 39:149–70

19. Staal R, Sonsalla P. (2000) Inhibition of brain vesicular monoamine transporter (VMAT2) enhances 1-methyl-4-phenylpyridinium neurotoxicity in vivo in rat striata. J Pharmacol Exp Ther. 293:336–42

20. Jakowec MW, Nixon K, Hogg L, McNeill T, Petzinger GM. (2004) Tyrosine hydroxylase and dopamine transporter expression following 1-methyl-4-phenyl-1,2,3,6-tetrahydropyridine-induced neurodegeneration in the mouse nigrostriatal pathway. J. Neurosci. Res. 76:539–50

21. Kilpatrick IC, Jones MW, Phillipson OT. (1986) A semiautomated analysis method for catecholamines, indoleamines, and some prominent metabolites in microdissected regions of the nervous system: an isocratic HPLC technique employing coulometric detection and minimal sample preparation. J. Neurochem. 46:1865–76.

22. Petzinger GM, Fisher BE, Hogg E, Abernathy A, Arevalo P, et al. (2006) Behavioral Recovery in the MPTP (1-methyl-4-phenyl-1,2,3,6-tetrahydropyridine)-lesioned Squirrel Monkey (Saimiri sciureus): Analysis of Striatal Dopamine and the Expression of Tyrosine Hydroxylase and Dopamine Transporter Proteins. J. Neuorsci. Res. 83:332–47

23. Paxinos G, Franklin KBJ. (2001). The Mouse brain in Stereotaxic Coordinates. (eds) Second Academic Press New York.

24. Gundersen HJ, Jensen EB. (1987) The efficiency of systematic sampling in stereology and its prediction. J Microsc 147:229–63

25. Jackson-Lewis V, Jakowec M, Burke RE, Przedborski S. (1995) Time course and morphology of dopaminergic neuronal death caused by the neurotoxin 1-methyl-4-phenyl-1,2,3,6-tetrahydropyridine. Neurodegen. 4:257–69

26. Petzinger GM, Walsh JP, Akopian G, Hogg E, Abernathy A, et al. (2007) Effects of treadmill exercise on dopaminergic transmission in the 1-methyl-4-phenyl-1,2,3,6-tetrahydropyridine-lesioned mouse model of basal ganglia injury. J Neurosci 27:5291–300

27. Sauer H, Oertel W. (1994) Progressive degeneration of nigrostriatal dopamine neurons following intrastriatal terminal lesions with 6-hydroxydopamine: a combined retrograde tracing and immunocytochemical study in the rat. Neurosci. 59:401–15

28. Nithianantharajah J, Hannan AJ. (2006) Enriched environments, experience-dependent plasticity and disorders of the nervous system. Nat Rev Neurosci 7:697–709

29. Ferro MM, Bellissimo MI, Anselmo-Franci JA, Angellucci ME, Canteras NS, Da Cunha C. (2005) Comparison of bilaterally 6-OHDA- and MPTP-lesioned rats as models of the early phase of Parkinson's disease: histological, neurochemical, motor and memory alterations. J Neurosci Methods 148:78–87

30. Shiotsuki H, Yoshimi K, Shimo Y, Funayama M, Takamatsu Y, et al. (2010) A rotarod test for evaluation of motor skill learning. J Neurosci Methods

31. Gorton LM, Vuckovic MG, Vertelkina N, Petzinger GM, Jakowec MW, Wood RI. (2010) Exercise Effects on Motor and Affective Behavior and Catecholamine Neurochemistry in the MPTP-Lesioned Mouse. Behav Brain Res 213:253–62

32. Vuckovic MG, Wood RI, Holschneider DP, Abernathy A, Togasaki DM, et al. (2008) Memory, mood, dopamine, and serotonin in the 1-methyl-4-phenyl-1,2,3,6-tetrahydropyridine-lesioned mouse model of basal ganglia injury. Neurobiol Dis 32:319–27

33. Akopian G, Walsh JP. (2007) Reliable long-lasting depression interacts with variable short-term facilitation to determine corticostriatal paired-pulse plasticity. J Physiol 580:225–40

34. Akopian G, Walsh JP. (2002) Corticostriatal paired-pulse potentiation produced by voltage-dependent activation of NMDA receptors and L-type Ca(2+) channels. J Neurophysiol. 87:157–65

35. Liu SJ, Cull-Candy SG. (2005) Subunit interaction with PICK and GRIP controls Ca2+ permeability of AMPARs at cerebellar synapses. Nat Neurosci 8:768–75

36. Shin J, Shen F, Huguenard J. (2007) PKC and polyamine modulation of GluR2-deficient AMPA receptors in immature neocortical pyramidal neurons of the rat. J Physiol 581:679–91

37. Patel J, Rice ME. (2006). Dopamine release in brain slices. In: Grimes CA, Dickey EC, Pishko MV (eds) Encyclopedia of Sensors, 1st ed. American Scientific Publishers, Stevenson Ranch, CA.

38. Jaffe EH, Marty A, Schulte A, Chow RH. (1998) Extrasynaptic vesicular transmitter release from the somata of substantia nigra

neurons in rat midbrain slices. J Neurosci 18:3548–53

39. Miles PR, Mundorf ML, Wightman RM. (2002) Release and uptake of catecholamines in the bed nucleus of the stria terminalis measured in the mouse brain slice. Synapse 44:188–97

40. Mosharov EV, Sulzer D. (2005) Analysis of exocytotic events recorded by amperometry. Nat Methods 2:651–8

41. Jakowec MW, Petzinger GM. (2004) 1-methyl-4-phenyl-1,2,3,6-tetrahydropyridine-lesioned model of parkinson's disease, with emphasis on mice and nonhuman primates. Comp Med 54:497–513

42. Przedborski S, Tieu K, Perier C, Vila M. (2004) MPTP as a mitochondrial neurotoxic model of Parkinson's disease. J Bioenerg Biomembr. 36:375–9

Chapter 20

The MPTP-Treated Primate, with Specific Reference to the Use of the Common Marmoset (*Callithrix jacchus*)

Michael J. Jackson and Peter Jenner

Abstract

For investigations into treatment of Parkinson's disease, the 1-methyl-4-phenyl-1,2,3,6 tetrahydropyridine (MPTP)-treated primate has become one of the most important and established animal models. Various species of both Old and New world primate have been used, employing differing MPTP-treatment regimes and methods for the assessment of motor deficits and motor complications. In this chapter, we briefly review the use of these varying models and treatment regimes, the effects of MPTP treatment and the general characteristics and drug response of the primate model. In detail, we describe the procedures we employ for MPTP treating the common marmoset including the specific welfare, husbandry and after-care protocols that are essential to produce a successful model. We also describe the methods for inducing the expression of persistent dyskinesia and how to accurately assess changes in locomotor activity, motor disability and the severity of dyskinesia.

Key words: MPTP, Primate, Common marmoset, Motor activity, Motor disability, Dyskinesia, Drug responsiveness

1. Introduction

Modelling the symptoms of Parkinson's disease (PD) in animals has largely focussed on the use of rodents. The unilaterally 6-OHDA-lesioned rat and the MPTP-treated mouse are the most commonly employed models, and these are extremely useful in early stages of drug development programmes and in replicating some of the motor components of PD exhibited in man and for examining the potential drug responsiveness of the disease to both dopaminergic and non-dopaminergic agents (1, 2). However, they have limitations with respect to the unilateral nature of nigro-striatal dopamine loss and in the resemblance of the motor deficits to those

Emma L. Lane and Stephen B. Dunnett (eds.), *Animal Models of Movement Disorders: Volume I*, Neuromethods, vol. 61,
DOI 10.1007/978-1-61779-298-4_20, © Springer Science+Business Media, LLC 2011

that characterise PD. For these reasons, a primate model of PD could offer significant advantages, but prior to the MPTP era, these had been restricted to radiofrequency or electrolytic lesions of basal ganglia or to the use of unilateral lesions involving toxins such as 6-OHDA (3). For this reason, the discovery of the selective toxicity of MPTP to dopaminergic neurones in man and the realisation that this could be reproduced in primate species completely revolutionised the approach to modelling PD (4, 5). The MPTP-treated primate has become an almost essential step for the development of new therapeutic approaches to the treatment of PD because of the high degree of translation of symptomatic drug action in man, and its use has been extensively reviewed elsewhere.

This chapter considers practical issues related to the use of MPTP in primate species and the ways in which the model can be used to assess the potential effects of therapeutic approaches to the treatment of PD. Initially, we examine the range of species to which MPTP treatment has been applied, the dosing schedules used and the motor and non-motor symptoms induced. We then provide detailed protocols for the use of MPTP in the common marmoset and the assessment of motor function and drug responsiveness relative to efficacy in man.

1.1. General Overview of the Model in Relation to Parkinson's Disease

The normal perception of the MPTP-treated primate is that it is a close mimic of what occurs in PD in man. Certainly, as will be discussed later, there are features of the model that closely resemble the primary motor deficits that occur in man as well as drug responsiveness and motor complications of treatment that are clinically relevant. But in many other respects, the MPTP-treated primate has limitations in relation to its similarity to the human disease.

In the doses normally used MPTP treatment causes very marked loss of nigral dopaminergic cells and this resembles the pathological change seen in the later stage of PD rather than in the early phases of the illness (6–9). Most treatment regimens utilised are acute and this means cell death and motor deficits appear with a time course far shorter than seen in PD. Some treatment schedules are varied to produce partial lesions of the substantia nigra, but the difficulty is that altered motor function and drug responsiveness are far more difficult to observe under these circumstances (10). The normal extensive loss of nigral cells explains why these animals develop dyskinesia so rapidly in response to levodopa administration (11). Because of the acute nature of the treatment, there is no progression of the effects of MPTP as would occur in PD and despite attempts to stretch the period of MPTP treatment, this largely results in a stepwise loss of dopaminergic neurones rather than a truly progressive period of neurodegeneration.

The effects of MPTP are also most entirely limited to the loss of nigral dopaminergic cells and even mesolimbic and mesocortical dopamine tracts remain largely intact (12). There is no hard evidence for a consistent effect on other non-dopaminergic nuclei that die in PD and that are outside of the basal ganglia. MPTP treatment has been associated with the loss of noradrenergic neurones in the locus coeruleus, but this does not occur in all laboratories and pathology in areas such as the raphe nuclei, dorsal motor nucleus of the vagus and substantia innominata has not been reported (13, 14). Acutely, MPTP can produce reserpine-like depletions of monoamines and it is necessary to distinguish these from those relating to neuronal destruction (6, 15).

MPTP treatment of primates leads to an inflammatory response in the substantia nigra that involves both astrocytes and microglia. This gliosis is present months or years after the termination of MPTP administration (16–18). However, the other pathological marker of PD, the Lewy body has not been reported after MPTP treatment although proteinous inclusions may be present in the substantia nigra (14). This presumably reflects the difference in pathogenic processes produced by MPTP compared to PD and also the short time course over which cell death occurs on MPTP treatment compared to the slow onset and progression of pathology in PD. Another difference between the primate and man is age. Invariably, young primates are used for MPTP treatment, whereas PD is a clearly age-related disorder. It appears necessary to destroy more dopaminergic cells with MPTP in a young primate to produce motor deficits than in older animals presumably reflecting the lesser plasticity occurring with age (19).

1.2. Primate Species and Treatment Protocols

A wide range of Old World and New World primates have been employed as detailed in Table 1. The range of species usually reflects the availability of animals to a research centre, and we are not aware of particular advantages between species for the onset of motor deficits. There are some differences relating to ease of treatment and specific symptom production. All show decreased motor function and the onset of motor disability following MPTP administration. Larger primates need to be anaesthetised for MPTP treatment, and this can be problematic when using repeated dose-treatment regimens. Smaller species such as the common marmoset can be manually restrained for toxin administration.

What is striking is the wide range of doses of MPTP utilised and the time periods of administration (Table 2). These are difficult to explain as many of the protocols have arisen through the experience gained at the centre using MPTP. They may reflect differences between species in terms of the extent of the behavioural response to MPTP and the general toxicity of the compound. Some species appear to be more tolerant to the acute effects of MPTP treatment than others in respect of the level of motor disability, difficulties with eating and drinking, decreased body temperature and depletion of cardiac noradrenaline content impairing cardiovascular function. It should also be noted that

Table 1
Non-human primate species that have been
treated with MPTP to produce models
of Parkinson's disease

Common name	Scientific classification
New world primates	
Common marmoset	*Calithrix jacchus*
Squirrel monkey	*Saimiri sciureus*
Capuchin	*Cebus apella*
Owl monkey	*Aotus trivirgatus*
Old world primates	
Cynomolgus	*Macaca fascicularis*
Rhesus	*Macaca mulatta*
Vervet	*Chlorocebus sabaeus*
Guinea Baboon	*Papio papio*
Olive Baboon	*Papio anubis*

there is recovery of motor function following the cessation of MPTP treatment that reflects both the onset of compensatory mechanisms and also the disappearance of the reserpine-like actions of MPTP. It is necessary to assess the extent to which treatment can be administered to an individual animal from the acute response to judge how pronounced motor deficits are likely to be on recovery and much of this can only be gained by experience. In some settings, MPTP treatment renders animals so parkinsonian that they require dopamine replacement therapy both during the recovery phase and then daily on a continuous basis. Our own experience is to ensure that we allow animals to resume relatively normal behavioural and feeding patterns while having clear motor deficits (see later).

The treatment regimes utilised also reflect other factors specific to individual laboratories and to regulations governing primate use. Long slow low dose administration is utilised to minimise the acute effects of MPTP treatment and to mimic a more prolonged period of neuronal loss as occurs in PD. However, care should be taken to ensure that low dose regimens are effective as MPTP is extensively metabolised by MAO-B in the periphery to MPP+ that does not then penetrate in to brain (20–22). Hence, the need to use boluses of MPTP that deliver sufficient MPTP to the basal ganglia to achieve dopaminergic cell loss.

Most MPTP-treatment regimes involve systemic administration resulting in bilateral lesioning of the substantia nigra and bilateral motor deficits that resemble what occurs in man. However, a few laboratories utilise intra-carotid administration of MPTP to

Table 2
Varying MPTP-treatment regimes utilised to produce non-human primate models of Parkinson's disease

Dose (mg/kg)	Route	Regime	Species	Reference
2.0	s.c.	5 Injections; once daily for 5 days	Marmoset	(32, 33)
3.0	i.p.	3 Injections; once daily for 3 days	Squirrel	(13)
1.0	i.p.	10 Injections; twice daily for 5 days	Squirrel	(13)
2.0	s.c.	6 Injections; 2 weeks between treatments	Squirrel	(83)
2.0	s.c.	2 Injections; 2 weeks between treatments	Squirrel	(83)
1.0–2.0	s.c.	Total 3.5–13.5 mg/kg	Squirrel	(84)
1.5	s.c.	3 Injections; over 6 months, 2 months between treatments	Squirrel	(85)
2 mg/dose	i.p.	Once weekly; total MPTP 6–44.5 mg cumulative dose	Squirrel	(86)
0.2	i.v.	Once daily until parkinsonian 13.7 ± 2.1 mg cumulative dose	Cynomolgus	(87)
2.0–3.0	s.c.	Once weekly for 4–32 weeks 9–23.5 mg cumulative dose	Cynomolgus	(88)
0.03; 0.3–0.967	i.v.	1 Initial dose of 0.03; further doses 0.3–0.967 once weekly for 2–12 weeks	Rhesus	(89)
0.3–0.7	i.v.	Once weekly for between 6 and 11 months	Rhesus	(90)
0.4	i.m.	5 Injections; given over 4 days	Vervet	(91)
0.4–0.5	i.m.	4 Injections; once daily for 4 days	Vervet	(92)
0.45	i.m.	5 Injections; once daily for 5 days	Vervet	(93)
0.2–0.5	i.v.	Once weekly for 5–21.5 months 11–37.6 mg cumulative dose	Baboon	(94)
0.4	i.m.	Once or twice daily for 6 days followed by 1 dose of 0.27 mg/kg on day 7	Baboon	(95)
1.0–2.0 mg/ monkey	i.v.	Once daily for 4–5 days	Owl monkey	(96)
0.5	i.m.	Once a day for 4 weeks	Cebus	(97)

produce unilateral dopaminergic neurone destruction (23–25). From a husbandry perspective, this appears less severe than systemic administration and results in animals that have normal motor function on one side that aids normal feeding and grooming. Whether these lesions are truly unilateral is an interesting question because there will be mixing of blood supply to both hemispheres through the Circle of Willis. From a behavioural perspective, these animals will show spontaneous unilateral motor deficits, but rotational behaviour appears on systemic treatment with dopaminergic drugs; and in some respects, this seems more difficult to rate as a mimic of PD than bilateral deficits occurring after systemic administration of MPTP.

1.3. General Characteristics of the Model and Drug Responsiveness

The general characteristics observed in MPTP-treated primates and those induced by subsequent drug treatment are summarised in Table 3. All primate species treated with MPTP develop slowness of movement, impairment of normal motor tasks (motor disability) and postural abnormalities. The characteristic rest tremor of PD is not seen in all laboratories or in all primate species. While rest

Table 3
Characteristics exhibited by MPTP-treated primates

	Marmoset	Cynomolgus	Vervet
Locomotor activity	Decreased	Decreased	Decreased
Motor disability	Present	Present	Present
Postural abnormalities	Yes	Yes	Yes
Rigidity	Yes	Yes	Yes
Resting tremor	No	No	Yes
Postural tremor	Infrequent	Infrequent	Infrequent
Action tremor	Yes	Yes	Yes
Balance	Impaired	Impaired	Impaired
Visual-spatial deficits	Yes	Yes	Yes
Cognitive deficits	Yes	Yes	Yes
Sleep irregularities	Yes	Yes	Yes
Autonomic dysfunction	Yes	Undefined	Undefined
Dopaminergic drug response	Yes	Yes	Yes
Dyskinesia induction	Yes	Yes	Yes
Wearing-off	Undefined	Indicated, but undefined	Undefined
On-off/freezing	Indicated, but undefined	Undefined	Undefined

tremor is reported in vervet monkeys, in other species it is often postural in nature or appears on intention of movement (26). The onset of motor disability can affect balance and cause postural instability that is very apparent in species that move between perches. Freezing or impairment of gait can also be present. The alterations in motor function may impair feeding and drinking and these need to be carefully monitored through direct measurement and by monitoring body weight. Associated with MPTP treatments are reports of alterations in visual fields, cognitive change and sleep disturbance, all of which are also features of PD (27–30). Autonomic function may be impaired, but this has not been extensively investigated.

Reversal of motor abnormalities is produced by the administration of dopamine replacement therapy, and this may lead to hyperactivity and stereotyped behaviour at higher dosages. Repeated treatment with L-dopa has been consistently found to induce dyskinesia consisting of chorea, dystonia and athetosis that closely resembles involuntary movements observed in PD (11, 31–33). Chronic treatment has also been reported to cause other motor complications and fluctuations, including "wearing off" and "on-off" (34). However, with the exception of dyskinesia, these have not been extensively studied or characterised. There are reports of hallucinations/psychosis in response to dopaminergic treatment of MPTP-treated primates, but assessment of this requires further validation (35).

In general, motor dysfunction responds to dopaminergic therapy in a manner that is highly predictive of the drug effect in man. Most studies of drug effect have been undertaken in relatively few laboratories in a few species, so it is difficult to be certain that there is consistency of drug response or of drug dosage or duration of drug effect. Certainly, those drugs used clinically to treat PD are effective – these include L-dopa and the ergot and non-ergot dopamine agonists. Amantadine is used to suppress dyskinesia in man and it can block dyskinetic movements in MPTP-treated primates, but the dosages required are relatively high and these effects are not easy to see when high doses of L-dopa are employed. The value of the MPTP-treated primate in assessing non-dopaminergic approaches to the treatment of PD remains to be validated by iteration from subsequent clinical trials. A number of pharmacological classes have been reported to produce symptomatic improvement of motor function and to prevent or suppress dyskinesia, but there have been gaps in attempting to translate these in to the clinic either because of lack of efficacy, problems with bioavailability or the occurrence of side-effects. The ability of the MPTP-treated primate to successfully predict a neuroprotective or neurorestorative effect is so far unproven and there have been disappointing results on subsequent clinical investigation for disease modification. However, the model should be viable for assessing the effects of neurotrophic factors, viral vectors and other gene therapies and for cell-based approaches to treatment (see later). Indeed, some approaches that have been demonstrated to be effective in the MPTP-treated primate have now been advanced to the clinic with some degree of both success and failure.

1.4. Rating Systems for Motor Deficits

The way in which motor dysfunction is assessed in MPTP-treated primates varies between the laboratories undertaking these studies (Table 4). Many laboratories use the automated rating of locomotor activity with technologies ranging from the use of photocells through actimeters and telemetry to computerised positional monitoring through video linkage. Visual rating is also used, but this can be subjective and only semi-quantitative in nature.

When considering components of motor disability, then invariably a rating scale is employed as the aspects of behaviour that are altered do not lend themselves to any type of automated recording. Ratings scales range from basic descriptions of the extent of motor disability through to those that break motor impairment down in to discrete categories and reflect worsening movement. In some laboratories, the rating of animals is undertaken in real time, whereas others utilise the subsequent rating of video recordings. Our own view is that the behaviour of the animals cannot be adequately assessed from a single angle and that a trained observer is more likely to correctly rate all behavioural components. Even so, ratings of motor disability are only made at intervals during the course of a study and as such are merely snapshots of events that occur.

Dyskinesia also does not lend itself to automated assessment, and rating scales are invariably employed. The involuntary movements can be focal, segmental or generalised and involve different body parts. Dyskinesia is a general term covering all involuntary movement, but there are clearly choreic, dystonic and athetotic components. Some scoring systems do not distinguish between these components, whereas others measure chorea and dystonia as being distinct. Rating scales are semi-quantitative and vary in the depth of assessment made as for motor disability. There have been attempts to produce a unified rating scale, but this has not been widely adopted (36). This probably reflects differences in opinion over what constitutes dyskinesia in different primate species. One clear distinction that must be made is between dyskinesia and stereotypy. We would describe stereotypy as the onset of repetitive purposeless movements that remain within the normal repertoire of voluntary movement, whereas dyskinesia is characterised by its involuntary, abnormal and fleeting expression.

There has been an attempt to rate psychosis and hallucinations induced by dopaminergic drugs in MPTP-treated primates (35). Our own view is that while strange behaviours are observed where animals are gazing in to the distance and appear startled by unseen events, it is difficult to relate this to the accepted features of a psychotic state.

Table 4
Methods of assessing behavioural changes in the MPTP-treated primate model

Species	Locomotor activity	Motor disability	Dyskinesia	References
Squirrel monkey	Visual rating scale −1 = hypoactive/sedated 0 = absent 1 = intermittent 2 = continuous 3 = hyperactive/hyper-reactive to external stimulae	–	Visual rating 0 = absent 1 = occasional/mild 2 = intermittent/moderate 3 = frequent/marked 4 = continuous/severe	(86)
Squirrel monkey	–	Visual rating A modified Parkinson's rating scale for the squirrel monkey	–	(98)
Squirrel monkey	Automated Photocell beam interruption monitoring	–	–	(99)
Squirrel monkey	–	Visual rating PPRS modified for the squirrel monkey Spatial hypokinesia (0–4), Body bradykinesia (0–4), Manual dexterity (right and left arm, 0–4 each), Balance (0–4), Freezing over a 4-min clinical observation period	Visual rating 1 = present 0 = absent	(100)

(continued)

Table 4
(continued)

Species	Locomotor activity	Motor disability	Dyskinesia	References
Cynomolgus	Automated electronic monitoring system (Datascience, St. Paul, MN) fixed in the cage of each animal; animals wear collars, which transmit a radiowave signal to the monitoring system and count locomotor movements	Visual rating Posture (normal: 0; flexed intermittent: 1; flexed constant: 2; crouched: 3) Mobility (normal: 0; mild reduction: 1; moderate reduction:2; very slow with freezing: 3) Climbing (normal: 0; absent: 1) Gait (normal: 0; slow: 1; very slow: 2; very slow with freezing: 3) Grooming (present: 0; absent: 1) Vocalisation (present: 0; absent: 1) Social interaction (present: 0; absent: 1) Tremor: (absent: 0; mild action tremor: 1; moderate action tremor: 2; resting tremor: 3)	Visual rating 0 = absent 1 = mild 2 = moderate 3 = severe	(101)
Rhesus		Visual rating Akinesia, Hunched posture, Tremor, Functionally disabled requiring feeding (0 = no disability; 1 = minimal disability; 2 = mild; 3 = moderate; 4 = severe; 5 = severe)	Visual rating Primate dyskinesia disability rating scale: 0 = absent, 1 = mild, 2 = moderate, 3 = severe, but not interfering with function, 4 = severe	(89)
Rhesus	Automated	Visual rating Overall score composed of subscores for assessing: Posture (0–2), Gait (0–4), Bradykinesia (0–4), Balance (0–2), Gross motor skills (0–3) and Defence reactions (0–2)	–	(90)
Vervet		Visual rating Modified primate parkinsonism and dyskinesia scale	Visual rating Modified primate parkinsonism and dyskinesia scale	(91)

			(49)
Marmoset	Automated monitoring Photocell beam interruption monitoring	Visual rating Alertness (normal 0, reduced 1, sleepy 2); checking movements (present 0, reduced 1, absent 2) Posture (normal 0, abnormal trunk +1, abnormal limbs +1, abnormal tail +1 or grossly abnormal 4); balance/co-ordination (normal 0, impaired 1, unstable 2, spontaneous falls 3) Reaction (normal 0, reduced 1, slow 2, absent 3) Vocalization (normal 0, reduced 1, absent 2); motility (normal 0, bradykinesia 1, akinesia 2)	Visual rating 0 = absent 1 = mild, fleeting and rare dyskinetic postures and movements 2 = moderate: more prominent abnormal movements, but not significantly affecting normal behaviour 3 = marked, frequent and at times continuous dyskinesia affecting the normal pattern of activity 4 = severe, virtually continuous dyskinetic activity, disabling to the animal and replacing normal behaviour

		(102)
Marmoset	Visual rating Motor activity measured by the animal's movements across either the four base segments (22×22 cm) of the floor of the test cage, or across the four vertical segments (22×20 cm) between the floor and perches of the test cage	Visual rating 0 = normal behaviour 1 = the animal appears quiet, but shows a normal repertoire of movements 2 = the animal can move freely, but is uncoordinated when making complicated movements, such as climbing down the cage wall 3 = the animal makes fewer and slower movements and is obviously uncoordinated in executing complex movements, such as jumping up to a perch or moving on a perch 4 = the animal makes few movements unless disturbed, and these are slow and limited to a small region of the cage 5 = the animal is akinetic and does not move even when disturbed
		–

(continued)

Table 4
(continued)

Species	Locomotor activity	Motor disability	Dyskinesia	References
Marmoset	Automated monitoring for each experimental cage, activity monitors consisted of a pair of externally located passive infrared volumetric detectors, each with 17 sensors, combined with a lens to diverge the passive sensor into the experimental cage	Visual rating 0 = no movement; 1 = movement of head, on the floor of the cage; 2 = movement of limbs, but no locomotion, on the floor of the cage; 3 = movement of head, on wall of cage or on perch; 4 = movement of limbs, but no locomotion, on wall of cage or perch; 5 = walking around floor of cage; 6 = hopping on floor of cage; 7 = climbing onto the wall of cage/onto the perch; 8 = climbing up and down walls, or along perch; 9 = running, jumping between roof, walls, perch, uses limbs through a wide range of activity Posture was rated from 0 to 1: 0 = normal, upright, holds head up, normal balance; 1 = abnormal, crouched, face down, may lose balance Bradykinesia was rated on a scale from 0 to 3: 0 = normal speed and initiation of movement; 1 = mild slowing, occasional hesitation and freezing; 2 = moderate slowing of movement, difficulty initiating and maintaining movement, marked freezing; 3 = akinetic, unable to move, prolonged freezing episodes	Visual rating Disability scale: 0 = absent 1 = mild, fleeting, rare, present less than 30% of the observation period 2 = moderate, present more than 30% of the observation period, but not interfering with normal activity 3 = marked, at times interfering with normal activity 4 = severe, disabling, replacing normal activity	(103)

1.5. MPTP Use in the Common Marmoset (Callithrix jacchus)

The common marmoset (*C. jacchus*) is commonly utilised for MPTP treatment and for the evaluation of potential antiparkinsonian therapies. This small, laboratory bred species has many advantages as the animal of choice for this type of investigation. The advantages of the common marmoset are listed in Table 3. We have utilised the common marmoset for 25 years and have found it to be highly suited to studies involving MPTP treatment and very predictive of drug effect subsequently uncovered in PD in man.

In the following sections, we describe in detail the exact procedure that is utilised in our laboratories for MPTP treatment that is necessary to ensure both appropriate animal welfare and the production of motor deficits closely resembling those that occur in PD. We also describe the procedure, which we use to subsequently evoke consistent dyskinesia. The methods for the assessment of locomotor activity, motor disability and dyskinesia are also described.

2. Methods

2.1. Husbandry and MPTP Treatment

2.1.1. Animal Selection and Acclimatisation Process

1. Animals of either sex must be in good health and have a minimum body weight of 320 g (see Note 1) and be at least 18 months of age prior to be being treated with MPTP (see Note 2).

2. If animals are sourced from an external supplier, they should be given a minimum of 2 weeks to acclimatise to their new environment before any regulated procedure is carried out.

3. During the first week of the acclimatisation period, the animals are observed in the home cages, but are not unduly disturbed or handled.

4. During the second week, the animals are weighed and examined by the veterinary surgeon and a member of the research team to establish their general health status. A record is made of any specific abnormalities (e.g. missing or malformed teeth, signs of previous, but healed injuries) (see Note 3). They are also accustomised to being handled and offered liquid treats orally using a 10-ml syringe.

5. Animals that fail to reach the correct body weight or have any indication of a health problem do not undergo MPTP-treatment until these issues have been resolved (see Note 4).

2.1.2. Safe Handling and Use of MPTP

Full PPE must be worn when working with MPTP, or animals that are being treated, or have recently been treated. A bleach solution (1%) is used to de-contaminate any equipment, or surfaces that may have been in contact with MPTP. A full review of the safe handling and use of MPTP is provided by Przedborski et al. (37).

2.1.3. MPTP Treatment

1. Sufficient MPTP for treatment at 1 mg/kg (partial lesion) or 2 mg/kg (full lesion) is weighed into five sealed glass vials to provide the correct amount of drug for each of the 5 treatment days.

2. On each day of treatment, one vial of the pre-weighed MPTP is dissolved in sterile saline at a concentration of 1 or 2 mg/ml to be administered at a volume of 1 ml/kg subcutaneously (s.c.)

3. On day 1 of MPTP-treatment, prior to administration, the home cage is modified by reducing the height of the cage, removing cage furniture and using dust-free tray liners for bedding instead of wood chips, etc. (see Note 5).

4. Animals are weighed and injected s.c. with MPTP once daily consecutively for up to 10 days (see Note 6).

2.1.4. Behavioural Changes and Specialised Care Regimes During and Following MPTP Treatment

The immediate acute effect of MPTP administration may produce mydriasis, impairment of balance and co-ordination of movement and sedation and an increased startle response to sudden loud noises. This syndrome may last up to 20-min post-treatment and is managed by ensuring that the animal is placed in a secure position when returned to the home cage after treatment (preferably into a nest box) and by minimising noise disturbance. Although persistent non-motor deficits have not generally been quantified observations indicate that they include increased urination due to a hyper reflexive bladder, decreased vocalisation, altered eye blink rate and blink response, increased day time somnolence and sleep disturbance.

1. Day 1 of treatment: There are no obvious signs of motor deficits and the animals will generally be alert and able to feed (see Note 7).

2. Day 2 of treatment: Some slowness of movement develops and there is reduced spontaneous vocalisation. Animals may have reduced ability for self-feeding and drinking. Hand feeding of liquidised food and fluids should be started (see Note 8).

3. Day 3 of treatment: Animals are noticeably bradykinetic, or akinetic and not self-feeding (see Note 9). Body weight may be reduced (see Note 10).

4. Day 4 of treatment: Animals are markedly bradykinetic, or akinetic and not self-feeding (see Note 11). Body weight may be reduced.

5. Day 5 of treatment: Animals are generally akinetic and not self-feeding. Body weight may be reduced.

6. Approximately 2 weeks after the end of treatment, 10–20% of the animals may develop a hyperactivity and obstinate progression syndrome (see Note 12) (38).

7. Daily weighing, hand feeding is continued until the animals have stabilised in regard to self-feeding and maintaining a stable body weight (see Note 13).

**2.2. Assessment
of Locomotor Activity
and Motor Disability**

Locomotor activity is assessed using automated test units, which are approximately 84 cm high × 60 cm wide and 70 cm deep. Each unit is fitted with two horizontal wooden perches, a water bottle and a clear Perspex door. Externally mounted are eight photoelectric transmitter/receivers with the light beams arranged so as to detect floor, perch and climbing activity. Interruption of a light beam, by a moving marmoset is automatically recorded as a single locomotor count by a computer using DasyLab acquisition software and collated in Microsoft Excel. Six individual test units are mounted in a rack allowing up to six animals to be individually assessed at one time. The test rack is positioned in a room that is viewed by a one-way mirror and is separate from the main housing room. Generally, animals are placed on test at the start of the day, before they are given food. Therefore, although pre-experiment they are not fasted as such, their last main meal would have been eaten the previous evening. Free access to drinking water is provided in the test cages and the animals are fed with their usual diet when returned to their home cages at the end of the test.

Motor disability is assessed simultaneously with assessment of locomotor activity through the one-way mirror by experienced observers blinded to treatment using an established motor disability rating scale; alertness (normal 0, reduced 1 and sleepy 2); checking movements (present 0, reduced 1 and absent 2); posture (normal 0, abnormal trunk +1, abnormal limbs +1, abnormal tail +1, or grossly abnormal 4); balance/co-ordination (normal 0, impaired 1, unstable 2 and spontaneous falls 3); reaction (normal 0, reduced 1, slow 2 and absent 3); vocalisation (normal 0, reduced 1 and absent 2); and motility (normal 0, bradykinesia 1 and akinesia 2). A total motor disability score of 0/10 min indicates normal behaviour and a score of 18/10 min indicates a high degree of behavioural deficits.

An average experimental group size is six and this has been calculated using power analysis to be the optimum size to achieve statistical significance.

1. Animals are weighed and placed into the test units.

2. The locomotor activity-monitoring program is stared.

3. Basal locomotor activity is assessed for 60 min prior to any treatment.

4. Basal disability and dyskinesia is assessed during the final 10 min of the 60 min basal locomotor activity period.

5. At time = 60 min, the animals are given the required treatment.

6. Locomotor activity monitoring is continued for the required time after drug administration.

7. Motor disability is assessed for 10 min once every 30 min (e.g. 20–30 and 50–60 min) for the required time after drug administration.

8. At the end of the test, the animals are returned to their home cages; total test duration is usually between 5 and 8 h.

2.3. Dyskinesia

2.3.1. Induction of Dyskinesia

Following MPTP-treatment and once animals have stabilised with regard to self-feeding and maintaining a stable body weight then priming for the expression of dyskinesia can be performed.

1. Animals are treated twice daily with L-dopa (12.5 mg/kg, p.o.) in combination with either carbidopa (12.5 mg/kg, p.o.) or benserazide (10 mg/kg, p.o.) at an interval of 3.5–4 h.

2. Twice daily treatment is performed on consecutive days for up to 56 days (see Note 14).

3. Behavioural assessments using the automated test units are performed twice weekly with at least 2 days between assessments. Motor disability and dyskinesia are assessed simultaneously with locomotor activity.

4. On non-behavioural assessment days, the animals are treated in their home cages and observed periodically after dosing for adverse effects and the severity of dyskinesia (see Note 15).

5. When an animal has displayed marked to severe dyskinesia on several consecutive behavioural assessment days, it can be considered to be "primed" to express dyskinesia (see Note 16).

2.3.2. Assessment of Dyskinesia

Dyskinesia is assessed simultaneously with motor disability for the same scoring periods. The severities of chorea and dystonia are assessed as individual components of dyskinesia and an overall dyskinesia rating is made using an established rating system. Chorea is composed of rapid random flicking and waving movements of the fore and hind limbs and may be associated with athetosis, a sinuous writhing movement of the limbs, or body. Dystonia is composed of sustained abnormal posturing (see Note 17).

- 0 = absent;
- 1 = mild, fleeting and rare dyskinetic postures and movements;
- 2 = moderate: more prominent abnormal movements, but not significantly affecting normal behaviour;
- 3 = marked, frequent and at times continuous dyskinesia affecting the normal pattern of activity;
- 4 = severe, virtually continuous dyskinetic activity, disabling to the animal and replacing normal behaviour.

3. Notes

1. Preferably, animals will have a body weight between 350 and 450 g. Animals below 320 g are highly susceptible to the effects of MPTP and are not used. Conversely, animals with a high body weight (above 500 g) can prove resistant to the effects of MPTP and may require an extended treatment period.

2. Animals below 18 months of age may prove to be either highly susceptible to the effects of MPTP or may recover rapidly and show minimal motor deficits.

3. MPTP-treated animals are more likely to injure themselves due to poor co-ordination and balance than naïve animals and having a pre-MPTP-treatment record of dentition and sites or indications of previous injury, such as broken tails/digits can prove invaluable in assessing any future suspected health problems.

4. Although it may seem obvious that an animal must have a good health status prior to any procedure, it is vital for successful MPTP-treatment. Conditions that may appear innocuous (e.g. tooth abscesses, minor superficial wounds) can substantially increase the mortality risk.

5. Due to the animals reduced mobility and impaired balance and co-ordination, it is advisable to reduce the cage height to minimise risk from falls and to provide easy access to nest areas. Similarly, furniture such as ladders and hoppers may also present an injury risk. MPTP-contaminated dust from bedding poses a health risk and this should be minimised by the use of dust-free tray liners. Cardboard boxes make useful nest boxes that can be disposed of when soiled.

6. The MPTP dose and the number of treatment days are dependent on the degree of lesion that is required. Although there is provision for up to 10 days of MPTP-treatment, to minimise the mortality risk and provide a more standardised treatment regime, up to 5 days of treatment at 2 mg/kg is generally used to create a full lesion (~90% loss of TH positive neurones). A small area on the lower back of the animal is shaved to facilitate the injection; any leakage of MPTP from the site is absorbed with a swab of tissue paper and discarded with clinical waste for de-contamination and disposal.

7. Although motor deficits will not be apparent at this stage, it is advisable to begin rating the degree of motor disability and general health condition from day 1 of MPTP-treatment.

8. To prevent malnutrition and dehydration, the hand feeding of liquidised food and fluids is essential. Various mixtures of commercially available high-protein/carbohydrate diets are used (e.g. Complan and Whey powder) combined with pureed bananas, vitamin supplements, etc. During the MPTP-treatment phase, the animals are weighed and then hand-fed prior to the injection being given. A second feeding is then carried out approximately 4 h later, with 20–30 ml of food/fluid being given using a 10-ml syringe at each feed. Larger primates (e.g. Squirrel monkey, Cynomolgus) are more difficult to hand feed and may require dopamine replacement therapy to encourage feeding. The benzodiazepine, Midazolam, has also been shown to encourage feeding in primates (39, 40).

9. At this stage, the animal may sit on the balance unrestrained and make little or no attempt to move, and this is a good indication of the degree of bradykinesia.

10. A weight loss of 10% is to be expected and this may increase to 20%. If weight loss does become excessive, then it will be necessary to change the feeding regime by increasing the number of feeding sessions and/or the addition of different and extra supplements (e.g. Calopet). However, percentage weight loss is not the overriding factor with regard to successful MPTP-treatment, as a very large animal may lose a higher proportion of body weight without ill-effects. If body weight falls below 300 g, regardless of the initial weight, this is an indicator that an animal may not survive the procedure.

11. The accurate assessment of motor deficits and general health condition on day 4, both before and after the fourth MPTP injection, is vital to the decision of whether a fifth MPTP injection is required. Body temperature, the degree of motor disability, lethargy, muscle tone, weight loss and difficulty with hand feeding should all be considered before proceeding to day 5 of MPTP-treatment. If there is any doubt regarding the condition of the animals, this fifth injection should be delayed or omitted completely.

12. This syndrome is sporadic and episodes may occur over a 2- to 3-week period. The risk of injury is substantial and we use a padded recovery housing unit (RHU) to house the animals when this activity is identified. Other groups use a further single dose of MPTP to reduce this syndrome (personal communication). The RHU, fitted with an electric warming blanket is also used for housing any animals that may become excessively lethargic and hypothermic following MPTP-treatment.

13. The time to stabilisation following MPTP-treatment is variable with some animals taking up to 12 weeks before hand feeding is stopped. It is necessary during the later stage of this period to "wean" the animals from being hand-fed by having weighing, but non-feeding days, to check whether they will self-feed. Some animals will not start to self-feed whilst being hand-fed.

14. Generally, 30-day treatment is sufficient to induce marked to severe dyskinesia. Some animals may develop dyskinesia within 1 or 2 days of treatment, and some animals may fail to respond to L-dopa and do not develop dyskinesia.

15. Observations are best made directly after treatment to check for signs of vomiting and at the time of peak dose effect, approximately 60–90 min after treatment.

16. At this stage to avoid the risk of any injury due to hyperactive and dyskinetic behaviour, it will be necessary to reduce, or stop L-dopa treatment, particularly for home cage dosing.

17. Choreic movements are generally very chaotic and tend to be associated with hyperactivity. The movements can range from mild hand or foot tapping akin to piano playing, to large limb waving movements, occasionally very rapid and repeated head bobbing are also observed. Athetosis is often exhibited as a rapid twisting or swaying body movement. Dystonia may be associated with a lower level of locomotor activity, and if severe can cause immobility. Movements include sustained limb extensions, neck twisting (torticollis and retrocollis), body twisting, elevation of the lower body and hip which may lead to handstands and moving backwards (retropulsion).

4. Characteristics of the MPTP-Treated Common Marmoset

The MPTP-treated common marmoset has well-documented motor deficits after toxin treatment, but also exhibits other symptomatology that is both neurological and autonomic in nature and that requires more detailed description. The key characteristics of the model are shown in Table 5. As with other primate species, MPTP treatment leads to a slowness of movement and postural abnormalities, but interestingly rigidity of the limbs and trunk is reported, but does not seem to have been formally demonstrated or quantified. Postural tremor affecting the hind quarters is commonly seen, but rest tremor is less common and it is more likely that intention tremor appears. Again, there do not seem to have been any formal studies of the tremor characteristics. Other neurological symptoms can be seen in these animals, but their existence is almost entirely based on casual observation. Myoclonic jerking, exaggerated startle responses, blepharospasm and spontaneous torticollis and retrocollis can present themselves. None of these have been analysed in detail and they do not affect every animal.

Autonomic dysfunction is clearly evident in the marmoset following MPTP treatment. Gastro-intestinal disturbances are not uncommon including both constipation and diarrhoea, and the animals are constantly wet in the groin region as a result of hyperreflexia of the bladder (41). The skin becomes scaly and the fur appears greasy and lank akin to seborrhea. All these signs are components of PD as it occurs in man. Sleep disturbance may occur as preliminary infrared video recordings during the night show MPTP-treated animals to be noticeably restless compared to normal common marmosets and day time somnolence can be observed.

Dopaminergic drug-related components of motor behaviour include the rapid onset of dyskinesia in response to L-dopa treatment and the appearance of stereotyped behaviours at high drug dosages (11). Post-treatment worsening is also observed, whereby as the drug effect subsides the animals appear to be more disabled

Table 5
Advantages of the use of the common marmoset

Category	Advantages
Husbandry	Small size Robust health with low incidence of disease Established veterinary care priciples Laboratory purpose bred Breeding records available Health status/records available Established husbandry protocols Established environmental enrichment protocols Adequate numbers can be used for complex studies Ease of dosing; p.o. (liquid and tablet),s.c., i.m., i.v., i.n., transdermal, buccal Ease of blood sampling from superficial vessels
MPTP treatment	Response to MPTP well documented Standardised MPTP-treatment protocols MPTP treatment is given s.c. without the need for anaesthesia Proven MPTP-treatment aftercare protocols established Low mortality rate from MPTP toxicity Stabilisation following MPTP treatment and no need for maintenance dopaminergic therapy Bilateral lesion tolerated well after recovery period
Drug response	Well defined Validated scoring of motor function and dyskinesia Quantitative/automated assessment of locomotor activity Widely published in peer reviewed journals Translation from marmoset to human well established
Surgery	Small size facilitates surgery Stereotaxic frames and brain atlas available Good and rapid recovery from anaesthesia Excellent recovery from surgical procedures Small volume of striatum (caudate-putamen) facilitates accurate localisation of implants Small volume of the striatum makes spread of injected material effective from minimum number of sites Established history of surgery for implantation of neuroprotective factors Implantation (s.c.) of osmotic minipumps well documented
Dyskinesia	Induction well established and reported Short duration of treatment required for priming Persistent and reproducible Established and validated rating system Major features of dyskinesia identifiable; chorea, dystonia, athetosis

and akinetic than at their pre-treatment baseline (42). In some instances, this may occur as a sudden and immediate cessation of activity with the animals sleeping and displaying negligible response to external stimuli. Anecdotally, changes in gait and balance occur as do episodes of freezing, but no serious investigation has been carried out. Analysis of the animal's response to repeated L-dopa treatment has not revealed the occurrence of "wearing off" and this differs from the experience in man where the disease process is progressive and motor deficits continue to worsen with time.

5. A Comparison Between Laboratories Using the MPTP-Treated Common Marmoset

Overall, investigations using MPTP-treated common marmosets have been undertaken at four major institutions and there is considerable commonality between the protocols employed and the assessment measures utilised (Table 6). This is important because majority of the pharmacological analysis of drug treatments using MPTP-treated primates has been carried out only in two species – the cynomolgus monkey and the common marmoset. Whereas, MPTP-related protocols and assessments of motor function are fairly variable between laboratories employing cynomolgus monkeys, the similarity in marmosets allows an assessment between centres that shows consistency of drug response. However, it should not be presumed that differences will not arise as colonies are commonly closed and in-bred genetic traits may influence drug responsiveness. Our experience has been that common marmosets sourced from one of the other centres, were not as responsive to L-dopa as we would have expected and that subcutaneous administration at relatively high doses was required for consistent motor responses.

6. Drug Responsiveness

The MPTP-treated common marmoset has probably been used more extensively for the assessment of novel therapeutic approaches to PD than any other primate species. Table 7 shows a selection of the drugs that we have tested in the marmoset, the doses employed and the therapeutic outcomes that have been achieved.

Not surprisingly, MPTP-treated common marmosets show an excellent reversal of motor disability and increase in locomotor activity when treated with L-dopa or with those dopamine agonists used to treat PD in man, for example the effects of ropinirole, pramipexole, rotigotine and piribedil are seen after oral, s.c. or transdermal administration (33, 43–49), appropriate delivery being determined by the characteristics of the drug molecule. Excellent effects can also

Table 6
A comparison of protocols and procedures between laboratories using the MPTP-treated common marmoset

| Parameter | Research group | | | |
	Jenner	Brotchie	Crossman	Phillipens
MPTP treatment	2.0 mg/kg/day s.c. for 5 consecutive days	2.0 mg/kg/day s.c. for 5 consecutive days	2.0 mg/kg/day s.c. for 5 consecutive days	Total of 6.0–8.75 mg/kg once daily over 10 days
MPTP after care	Published protocols			
Locomotor activity	Quantitative automated monitoring	Quantitative automated monitoring	Quantitative automated monitoring	Bungalow test Tower test
Motor disability	Visual rating scales Observations in real time	Visual rating scales Video recorded for post-hoc analysis	Visual rating scales Video recorded for post-hoc analysis	Visual rating scales Hand eye co-ordination test Hourglass test
Dyskinesia	Visual rating scales Observations in real time	Visual rating scales Video recorded for post-hoc analysis	Visual rating scales Video recorded for post-hoc analysis	Visual rating scales
Tremor assessment	Recorded as present or absent	Undefined	Undefined	Undefined
Video recording	Sample only	Full recording	Full recording	Full recording
Isolated test assessment	Yes	Yes	Yes	Yes

be achieved using continuous drug delivery through the subcutaneous implantation of Alzet osmotic minipumps in to the interscapular, or lower flank region (46–48). The effects of L-dopa are potentiated by decarboxylase inhibitors, such as carbidopa and benserazide, COMT inhibitors, such as entacapone, and MAO-B inhibitors, including selegiline in line with the effects in man (50, 51).

Other classes of dopaminergic drugs are also effective in MPTP-treated common marmosets. Notably, there is a good response to D_1 dopamine agonists, such as ABT-431, although it is necessary to use subcutaneous administration because of bioavailability problems (52, 53). The effects are robust and behaviour appears more normal than for $D_{2/3}$ agonists. These compounds are also effective in man, but there has been relatively little clinical evaluation because

Table 7
Examples of drug responsiveness of MPTP-treated common marmoset and dose ranges employed

Drug	Dose (mg/kg)	Increase in locomotor activity	Reversal of motor disability	Reference	Response in man
L-dopa	1.56–12.5	+++	+++	(104)	Improves motor function
Ropinirole	0.1–0.5	+++	+++	(32)	Improves motor function
Pramipexole	0.04–0.3	+++	+++	(49)	Improves motor function
Piribedil	1.25–12.5	+++	+++	(33)	Improves motor function
Bromocriptine	0.5–1.0	+++	+++	(105)	Improves motor function
Quinpirole	0.15–0.6	+++	+++	(106)	Improves motor function
Pergolide	0.4–0.5	+++	+++	(107)	Improves motor function
Rotigotine	0.019–0.3	+++	+++	(44)	Improves motor function
Apomorphine	0.35–1.5	+++	+++	(106)	Improves motor function
Pardoprunox	0.01–0.3	++	+++	(57)	Improves motor function
Aplindore	0.05–1.0	+++	+++	(55)	APLIED study in progress
A-77636	0.5–2.0 µmol/kg	+++	+++	(108)	Undefined
A-86929	0.1–3.0 µmol/kg	+++	+++	(52)	Undefined
ABT-413	0.03–1.0 µmol/kg	+++	+++	(52)	Improves motor function
SKF38393	1–20	---	---	(109)	Undefined
Istradefylline	0.5–100.0	++	+++	(64)	Improves motor function
ST1535	10–40	+++	++	(65)	Undefined
Fipamezole	1–1.0	Reduces L-dopa dyskinesia, without loss of motor improvement		(66)	Undefined
BTS 74398	2.5–10.0	++	++	(62)	Undefined
GBR 12909	5–10	++	++	(61)	Undefined

of bioavailability problems (54). Similarly, partial agonists, such as aplindore and pardoprunox, show good efficacy in this model, but their clinical advantages are not known (55–58).

The motor deficits seen in MPTP-treated common marmosets are also reversed by monoamine reuptake blockers. These are non-specific drugs that affect dopamine, noradrenaline and 5-HT reuptake, such as NS-2214, BTS 74–398, and more specific agents, such as GBR12909 and bupropion (59–62). However, this is an area where caution is required as these effects did not translate in to an effect on PD in subsequent clinical investigation. The reasons for this are not clear, but the measures of motor function used are relatively crude as behavioural measures go and they may respond to a general increase in mood and activation.

An increasing number of non-dopaminergic drugs have been examined in the MPTP-treated common marmoset for both their potential symptomatic effects on motor signs related to PD and for their ability to suppress established dyskinesia (63–66). Whether the effects seen in the marmoset translate in to a clinical benefit is not yet clear as too few non-dopaminergic drugs have undergone full clinical evaluation. Adenosine A2a antagonists, such as istradefylline, have been most fully evaluated and as a class, they appear to be able to enhance on-time in patients receiving dopaminergic medication without increasing troublesome dyskinesia (67, 68). However, the effects so far have been modest compared to the effects in common marmosets, but there are marked differences in the treatment protocols used. Serotoninergic drugs, particularly 5-HT$_{1A}$ agonists, are able to suppress dyskinesia in the marmoset, but at the expense of a worsening of motor function and this appears also to be the case in man (69, 70).

7. Use in Neuroprotection, Neurorestoration and Surgical Approaches to Treatment

The MPTP-treated common marmoset also needs to be thought of as a test bed for approaches to the treatment of PD other than those employing pharmacological manipulation of motor symptoms. This particularly applies to novel therapeutic strategies aimed at neuroprotection and neurorestoration. Studies of the potential neuroprotective actions of small molecules have been undertaken and shown drugs such as pramipexole and modafinil to reduce MPTP-induced nigral dopaminergic neurone loss (71, 72). How this translates to potential effects in PD remains unknown as insufficient studies and clinical trials have been undertaken to allow iteration.

The common marmoset is a very useful species in which to assess the effectiveness of surgical invasive technologies that are to be applied to PD (73–77). The ease of undertaking stereotaxic surgery and the excellent recovery from surgical procedures are major advantages as are the existence of a stereotaxic atlas, the small volume of the basal ganglia making target direction easy along with the small size and number of sites for injection that are required to cover this space (see Table 5). The small size of the caudate-putamen has also made the common marmoset an amenable species for undertaking in vivo microdialysis studies (78–80).

The common marmoset has been used for stereotaxic injection of toxins other than MPTP, for example 6-OHDA, and combinations of toxins, such as MPTP and 3-nitropropionic acid, to mimic either Huntington's chorea or multiple system atrophy (81, 82). But it has been in the neurorestorative arena that this species has had most use with studies of the actions of neurotrophic factors such as GDNF or BDNF that require direct injection in to the ventricular system or in to the basal ganglia (74–77). Similarly, the MPTP-treated common marmoset and other primate species form a viable test bed for assessing the effects of viral vector technologies for replacing dopamine production or for introducing genes for trophic factor production. Lastly, the MPTP-treated common marmoset is also valuable for assessing ablative approaches to restoring normal basal ganglia function and movement through the use of pallidotomy or subthalamotomy (73).

References

1. Bove J, Prou D, Perier C, Przedborski S (2005) Toxin-induced models of Parkinson's disease. NeuroRx 2:484–494

2. Lundblad M, Andersson M, Winkler C, Kirik D, Wierup N, Cenci MA (2002) Pharmacological validation of behavioural measures of akinesia and dyskinesia in a rat model of Parkinson's disease. Eur J Neurosci 15:120–132

3. POIRIER LJ (1960) Experimental and histological study of midbrain dyskinesias. J Neurophysiol 23:534–551

4. Davis GC, Williams AC, Markey SP, Ebert MH, Caine ED, Reichert CM, Kopin IJ (1979) Chronic Parkinsonism secondary to intravenous injection of meperidine analogues. Psychiatry Res 1:249–254

5. Langston JW, Ballard P, Tetrud JW, Irwin I (1983) Chronic Parkinsonism in humans due to a product of meperidine-analog synthesis. Science 219:979–980

6. Burns RS, Markey SP, Phillips JM, Chiueh CC (1984) The neurotoxicity of 1-methyl-4-phenyl-1,2,3,6-tetrahydropyridine in the monkey and man. Can J Neurol Sci 11:166–168

7. Jenner P, Rupniak NM, Rose S, Kelly E, Kilpatrick G, Lees A, Marsden CD (1984) 1-Methyl-4-phenyl-1,2,3,6-tetrahydropyridine-induced parkinsonism in the common marmoset. Neurosci Lett 50:85–90

8. Crossman AR, Clarke CE, Boyce S, Robertson RG, Sambrook MA (1987) MPTP-induced parkinsonism in the monkey: neurochemical pathology, complications of treatment and pathophysiological mechanisms. Can J Neurol Sci 14:428–435

9. Langston JW, Forno LS, Rebert CS, Irwin I (1984) Selective nigral toxicity after systemic administration of 1-methyl-4-phenyl-1,2,5,6-tetrahydropyrine (MPTP) in the squirrel monkey. Brain Res 292:390–394

10. Iravani MM, Syed E, Jackson MJ, Johnston LC, Smith LA, Jenner P (2005) A modified MPTP treatment regime produces reproducible partial nigrostriatal lesions in common marmosets. Eur J Neurosci 21:841–854

11. Pearce RK, Jackson M, Smith L, Jenner P, Marsden CD (1995) Chronic L-DOPA administration induces dyskinesias in the 1-methyl-4-phenyl-1,2,3,6-tetrahydropyridine-treated common marmoset (Callithrix Jacchus). Mov Disord 10:731–740

12. Jacobowitz DM, Burns RS, Chiueh CC, Kopin IJ (1984) N-methyl-4-phenyl-1,2,3,6-tetra-hydropyridine (MPTP) causes destruction of the nigrostriatal but not the mesolimbic dopamine system in the monkey. Psychopharmacol Bull 20:416–422

13. Forno LS, Langston JW, DeLanney LE, Irwin I, Ricaurte GA (1986) Locus ceruleus lesions and eosinophilic inclusions in MPTP-treated monkeys. Ann Neurol 20:449–455

14. Forno LS, Langston JW, DeLanney LE, Irwin I (1988) An electron microscopic study of MPTP-induced inclusion bodies in an old monkey. Brain Res 448:150–157

15. Schmidt CJ, Matsuda LA, Gibb JW (1984) In vitro release of tritiated monoamines from rat CNS tissue by the neurotoxic compound 1-methyl-phenyl-tetrahydropyridine. Eur J Pharmacol 103:255–260

16. Barcia C, Sanchez BA, Fernandez-Villalba E, Bautista V, Poza YP, Fernandez-Barreiro A, Hirsch EC, Herrero MT (2004) Evidence of active microglia in substantia nigra pars compacta of parkinsonian monkeys 1 year after MPTP exposure. Glia 46:402–409

17. Waters CM, Hunt SP, Jenner P, Marsden CD (1987) An immunohistochemical study of the acute and long-term effects of 1-methyl-4-phenyl-1,2,3,6-tetrahydropyridine in the marmoset. Neuroscience 23:1025–1039

18. McGeer PL, Schwab C, Parent A, Doudet D (2003) Presence of reactive microglia in monkey substantia nigra years after 1-methyl-4-phenyl-1,2,3,6-tetrahydropyridine administration. Ann Neurol 54:599–604

19. Rose S, Nomoto M, Jackson EA, Gibb WR, Jaehnig P, Jenner P, Marsden CD (1993) Age-related effects of 1-methyl-4-phenyl-1,2,3,6-tetrahydropyridine treatment of common marmosets. Eur J Pharmacol 230:177–185

20. Bartlett RM, Holden JE, Nickles RJ, Murali D, Barbee DL, Barnhart TE, Christian BT, DeJesus OT (2009) Paraquat is excluded by the blood brain barrier in rhesus macaque: An in vivo pet study. Brain Res 1259:74–79

21. Jenner P (1989) Clues to the mechanism underlying dopamine cell death in Parkinson's disease. J Neurol Neurosurg Psychiatry Suppl:22–28

22. Maret G, el TN, Carrupt PA, Testa B, Jenner P, Baird M (1990) Toxication of MPTP (1-methyl-4-phenyl-1,2,3,6-tetrahydropyridine) and analogs by monoamine oxidase. A structure-reactivity relationship study. Biochem Pharmacol 40:783–792

23. Clarke CE, Boyce S, Robertson RG, Sambrook MA, Crossman AR (1989) Drug-induced dyskinesia in primates rendered hemiparkinsonian by intracarotid administration of 1-methyl-4-phenyl-1,2,3,6-tetrahydropyridine (MPTP). J Neurol Sci 90:307–314

24. Guttman M, Fibiger HC, Jakubovic A, Calne DB (1990) Intracarotid 1-methyl-4-phenyl-1,2,3,6-tetrahydropyridine administration: biochemical and behavioral observations in a primate model of hemiparkinsonism. J Neurochem 54:1329–1334

25. Andringa G, Vermeulen RJ, Drukarch B, Renier WO, Stoof JC, Cools AR (1999) The validity of the pretreated, unilaterally MPTP-treated monkeys as a model of Parkinson's disease: a detailed behavioural analysis of the therapeutic and undesired effects of the D2 agonist quinpirole and the D1 agonist SKF 81297. Behav Pharmacol 10:163–173

26. Guehl D, Pessiglione M, Francois C, Yelnik J, Hirsch EC, Feger J, Tremblay L (2003) Tremor-related activity of neurons in the "motor" thalamus: changes in firing rate and pattern in the MPTP vervet model of parkinsonism. Eur J Neurosci 17:2388–2400

27. Barraud Q, Lambrecq V, Forni C, McGuire S, Hill M, Bioulac B, Balzamo E, Bezard E, Tison F, Ghorayeb I (2009) Sleep disorders in Parkinson's disease: the contribution of the MPTP non-human primate model. Exp Neurol 219:574–582

28. Bodis-Wollner I (1990) Visual deficits related to dopamine deficiency in experimental animals and Parkinson's disease patients. Trends Neurosci 13:296–302

29. Decamp E, Schneider JS (2004) Attention and executive function deficits in chronic low-dose MPTP-treated non-human primates. Eur J Neurosci 20:1371–1378

30. Schneider JS, Kovelowski CJ (1990) Chronic exposure to low doses of MPTP. I. Cognitive deficits in motor asymptomatic monkeys. Brain Res 519:122–128

31. Bedard PJ, Di PT, Falardeau P, Boucher R (1986) Chronic treatment with L-DOPA, but not bromocriptine induces dyskinesia in MPTP-parkinsonian monkeys. Correlation with (3H)spiperone binding. Brain Res 379:294–299

32. Jackson MJ, Smith LA, Al-Barghouthy G, Rose S, Jenner P (2007) Decreased expression of l-dopa-induced dyskinesia by switching to ropinirole in MPTP-treated common marmosets. Exp Neurol 204:162–170

33. Smith LA, Jackson MJ, Johnston L, Kuoppamaki M, Rose S, Al-Barghouthy G, Del SS, Jenner P (2006) Switching from levodopa to the long-acting dopamine D2/D3 agonist piribedil reduces the expression of dyskinesia while maintaining effective motor activity in MPTP-treated primates. Clin Neuropharmacol 29:112–125

34. Blanchet PJ, Grondin R, Bedard PJ (1996) Dyskinesia and wearing-off following dopamine D1 agonist treatment in drug-naive 1-methyl-4-phenyl-1,2,3,6-tetrahydropyridine-lesioned primates. Mov Disord 11:91–94

35. Visanji NP, Gomez-Ramirez J, Johnston TH, Pires D, Voon V, Brotchie JM, Fox SH (2006) Pharmacological characterization of psychosis-like behavior in the MPTP-lesioned nonhuman primate model of Parkinson's disease. Mov Disord 21:1879–1891

36. Petzinger GM, Quik M, Ivashina E, Jakowec MW, Jakubiak M, Di MD, Langston JW (2001) Reliability and validity of a new global dyskinesia rating scale in the MPTP-lesioned nonhuman primate. Mov Disord 16:202–207

37. Przedborski S, Jackson-Lewis V, Naini AB, Jakowec M, Petzinger G, Miller R, Akram M (2001) The parkinsonian toxin 1-methyl-4-phenyl-1,2,3,6-tetrahydropyridine (MPTP): a technical review of its utility and safety. J Neurochem 76:1265–1274

38. Jackson MJ (2001) Environmental enrichment and husbandry of the MPTP-treated common marmoset. Animal Technology 52:21–27

39. Foltin RW, Fischman MW, Byrne MF (1989) Food intake in baboons: effects of diazepam. Psychopharmacology (Berl) 97:443–447

40. Locke KW, Brown DR, Holtzman SG (1982) Effects of opiate antagonists and putative mu- and kappa-agonists on milk intake in rat and squirrel monkey. Pharmacol Biochem Behav 17:1275–1279

41. Albanese A, Jenner P, Marsden CD, Stephenson JD (1988) Bladder hyperreflexia induced in marmosets by 1-methyl-4-phenyl-1,2,3,6-tetrahydropyridine. Neurosci Lett 87:46–50

42. Kuoppamaki M, Al-Barghouthy G, Jackson M, Smith L, Zeng BY, Quinn N, Jenner P (2002) Beginning-of-dose and rebound worsening in MPTP-treated common marmosets treated with levodopa. Mov Disord 17:1312–1317

43. Maratos EC, Jackson MJ, Pearce RK, Jenner P (2001) Antiparkinsonian activity and dyskinesia risk of ropinirole and L-DOPA combination therapy in drug naive MPTP-lesioned common marmosets (Callithrix jacchus). Mov Disord 16:631–641

44. Rose S, Scheller DK, Breidenbach A, Smith L, Jackson M, Stockwell K, Jenner P (2007) Plasma levels of rotigotine and the reversal of motor deficits in MPTP-treated primates. Behav Pharmacol 18:155–160

45. Smith LA, Jackson MG, Bonhomme C, Chezaubernard C, Pearce RK, Jenner P (2000) Transdermal administration of piribedil reverses MPTP-induced motor deficits in the common marmoset. Clin Neuropharmacol 23:133–142

46. Stockwell KA, Virley DJ, Perren M, Iravani MM, Jackson MJ, Rose S, Jenner P (2008) Continuous delivery of ropinirole reverses motor deficits without dyskinesia induction in MPTP-treated common marmosets. Exp Neurol 211:172–179

47. Stockwell KA, Scheller D, Rose S, Jackson MJ, Tayarani-Binazir K, Iravani MM, Smith LA, Olanow CW, Jenner P (2009) Continuous administration of rotigotine to MPTP-treated common marmosets enhances antiparkinsonian activity and reduces dyskinesia induction. Exp Neurol 219:533–542

48. Stockwell KA, Scheller DK, Smith LA, Rose S, Iravani MM, Jackson MJ, Jenner P (2010) Continuous rotigotine administration reduces dyskinesia resulting from pulsatile treatment with rotigotine or L-DOPA in MPTP-treated common marmosets. Exp Neurol 221:79–85

49. Tayarani-Binazir KA, Jackson MJ, Rose S, Olanow CW, Jenner P (2010) Pramipexole combined with levodopa improves motor function but reduces dyskinesia in MPTP-treated common marmosets. Mov Disord

50. Smith LA, Jackson MJ, Al-Barghouthy G, Rose S, Kuoppamaki M, Olanow W, Jenner P (2005) Multiple small doses of levodopa plus entacapone produce continuous dopaminergic stimulation and reduce dyskinesia induction in MPTP-treated drug-naive primates. Mov Disord 20:306–314

51. Tayarani-Binazir KA, Jackson MJ, Fisher R, Zoubiane G, Rose S, Jenner P (2010) The timing of administration, dose dependence and efficacy of dopa decarboxylase inhibitors on the reversal of motor disability produced by L-DOPA in the MPTP-treated common marmoset. Eur J Pharmacol 635:109–116

52. Shiosaki K, Jenner P, Asin KE, Britton DR, Lin CW, Michaelides M, Smith L, Bianchi B, Didomenico S, Hodges L, Hong Y, Mahan L, Mikusa J, Miller T, Nikkel A, Stashko M, Witte D, Williams M (1996) ABT-431: the diacetyl prodrug of A-86929, a potent and selective dopamine D1 receptor agonist: in vitro characterization and effects in animal models of Parkinson's disease. J Pharmacol Exp Ther 276:150–160

53. Smith LA, Jackson MJ, Al-Barghouthy G, Jenner P (2002) The actions of a D-1 agonist in MPTP treated primates show dependence on both D-1 and D-2 receptor function and tolerance on repeated administration. J Neural Transm 109:123–140

54. Rascol O, Blin O, Thalamas C, Descombes S, Soubrouillard C, Azulay P, Fabre N, Viallet F, Lafnitzegger K, Wright S, Carter JH, Nutt JG (1999) ABT-431, a D1 receptor agonist prodrug, has efficacy in Parkinson's disease. Ann Neurol 45:736–741

55. Jackson MJ, Andree TH, Hansard M, Hoffman DC, Hurtt MR, Kehne JH, Pitler TA, Smith LA, Stack G, Jenner P (2010) The dopamine D(2) receptor partial agonist aplindore improves motor deficits in MPTP-treated common marmosets alone and combined with L-dopa. J Neural Transm 117:55–67

56. Johnston LC, Jackson MJ, Rose S, McCreary AC, Jenner P (2010) Pardoprunox reverses motor deficits but induces only mild dyskinesia in MPTP-treated common marmosets. Mov Disord 25:2059–2066

57. Jones CA, Johnston LC, Jackson MJ, Smith LA, van SG, Rose S, Jenner PG, McCreary AC

(2010) An in vivo pharmacological evaluation of pardoprunox (SLV308)--a novel combined dopamine D(2)/D(3) receptor partial agonist and 5-HT(1A) receptor agonist with efficacy in experimental models of Parkinson's disease. Eur Neuropsychopharmacol 20:582–593

58. Tayarani-Binazir K, Jackson MJ, Rose S, McCreary AC, Jenner P (2010) The partial dopamine agonist pardoprunox (SLV308) administered in combination with l-dopa improves efficacy and decreases dyskinesia in MPTP treated common marmosets. Exp Neurol 226:320–327

59. Pearce RK, Smith LA, Jackson MJ, Banerji T, Scheel-Kruger J, Jenner P (2002) The monoamine reuptake blocker brasofensine reverses akinesia without dyskinesia in MPTP-treated and levodopa-primed common marmosets. Mov Disord 17:877–886

60. Hansard MJ, Smith LA, Jackson MJ, Cheetham SC, Jenner P (2002) Dopamine, but not norepinephrine or serotonin, reuptake inhibition reverses motor deficits in 1-methyl-4-phenyl-1,2,3,6-tetrahydropyridine-treated primates. J Pharmacol Exp Ther 303:952–958

61. Hansard MJ, Smith LA, Jackson MJ, Cheetham SC, Jenner P (2002) Dopamine reuptake inhibition and failure to evoke dyskinesia in MPTP-treated primates. Eur J Pharmacol 451:157–160

62. Hansard MJ, Smith LA, Jackson MJ, Cheetham SC, Jenner P (2004) The monoamine reuptake inhibitor BTS 74 398 fails to evoke established dyskinesia but does not synergise with levodopa in MPTP-treated primates. Mov Disord 19:15–21

63. Hill MP, Ravenscroft P, Bezard E, Crossman AR, Brotchie JM, Michel A, Grimee R, Klitgaard H (2004) Levetiracetam potentiates the antidyskinetic action of amantadine in the 1-methyl-4-phenyl-1,2,3,6-tetrahydropyridine (MPTP)-lesioned primate model of Parkinson's disease. J Pharmacol Exp Ther 310:386–394

64. Kanda T, Jackson MJ, Smith LA, Pearce RK, Nakamura J, Kase H, Kuwana Y, Jenner P (1998) Adenosine A2A antagonist: a novel antiparkinsonian agent that does not provoke dyskinesia in parkinsonian monkeys. Ann Neurol 43:507–513

65. Rose S, Jackson MJ, Smith LA, Stockwell K, Johnson L, Carminati P, Jenner P (2006) The novel adenosine A2a receptor antagonist ST1535 potentiates the effects of a threshold dose of L-DOPA in MPTP treated common marmosets. Eur J Pharmacol 546:82–87

66. Savola JM, Hill M, Engstrom M, Merivuori H, Wurster S, McGuire SG, Fox SH, Crossman AR, Brotchie JM (2003) Fipamezole (JP-1730) is a potent alpha2 adrenergic receptor antagonist that reduces levodopa-induced dyskinesia in the MPTP-lesioned primate model of Parkinson's disease. Mov Disord 18:872–883

67. Hauser RA, Shulman LM, Trugman JM, Roberts JW, Mori A, Ballerini R, Sussman NM (2008) Study of istradefylline in patients with Parkinson's disease on levodopa with motor fluctuations. Mov Disord 23:2177–2185

68. Mizuno Y, Hasegawa K, Kondo T, Kuno S, Yamamoto M (2010) Clinical efficacy of istradefylline (KW-6002) in Parkinson's disease: a randomized, controlled study. Mov Disord 25:1437–1443

69. Iravani MM, Tayarani-Binazir K, Chu WB, Jackson MJ, Jenner P (2006) In 1-methyl-4-phenyl-1,2,3,6-tetrahydropyridine-treated primates, the selective 5-hydroxytryptamine 1a agonist (R)-(+)-8-OHDPAT inhibits levodopa-induced dyskinesia but only with\ increased motor disability. J Pharmacol Exp Ther 319:1225–1234

70. Olanow CW, Damier P, Goetz CG, Mueller T, Nutt J, Rascol O, Serbanescu A, Deckers F, Russ H (2004) Multicenter, open-label, trial of sarizotan in Parkinson disease patients with levodopa-induced dyskinesias (the SPLENDID Study). Clin Neuropharmacol 27:58–62

71. Iravani MM, Haddon CO, Cooper JM, Jenner P, Schapira AH (2006) Pramipexole protects against MPTP toxicity in non-human primates. J Neurochem 96:1315–1321

72. van Vliet SA, Blezer EL, Jongsma MJ, Vanwersch RA, Olivier B, Philippens IH (2008) Exploring the neuroprotective effects of modafinil in a marmoset Parkinson model with immunohistochemistry, magnetic resonance imaging and spectroscopy. Brain Res 1189:219–228

73. Iravani MM, Costa S, Al-Bargouthy G, Jackson MJ, Zeng BY, Kuoppamaki M, Obeso JA, Jenner P (2005) Unilateral pallidotomy in 1-methyl-4-phenyl-1,2,3,6-tetrahydropyridine-treated common marmosets exhibiting levodopa-induced dyskinesia. Eur J Neurosci 22:1305–1318

74. Costa S, Iravani MM, Pearce RK, Jenner P (2001) Glial cell line-derived neurotrophic factor concentration dependently improves disability and motor activity in MPTP-treated common marmosets. Eur J Pharmacol 412:45–50

75. Dass B, Iravani MM, Jackson MJ, Engber TM, Galdes A, Jenner P (2002) Behavioural and immunohistochemical changes following supranigral administration of sonic hedgehog in 1-methyl-4-phenyl-1,2,3,6-tetrahydropyridine-treated common marmosets. Neuroscience 114:99–109

76. Iravani MM, Costa S, Jackson MJ, Tel BC, Cannizzaro C, Pearce RK, Jenner P (2001) GDNF reverses priming for dyskinesia in MPTP-treated, L-DOPA-primed common marmosets. Eur J Neurosci 13:597–608

77. Pearce RK, Costa S, Jenner P, Marsden CD (1999) Chronic supranigral infusion of BDNF in normal and MPTP-treated common marmosets. J Neural Transm 106:663–683

78. Kaseda S, Nomoto M, Iwata S (1999) Effect of selegiline on dopamine concentration in the striatum of a primate. Brain Res 815:44–50

79. Nomoto M (1995) (Application of the common marmoset to pharmacological studies). Nippon Yakurigaku Zasshi 106:11–18

80. Nomoto M, Kaseda S, Iwata S, Shimizu T, Fukuda T, Nakagawa S (2000) The metabolic rate and vulnerability of dopaminergic neurons, and adenosine dynamics in the cerebral cortex, nucleus accumbens, caudate nucleus, and putamen of the common marmoset. J Neurol 247 Suppl 5:V16-V22

81. Eslamboli A, Baker HF, Ridley RM, Annett LE (2003) Sensorimotor deficits in a unilateral intrastriatal 6-OHDA partial lesion model of Parkinson's disease in marmoset monkeys. Exp Neurol 183:418–429

82. Svenningsson P, Arts J, Gunne L, Andren PE (2002) Acute and repeated treatment with L-DOPA increase c-jun expression in the 6-hydroxydopamine-lesioned forebrain of rats and common marmosets. Brain Res 955:8–15

83. Petzinger GM, Fisher B, Hogg E, Abernathy A, Arevalo P, Nixon K, Jakowec MW (2006) Behavioral motor recovery in the 1-methyl-4-phenyl-1,2,3,6-tetrahydropyridine-lesioned squirrel monkey (Saimiri sciureus): changes in striatal dopamine and expression of tyrosine hydroxylase and dopamine transporter proteins. J Neurosci Res 83:332–347

84. Quik M, Cox H, Parameswaran N, O'Leary K, Langston JW, Di MD (2007) Nicotine reduces levodopa-induced dyskinesias in lesioned monkeys. Ann Neurol 62:588–596

85. Quik M, Parameswaran N, McCallum SE, Bordia T, Bao S, McCormack A, Kim A, Tyndale RF, Langston JW, Di Monte DA (2006) Chronic oral nicotine treatment protects against striatal degeneration in MPTP-treated primates. J Neurochem 98:1866–1875

86. Boyce S, Rupniak NM, Steventon MJ, Iversen SD (1990) Characterisation of dyskinesias induced by L-dopa in MPTP-treated squirrel monkeys. Psychopharmacology (Berl) 102:21–27

87. Visanji NP, Fox SH, Johnston TH, Millan MJ, Brotchie JM (2009) Alpha1-adrenoceptors mediate dihydroxyphenylalanine-induced activity

in 1-methyl-4-phenyl-1,2,3,6-tetrahydropyridine-lesioned macaques. J Pharmacol Exp Ther 328:276–283

88. Hadj TA, Ekesbo A, Gregoire L, Bangassoro E, Svensson KA, Tedroff J, Bedard PJ (2001) Effects of acute and repeated treatment with a novel dopamine D2 receptor ligand on L-DOPA-induced dyskinesias in MPTP monkeys. Eur J Pharmacol 412:247–254

89. Hallett PJ, Brotchie JM (2007) Striatal delta opioid receptor binding in experimental models of Parkinson's disease and dyskinesia. Mov Disord 22:28–40

90. Raju DV, Ahern TH, Shah DJ, Wright TM, Standaert DG, Hall RA, Smith Y (2008) Differential synaptic plasticity of the corticostriatal and thalamostriatal systems in an MPTP-treated monkey model of parkinsonism. Eur J Neurosci 27:1647–1658

91. Heimer G, Bar-Gad I, Goldberg JA, Bergman H (2002) Dopamine replacement therapy reverses abnormal synchronization of pallidal neurons in the 1-methyl-4-phenyl-1,2,3,6-tetrahydropyridine primate model of parkinsonism. J Neurosci 22:7850–7855

92. Raz A, Feingold A, Zelanskaya V, Vaadia E, Bergman H (1996) Neuronal synchronization of tonically active neurons in the striatum of normal and parkinsonian primates. J Neurophysiol 76:2083–2088

93. Bjugstad KB, Teng YD, Redmond DE, Jr., Elsworth JD, Roth RH, Cornelius SK, Snyder EY, Sladek JR, Jr. (2008) Human neural stem cells migrate along the nigrostriatal pathway in a primate model of Parkinson's disease. Exp Neurol 211:362–369

94. Varastet M, Riche D, Maziere M, Hantraye P (1994) Chronic MPTP treatment reproduces in baboons the differential vulnerability of mesencephalic dopaminergic neurons observed in Parkinson's disease. Neuroscience 63:47–56

95. Kowall NW, Hantraye P, Brouillet E, Beal MF, McKee AC, Ferrante RJ (2000) MPTP induces alpha-synuclein aggregation in the substantia nigra of baboons. Neuroreport 11:211–213

96. Collins MA, Neafsey EJ (1985) Beta-carboline analogues of N-methyl-4-phenyl-1,2,5,6-tetrahydropyridine (MPTP): endogenous factors underlying idiopathic parkinsonism? Neurosci Lett 55:179–184

97. Lipina SJ, Colombo JA (2007) Premorbid exercising in specific cognitive tasks prevents impairment of performance in parkinsonian monkeys. Brain Res 1134:180–186

98. Jakowec MW, Petzinger GM (2004) 1-methyl-4-phenyl-1,2,3,6-tetrahydropyridine-lesioned

model of parkinson's disease, with emphasis on mice and nonhuman primates. Comp Med 54:497–513

99. Moratalla R, Quinn B, DeLanney LE, Irwin I, Langston JW, Graybiel AM (1992) Differential vulnerability of primate caudate-putamen and striosome-matrix dopamine systems to the neurotoxic effects of 1-methyl-4-phenyl-1,2,3,6-tetrahydropyridine. Proc Natl Acad Sci USA 89:3859–3863

100. Stephenson DT, Meglasson MD, Connell MA, Childs MA, Hajos-Korcsok E, Emborg ME (2005) The effects of a selective dopamine D2 receptor agonist on behavioral and pathological outcome in 1-methyl-4-phenyl-1,2,3,6-tetrahydropyridine-treated squirrel monkeys. J Pharmacol Exp Ther 314:1257–1266

101. Hodgson RA, Bedard PJ, Varty GB, Kazdoba TM, Di PT, Grzelak ME, Pond AJ, Hadjtahar A, Belanger N, Gregoire L, Dare A, Neustadt BR, Stamford AW, Hunter JC (2010) Preladenant, a selective A(2A) receptor antagonist, is active in primate models of movement disorders. Exp Neurol 225:384–390

102. Nomoto M, Kita S, Iwata SI, Kaseda S, Fukuda T (1998) Effects of acute or prolonged administration of cabergoline on parkinsonism induced by MPTP in common marmosets. Pharmacol Biochem Behav 59:717–721

103. Fox SH, Henry B, Hill M, Crossman A, Brotchie J (2002) Stimulation of cannabinoid receptors reduces levodopa-induced dyskinesia in the MPTP-lesioned nonhuman primate model of Parkinson's disease. Mov Disord 17:1180–1187

104. Kuoppamaki M, Al-Barghouthy G, Jackson MJ, Smith LA, Quinn N, Jenner P (2007) L-dopa dose and the duration and severity of dyskinesia in primed MPTP-treated primates. J Neural Transm 114:1147–1153

105. Pearce RK, Banerji T, Jenner P, Marsden CD (1998) De novo administration of ropinirole and bromocriptine induces less dyskinesia than L-dopa in the MPTP-treated marmoset. Mov Disord 13:234–241

106. Loschmann PA, Smith LA, Lange KW, Jahnig P, Jenner P, Marsden CD (1992) Motor activity following the administration of selective D-1 and D-2 dopaminergic drugs to MPTP-treated common marmosets. Psychopharmacology (Berl) 109:49–56

107. Maratos EC, Jackson MJ, Pearce RK, Cannizzaro C, Jenner P (2003) Both short- and long-acting D-1/D-2 dopamine agonists induce less dyskinesia than L-DOPA in the MPTP-lesioned common marmoset (Callithrix jacchus). Exp Neurol 179:90–102

108. Kebabian JW, Britton DR, DeNinno MP, Perner R, Smith L, Jenner P, Schoenleber R, Williams M (1992) A-77636: a potent and selective dopamine D1 receptor agonist with antiparkinsonian activity in marmosets. Eur J Pharmacol 229:203–209

109. Nomoto M, Jenner P, Marsden CD (1985) The dopamine D2 agonist LY 141865, but not the D1 agonist SKF 38393, reverses parkinsonism induced by 1-methyl-4-phenyl-1,2,3,6-tetrahydropyridine (MPTP) in the common marmoset. Neurosci Lett 57:37–41

Chapter 21

Behavioral Assessment in the African Green Monkey After MPTP Administration

D. Eugene Redmond Jr.

Abstract

This chapter describes the behavioral methods of our research program for assessing the effects of 1-methyl-4-phenyl-1,2,3,6-tetrahydropyridine (MPTP) intoxication in the African green monkey as well as various strategies for reversing those effects. MPTP intoxication has been extensively described as a model for Parkinson's disease. I will describe how and why our particular methods were developed, how they are done, and what advantages and disadvantages that they have, with illustration using data collected from our program over a number of years. Analysis of these data replicated earlier findings as well as revealed new information about the effects of MPTP in primates. These data also show the validity, sensitivity, and utility of these methods.

Key words: MPTP, African green monkey, Ethological, Ratings, Parkinson's, L-DOPA, Long-term effects

1. Introduction

The discovery that MPTP induces parkinsonism in humans [1, 2] and in nonhuman primates [3–5] via a highly selective destruction of dopamine neurons led to my interest in studying whether these cells could be restored by transplantation of fetal mesencephalic precursor cells [6]. Since little had been published at that time in primates, my group determined to characterize the histological, biochemical, cognitive, and motor/behavioral effects in a primate species that had not been previously studied, the African green monkey.

The aim of these studies was to determine whether MPTP intoxication reproduced the signs, characteristics, and complications of Parkinson's disease in primates as well as it did in humans,

Emma L. Lane and Stephen B. Dunnett (eds.), *Animal Models of Movement Disorders: Volume I*, Neuromethods, vol. 61, DOI 10.1007/978-1-61779-298-4_21, © Springer Science+Business Media, LLC 2011

and equally important, whether the model could predict success of potential new treatments in patients.

Accurate and comprehensive characterization of the model was therefore essential using histological and biochemical methods (7–14). Initially, we made no assumptions regarding behavioral and functional manifestations of parkinsonism in the monkey, but wanted a quantitative system that was selective for parkinsonism while also providing information about other behaviors and level of function. Options, which were available and have been used by others, included various operant activities (15), lever pressing (16), directional rotation (17), motor activity monitors (18), video assessment (19), and the application of qualitative/semiquantitative rating scales almost directly from the neurology clinic (20–29).

We were concerned that training and practice effects could confound operant methods and though they would reflect motor changes, they would not do so selectively. In addition, they would provide data only up to a moderate level of parkinsonism after which the animals would not or could not perform the task at all. Motor activity alone, measured by an accelerometer or automatic movement detectors, while quantitative, was felt to be too nonselective, unable to distinguish parkinsonism from sedation or other illnesses. After some preliminary studies with MPTP, we decided to try a combination of an ethologically based time sampling behavioral observation method with rated items taken from neurological rating scales.

2. Development of Behavioral Methods

2.1. Selection of Behavioral Items

We compiled a list of 200 behaviors of the African green monkey which had been described from the existing primatology literature and defined them as precisely as possible. With a group of observers who were experienced in the observation of this species, these behaviors were quantitated using a one-zero time sampling method (30) which counts the presence or absence of a behavior during a 5-s interval, which is counted audibly by a tape recorder. The behavior cannot exceed more than one count each 5 s, with a maximum of 60 counts in a 5-min observation period. Many items were dropped from our list of behaviors, which could not be performed by a single monkey in a standard size cage (social grooming, aggression, etc.). Spontaneous undisturbed behavior was observed for a 5-min period that was rated one-zero or by 5-s duration, followed by a mild verbal and facial threat gesture by the observer and the presentation of a piece of fruit. These "challenges" were used to elicit behavioral responses and to insure that certain behaviors were likely to occur, which might not otherwise be observed during a random 5-min period. Food presentation was used to observe the speed and quality of the subject's response, the effect of voluntary movement on tremor, or any difficulties in handling food, eating, or swallowing. Mild threat gestures were given to be able to assess motor speed, spontaneous freezing, and aggressive responses. Items that could not be quantitated one-zero were "rated" at the end of the observation period, and included the effects of the "challenges," on a 5-point scale to provide a larger range of behavioral variability.

2.2. Selection Based Upon Reliability and Frequency

Based upon a significant amount of data, more behaviors were dropped due to very low frequencies or the inability of the observers to achieve adequate inter-rater reliability of 95% on Kendall's coefficient of concordance (31, 32). Notably dropped were two behaviors, "facial expression/masked facies" and "hunched posture" which many of the other scales have incorporated from clinical rating scales. Our observers were unable to reliably rate these items, perhaps because green monkeys have fairly blank facial expressions and hunched postures normally. The remaining behaviors and their definitions are shown in Tables 1–3. With

Table 1
Rated and scored behavioral measures: definitions of scored behaviors or signs for assessment of MPTP[a]

Bipedal lookout	Standing on legs with hands not touching the floor and gazing outside of cage. Legs may be standing on perch or other object, and hands may be resting against the cage
Cage pick	Manual exploration or manipulation of any part of the cage. Eating (if food is clearly apparent or can be reliably inferred) is not scored as cage pick. Observers should note that monkeys almost always appear to introduce something into their mouths after grooming and cage picking, but this should not be scored as eating unless it is clearly food
Chew/bruxism	Alternating mandibular-maxillary apposition without previous introduction of food or objects into mouth. May occur with or without audible grinding of the teeth. Yawns, grooms, cage picks, and facial threat gestures are excluded and should be scored as those behaviors if their defined criteria are met
Drinking	Contact between mouth and drinking spout. It is not necessary that you see the monkey swallowing the water
Eating	Actively gathering, manipulating, introducing into the mouth, chewing, and swallowing of food. This includes holding of food in the hands or feet, but must include some element of activity. For example, a monkey might freeze while holding food in its hands, or it might have visible food in its cheek pouches which would not score as eating because there is no activity present
Eyes closed	Motionless (as defined below) with eyelids closed for 5 s, or motionless for 5 s in the traditional sleep posture-sitting hunched over with head down. Scored with priority over motionless, per 5-s duration (Since each count of eyes closed must also meet the definition of motionless, score only eyes closed, not both behaviors)

(continued)

Table 1
(continued)

Facedown	Lying down with ventral or lateral surface of trunk and abdomen in contact with the cage floor or perch. Scored once per 5-s duration
Freeze/motionless	Remaining motionless for 5-s duration, excluding eyelid movement or blinks. Some very small eye movement can be tolerated, but if monkey actively changes its field of vision that will break the motionless/freeze. A minimal period of 5 s must occur before any motionless would be scored in the next interval. For example, 8 full seconds of motionless would receive a score of 0 (5 s to meet criteria and only 3 s in the second period, not meeting criteria). 12 s would receive a score of 1 (5 s initial criteria, 5-s duration = 1 count, and 2 s in the third period, not meeting criteria)
Masturbation	Any rubbing or manipulation of the genital area which does not meet the criteria for self-grooming. Penile erection is not necessary for a male for this to be scored
Penile erection	Protrusion or raising of the penis. This should be scored if the glans of the penis is fully visible, even if not fully protruded or erect
Scratch	Rapid and repeated rubbing of any body part with fingers, hand, or foot. This behavior is distinguished from self-grooming
Self-groom	Gentle manipulation of the monkey's own hair or skin with hands or mouth, usually accompanied by introduction of particles into the mouth
Shift	Pacing or walking about the cage. Must be on all 4 feet and must take at least two steps for the behavior to be scored. The behavior is scored a maximum of 1 count per 5-s interval during which it occurs
Tail flag	While standing or walking, scored if any part of the tail passes forward of the plane which is perpendicular to the longitudinal axis of the monkey and passing through the point at the base of the tail
Threaten outside	Prolonged stare at observer or monkeys in other cages usually accompanied by head bobbing, raising of eyebrows, flattening of ears, or a "square-mouth" face
Vertical climb	Any locomotion on sides or top of the cage in which the monkey's feet do not touch the cage floor or perch, scored in frequency per 5-min period. A monkey may score shift, while on a perch, but vertical climb only when going or coming from the floor to the perch
Vocalization	Any pharyngeal or laryngeal sounds
Stereotypy	Repeated similar or identical patterns of movement with any part of the body which do not fit any other definition, or which are repetitions of behaviors which are scored and recognized. Examples, head movements back and forth, repeated pacing, etc. If the specific movements fit usual definitions, they should also be scored – for example, shift
Dyskinesia	Abnormal uncontrolled movements of body torso, head, neck, arms, or legs. Often has a writhing quality or is particularly uncontrolled. Tremor is excluded and should be scored separately
Dystonia	Abnormal posturing that appears to be caused by abnormal contractions of muscles. Different from dyskinesia that involves *movements*

[a] Extended operational definitions. Individual observed behaviors score 1 or 0 per 5 s or per 5-s duration

Table 2
Rated and scored behavioral measures: behaviors scored at end of observation period after behavioral challenge[a]

Food response	Speed with which animal reaches for, handles, and eats segment of banana or other highly prized fruits (scored from 0 to 5, normal response scores 0, no response scores 5)
Delayed movement	Degree to which motor movement appears to be delayed although eventually carried out by the monkey (scored 0 = no delay to 5 severe inhibition). If a monkey is unable to move at all, he should be scored 5
Difficulty eating	Physical difficulty in handling, biting, chewing, or swallowing food or liquids (scored 0–5). If he cannot accomplish every aspect of feeding himself, he should be scored 5, i.e., he has sufficient difficulty in eating that he would die without an intervention, he should get the highest rating
Appearance	General condition of grooming and appearance (scored 0 for normal to 5 very abnormal). This is not a measure of "blank" faces seen in Parkinson's patients
Poverty of movement	Slowness, decreased complexity, and small quantity of movement (scored from 0 absent to 5 severe)
Threat response	Motor, facial, and vocal responses to postural and vocal threats from humans (scored 0 normal to 5 absent response)
Head tremor	Oscillating movement of head at rest or during "voluntary" movement (scored 0–5 as follows: 0 = absent; 1 = fine tremor, as defined, seen intermittently, or under special situations; 2 = fine tremor seen more often; 3 = larger amplitude tremor seen occasionally or fine tremor seen most of the time; 4 = constant fine tremor or large amplitude tremor most of the time; 5 = large amplitude tremor seen almost constantly)
Limb tremor	Oscillating movement of any limbs at rest or during "voluntary" movement (scored 0 absent to 5 severe, as defined for head tremor)
Effect of "intention"	Tremor observed and scored either decreases (negative value from 0 to 5) or increases (positive value from 0 to 5) with "voluntary" movements
Dyskinesia	Any abnormal or unusual movement of face, lips, tongue, head, or limbs, besides tremor as defined (scored 0 absent to 5 severe and persistent). If such movements are identified during any observation session, an effort should be made to videotape them for later more definitive identification
Spontaneous freeze	Interruption of an on-going motor movement pattern of any type, which lasts at least 5 s (scored 0 absent to 5 severe, as defined for head tremor.) Some movement must occur for this to be scored
Left < right	If any of the above rated behaviors affect one side of the body more than the other, this item should be scored to indicate the most affected side. 0 indicates no lateralization. Scores may range from –5 (unilateral symptoms on the left only) to +5 (unilateral symptoms on the right)
Hypervigilance	Manifested by rapid head and eye movements back and forth and hyper-responsiveness to stimuli such as noises or movements in the environment. Score 0 (none) to 5 (extreme) for the entire observation period

[a] A behavioral challenge involves a presentation of food and vigorous postural and vocal threat from humans; these items are to be scored based upon the entire period of observation including the special challenges which occur at the end of the session

Table 3
Rated and scored behavioral measures: behavior summaries from factor analyses[a]

Parkscore	The sum of the following individual scores, as defined: head tremor, limb tremor, appearance, freeze/motionless/5, difficulty eating, delayed initiation of movement, poverty of movement, response to threat, facedown
Tremor	Head tremor, limb tremor
Anxiety	Yawn, chew, scratch, self-groom, penile erection
Arousal	Shift, tail flag, bipedal lookout, vertical climb
Sedation	Eyes closed and freeze
Quiet OK	Self-groom, cage pick, eating, drinking
Healthy	Arousal, anxiety, quiet okay

[a] These behaviors are derived by summing the primary behaviors scored

time-sampled behaviors, only one behavior can be scored during the 5-s period, with a hierarchy of behaviors also defined in the table. The hierarchy was generally determined by the level and uniqueness of purposeful activity that the behavior required. These behaviors were used to collect thousands of observations of 11 untreated normal adult male monkeys and 66 similar monkeys after the administration of MPTP.

2.3. Original Factor Analysis

Rather than assuming which behaviors represented specific internal behavioral states, we used factor analysis to determine whether groups of behaviors clustered together and whether such factors might provide a more empirical method of assessing the behavioral state of the animal. The principal component factor analysis (32) identified five factors that accounted for approximately 50% of the standardized variance. In a series of experiments, we identified and validated these factors under a number of conditions and constructed the factor scores by summing the individual items. The items included in each of the factors are shown in Table 4. This factor analysis was in agreement with previous summary factors that were based on a small number of MPTP-treated monkeys (6). The derivation of the factor scores that we have used in many studies was thoroughly described, along with extensive information about the MPTP model in African green monkeys, by Taylor et al. (33).

A factor score provides broader measurement of the underlying behavioral state than an individual item would do. For example, early after MPTP, a monkey might show tremor that disappears as it becomes more frozen (and in fact incapacitated). Plotting tremor alone, however, might make it appear that the animal had improved. Making sure that items are properly weighted is another

Table 4
Behavioral item loading for each of five factors at 1–4 months after MPTP

Behavior	Factor 1	Factor 2	Factor 3	Factor 4	Factor 5
Parkscore	*Anxiety*	*Arousal*	*Tremor*	*Quiet*	*OK*
Shift	−0.20780	0.16592	0.57029	0.17455	0.06405
Tlflag	−0.11823	0.23215	0.49656	0.31581	−0.00451
Yawn	−0.36566	0.60287	−0.26483	0.15457	0.13955
Chew	−0.33983	0.54192	−0.29245	0.03769	0.21236
Scratch	−0.17076	0.29693	−0.24700	−0.05239	0.21610
Vocaliz	−0.02991	0.06162	0.08250	0.07875	0.05563
Slfgrm	−0.26362	0.19277	−0.08267	−0.35691	0.28735
Cagepck	−0.26278	0.20757	0.07055	−0.34943	0.40858
Eating	−0.14449	−0.12640	−0.05376	−0.18520	−0.27209
Drink	−0.20381	0.01419	0.10231	−0.27089	0.31156
Threat	−0.28802	0.30130	0.26743	0.27363	−0.19385
Lookout	−0.20293	0.16341	0.37217	0.08092	0.11708
Vertclm	−0.12439	0.10452	0.45834	0.09032	0.17433
Penerct	−0.18023	0.44736	−0.25774	0.18611	−0.35043
Mastrb	−0.09092	0.31242	−0.21401	0.16510	−0.34580
Freeze	0.41356	0.06623	0.03748	−0.03158	0.01183
Eyescls	0.10647	−0.26524	−0.13078	*0.34802*	0.14449
Foodres	0.90239	0.19032	0.11610	−0.15817	−0.07331
Delay	0.91334	0.18948	0.10636	−0.14166	−0.05589
Eatprobs	0.92779	0.17797	0.09210	−0.11904	−0.04049
Appear	0.90115	0.10896	−0.00467	−0.00703	0.01481
Poverty	0.94346	0.06611	0.00167	−0.00921	0.00444
Thrtres	0.88619	0.07747	−0.00756	−0.00861	0.01797
Trintent	0.70594	−0.03828	−0.10782	0.24991	0.20098
Tremhead	0.81653	−0.03668	−0.12533	0.24307	0.18577
Tremlimb	0.81871	−0.00771	−0.12622	0.23228	0.18554
Sponfrz	0.15882	−0.11789	−0.11810	*0.37928*	0.34756
Facedown	0.71602	0.27201	0.16163	−0.30809	−0.17148

(continued)

Table 4
(continued)

Behavior	Factor 1	Factor 2	Factor 3	Factor 4	Factor 5
Parkscore	*Anxiety*	*Arousal*	*Tremor*	*Quiet*	*OK*
Eigenvalue	8.27	1.64	1.48	1.26	1.15
% Variance	29.5	5.8	5.3	4.5	4.1

Components identified and summed for each factor score are underlined based upon the prior factor analysis, and were used also to analyze data in this chapter. The loadings shown are from the repeated factor analysis of the new 1 month data presented here. Differences from the prior analysis are shown in italics if an item showed a significantly different loading or was not included in the summary factor. Whether a behavior was included in the summary factors was also influenced by experimental conditions to test the factors, for example, did the behaviors and factor for anxiety move appropriately in response to fear stimuli

goal of a useful behavioral assessment. One would want an animal that is unable to walk to have a higher score of parkinsonism than one that has a combination of less incapacitating signs. In a group of experimentally exposed animals, there are advantages if scores on a scale or measurement are distributed across a broad range rather than be clustered together if there is real variability.

Factor 1 identified behaviors that were increased after MPTP administration, with high positive correlations (0.6–0.9), and accounting for 23% of the variance. Based on these behaviors, we created our Parkinson's score or *Parkscore*, which includes *difficulty eating, facedown, motionless/freeze, head and limb tremor, abnormal response to threat, poverty of movement, delayed initiation of movement, poor appearance, abnormal response to food presentation.* Although all other behaviors were summed to make up the score, we gave *freeze* a diminished weight (divided by 5, since it is a part of a subject's normal behavioral repertoire). It should be noted that this normal behavior, remaining motionless for at least 5 s, is different from *spontaneous freezing*, which was not identified as part of the factor.

The behaviors with negative loading in factor 1 were the behaviors that had high loadings in factors 2, 3, and 5. These were also identified by their appropriate responsiveness to conditions of anxiety, arousal, or quiet normal behaviors, so were named *anxiety, arousal,* and *quiet OK* behavior. To represent a significant proportion of nonpathological behaviors exhibited by the monkeys, we summed these three to form a super-factor, *Healthy Behavior.*

2.4. Variability Assessed by Quintiles of 1 Month Outcome

In those early experiments, we showed the effects of MPTP on these behaviors, noting, as expected that MPTP increases *Parkscore* and decreases *Healthy Behavior* (33). We noticed considerable variability in response to the same doses of MPTP, as well as recovery that appeared greater in mildly affected monkeys. To study this, the monkeys were divided into quintiles based upon their *Parkscore* at 1 month after MPTP. A wide range of outcomes was apparent, from one-fifth that were essentially asymptomatic to one-fifth that were severely incapacitated and unable to ambulate. Based upon the individual signs being expressed in each of these quintiles, this severity classification corresponds remarkably well to the 5-point Hoehn and Yahr clinical rating scale (34). Severity Group 0 included all of the untreated animals, as well as several of the MPTP-treated animals that remained asymptomatic. They had the lowest *Parkscores*, and the highest *Healthy Behavior* scores. The highest quintile animals were severely impaired, and had the highest scores of *facedown* (largely could not walk, similar to the highest rank of Hoehn and Yahr). Along with the highest *Parkscores*, they had the lowest *Healthy Behavior* scores. Being in the highest quintile, the month after treatment largely predicted whether the subject died or was sacrificed prior to the end of the experiment ($\chi^2(4) = 27.54$, $p < 0.01$). Fifty percent of subjects in category 4 (severe) and 25% in category 3 (moderate) died or were sacrificed for humane reasons within 4 months of MPTP treatment despite constant nursing care and medical treatments (33). No deaths occurred in the other severity categories.

Factors 2, 3, and 5 represent all of the other behavioral states that can be identified by this particular very comprehensive set of behavioral categories and ratings. Not surprisingly, there is an inverse correlation between increasing *Parkscore* and the sum of these other non-Parkinson's factors (*Healthy Behavior*). As might be expected, *Healthy Behavior* is not selective for parkinsonism and can be reduced by sedative and tranquilizing drugs or by other physical illnesses, such as pneumonia. Nonetheless, this measure might be useful and is similar, for monkey behavior, to the "activities of daily living" component of the Unified Parkinson's Disease Rating Scale (UPDRS) (35). Comprehensive assessment of other behavioral activities of monkeys is missing from many of the clinical scales applied to monkeys (29). A final advantage of the *Parkscore* factor is that a single outcome variable in a therapeutic experiment eliminates the need for multiple independent statistical analyses that make it difficult to predict expected probabilities.

2.5. Assessing the Required Behavioral Sample Size

We also studied other methodological questions that would affect reliability and reproducibility of the data. We determined how behaviors of the animals varied during a 24-h period by doing observations during each 2-h period every day. We also studied how much time was necessary for each observation period to give a reasonable sample of the behavior. We found, not surprisingly, that animals' behavioral activities are different throughout the 24-h period (36). The duration of the observation necessary for a reliable sample depended upon the behaviors of interest. If you were interested in a short duration and infrequent behavior, a much longer observation period would be required. Parkinsonism, although somewhat variable, is fairly stable over time, so that a shorter sample of behavior, repeated over a week, will give a reasonably reliable estimate of parkinsonian disability. Just as adding "cost" factors to sample size calculations, the labor costs of doing observations need to be considered and adjusted to what is otherwise known about experimental interventions that are being evaluated. A cell transplant, even if successful, is likely to take a matter of weeks or months to improve behavior, whereas an intravenous injection of a dopamine agonist would have to be evaluated over the time course of its pharmacokinetics and physiological activity.

For our standard observations, two 5-min observation periods using this scale were made of each animal each day, by an observer seated in plain view in front of the cage, and repeated 2–5 times per week. These observations were done at a rotating time within a 2-h period between 9:30 and 11:30 a.m., so that they would be distributed throughout that period (some animals not always observed at one time and others at another time). This particular period was chosen by analysis of data from four different 2-h periods during the day. This particular period was statistically not different from any of the other three periods for any behavior, with the exception of *chew*, which differed between the first period (after feeding) and the last period of the day (33). We found that behavioral changes in the animals in response to the observers, after some preliminary conditioning, returned to normal 5 min after the rater was seated. The monkeys seem to have learned that seated humans are unlikely to do them any harm. No behaviors were formally rated again on any monkey after the "challenges" until later in the observation schedule on that day. There were usually multiple observers, each observing one monkey at a time, but their sequence of sitting, observing, and "challenging" was synchronized by the tape recorder timing signal.

2.6. Controlling for Observer Bias

Another important consideration in the use of ratings, which involve some subjective judgments, is that the ratings be done without knowledge of the experimental manipulations. We have attempted to do this, but it requires the use of control vehicle injections and surgeries. Some outcomes are obvious and will "unblind" the observers of those animals. Inter-rater reliability is also a challenge and is achieved by weekly inter-rater reliability tests (Kendall's concordance at >95%), with discussion of differences to maintain consistency and to prevent experimental drift. In addition, in preliminary data analyses, observer differences were analyzed by ANOVA, and all observations from any observers that were significant outliers from the other observers were dropped from the dataset (this is not a problem with very large datasets, but might be with smaller ones and few observers).

Since the Taylor et al. (33) chapter in 1994, we have accumulated a significant amount of data on many monkeys treated for a variety of experiments. This earlier chapter remains a comprehensive description of the MPTP model, the management of the animals, and the side effects and medical complications, which I have not attempted to repeat. In the present chapter, I will illustrate the behavioral methods by re-examining some of the key questions with a substantially larger dataset, which covers a longer period of time, more monkeys, and may answer some long-standing questions about the MPTP model.

3. Illustrative Use of the Methods with New Unpublished Data

3.1. Description of Animals and MPTP Treatments

All monkeys studied were male African green monkeys (*Chlorocebus sabaeus*), weighing from 5 to 8 kg at the beginning of the experiments. Note that these animals from St. Kitts, West Indies, have had several official taxonomic name changes during the course of our studies and are designated in various reports as *Cercopithecus aethiops sabaeus*, *Chlorocebus aethiops sabaeus*, and other variations, all being exactly the same type of monkey. The animals were given free access to water and fed standard daily rations of Harland Teklad monkey chow (20% protein and without isoniazide or other drugs or contaminants), and were hand fed or tube fed if necessary to ensure minimal nutrition and fluid requirements. They were sometimes given supplemental fresh tropical fruits, such as banana, soursop, starfruit, and guava. They were held in roofed enclosures in standard primate squeeze cages exposed to natural daylight at 17° North latitude with natural breezes and ambient temperatures. Care and treatment was in accordance with the US Public Health Service *Guide for the Care and Use of Animals* (37). All of the relevant protocols were approved by the institutional review committees. Antiparkinsonian drugs were not administered so as to avoid confounding results, except for L-DOPA administration to determine its effects on the model. Although most of the animals went on to participate in experiments for potential anti-Parkinson's treatments, only baseline or pre-MPTP exposure data were included in the present analyses.

MPTP was administered intramuscularly over a 5-day period in most of the monkeys, with the total dose divided into five treatments. Four doses were administered at approximately 4 p.m., the following 8 a.m. and 4 p.m., the next 8 a.m. The fourth day was skipped, and the fifth dose administered at 8 a.m. on the fifth day. When an animal showed a significant early toxic effect, the fifth dose was skipped. The monkeys included in these analyses, however, all received either a total of 2 or 2.25 mg/kg. Some animals were re-dosed with various small repeated doses over a period of months, but these animals were analyzed separately to evaluate these different injection schedules.

3.2. Data Collection and Statistical Analyses

Behavioral data were collected by trained observers as described earlier and in detail by Taylor et al. (33). Regular inter-rater reliability testing was done, with discussion to minimize and reconcile differences and to prevent observer drift. These data were entered into a computer at the observation site from paper data recording sheets, which were then checked for accuracy. Data were analyzed using the Statistical Analysis System. After the assumptions for the use of parametric tests were confirmed (normality and lack of heteroscedasticity), the effects of observers, individual differences between monkeys, and repeated measures over time, dose response to L-DOPA were made using analysis of variance (38). Data from any observers that were significantly different from the other observers were dropped from the dataset. Post hoc tests, after ANOVA, used Student–Newman–Keuls test (at $p < 0.05$) (38). Linear regressions were determined using PROC GLM from SAS (32). Approximately 33,000 observations entered into the different analyses.

3.3. How Does MPTP Treatment Change Motor and Behavioral Function?

Two hundred and eighteen monkeys were studied during a baseline period of 30–90 days before treatment and then the month following MPTP administration. From their normal baseline state, the monkeys became progressively parkinsonian during the first month. After some acute reactions during treatment, such as occasional vomiting, monkeys became progressively slower in their motor movements, showed decreased normal behaviors and activities, ate and drank less, and then developed more characteristic signs of parkinsonism such as bradykinesia, incoordination, delayed initiation of movement, freezing, and muscle rigidity, tremor with intention and at rest, and difficulties with ambulation and eating and swallowing.

3.4. Behavioral Differences 1 Month After MPTP Administration

Every scored and rated behavior changed during the first month. The *Parkscore* changed from a baseline of 0.97–29.47 during the first month ($F = 305$, df = 1, 6910, $p < 0.0001$). There were significant differences between monkeys, however ($F = 13.41$, df = 217, $p < 0.0001$). Conversely, normal behaviors, as measured by *Healthy Behavior*, were significantly reduced from 21 during baseline to 4.55 in month 1 ($F = 1515$, df = 1, 6886, $p < 0.0001$), also with significant differences among monkeys ($F = 9.29$, df = 217, 6886, $p < 0.0001$). The components which contributed to the *Parkscore* are shown in Fig. 1. Every item changed ($p < 0.0001$, for all except *vocalize*, which was $p < 0.0004$). Parkinsonian behaviors increased, and other normal activities decreased. Effects of MPTP administration on the individual non-Parkinson's behaviors were detailed previously (33).

3.5. New Factor Analysis and Loading of Behaviors

Since factor analyses are sometimes dependent upon the specific sample of data, the same type of factor analysis was run again on these newer data. The five principal factors were replicated, and the

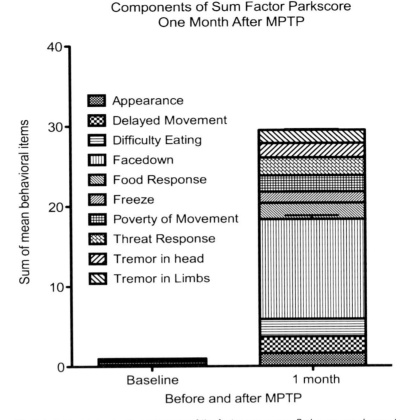

Fig. 1. Individual behavioral components of the factor sum score, *Parkscore*, are shown at 1 month after MPTP. The standard error of the means is plotted, but is so small that they are almost invisible. All of these behaviors changed significantly in the first month as determined by analysis of variance. *Parkscore* would equal the sum of those behaviors, except freeze divided by 5.

component individual behaviors are shown in Table 4. Many of the loadings, which are shown from the new analysis, are almost identical to the previous analysis (see Table II in Taylor et al. (33)), but some of the parkinsonian behaviors show higher loadings, consistent with the fact that, as a whole, the monkeys in this sample were more severe. There was more variability in the nonparkinsonian behaviors, probably because this dataset included more data from animals with significant parkinsonism. There were no monkeys in the present data that had not received MPTP (normal monkeys), but there were extensive baseline/normative data from the same monkeys prior to MPTP administration.

3.6. Response Variability in Response to MPTP

We repeated the same approach described by Taylor et al. (33) initially to examine the degree of variability. The animals were categorized into five severity scores (0, 1, 2, 3, and 4) by rank of *Parkscore* at 1 month. It was apparent from this analysis that some animals showed much greater sensitivity to MPTP, becoming more parkinsonian more rapidly in the same fashion. Approximately 44 monkeys were in each severity group. Group 0 had values of *Parkscore* from 0.45 to 7.3. Since there were no normal controls in this dataset, all of these monkeys were treated at least with 2.0 mg/ kg of MPTP, and some with more. The next groups had *Parkscore* ranges as follows: Group 1, from 7.3 to 14.3; Group 2, from 14.3 to 26.4; Group 3, from 27 to 60.6; and Group 4, from 60.7 to 98.3. The animals that became more parkinsonian in the first month were also more likely to die from some complication of the disease. To determine the impact of severity on the rate of progression or possible recovery, we analyzed these animals during the first 4 months after MPTP treatment, computing slopes by each severity grouping for each month. Groups 0, 1, and 2 remained fairly flat, with some suggestion of increases after 3 months. Based on their classifications of severity after MPTP, none of the monkeys in these severity groups were different during the baseline period, although some significant slopes in opposite directions were identified (see Fig. 2). After MPTP, Groups 0, 1, 3, and 4 showed highly significant increasing slopes in the first month, and Groups 3 and 4 showed significant decreasing slopes of *Healthy Behavior* at that time. Of course, mean values of *Parkscore* during the month

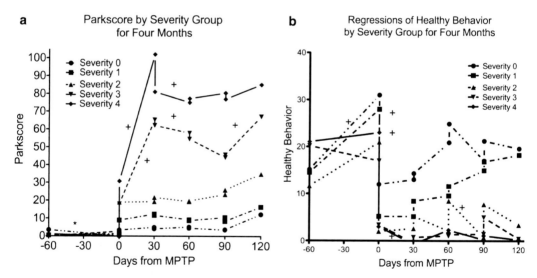

Fig. 2. Does severity at 1 month predict outcome and recovery? To minimize the impact of varying numbers of monkeys to be analyzed after month 1, each animal was categorized by its severity category (0–4). Then the data were analyzed using a linear regression analysis separately during each of the first 4 months, with the predicted value plotted for *Parkscore* () or *Healthy Behavior* () regression over time. Each severity group was plotted with a different style line, and the points were marked with different symbols. Significantly positive or negative slopes were designated with "+". All groups had very low baseline *Parkscores*, and a random pattern of *Healthy* during the baseline period. In the second month, both Severity Group 3 and 4 showed a significant decreasing slope, but this leveled out in months 3 and 4 for Severity Group 4. Severity Group 3 had a significant increasing slope. None of the other groups had significant nonzero slopes, with the interpretation that they were stable during this period. *Healthy* behavior was highest in Severity Group 0 and lowest in Group 4.

ranked perfectly with severity classification (since the groupings were based on these 1 month data). Thereafter, both *Parkscore* and *Healthy Behavior* remained stable (no significant increasing slopes for *Parkscore* or evidence of recovery from decreasing slopes). The only exception was Group 3 in the fourth month, which showed an increasing slope for *Parkscore* ($p < 0.003$) and decreasing *Healthy Behavior* ($p < 0.03$). Group 4 animals showed the steepest increases in the first month and remained highly parkinsonian during this time period, although, as with initial sensitivity to MPTP, there were some individuals that did better. The *Healthy Behavior* scores were always inverted, with the highest scores in Groups 0 and 1. The numbers of observations per period analyzed ranged from 1,015 in the baseline to 160 in one of the groups in month 4. Because the number of animals was not constant over the period, the slopes were analyzed independently for each month so that more severe monkeys that died in a previous month would not affect the slopes in the next month, although these could have influenced changes in the mean scores. From these data, the previous report, that severity ranking in the first month would predict the outcome, continued to be true for this time period, although with significantly larger numbers of animals there are clearly some exceptions. In addition, 4 months is clearly not long enough to assume that there is no recovery, as we have learned.

3.7. Stability Over a 2-Year Period

We had many more animals that entered into therapeutic experiments after 4 months that confound a longer analysis with those treatments. To look at a longer period in the same monkeys, we investigated a group of seven monkeys that were sham operated or had MPTP injections only as controls in other experiments, but were observed and scored for 24 months or longer (Fig. 3). These animals were all given a single acute series with 2.25 mg/kg of MPTP. The total number of observations on these monkeys was approximately 350 each, for a total of 2,390. These animals showed a pattern of recovery after the first month. After month 5, they appeared quite stable, with gradual decreases in *Parkscores*, to the level of Severity Group 1, but not decreasing to the status of fully normal monkeys. In a therapeutic experiment, this result, without matching controls, would likely be interpreted as showing a dramatic and significant effect, but the monkeys continued to be mildly parkinsonian. In this group, the *Parkscores* in the first month were from 17 (Y079) to 82 (X391). Testing the prediction that the most severe monkey (monkey X391, Severity Group 4) would remain the most parkinsonian, at 24 months, this was not the case. After 24 months, both were essentially asymptomatic with scores of 5 (Y079) and 2.2 (X391). So, clearly, in this example, the prediction was not supported. This finding, however, is consistent with variability in susceptibility and recovery from MPTP. Compared with the published literature, a group of seven monkeys

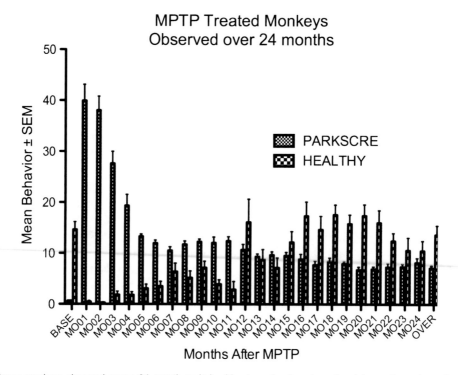

Fig. 3. Seven monkeys observed over a 24-month period, without any treatment or other interventions showed apparent stabilization to a mild level of parkinsonism after 4 months. There was significant variability in individual monkeys, with some monkeys with high *Parkscores* that returned almost to normal after 2 years, compared with a more stable pattern that was seen in the larger number of monkeys in the previous figure.

seems like a reasonably large number. However, selected from several hundred, X391 is likely to be an extraordinary survivor which was included in this analysis of 2-year data because he had survived, when in fact the majority of animals at that level of severity had died or had been sacrificed. As we previously reported (33), 50% of the animals at that level of severity would be likely to die. This animal indicates that there can be some remarkable exceptions, and potential resiliency factors in animals would be valuable to investigate further.

3.8. Is Variability Due to Different Doses of MPTP?

In the data analyzed by Taylor et al., there was some small variability in the doses of MPTP, which did not correlate with behavioral outcome, but the variations might have been too small for the sample size to detect. In the present data, the MPTP doses were more consistent for large numbers of monkeys, and some large variations in MPTP dose provide an opportunity to investigate the effect of MPTP dosing as a cause of this variability. In this sample, 86 monkeys were injected with cumulative doses of 2.0 mg/kg and 136 monkeys were given 2.25 mg/kg. Another 48 monkeys were injected with increasing doses over weeks to months. The two lower doses (with identical amounts and administration schedules)

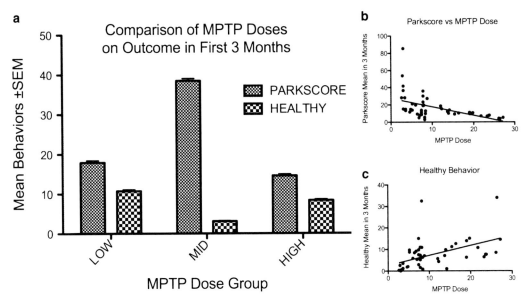

Fig. 4. The effects of doses of MPTP on *Parkscore* and *Healthy Behaviors* were compared, with low dose = 2.0 mg/kg, mid dose = 2.25 mg/kg, and high dose being greater than 2.25 mg/kg, with a mean of 11.5 mg/kg. The dose groups were compared in panel **a** with the mean behavioral outcome for *Parkscore* and *Healthy* in (**a**). As expected, there was a significant increase in *Parkscore* from 2.0 to 2.5 mg/kg. Unexpectedly, the higher dose showed significantly lower *Parkscores* and higher *Healthy Behaviors*. Because there was a much larger variation in doses in this group, *Parkscore* is plotted against MPTP in (**b**) and *Healthy Behavior* in (**c**). The regression line is plotted, showing a significant correlation in the opposite direction that might be expected, as increasing doses are associated with lower *Parkscores* ($F = 14.9$, df = 1, 46, $p < 0.0003$) and higher *Healthy Behaviors* ($F = 14.1$, df = 1, 46, $p < 0.0005$). Since animals were automatically selected for higher doses by failing to become parkinsonian on the standard doses, these MPTP-resistant monkeys continue to be resistant to higher and higher doses.

were designated as "Low" and "Mid" and compared with the "High" dose group, which was more variable, with total doses from 2.6 to 27 mg/kg and a mean dose of 11.5 mg/kg (see Fig. 4a). The higher doses were given with the hope that monkeys that did not become symptomatic with the initial 2.0 or 2.25 mg/kg dose would become symptomatic. This led to a wide range of doses and durations, with some animals succumbing to fatal toxic levels, and some remaining asymptomatic after very large doses (Y233 received a total of 27 mg/kg over a period of several months, and remained at a *Parkscore* of 5, with *Healthy Behavior* of 14.5). The differences between the three dose groups were highly significant and in unexpected directions. The overall effect of MPTP dose on *Parkscore* was significant ($F = 886$, df = 2, 9306, $p < 0.001$). Post hoc tests showed the "Mid" dose (2.25 mg/kg) to have a significantly greater effect on Parkscore than both the "Low" (2.0) and the "High" doses. *Healthy Behavior* was also significantly different ($F = 119$, df = 2, 8872, $p < 0.0001$). All three doses were significantly different from each other (Fig. 4a). The surprising result is that the middle dose produced the highest *Parkscores* and the lowest *Healthy Behaviors*. There was no difference between several different production lots of the MPTP that were used over this time period (Sigma-Aldrich).

To confirm and examine the failure of repeated dosing to induce parkinsonism in the higher dose group, because of the variability of the dosing in that group, we plotted the relationship between dose and *Parkscore* in the high dose group separately (see Fig. 4b). This analysis showed that there is a significant relationship between MPTP dose and *Parkscore* in these animals (slope is nonzero, $F = 14.96$, df = 1, 46, $p < 0.0003$), but opposite from the predicted direction (higher doses with lower scores). A similar, but opposite effect was found on *Healthy Behavior* (slope is nonzero, $F = 14.09$, df = 1, 46, $p < 0.0005$) (Fig. 4c).

In summary, there was a highly significant effect to increase parkinsonism and to decrease healthy behaviors of a small increment in the MPTP dose from 2.0 to 2.25 mg/kg. The explanation for the absence of an increase in parkinsonism between the "Mid" and "High" doses probably relates to the notable resistance of some monkeys to the drug. Thus, those treated with the "High" dose were, in effect, preselected for relative insensitivity to MPTP, such that repeated chronic dosing and higher doses of MPTP had little effect. A similar finding was reported in marmosets (39). It is an intriguing possibility that the factor(s) responsible for marked sensitivity to MPTP in monkeys is/are the same as those responsible for the exaggerated nigrostriatal dopamine system degeneration observed in Parkinson's disease. Alternatively, the mechanism behind the resistance to MPTP in some monkeys might be exploited for its therapeutic potential.

3.9. What Are the Longer-Term Effects of MPTP?

Are monkeys actually different from people by sometimes recovering spontaneously and never progressing the way people have done after accidental MPTP exposure? It is very difficult to find out experimentally. Grant committees generally prefer not to spend money on monkey studies at all, especially when rodents and fruit flies are cheaper, much less to fund studies over a period of many years. Carrying out such studies could also be difficult, with turnover of observers and other uncontrollable factors. In addition, we already know the answer in humans. Nonetheless, it might be valuable to know if and how the "model" differs between monkeys and humans. These caveats, in advance, are intended as an apology for the fact that the next group of monkeys were not studied consistently or with proper controls, but nonetheless might be worthwhile to consider since this same observation method has been used over a long period and has been successfully taught to many observers. We have now analyzed data from seven monkeys, which were "rejected" from therapeutic experiments because they were not sufficiently symptomatic to be put into a therapeutic study. We hesitated to re-expose them to additional doses of MPTP, not

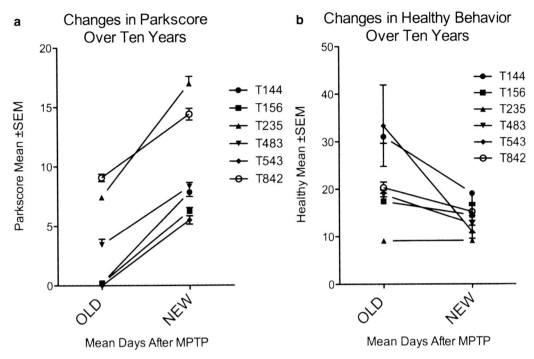

Fig. 5. Long-term changes in *Parkscore* or *Healthy Behavior* over a 10-year period after MPTP treatment. Mean values are shown for the earlier data (100 days or more after MPTP) compared with data from more recent observations. (**a**) All six monkeys showed increases in *Parkscore* (old vs. new period – $F = 27.03$, df = 1, 2816, $p < 0.0001$; however, interactions between observer or monkey and the two periods were not significant, $p < 0.19$ and $p < 0.06$). When the slopes of the more recent observations were analyzed separately, to eliminate early effects that might have biased the long-term data, five of the seven monkeys showed significant nonzero slopes to higher *Parkscores* (see Table 5) during the recent data collection period alone. (**b**) Although five of six showed decreases in *Healthy Behavior* over this time period, these were not significant, according to the ANOVA, which also examined factors due to monkey and observer. These interactions with the old vs. new period were not significant ($p < 0.14$ and $p < 0.93$). Overall difference between the two periods was not significant ($F = 3.12$, df = 1, 2846, $p < 0.07$).

knowing what longer-term (biological or selection) effects that might have on a future experiment. We believed that they might someday be useful for determining effects of MPTP and also interactions with age, so they were allowed to "retire." They were originally given the standard 2.0 or 2.25 mg/kg MPTP doses. We compared prior identically collected behavioral data on six of these monkeys with a series of observations taken a mean of 10 years later ("new" data). The first 100 days after MPTP were dropped in order to compare the later more stable effects and to eliminate any bias from initial acute responses to MPTP. The "old" means were plotted against the "new" ones in Fig. 5a, b (*Parkscore* and *Healthy*). All six monkeys showed increases in mean *Parkscores* and five of six showed decreases in *Healthy Behaviors* over that time period. The earliest were given MPTP more than 10 years before the "new" observations, so many of them added advanced age to their MPTP effects (shown in Table 5). Monkey ages were calculated from

Table 5
Long-term MPTP changes in *Parkscore* by individual monkey

Monkey	Age now	MPTP dose	*Parkscore* (mean±SEM)	N (obs)	Years after MPTP	*Parkscore* (mean±SEM)	N (obs)	Years after MPTP	New slope	Intercept	p<0.05
T144	19.7	2.00	0.21±0.15	442	0.75	7.85±0.37	63	10.97	0.0063	−27.6	0.0021
T156	21.6	2.00	0.07±0.05	441	0.74	6.26±0.29	58	11.92	0.0033	−7.9	0.0600
T235	21.7	2.50	6.37±4.03	1340	2.59	17.00±0.59	83	10.02	0.0145	−36.0	0.0001
T483	21.2	2.00	4.00±2.83	391	2.90	8.31±0.35	63	14.99	0.0074	−32.7	0.0001
T543	21.2	2.00	0.00±0.00	9	0.28	5.51±0.35	83	14.91	0.0048	−20.6	0.0210
T842	16.4	2.50	9.19±5.81	265	0.65	14.40±0.49	60	6.15	0.0020	10.0	0.4600
T867	16.1	2.25		0		5.20±0.30	61	10.14	0.0037	−8.1	0.0300
Mean	19.7		3.31±2.14		1.3	9.89±0.41		11.5			

The mean values and number of observations during the old and new periods are shown. Significant slopes for *Parkscore* during this period for five of the seven monkeys. Age shown is the age at the end of the recent data collection period. (These monkeys show stigmata of aging in their 20 s and live to about 30 years.) The individual item *tremor* (not shown) also increased in five of the six monkeys, suggesting that the overall increases in *Parkscore* ($F = 16.90$, df = 1, 2862, $p < 0.0001$) were not due merely to slow movement during advancing age. There were no significant interactions between observer and the analysis period ($p < 0.32$), although there were significant interactions between monkey and period ($p < 0.005$)

known birth dates or estimated from the time since trapping plus 5 years if an animal appeared to be fully adult at the time of trapping (juvenile and infant ages were estimated originally using body weight).

In addition to the mean comparisons, the "new" sample of behaviors were reanalyzed to determine whether the monkeys had increasing *Parkscore* based upon the recent measurements. Six of the seven monkeys showed nonzero slopes (increasing *Parkscore*) during this more recent observation period. The mean *Parkscores* and the slopes are shown in Table 5. The results are now confounded, perhaps similarly to human Parkinson's disease, with the effects of advancing age. We have recently documented a group of aged female monkeys (over 20 years of age), never exposed to MPTP or known neurotoxic drugs, which also showed elevated *Parkscores* compared with younger animals (40). But generally, such classic signs of Parkinson's disease, such as tremor, have not been reported in aged monkeys (41–44).

3.10. Do the Behavioral Measures Change with Known Treatments for Parkinson's Disease

Many studies show the anti-Parkinson's effect of L-DOPA in MPTP exposed primates. To illustrate the sensitivity of our observation method to the effects of L-DOPA, we studied 22 adult male monkeys 2 months or more after receiving MPTP and reaching a criterion *Parkscore* of 10 or more (mildly Parkinsonian). L-DOPA/carbidopa (4:1) was administered orally in several doses from a syringe in a sweetened orange concentrate (Sunquick). Monkeys were observed before the drug administration, and 1–2 h after the drug and during a period that would minimize disturbances to the animals. A saline administration (dose 0 mg/kg) was studied to control for the disturbance that was associated with drug administration and injections and was found not to be different from other baseline observations during the same period. Doses were 0, 30, 60, and 90 mg/kg, large doses that were aimed to induce dyskinesia, if possible. There was a significant effect of the drug administrations on *Anxiety* factor ($F = 5.74$, df = 4, 12133, $p < 0.0001$), with the highest levels after 60 mg, followed by the saline/control periods. The 90 mg dose had the lowest levels of *Anxiety* factor behaviors and was significantly different from all of the other doses, based upon ANOVA, with post hoc test using the critical range of Student–Newman–Keuls test at $p < 0.05$. *Parkscore* ($F = 18$, df = 4, 12134, $p < 0.0001$), *Healthy* ($F = 8.6$, df = 4, 12090, $p < 0.0001$), *Tremor* ($F = 110.3$, df = 4, 12174, $p < 0.0001$), and *Arousal* ($F = 50.9$, df = 4, 12131, $p < 0.0001$) all showed the expected responses for an anti-Parkinson effect and are shown in Fig. 6. *Appearance, threat response, facedown*, and *sedation* (*eyes closed* and *freeze*) also showed effects at $p < 0.0001$ (*food response, delayed movement, eating problems, poverty of movement, spontaneous freeze*, and *Quiet OK* factor did not change significantly).

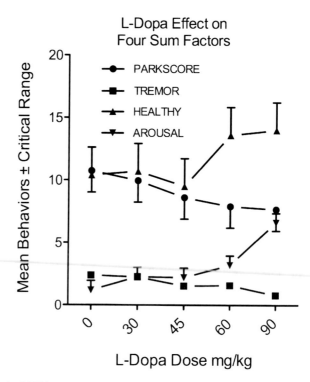

L-Dopa Effect on Four Sum Factors

Fig. 6. Oral L-DOPA/carbidopa administration (0, 30, 45, 60, and 90 mg/kg L-DOPA) showed consistent dose–response effects to decrease *Parkscore* and to increase *Healthy Behaviors*. Mean values are plotted for each dose, with the critical range for 5 means from two-tailed Newman–Keuls test at $p < 0.05$. There was a significant reduction in tremor, and increases in the *Arousal* factor at the highest dose.

3.11. Does the Model Reflect the Biology of Parkinson's Disease?

Most relevant for this chapter is how well the behavioral method reflects the biology of Parkinson's disease. To make this determination, we have examined the relationship of measures of dopamine function in multiple sites in the nigrostriatal system to assess what relationships might exist with the various behavioral measures. We found a highly significant relationships between our *Parkscore* or the Severity Group (0–4) and dopamine depletion in several nigrostriatal brain areas measured postmortem (33, 45). Parkscore correlated highly with measures of dopamine transporters in vivo and dopamine concentrations postmortem in the nigrostriatal region. Studies of dopamine transporters in vivo using SPECT imaging with (^{123}I)beta-CIT in the striatum of MPTP monkeys have included determination of the reproducibility of the imaging measures, optimization of the SPECT outcome measure, and longitudinal assessment of MPTP animals with SPECT. A study involving 148 SPECT scans in 74 monkeys was performed, with John Seibyl and Paul Hoffer, in which the percent striatal uptake measured by SPECT was correlated with *Parkscore* ($r = -0.67$, $p < 0.0001$). Postmortem biochemical correlations were studied by John Elsworth in 18 monkeys with prior SPECT imaging for quantitation of DA content in striatal punch samples. Mean dopamine punch biopsy measures were significantly correlated with SPECT striatal uptake ($r=0.80$, $n=18$, $p<0.015$). Correlation of dopamine concentrations in a punch with both the *Parkscore* ($r=-0.72$, $p<0.006$) and *Healthy Behavior* Score ($r=+0.84$, $p=0.0003$) were also statistically significant. This finding is similar to the finding that UPDRS correlates significantly with SPECT beta-CIT in patients with Parkinson's disease ($r=-0.47$, $n=28$) (46–49), although the *Parkscore* from our behavioral method accounts for a larger percentage of the variance than the clinical UPDRS score (35). This difference might be due to the fact that the patients were medicated, and the monkeys were not, or our behavioral method may be more precise than the UPDRS clinical rating scale.

There are certainly many other abnormalities in Parkinson's disease at the genetic, cellular, and biochemical level. Yet there is a remarkably good relationship between the dopamine depletion induced by MPTP and the group of behaviors and physical signs identified as Parkinson's disease, including nearly every medical complication of the disease. And there is a strong relationship between our *Parkscore* measure and dopamine in the nigrostriatal region.

4. Discussion

4.1. Other Means of Behavioral Assessment in Primates After MPTP

4.1.1. Cognitive Testing

We have used our behavioral method in a number of published studies (6, 11, 50–67), but we have used other methods to describe the broader effects of MPTP besides "motor" effects. Cognitive function is impaired in Parkinson's disease, as well as after MPTP (68–72). The tasks we have used involved both the learning and the performance of an object retrieval task that involved reaching around a transparent barrier (a detour) in order to retrieve a reward. The task not only involves direct motor control but also tests planning, the association of visual information in space, the recognition of object permanence, and other features known to reflect mental computations or cognition (73–75). This task also measures perseverative responses, requires complex sequential motor planning, and measures motor response biases, such as hand preference. It was described in detail and used in several studies in African green monkeys to study the effects of MPTP and fetal ventromesencephalic tissue transplantation (76–78). In addition to cognition, this method does provide an assessment of motor difficulties that is very precise, as well as measuring cognition that is not captured by the behavioral scoring and rating system. Cognitive deficits are more subtle and appear in monkeys with known dopamine deficits, but without observable motoric signs of PD. The disadvantage of these tests is that they are more time consuming to perform and can only be done with normal to moderately affected animals. The more severe ones, similar to patients with the fewest treatment options, are unable to perform the task at all.

4.1.2. Eyeblink Rate Measurements

In addition to cognitive testing, the simple measurement of eyeblink rate provides an extraordinarily simple and accurate window on nigrostriatal dopamine function, and moves in the predicted directions with dopamine agonists and antagonists, as well as also being reduced in Parkinson's disease, although the mechanisms that underlie the phenomenon are complicated (79, 80). We have assessed eyeblink rate after MPTP administration and shown the effects of drugs that act via the dopamine system (81–83).

4.1.3. Dyskinesia Assessment

Finally, if dyskinesia is a possibility, such as after repeated administration of L-DOPA, dopamine agonists, or neural transplantation, we have used a separate scale for rating dyskinesia that was derived from Di Monte et al. (84–86). We rated the intensity and frequency of four different categories of dyskinetic movement in six different regions of the body. The four movements were: stereotypy (repetitive, controlled, purposeless behavior); chorea (rapid random flicking movements); athetosis (sinuous writhing limb movements); and dystonia (sustained abnormal posturing). These behaviors may appear simultaneously and thus rating of the four categories was not mutually exclusive. The total of 48 different numerical variables offered the advantage of being able to determine how and where each of the four types of abnormal movements were changing with L-DOPA administration and prevented anomalies in one region or type of movement from interfering with the data from other areas. This assessment was utilized to study the potential induction of dyskinesia after the transplantation of fetal precursor dopamine cells into the nigrostriatal system (87, 88).

4.1.4. Videotaped Assessment

We also routinely videotaped some segments of behavior using a standardized protocol that allows review of the animal's status at a later time by individuals with specific perspectives, such as neurologists, or for testing the blind conditions, since videotape is easier to randomize and blind than is reality which has a regular time sequence. The same rating system can be applied to the videotape as with the live ratings, or a completely different scoring system can be applied. The disadvantage of videotape is that the job is only beginning after it is recorded – it still has to be rated, which will take at least as long as live ratings. Videotape also has the disadvantage that it is difficult to record some behaviors from the fixed position of the camera, whereas a live rater can often shift slightly to avoid an obstacle. Also, in general, the resolution is poorer than real life in spite of improving modern video equipment. For inter-rater reliability studies, however, it is superior since each rater can be presented with exactly the same images and perspective. It is also useful for data presentation, but must always be backed up by quantitative and statistically analyzed data. Videotape images can be easy to edit and manipulate and can therefore be significantly misleading.

4.2. Advantages and Disadvantages

What are the advantages of this observational method? (1) it provides a comprehensive assessment of the full behaviors and activities of each animal; (2) it quantitates a number of individual signs and manifestations of parkinsonism; (3) it is comparably scaled to severity as a well-known Parkinson's clinical scale (Hoehn–Yahr) (34); (4) it can be applied over a full range of signs and symptoms from asymptomatic to the most severe "end-stage"; (5) nonspecialists can learn the definitions and apply the scale reliably between raters and over time; (6) it is highly selective for parkinsonism in contrast to a number of major items that appear in other scales that are clearly nonselective; (7) it responds appropriately to drugs which are known to improve or worsen parkinsonism; and (8) to the extent that parkinsonism is reproduced by the loss of dopamine function in the nigrostriatal system, it has high validity and correlates with both in vivo SPECT measures of dopamine transporters or with postmortem dopamine concentrations measured directly.

The disadvantage is its complexity. An expert with Parkinson's disease in patients likely finds it easier to apply a human clinical scale to monkeys, with minor variations than to deal with the amount of defined primate behavior that this scale includes. Another major weakness is the limited attention to unilateral manifestations of motor problems, which are only captured by identifying whether manifestations are greater on one side or the other, but not quantitated. This would be especially important for induced hemiparkinsonism that has significant lateralization. When we have studied unilateral carotid injections, the scale has been modified to include additional items to identify unilateral deficits. Similarly, unilateral animals, without additional systemic or bilateral injections fail to show a number of the cardinal features of Parkinson's disease.

Another weakness is the fact that our raters are unable to reliably identify rigidity by observation alone, which a neurologist can easily identify by examination along with the classic cog-wheeling pattern. Actually handling the animals would be impossible without chemical restraint, which would also significantly interfere with the other measurements and likely the experiment.

4.3. Critique of the MPTP "Model" in Primates: Is It Useful for Advancing Our Knowledge of Human Parkinson's Disease?

4.3.1. Variability

The variability in response to identical doses of MPTP was initially somewhat surprising, although there is now overwhelming evidence of differences in response to MPTP among different species. Even among closely related nonhuman primates, there appear to be significant differences in responsiveness to MPTP. These differences might provide useful genomic data to understand the sensitivity of some individuals to environmental or genetic factors that induce the disease or, conversely, protect others from getting it. From an experimental perspective, however, this variability is a major problem.

Other investigators have reported in some primate species that they can always produce the same extent of severity by special treatment schedules, routes of administration, or chronic administration over time (17, 23, 89–93). We have been unable to do this. Whether given by intracarotid administration (we have studied about 30 monkeys after intracarotid administration) or systemically, the large variability of response remains. Some monkeys have received as much as 27 mg/kg of MPTP without becoming symptomatic, with others proceeding to die with severe parkinsonism with doses of 2.0 mg/kg. The response variability may be more pronounced in African green monkeys, since these other investigators are using different primate species. However, those other species do not show as many characteristic manifestations of Parkinson's disease, or the assessments are based on more indirect manifestations, such as motor performance or general activity.

For our studies, we have instead selected animals with restricted ranges of parkinsonism for our experimental studies, assuming that they are less likely to have some unknown mechanism for protection or recovery and that we can determine from prior similar animals what their levels of dopamine depletion are likely to be. This strategy is expensive because significantly more animals have to be given injections of MPTP than are required for the final study, as well as the fact that preexposure baseline measurements and workups have to include data from and care of more animals than are needed for the final experiments. The mild/asymptomatic animals can be used in biological studies that benefit from mild dopamine depletion but do not depend upon a behavioral outcome measure. The deliberately restricted range of severity of a group selected for therapeutic experiments does reduce the number of animals required to achieve adequate statistical power for analyses.

4.3.2. Recovery

Another problem with the MPTP model is the extent of recovery that often occurs independent of any experimental intervention. Again, various investigators have reported that their behavioral deficit is "stable" over varying periods of time, particularly after unilateral carotid artery injections (92, 93) and the stability, again, can be improved by special methods of administration (25). This recovery phenomenon is at variance from the observations of MPTP intoxicated patients, whose signs and symptoms have not improved or have worsened. There are numerous regulatory mechanisms that have been reported that could underlie both the variability in initial response as well as recovery and may account for the fact that signs of parkinsonism do not become apparent until after the majority of dopamine has been depleted. In addition, MPTP clearly diminishes dopamine function prior to the death of the dopamine neuron, which also has been shown to sprout new axons and eventually recover (94).

In addition, one criticism of the MPTP model itself is that it is primarily a presynaptic effect during the period of most therapeutic experiments, leaving the postsynaptic systems less affected or even hypersensitive until longer-term postsynaptic changes occur. Many of the effects of the 6-hydroxydopamine model are due to abnormal postsynaptic effects in response to drug challenges (amphetamine rotation) (95–97). Some have postulated that chronic MPTP exposures (presynaptic damage) over a longer period of weeks or months might do a better job of emulating the postsynaptic changes, such as losses of receptor density and sensitivity that would be expected to have occurred in the sporadic disease in humans over a period of years or decades (98, 99). One to 2 months after MPTP, only a small difference in striatal dopamine loss distinguishes monkeys with widely different parkinsonian behavior; however, over a 1-year time period this differential in dopamine levels becomes greater as monkeys with less pronounced signs recover striatal dopamine function, whereas those with more severe signs do not. For example, at 1–2 months after MPTP, the dopamine loss in caudate nucleus is 96% in Severity Group 1 monkeys and 99% in Group 4, but after 1 year the losses are 70% for Group 1 and 99% for Group 4 (11, 45).

We have previously reported that the only group of animals that does not recover over a period of 4–12 months are those that are rendered severely parkinsonian by MPTP (Severity Groups 3 and 4), although with extensive and effective nursing care, even some of these animals have been found to recover without other therapies. The most important lesson for the use of MPTP administration in primates is that experimental groups must be appropriately matched with controls, with the expectation that some of them may recover with proper nursing care.

4.3.3. Do the Effects of MPTP Progress Over Time?

I do not know of any data which show worsening of MPTP-induced parkinsonism in primates after the initial period and a phase of often significant recovery unless additional MPTP is administered. These new data presented in this chapter show a significant increase in *Parkscore* over a period of 10 years. Collection of sensitive data to evaluate this small effect required using an identical method by 46 observers over a 10 year period. 2975 observations were analyzed after we had dropped data from a few observers who watched very few monkeys or were identified by analysis as different from the rest (for Parkscore, Observer effect $F = 4.73, \mathrm{df} = 44, 2816, p < 0.0001$). There was no significant interaction between observer or monkey and the old vs. new period, so we do not believe that the worsening shown over the 10-year period or the continuing slopes in the recent period are due to observer differences or bias. Instead, our new data do suggest that, with longer periods of time that may more closely match human disease progression, monkeys do become more parkinsonian after a single course of MPTP administration.

As with humans, it is not clear whether this occurs completely independent of effects of aging (100–113), or whether the acute loss of dopamine neurons initiates other changes that are progressive and independent of the aging process, such as increased dopamine turnover, oxidative stress, and more cell death. In people over age 85, more than half display Parkinson's disease symptoms and signs whether or not they have been diagnosed with Parkinson's disease (114). Dopamine depletion occurs with normal aging in humans, but the key difference with Parkinson's disease may be the rate of decline, with major signs occurring after compensatory mechanisms have failed (103).

5. Summary: What Is the Future of the MPTP Model in Primates?

We know now that MPTP in monkeys does reproduce nearly all of the behavioral and functional aspects of Parkinson's disease, and from hundreds of studies, which have been extensively reviewed (115), that new methods of treatment have been successfully predicted (116, 117). After some early hope that it might provide more direct evidence of the pathophysiology of the disease, this looks not to be the case (118–122). An even better model would be derived from recently discovered genetic variations, was exacerbated by environmental factors (123), showed widespread Lewy bodies and activated oxidative and inflammatory effects, and showed rapid progression. In the meantime, there are other uses for the MPTP model that have yet to be tried (124) and new strategies for reconstruction and repair of dopamine damage that can still be studied and exploited using the MPTP primate model of Parkinson's disease.

Acknowledgments

I thank numerous collaborators and coworkers for collecting the data reported here and for the various studies and grants which made it possible. It would be impossible to rank the list in order of importance, but much thanks is owed to Jane R. Taylor, Ph.D., who developed the behavioral methods with me and has published previous details, as well as applications to studies of fetal precursor dopamine transplant studies. My long-time collaborators, Robert H. Roth, Ph.D., John D. Elsworth, Ph.D. and John R. Sladek, Jr., Ph.D. provided support for long-term projects and key biochemical data. Other key collaborators include Patrick Aebischer, M.D., Dr. Med., Kimberly Bjugstad, Ph.D., Barbara Blanchard, M.S., Jocelyne Bloch, M.D., Martha C. Bohn, Ph.D., Alfred L.M.

Bothwell, Ph.D., Jean Francois Brunet, Ph.D., Derek Choi-Lundberg, Ph.D., Timothy J. Collier, Ph.D., Ariel Y. Deutch, Ph.D., Joseph P. Glorioso, Ph.D., S.N. Haber, Ph.D., Paul Hoffer, M.D., Tamas L. Horvath, D.V.M., Ph.D., R.B. Innis, M.D., Ph.D., Ole Isacson, Ph.D., M.D., J. David Jentsch, Ph.D., Maryanne Johnson, Jeffrey Kordower, Ph.D., Matthew S. Lawrence, M.D., Ph.D., Csaba Leranth, M.D., Ph.D., Robert W. Makuch, Ph.D., Eleni Markakis, Ph.D., Frederick Naftolin, M.D., D.Phil., Peter Olausson, Ph.D., Richard J. Samulski, Ph.D., John Seibyl, M.D., Richard Sidman, M.D., Joanne Simiola, Eileen Smith, M.S., Evan Y. Snyder, M.D., Ph.D., Caryl E. Sortwell, Ph.D., Dennis Spencer, M.D., Ted Teng, M.D., Ph.D., Angel Vinuela, M.D., Kenneth Vives, M.D., and Dustin Wakeman, Ph.D.

Especially critical were the staff at St. Kitts Biomedical Research Foundation who actually cared for all of these parkinsonian monkeys, including on their weekends and holidays, as well as carrying out the experiments, doing the thousands of observations, and keeping the facility operating and meeting the highest standards for AAALAC accreditation. These again are alphabetical: Darnel Allen, Curtis Archibald, Verna Archibald, Zelia Archibald, Lashley Benjamin, Rohn Brookes, Dave Charles, Dudley Christopher, Dexter Collins, Franklyn Connor, Royden Conwell, Edwin Dietrich, Joseph E. Dunrod, Levi Estridge, Lawrence E. Evans, DVM, Shaied Fahie, Trevor Felle, Junie Fergus, Stokely Grey, Collin Hanley, Steve Henry, Loydon Henry-Phillip, Tasino Herbert, Patrick Hurley, Velda Irish, MBA, Angelle James, Joseph P. Kingsbury, Shervin Liddie, Akeba Matthew, Mavis Matthew, Chad Morton, Xavier Morton, Maurice Newton, Ernell Nisbett, Alexis Nisbett, MBA, Sean O'Loughlin, Tarik Phillip, Samuel Phipps, Wade Phipps, Ricaldo Pyke, DVMZ, Dwayne Rawlins, Andy J. Redmond, M.D., Kathlyn Reid Redmond, M.B.A., Gerard Saddler, Wentworth Sargeant, Terrence Scarborrough, Walter Streett, D.V.M., Mike Struharik, Wellington Sutton, M.S., Junior Swanston, Conrad Taylor, John Wharton, Leroy Whattley, O'Neal Whattley, Steve Whittaker, Milton Whittaker, Ph.D., Maurice Whyte, Sylvester Whyte, Kevin Wilkenson, Lewis Clifford Williams, Clive Audain Wilson, Clive R Wilson, and Brett Youngerman.

I thank the following for support: H. Roy Cullen, the Charles M. Solomon family, Project ALS, and the Michael J. Fox Foundation for Parkinson's Research for critical research facilities and support, Axion Research Foundation, and grants from NIH: K05-MH00643, RO1-NS24032, RO1-NS33909, RO1-NS35998, RO1-MH57958, RO1-NS40822, PO1-NS044281, RO1-NS40822.

When I use the first person plural in this chapter, I am speaking for the work done by this group of collaborators, and I thank them all for their outstanding contributions. Since I was unable to persuade any of them to co-author this chapter with me, I must bear full responsibility for any errors or omissions. I thank Jane Taylor,

John Elsworth, Matt Lawrence, Robert Roth, Tim Collier, John Sladek, Jeff Kordower, and Ole Isacson for helpful comments and improvements to the manuscript.

References

1. Davis GC, Williams AC, Markey SP, Ebert MH, Caine ED, Reichert CM, Kopin IJ (1979) Chronic parkinsonism secondary to intravenous injection of meperidine analogues. Psychiatric Research **1**: 249–154

2. Langston JW, Ballard P, Tetrud JW, Irwin I (1983) Chronic parkinsonism in humans due to a product of meperidine-analog synthesis. Science **219**: 979–980

3. Burns RS, Chiueh CC, Markey SP, Ebert MH, Jacobowitz DM, Kopin IJ (1983) A primate model of parkinsonism: Selective destruction of minergic neurons in the pars compacta of the substantia nigra by N-methyl-4-phenyl-1,2,3,6-tetrahydropyridine. Proceedings of the National Academy of Science, USA **80**: 4546–4550

4. Langston JW, Forno LS, Rebert CS, and Irwin I (1984) Selective nigral toxicity after systemic administration of 1-methyl-4- phenyl-1,2,5, 6-tetrahydropyrine (MPTP) in the squirrel monkey. Brain Research **292**: 390–394

5. Tetrud JW, Langston JW, Redmond DE, Jr., Roth RH, Sladek JR, Jr., Angel RW (1986) MPTP-induced tremor in human and non-human primates. Neurology **36**: 308

6. Redmond DE, Jr., Sladek JR, Jr., Roth RH, Collier TJ, Elsworth JD, Deutch AY, Haber S (1986) Fetal neuronal grafts in monkeys given methylphenyltetrahydropyridine. Lancet **1**: 1125–1127

7. Deutch AY, Elsworth JD, Goldstein M, Fuxe K, Redmond DE, Jr., Sladek JR, Jr., Roth RH (1986) Preferential vulnerability of A8 dopamine neurons in the primate to the neurotoxin 1-methyl-4-phenyl- 1,2,3,6-tetrahydropyridine. Neuroscience Letters **68**: 51–56

8. Elsworth JD, Deutch AY, Redmond DE, Jr., Sladek JR, Jr., Roth RH (1987) Effects of 1-methyl-4-phenyl-1,2,3,6-tetrahydropyridine (MPTP) on catecholamines and metabolites in primate brain and CSF. Brain Research **415**: 293–299

9. Elsworth JD, Deutch AY, Redmond DE, Jr., Sladek JR, Jr., Roth RH (1987) Differential responsiveness to 1-methyl-4-phenyl-1,2,3,6-tetrahydropyridine toxicity in sub-regions of the primate substantia nigra and striatum. Life Sciences **40**: 193–202

10. Elsworth JD, Leahy DJ, Roth RH, Redmond DE, Jr. (1987) Homovanillic acid concentrations in brain, CSF and plasma as indicators of central dopamine function in primates. Journal of Neural Transmission **68**: 51–62

11. Elsworth JD, Deutch AY, Redmond DE, Jr., Taylor JR, Sladek JR, Jr., Roth RH (1989) Symptomatic and asymptomatic 1-methyl-4-phenyl-1,2,3,6-tetrahydropyridine- treated primates: biochemical changes in striatal regions. Neuroscience **33**: 323–331

12. Elsworth JD, Taylor JR, Redmond DE, Jr., Collier TJ, Sladek JR, Roth RH (1989) Biochemical assessment of reversal of MPTP-induced parkinsonism following intrastriatal transplants of fetal substantia nigra in primates. Restorative Neurology and Neuroscience **1**: 59

13. Elsworth JD, Deutch AY, Redmond DE, Jr., Sladek JR, Jr., Roth RH (1990) MPTP reduces dopamine and norepinephrine concentrations in the supplementary motor area and cingulate cortex of the primate. Neuroscience Letters **114**: 316–322

14. Elsworth JD, Deutch AY, Redmond DE, Jr., Sladek JR, Jr., Roth RH (1990) MPTP-induced parkinsonism: relative changes in dopamine concentration in subregions of substantia nigra, ventral tegmental area and retrorubral field of symptomatic and asymptomatic vervet monkeys. Brain Research **513**: 320–324

15. Schneider JS, Unguez G, Yuwiler A, Berg SC, Markham CH (1988) Deficits in operant behaviour in monkeys treated with N-methyl-4-phenyl-1,2,3,6-tetrahydropyridine (MPTP). Brain **111** (Pt 6): 1265–1285

16. Brooks BA, Eidelberg E, Morgan WW (1987) Behavioral and biochemical studies in monkeys made hemiparkinsonian by MPTP. Brain Res **419**: 329–332

17. Bankiewicz KS, Oldfield EH, Chiueh CC, Doppman JL, Jacobowitz DM, Kopin IJ (1986) Hemiparkinsonism in monkeys after unilateral internal carotid artery infusion of 1-methyl-4-phenyl-1,2,3,6-tetrahydropyridine (MPTP). Life Sci **39**: 7–16

18. Togasaki DM, Hsu A, Samant M, Farzan B, DeLanney LE, Langston JW, Di Monte DA,

Quik M (2005) The Webcam system: a simple, automated, computer-based video system for quantitative measurement of movement in nonhuman primates. J Neurosci Methods **145**: 159–166

19. Saiki H, Hayashi T, Takahashi R, Takahashi J (2010) Objective and quantitative evaluation of motor function in a monkey model of Parkinson's disease. J Neurosci Methods **190**: 198–204

20. Schwab RS, England AC, Jr. (1969) Projection techniques for evaluating surgery in Parkinson's disease. In: Gillingham F, Donaldson I (eds) Third Symposium on Parkinson's Disease. E. and S. Livingstone, Edinburgh, Scotland, pp 152–157

21. Fahn S, Elton R, and members of the UPDRS Development Committee (1987) Unified Parkinson's disease rating scale. In: Fahn S, Marsden C, Goldstein M, Calne C (eds) Recent Developments in Parkinson's Disease. Macmillan, Florham Park, NJ, pp 153–163

22. Kurlan R, Kim MH, Gash DM (1991) Oral levodopa dose-response study in MPTP-induced hemiparkinsonian monkeys: assessment with a new rating scale for monkey parkinsonism. Mov Disord **6**: 111–118

23. Kurlan R, Kim MH, Gash DM (1991) The time course and magnitude of spontaneous recovery of parkinsonism produced by intracarotid administration of 1-methyl-4-phenyl-1,2,3,6-tetrahydropyridine to monkeys. Ann Neurol **29**: 677–679

24. Gomez-Mancilla B, Bedard PJ (1993) Effect of nondopaminergic drugs on L-dopa-induced dyskinesias in MPTP-treated monkeys. Clin Neuropharmacol **16**: 418–427

25. Smith RD, Zhang Z, Kurlan R, McDermott M, Gash DM (1993) Developing a stable bilateral model of parkinsonism in rhesus monkeys. Neuroscience **52**: 7–16

26. Benazzouz A, Boraud T, Dubedat P, Boireau A, Stutzmann JM, Gross C (1995) Riluzole prevents MPTP-induced parkinsonism in the rhesus monkey: a pilot study. Eur J Pharmacol **284**: 299–307

27. Schneider JS, Lidsky TI, Hawks T, Mazziotta JC, Hoffman JM (1995) Differential recovery of volitional motor function, lateralized cognitive function, dopamine agonist-induced rotation and dopaminergic parameters in monkeys made hemi-parkinsonian by intracarotid MPTP infusion. Brain Res **672**: 112–117

28. Papa SM, Chase TN (1996) Levodopa-induced dyskinesias improved by a glutamate antagonist in Parkinsonian monkeys. Ann Neurol **39**: 574–578

29. Imbert C, Bezard E, Guitraud S, Boraud T, Gross CE (2000) Comparison of eight clinical rating scales used for the assessment of MPTP-induced parkinsonism in the Macaque monkey. J Neurosci Methods **96**: 71–76

30. Altman J (1974) Observational study of behavior: Sampling methods. Behavior **49**: 222–267

31. SAS Institute I (1988) SAS/STAT User's Guide. SAS Institute, Inc., Cary, NC

32. SAS Institute (1988) SAS/STAT User's Guide. SAS Institute, Inc., Cary, NC

33. Taylor JR, Elsworth JD, Sladek JR, Jr., Roth RH, Redmond DE, Jr. (1994) Behavioral effects of MPTP administration in the vervet monkey: a primate model of Parkinson's disease. In: Woodruff ML, Nonneman AJ (eds) Toxin-Induced Models of Neurological Disorders. Plenum Press, New York, pp 139–174

34. Hoehn MM, Yahr MD (1967) Parkinsonism: onset, progression, and mortality. Neurology **17**: 427–442

35. Fahn S, Elton R, Committee amotUD (1987) Unified Parkinson's disease rating scale. In: Fahn S, Marsden C, Goldstein M, Calne C (eds) Recent Developments in Parkinson's Disease. Macmillan, Florham Park, NJ, pp 153–163

36. Baulu J, Redmond DE, Jr. (1978) Some sampling considerations in the quantitation of monkey behavior under field and captive conditions. Primates **19**: 391–399

37. Service USPH (1985) Guide for the Care and Use of Animals. U.S. Government Printing Office, Washington D.C.

38. Winer BJ (1971) Statistical Principles in Experimental Design. McGraw-Hill Book Company, New York

39. Ueki A, Chong PN, Albanese A, Rose S, Chivers JK, Jenner P, Marsden CD (1989) Further treatment with MPTP does not produce parkinsonism in marmosets showing behavioural recovery from motor deficits induced by an earlier exposure to the toxin. Neuropharmacology **28**: 1089–1097

40. Hurley PJ, Elsworth JD, Whittaker MC, Roth RH, Redmond DE, Jr (2011). Aged monkeys as a partial model for Parkinson's disease. Pharmacology, biochemistry, and behavior **99**: 324–332

41. Bachevalier J, Landis LS, Walker LC, Brickson M, Mishkin M, Price DL, Cork LC (1991) Aged monkeys exhibit behavioral deficits indicative of widespread cerebral dysfunction. Neurobiol Aging **12**: 99–111

42. Irwin I, DeLanney LE, McNeill T, Chan P, Forno LS, Murphy GM, Jr., Di Monte DA,

Sandy MS, Langston JW (1994) Aging and the nigrostriatal dopamine system: a non-human primate study. Neurodegeneration 3: 251–265

43. Emborg ME, Ma SY, Mufson EJ, Levey AI, Taylor MD, Brown WD, Holden JE, Kordower JH (1998) Age-related declines in nigral neuronal function correlate with motor impairments in rhesus monkeys. J Comp Neurol 401: 253–265

44. Zhang Z, Andersen A, Smith C, Grondin R, Gerhardt G, Gash D (2000) Motor slowing and parkinsonian signs in aging rhesus monkeys mirror human aging. J Gerontol A Biol Sci Med Sci 55: B473–480

45. Elsworth JD, Taylor JR, Sladek JR, Jr., Collier TJ, Redmond DE, Jr., Roth RH (2000) Striatal dopaminergic correlates of stable parkinsonism and degree of recovery in old-world primates one year after MPTP treatment. Neuroscience 95: 399–408

46. Innis RB, Seibyl JP, Scanley BE, Laruelle M, Abi-Dargham A, Wallace E, Baldwin RM, Zea-Ponce Y, Zoghbi S, Wang S, et a (1993) Single photon emission computed tomographic imaging demonstrates loss of striatal dopamine transporters in Parkinson disease. Proc Natl Acad Sci USA 90: 11965–11969

47. al-Tikriti MS, Zea-Ponce Y, Baldwin RM, Zoghbi SS, Laruelle M, Seibyl JP, Giddings SS, Scanley BE, Charney DS, Hoffer PB, et al. (1995) Characterization of the dopamine transporter in nonhuman primate brain: homogenate binding, whole body imaging, and ex vivo autoradiography using (125I) and (123I)IPCIT. Nucl Med Biol 22: 649–658

48. Seibyl JP, Marek KL, Quinlan D, Sheff K, Zoghbi S, Zea-Ponce Y, Baldwin RM, Fussell B, Smith EO, Charney DS, van Dyck C, et al. (1995) Decreased single-photon emission computed tomographic (123I)beta-CIT striatal uptake correlates with symptom severity in Parkinson's disease. Ann Neurol 38: 589–598

49. Marek K, Innis R, van Dyck C, Fussell B, Early M, Eberly S, Oakes D, Seibyl J (2001) (123I) beta-CIT SPECT imaging assessment of the rate of Parkinson's disease progression. Neurology 57: 2089–2094

50. Sladek JR, Jr., Redmond DE, Jr., Collier TJ, Haber SN, Elsworth JD, Deutch AY, Roth RH (1987) Transplantation of fetal dopamine neurons in primate brain reverses MPTP induced parkinsonism. Progress In Brain Research 71: 309–323

51. Redmond DE, Jr., Naftolin F, Collier TJ, Leranth C, Robbins RJ, Sladek CD, Roth RH, Sladek JR, Jr. (1988) Cryopreservation, culture,

and transplantation of human fetal mesencephalic tissue into monkeys. Science 242: 768–771

52. Elsworth JD, Redmond DE, Jr., Sladek JR, Jr., Deutch AY, Collier TJ, Roth RH (1989) Reversal of MPTP-induced parkinsonism in primates by fetal dopamine cell transplants. In: Franks AJ, Ironside JW, Mindham RHS, Smith RJ, Spokes EGS, Winlow W (eds) Function and Dysfunction of the Basal Ganglia. Manchester University Press, New York, pp 161–180

53. Taylor J, Elsworth J, Roth R, Sladek J, Jr., Collier T, Redmond D, Jr. (1991) Grafting of fetal substantia nigra to striatum reverses behavioral deficits induced by MPTP in primates: A comparison with other types of grafts as controls. Experimental Brain Research 85: 335–348

54. Redmond DE, Jr., Roth RH, Elsworth JD, Smith E, Al-Tikriti M, Taylor JR, Sladek JR, Neumeyer RB, Innis RB, Hoffer P (1993) SPECT studies with CIT provide reliable estimates of striatal dopamine (DA) depletion corresponding to the degree of parkinsonism and identify dopamine grafts in vivo. American College of Neuropsychopharmacology Abstracts: 131

55. Redmond DE, Jr., Sladek JR, Jr., Collier TJ, Elsworth JD, Spencer D, Naftolin F, Leranth C, Robbins RJ, Marek KL, Bunney BS, Roth RH (1993) Impact of fetal age on mesencephalic graft survival in monkeys and humans. Neuroscience Abstracts 19: 864

56. Sladek JR, Jr., Elsworth JD, Roth RH, Evans LE, Collier TJ, Cooper SJ, Taylor JR, Redmond DE, Jr. (1993) Fetal dopamine cell survival after transplantation is dramatically improved at a critical donor gestational age in nonhuman primates. Experimental Neurology 122: 16–27

57. Elsworth JD, Al-Tikriti MD, Sladek JR, Jr., Taylor JR, Innis RB, Redmond DE, Jr., Roth RH (1994) Novel radioligands for the dopamine transporter demonstrate the presence of intrastriatal nigral grafts in the MPTP-treated monkey: correlation with improved behavioral function. Exp Neurol 126: 300–304

58. Sladek J, Jr., Elsworth J, Taylor J, Roth R, Redmond D, Jr. (1995) Techniques for neural transplantation in non-human primates. In: Ricordi C (ed) Methods in Cell Transplantation. R.G. Landes, Austin, TX, pp 391–408

59. Taylor JR, Elsworth JD, Sladek JR, Jr., Collier TJ, Roth RH, Redmond DE, Jr. (1995) Sham surgery does not ameliorate MPTP-induced

behavioral deficits in monkeys. Cell Transplantation **4**: 13–26

60. Leranth C, Sladek J, Jr., Roth R, Redmond D, Jr. (1998) Efferent synaptic connections of dopaminergic neurons grafted into the caudate nucleus of experimentally induced parkinsonian monkeys are different from those of control animals. Experimental Brain Research **123**: 323–333

61. Sladek JR, Jr., Collier TJ, Elsworth JD, Roth RH, Taylor JR, Redmond DE, Jr. (1998) Intrastriatal grafts from multiple donors do not result in a proportional increase in survival of dopamine neurons in nonhuman primates. Cell Transplant 7: 87–96.

62. Leranth C, Roth RH, Elsworth JD, Naftolin F, Horvath TL, Redmond DE, Jr. (2000) Estrogen is essential for maintaining nigrostriatal dopamine neurons in primates: implications for Parkinson's disease and memory. J Neurosci **20**: 8604–8609.

63. Sladek JR, Collier T, Bundock E, Roth RH, Elsworth JD, Taylor J, Redmond DE, Jr (2001) Striatal grafts direct the growth of fibers from co-grafted and host dopaminergic neurons in monkey. American Society of Neural Transplantation and Repair **8**: 34

64. Collier TJ, Sortwell CE, Elsworth JD, Taylor JR, Roth RH, Sladek JR, Jr., Redmond DE, Jr (2002) Embryonic ventral mesencephalic grafts to the substantia nigra of MPTP-treated monkeys: feasibility relevant to multiple-target grafting as a therapy for Parkinson's disease. J. Comp. Neurol. **442**: 320–330

65. Elsworth JD, Redmond DE, Jr., Leranth C, Bjugstad KB, Sladek JR, Jr., Collier TJ, Foti SB, Samulski RJ, Vives KP, Roth RH (2008) AAV2-mediated gene transfer of GDNF to the striatum of MPTP monkeys enhances the survival and outgrowth of co-implanted fetal dopamine neurons. Exp Neurol **211**: 252–258

66. Brunet JF, Redmond DE, Jr., Bloch J (2009) Primate adult brain cell autotransplantation, a pilot study in asymptomatic MPTP treated monkeys. Cell Transplant

67. Redmond DE, Jr., Elsworth JD, Roth RH, Leranth C, Collier TJ, Blanchard B, Bjugstad KB, Samulski RJ, Aebischer P, Sladek JR, Jr. (2009) Embryonic substantia nigra grafts in the mesencephalon send neurites to the host striatum in non-human primate after overexpression of GDNF. J Comp Neurol **515**: 31–40

68. Reitan RM, Boll TJ (1971) Intellectual and cognitive functions in Parkinson's disease. J. Consulting and Clinical Psychology **37**

69. Loranger AW, Goodell H, McDownell FH, Lee JE, Sweet RD (1972) Intellectual impairment in Parkinson's syndrome. Brain **95**: 405–412

70. Lees AJ, Smith E (1983) Cognitive deficits in the early stages of Parkinson's disease. Brain **106**: 257–270

71. Taylor AE, Saint-Cyr JA, Lang AE (1986) Frontal dysfunction in Parkinson's disease. The cortical focus of neostriatal outflow. Brain **109**: 845–883

72. Sass KJ, Buchanan CP, Westerveld M, Marek KL, Farhi A, Robbins RJ, Naftolin F, Vollmer TL, Leranth C, Roth RH, et al (1995) General cognitive ability following unilateral and bilateral fetal ventral mesencephalic tissue transplantation for treatment of Parkinson's disease. Archives of Neurology **52**: 680–686

73. Piaget J (1954) Construction of reality in the child. Basic Books, New York

74. Diamond A (1989) Retrieval of an object from an open box: The development of visual-tactile control of reaching in the first year of life. Monographs of the Society for Research in Child Development

75. Diamond A (1989) Frontal Lobe Involvement in Cognitive Changes During the First Year of Life. Aldine, New York

76. Taylor JR, Elsworth JD, Roth RH, Collier TJ, Sladek JR, Redmond DE, Jr. (1990) Improvements in MPTP-induced object retrieval deficits and behavioral deficits after fetal nigral grafting in monkeys. In: Richards SJ, Dunnett SB (eds) Progress in Brain Research, vol 82. Elsevier, Amsterdam, pp 543–559

77. Taylor JR, Elsworth JD, Roth RH, Sladek JR, Jr., Redmond DE, Jr. (1990) Cognitive and motor deficits in the acquisition of an object retrieval/detour task in MPTP-treated monkeys. Brain **113**: 617–637

78. Taylor JR, Roth RH, Sladek JR, Jr., Redmond DE, Jr. (1990) Cognitive and motor deficits in the performance of an object retrieval task with barrier-detour in monkeys *(Cercopithecus aethiops sabaeus)* treated with MPTP: Long-term performance and effect of transparency of the barrier. Behavioral Neuroscience **104**: 564–576

79. Korosec M, Zidar I, Reits D, Evinger C, Vanderwerf F (2006) Eyelid movements during blinking in patients with Parkinson's disease. Mov Disord **21**: 1248–1251

80. Agostino R, Bologna M, Dinapoli L, Gregori B, Fabbrini G, Accornero N, Berardelli A (2008) Voluntary, spontaneous, and reflex blinking in Parkinson's disease. Mov Disord **23**: 669–675

81. Elsworth JD, Lawrence MS, Roth RH, Taylor JR, Mailman RB, Nichols DE, Lewis MH, Redmond DE, Jr. (1991) D1 and D2 dopamine receptors independently regulate spontaneous blink rate in the vervet monkey. J. of Pharmacology and Experimental Therapeutics **259**: 595–600

82. Lawrence MS, Redmond DE, Jr. (1991) MPTP lesions and dopaminergic drugs alter eye blink rate in African green monkeys. Pharmacology, Biochemistry and Behavior **38**: 869–874

83. Taylor JR, Elsworth JD, Lawrence MS, Sladek JR, Jr., Roth RH, Redmond DE, Jr. (1999) Spontaneous blink rates correlate with dopamine levels in the caudate nucleus of MPTP-treated monkeys. Exp Neurol **158**: 214–220

84. Di Monte DA, McCormack A, Petzinger G, Janson AM, Quik M, Langston WJ (2000) Relationship among nigrostriatal denervation, parkinsonism, and dyskinesias in the MPTP primate model. Mov Disord **15**: 459–466

85. Langston JW, Quik M, Petzinger G, Jakowec M, Di Monte DA (2000) Investigating levodopa-induced dyskinesias in the parkinsonian primate. Ann Neurol **47**: S79-89

86. Petzinger GM, Quik M, Ivashina E, Jakowec MW, Jakubiak M, Di Monte D, Langston JW (2001) Reliability and validity of a new global dyskinesia rating scale in the MPTP-lesioned non-human primate. Mov Disord **16**: 202–207

87. Redmond DE, Jr., Vinuela A, Kordower JH, Isacson O (2008) Influence of cell preparation and target location on the behavioral recovery after striatal transplantation of fetal dopaminergic neurons in a primate model of Parkinson's disease. Neurobiol Dis **29**: 103–116

88. Isacson O, Kordower, J., Vinuela, A, Redmond, DE, Jr. (2011) Cell transplant consortium studies of impact of fetal tissue grafting on the induction of dyskinesia in MPTP treated monkeys. In preparation.

89. Joyce JN, Marshall JF, Bankiewicz KS, Kopin IJ, Jacobowitz DM (1986) Hemiparkinsonism in a monkey after unilateral internal carotid artery infusion of 1-methyl-4-phenyl-1,2,3,6-tetrahydropyridine (MPTP) is associated with regional ipsilateral changes in striatal dopamine D-2 receptor density. Brain Res **382**: 360–364

90. Annett LE, Rogers DC, Hernandez TD, Dunnett SB (1992) Behavioural analysis of unilateral monoamine depletion in the marmoset. Brain **115** (Pt 3): 825–856

91. Schneider JS, McLaughlin WW, Roeltgen DP (1992) Motor and nonmotor behavioral deficits in monkeys made hemiparkinsonian by intracarotid MPTP infusion. Neurology **42**: 1565–1572

92. Eberling JL, Jagust WJ, Taylor S, Bringas J, Pivirotto P, VanBrocklin HF, Bankiewicz KS (1998) A novel MPTP primate model of Parkinson's disease: neurochemical and clinical changes. Brain Res **805**: 259–262

93. Oiwa Y, Eberling JL, Nagy D, Pivirotto P, Emborg ME, Bankiewicz KS (2003) Overlesioned hemiparkinsonian non human primate model: correlation between clinical, neurochemical and histochemical changes. Front Biosci **8**: a155-166

94. Bohn MC, Marciano F, Cupit L, Gash DM (1988) Recovery of dopaminergic fibers in striatum of the 1-methyl-4-phenyl-1,2,3,6-tetrahydropyridine-treated mouse is enhanced by grafts of adrenal medulla. Prog Brain Res **78**: 535–542

95. Gerlach M, Riederer P (1996) Animal models of Parkinson's disease: an empirical comparison with the phenomenology of the disease in man. J Neural Transm **103**: 987–1041

96. Glinka Y, Gassen M, Youdim MB (1997) Mechanism of 6-hydroxydopamine neurotoxicity. J Neural Transm Suppl **50**: 55–66

97. Blum K, Chen TJ, Downs BW, Bowirrat A, Waite RL, Braverman ER, Madigan M, Oscar-Berman M, DiNubile N, Stice E, Giordano J, Morse S, Gold M (2009) Neurogenetics of dopaminergic receptor supersensitivity in activation of brain reward circuitry and relapse: proposing "deprivation-amplification relapse therapy" (DART). Postgrad Med **121**: 176–196

98. Bezard E, Imbert C, Deloire X, Bioulac B, Gross CE (1997) A chronic MPTP model reproducing the slow evolution of Parkinson's disease: evolution of motor symptoms in the monkey. Brain Res **766**: 107–112

99. Bezard E, Dovero S, Prunier C, Ravenscroft P, Chalon S, Guilloteau D, Crossman AR, Bioulac B, Brotchie JM, Gross CE (2001) Relationship between the appearance of symptoms and the level of nigrostriatal degeneration in a progressive 1-methyl-4-phenyl-1,2,3,6-tetrahydropyridine-lesioned macaque model of Parkinson's disease. J Neurosci **21**: 6853–6861

100. Hiral S (1968) Ageing of the substantia nigra. Adv Neurol Sci **7**: 12845–12849

101. Adolfsson R, Gottfries CG, Roos BE, Winblad B (1979) Post-mortem distribution of dopamine and homovanillic acid in human brain, variations related to age, and a review of the literature. J Neural Transm **45**: 81–105

102. Calne DB, Peppard RF (1987) Aging of the nigrostriatal pathway in humans. Can J Neurol Sci **14**: 424–427

103. Fearnley JM, Lees AJ (1991) Ageing and Parkinson's disease: substantia nigra regional selectivity. Brain **114 (Pt 5)**: 2283–2301

104. Felten DL, Felten SY, Steece-Collier K, Date I, Clemens JA (1992) Age-related decline in the dopaminergic nigrostriatal system: the oxidative hypothesis and protective strategies. Ann Neurol **32 Suppl**: S133-136

105. Agid Y, Hirsch E, Anglade P, Michel P, Brugg B, Ruberg M (1996) Aging, disease and death of nerve cells. Acta Neurol Belg **96**: 219–223

106. Adler CH, Hentz JG, Joyce JN, Beach T, Caviness JN (2002) Motor impairment in normal aging, clinically possible Parkinson's disease, and clinically probable Parkinson's disease: longitudinal evaluation of a cohort of prospective brain donors. Parkinsonism Relat Disord **9**: 103–110

107. Chu Y, Kompoliti K, Cochran EJ, Mufson EJ, Kordower JH (2002) Age-related decreases in Nurr1 immunoreactivity in the human substantia nigra. J Comp Neurol **450**: 203–214

108. Bender A, Krishnan KJ, Morris CM, Taylor GA, Reeve AK, Perry RH, Jaros E, Hersheson JS, Betts J, Klopstock T, Taylor RW, Turnbull DM (2006) High levels of mitochondrial DNA deletions in substantia nigra neurons in aging and Parkinson disease. Nat Genet **38**: 515–517

109. Burton A (2006) mtDNA deletions associated with ageing and PD. Lancet Neurol **5**: 477

110. Chu Y, Kordower JH (2007) Age-associated increases of alpha-synuclein in monkeys and humans are associated with nigrostriatal dopamine depletion: Is this the target for Parkinson's disease? Neurobiol Dis **25**: 134–149

111. Collier TJ, Lipton J, Daley BF, Palfi S, Chu Y, Sortwell C, Bakay RA, Sladek JR, Jr., Kordower JH (2007) Aging-related changes in the nigrostriatal dopamine system and the response to MPTP in nonhuman primates: diminished compensatory mechanisms as a prelude to parkinsonism. Neurobiol Dis **26**: 56–65

112. Hawkes CH (2008) Parkinson's disease and aging: same or different process? Mov Disord **23**: 47–53

113. Banerjee R, Starkov AA, Beal MF, Thomas B (2009) Mitochondrial dysfunction in the limelight of Parkinson's disease pathogenesis. Biochim Biophys Acta **1792**: 651–663

114. Bennett DA, Beckett LA, Murray AM, Shannon KM, Goetz CG, Pilgrim DM, Evans DA (1996) Prevalence of parkinsonian signs and associated mortality in a community population of older people. N Engl J Med **334**: 71–76

115. Emborg ME (2007) Nonhuman primate models of Parkinson's disease. Ilar J **48**: 339–355

116. Jenner P (2003) The contribution of the MPTP-treated primate model to the development of new treatment strategies for Parkinson's disease. Parkinsonism Relat Disord **9**: 131–137

117. Jenner P (2003) The MPTP-treated primate as a model of motor complications in PD: primate model of motor complications. Neurology **61**: S4-11

118. Orr CF, Rowe DB, Halliday GM (2002) An inflammatory review of Parkinson's disease. Prog Neurobiol **68**: 325–340

119. Marchetti B, Serra PA, L'Episcopo F, Tirolo C, Caniglia S, Testa N, Cioni S, Gennuso F, Rocchitta G, Desole MS, Mazzarino MC, Miele E, Morale MC (2005) Hormones are key actors in gene x environment interactions programming the vulnerability to Parkinson's disease: glia as a common final pathway. Ann N Y Acad Sci **1057**: 296–318

120. Carvey PM, Punati A, Newman MB (2006) Progressive dopamine neuron loss in Parkinson's disease: the multiple hit hypothesis. Cell Transplant **15**: 239–250

121. Miller RL, James-Kracke M, Sun GY, Sun AY (2009) Oxidative and inflammatory pathways in Parkinson's disease. Neurochem Res **34**: 55–65

122. Olanow CW, Kordower JH (2009) Modeling Parkinson's disease. Ann Neurol **66**: 432–436

123. Horowitz MP, Greenamyre JT (2010) Gene-environment interactions in Parkinson's disease: the importance of animal modeling. Clin Pharmacol Ther **88**: 467–474

124. Jenner P (2009) From the MPTP-treated primate to the treatment of motor complications in Parkinson's disease. Parkinsonism Relat Disord **15 Suppl 4**: S18–23

INDEX

A

Abnormal involuntary movements (AIMs). *See* L-DOPA induced dyskinesia

ADC. *See* Apparent diffusion coefficient

AIF. *See* Arterial input function

Allatostatin receptor (AlstR) 227

Alpha-synuclein
  C. elegans 36, 45–47
  DA neuron analysis 46
  degenerating neurons 47
  dose-dependent neurodegeneration 46
  Parkinson's disease 6, 25, 244
  animal model 47, 321, 326, 330

Amphetamine rotation. *See* Rotation, amphetamine

Amytrophic Lateral Sclerosis. *See* Volume II (Neuromethods 62)

Anaesthesia
  fMRI, rodents 146
  MRI experiment
    α-chloralose and medetomidine 136–137
    halothane 136

Animal fixation 137

Apomorphine rotation.
  See Rotation, apomorphine

Apparent diffusion coefficient (ADC)
  maps 139, 140
  representative time course 139, 140
  thresholds, ischemic core and penumbra 138–139

Arterial input function (AIF)
  definition 167
  PET
    description 167
    less-invasive method 167
    TAC and image-derived 167

B

Balance beam test
  balance and motor coordination 79
  materials and methods 80–82, 118–119, 287–288, 328–329
  mouse 80–83, 118–119, 287–288, 329
  rat 328–329

Basal ganglia. *See also* Volume II (Neuromethods 62)
  BAC transgenic mice 197
  dendritic calcium imaging 202–207
  intensive treadmill exercise 360
  IR-DIC optics 196
  MPTP-lesioning 356
  nonlinear fluorescence microscopy
    confocal pinhole aperture 192
    high-resolution optical study 192
    patch-clamp techniques 191
    photo-stimulation and photolysis 195–196
    2PLSM 192
    system components 192–195
  optical voltage measurements
    FRET mechanism 216
    mammalian neurons 215
    Purkinje neurons 215
    strengths and limitations 215
  optogenetic approaches
    axons and synaptic terminals 214
    ChR2 213
    circuit analysis 213
    coherent (laser) light source 214
    corticostriatal and thalamostriatal
      microcircuits 213–214
    dendritic topography 214
    hyperpolarization 213
    2 photon microscopy 215
    somatic patch electrode 214
  organelle imaging, neurons 210–212
  patch-clamp experiment 196
  2PLU 208–209
  SPNs 196–202

Behavioral phenotyping 4

Behavioural testing, rodent. *See also* Volume II (Neuromethods 62)
  exercise paradigms
    Fast-Trac running wheel 359
    intensive treadmill exercise 360
    6-lane treadmill apparatus 359
    running incentive 359–360
  motor behavior
    accelerating rotarod 360

Behavioural testing, rodent *(continued)*
    corridor test 289
    cylinder test 288–289, 327–328
    ladder.. 119–120
    ledged beam test 328–329
    limb-use asymmetry................................ 327–328
    mid-field crossings and time spent 360–361
    motorized treadmill 360
    rotation *(see* Rotation)
    vibrissae-elicited limb placing.................... 330–331
  non-motor behavior
    cognitive function and mood disorders............. 361
    elevated plus maze 361–362
    object recognition task............................ 126–128
    sensory stimuli 330
    stimulus contact and removal............................ 330
    sucrose preference test 361
    tactile stimulation test............................ 32
BAC transgenic mice................................ 197

C

Caenorhabditis elegans models
  behaviors.. 33
  definition 32
  genome and shares................................ 33
  human movement disorders.................... 33
  large-scale RNAi screening 40–42
  nervous system
    anatomy and complete connectivity.................... 34
    anterior deirid neurons (ADEs)........................... 34
    basal slowing response 36
    DA neurons 34–35
    mammalian neuronal function............................ 34
  online resources 32
  Parkinson's disease
    LRRK2 and PINK1 36
    orthologs and mutants 36–37
    susceptibility genes 36–38, 43–44
    therapeutic targets and chemicals 38–39
  6-OHDA-induced dopaminergic toxicity........... 44–45
  reverse genetic screens 33
  RNAi, neuronal cells.................................... 47–50
  α-syn proteotoxicity 45–47
  959 somatic cells.. 32
Carbon–11, radiochemistry
  primary precursor
    description 157–158
    methane 158
  14 N(p, α) 11C nuclear reaction.............................. 156
  reaction types............................ 156, 157
Cat–2 gene 45
Cerebellar and brain stem systems. *See* Volume II
    (Neuromethods 62)
Channelrhodopsin–2 (ChR2)
  activation 215

  advantages.................................... 214
  brain slices 228
  description 213
  functional synaptic inputs 214
  presynaptic terminals 214
Common marmoset. *See* MPTP-treated primate,
    common marmoset
Conditioning
  associative conditioning........................... 313
  classical *(see* volume II (Neuromethods 62))
  6-OHDA lesioned rat 313, 339
  operant conditioning 313
Corridor test.................................... 289
Cylinder test
  paw preference 288
  rat protocol 289, 327–328, 331
  mouse protocol 288, 327–328, 331
  dopamine depletion 327
  stroke .... 328 *(see* volume II (Neuromethods 62))

D

Danio rerio. See Zebrafish
Dendritic calcium imaging
  calcium dye selection 203–204
  direct patch clamping 202
  line scans................................ 204–207
  neurons 207
  voltage-dependent channels....................... 202
Dendritic spines
  density determination, biocytin-HCl 365
  recording electrodes................................ 363
Diffusion tensor imaging (DTI)
  fibre tracking 142–143
  myelinated fibre bundles........................ 142
Dopamine (DA)
  active drugs
    amphetamine 269–270
    apomorphine.................................... 270
  dopamine transporter (DAT) 356
  fast-scan cyclic voltammetry...................... 365–366
  lesion
    6-OHDA....................267–9, 281–283
    MPTP 353, 371–373, 401
  neurons
    Caenorhabditis elegans 34
    cellular aspects 35
    dose-dependent degeneration 43
    human α–syn 39
    neurodegeneration 43
    normal and degenerative states 35
    α-synuclein proteotoxicity............................ 45–47
Drosophila
  anti-Parkinson drugs 63
  behavioural activity
    geotaxis assay...................................... 59–60

locomotor assay 59, 61
phototaxis assay 59, 61
sensory function 55–56
random locomotor activities, Canton-S
 and stocks 56
Canton-S flies, UAS transgene 60–62
human coding gene 56
in vitro assays
 dot blot analysis 60
 immunoblot assays 58
 western blot assays 58–59
maintenance, stock cultures
 and experimental flies 60
media preparation
 procedure 56, 58
 recipe 56, 57
α–synuclein 62
DTI. *See* Diffusion tensor imaging
Dyskinesia. *See* L-dopa-induced dyskinesia

E

Electrophysiological effects
 electrotonic potentials and dye-coupling 226
 HD and PD 228
 ionic currents 230
 optogenetics 227–228
 passive membrane properties 228
 patch-clamp
  and optical imaging techniques 226–227
  recordings 225–226
 preparations
  acute brain slices 222
  acutely dissociated neurons 223
  in vivo recordings 223, 224
  neuronal cultures and networks 222–224
  selective neuronal ablation 227
  sharp electrode recordings 225
 synaptic activity 230–235
Elevated beam test. *See* Balance beam test
EPSCs. *See* Excitatory postsynaptic currents
Eshkol-Wachmann movement notation (EWMN)
 limb 96
 movement elements 95–96
Ethological ratings 402
European Mouse Disease Clinic
 (EUMODIC) 111–112
European Mouse Phenotyping Resource
 of Standardised Screens, (EMPReSS) 67
European Union Mouse Research
 for Public Health and Industrial
 Applications (EUMORPHIA) 111
EWMN. *See* Eshkol-Wachmann movement notation
Excitatory postsynaptic currents (EPSCs)
 amplitude and frequency 364
 changes, amplitude and frequency 364

MSSNs 234
physiological readouts and properties 362
pyramidal neurons 231
Exercise
motorized treadmill 360
paradigms
 Fast-Trac running wheel 359
 intensive treadmill exercise 360
 6-lane treadmill apparatus 359
 running incentive 359–360
repair processes, striatum and motor learning 358

F

Fast-scan cyclic voltammetry
 changes, extracellular dopamine 366
 coronal corticostriatal slices 365–366
 electrical stimulation 366
fcMRI. *See* Functional connectivity MRI
Fibre tracking
 DTI 142
 thalamo-cortical connectivity, changes 142
Fluorine–18, radiochemistry
 vs. carbon–11 158
 description, reaction types 158
 electrophilic radiofluorination 158–159
 nucleophilic radiofluorination
  aliphatic fluorination, indirect 160
  aliphatic substitution, direct 159
  aromatic fluorination, indirect 160–161
  fluorination, direct 159–160
  prosthetic groups 161
  structure, direct and indirect 159, 160
Forepaw stimulation 143, 144
protocol
 α2-adrenoreceptor 143
 α–chloralose 143
SSEPs signals 143, 144
therapeutic effect, stem cell 145
Functional assessment, neuroplasticity.
 See also Motor behaviour, rodent
 AMPA receptor frequency 364
 amplitude and electrophysiology studies
  and biocytin labeling
 biocytin injection, MSNs 36
 current-voltage relationship (I-V) curve 364
 mediated synaptic responses 364–365
 MSNs and EPSCs 362
 paired-pulse ratio (PPR) 364
 preparation, brain and coronal
  corticostriatal slices 362
 rectangular current pulses 364
 series resistance (Rs) 363
 spermine concentration 363
 voltage clamp and electrical stimulation
  methods 363

Functional connectivity
  brain networks
    fcMRI 143
    rs-fMRI 142–143
  tracing
    axonal 141
    fibre 142

G

Gait analysis. *See also* footprint and gait analysis
  method 77–78,115–117
  pitfalls 77–78, 117–118
  rationale 112, 115
    digital systems 76
  runways 76, 77, 112
  video imaging 78
Geotaxis assay 59–60
German mouse clinic (GMC) 109
  Web pages 111
Grip strength test 123–124
GMC. *See* German mouse clinic

H

Handedness 317
Hand shaping movement 98–99
  rating scale 98
  transitions 98–99
HD. *See* Huntington's disease
High-performance liquid chromatography (HPLC)
  gold standard method 168
  radiosynthesis 155
  ultra-high PLC 168
High-throughput mouse phenotyping
  comprehensive analysis approach 110
  EUMORPHIA and d EUMODIC 111–112
  GMC (*See* German mouse clinic)
  IMPC 112
  large-scale systematic mutagenesis 109
HPLC. *See* High-performance liquid chromatography
Huntington's disease (HD)
  chorea treatment 234
  electrophysiological outcomes 229
  motor function 232
  mouse models 79–83, 228, 231
    and Volume II (Neuromethods 62)
  neuron loss 221
    and Volume II (Neuromethods 62)
  pathogenesis 234
    and Volume II (Neuromethods 62)
  synaptic plasticity 233
6-Hydroxydopamine (6-OHDA)
  advantages 282
  amphetamine 269–270, 309–310, 312
    methamphetamine 275
  apomorphine 270

conditioning
  associative conditioning 313
  operant conditioning 313
C. elegans 44–45
dopaminergic toxicity 44–45
dopamine receptors and postsynaptic
  signaling 301
general surgical procedure
  cannula and tubing 271, 273–274, 283
  hydrobromide salt and ascorbic acid 269
  inhalation, neurotoxin 270–271
  post-operative care 276, 285
  ten-microlitre microsyringe setup ... 271, 273–274
  surgery considerations
    rats 290–291, 312
    mice 284
gene expression and receptor binding 307–308
lesion
  MFB 267–268, 272–273, 277, 306
  striatal 273–275
  nigral 276
mice 281–296
motor behaviour
  automated freely moving systems 305–306
  elevated beam 293
  cylinder test 288–289, 327–328, 331
  rotation (*see also* Rotation)
  spontaneous turning 307
  drug-induced rotation 269–270, 285,
    309–310, 312
  manual observation 300
  rotometer 275, 301–305
  recording rotational behavior 286
  motor and sensory motor deficits 287–290
  torsional and postural biases 307
  rotarod test 71–75, 122–123, 289–90, 294, 360
  staircase test 276, 288–289, 293–294
  motorized treadmill 360
neurotoxic effects 44, 267
non-motor behavior
  cognitive function and mood disorders 361
  elevated plus maze 361–362
  sucrose preference test 361
  object recognition task 126–128
post-mortem assessment/group
  allocations 276, 311
postsynaptic activation 311
rats 267, 299, 317–318, 325
SHSY5Y cells 38
video recording 275

I

Imaging
  calcium imaging (*See* Calcium imaging)
  DTI (*See* Diffusion tensor imaging)

light and EM imaging Volume II (Neuromethods 62)

MRI (*See* Magnetic resonance imaging and Volume II (Neuromethods 62))

PET (*See* Positron emission tomography)

SPECT (*See* Single positron emission computed tomography)

Immunoblot assays 58

International Mouse Phenotyping Consortium (IMPC) 112

IR-DIC optics 196

K

Knockout. *See also* Tamoxifen, inducible gene knockout

disease-associated alleles 246

gene inactivation 246

protein expression 245

targeting vectors 256

Kruskal-Wallis non-parametric analysis of variance 71

L

L-dopa-induced dyskinesia

AIMs/dyskinetic movements

AIM subtype 344

amplitude scale 42–344

basic rating scale 341–342

rating scales, training 348

data analysis

ANOVA 345

dyskinesia 424

non-parametric statistics 344–345

theoretical maxima.................... 348–349

experimental design

AIMs 338

chronic treatment period 338–339

putative anti-dyskinetic agents 338, 340

6-OHDA lesion types and treatment groups 344, 346–347

methodology 340–341

strain and species 345

testing environment 347–348

L-dopa

administration route 348

anti-Parkinson's effect 421, 422

behavioral activation 345

benserazide dosage.................... 347

MPTP-treated primate

African Green Monkey 424

common marmoset

dyskinesia protocols 384, 387, 392

drug testing.................. 391, 393–394

dyskinesia rating scales 379–382

primate species................... 378

Ladder 119–120

Learned nonuse 91, 92, 103

Limb-use asymmetry. *See* Cylinder test

Locomotion 20, 23

Locomotor

activity and motor disability

automated test units................... 386

DasyLab acquisition software 386

rating scale 386

statistical significance, power analysis 386

larval and adult zebrafish

anxiolytic effect, drugs 20

functions 19

place preference behaviour 19–20

video tracking 19

well plates 19

Long-term depression (LTD)................. 225, 229, 232, 233, 270

Long-term potentiation (LTP)................. 225

M

Magnetic resonance imaging (MRI)

description 135–136

neurological damage, rats and mice

anaesthesia.................... 136–137, 146

animal fixation 137

functional activation and connectivity 141–145

movement artefacts............... 145–146

shine-through effect 146

structural damage.............. 138–141

physical variables and in vivo imaging 137–138

RF signals 136

Mann-Whitney U–test................... 71

MAO genes. *See* Monoamine oxidase genes

MCAO (middle cerebral artery occlusion)....... 140, 144, 326 and Volume II (Neuromethods 62)

Median forebrain bundle (MFB)

intracerebral injection 268

lesion coordinates 277

ten-microlitre microsyringe setup........... 271

unilateral lesion.............. 272–273

Medium spiny neurons (MSNs)

biocytin injection 365

physiological properties 362

whole-cell recordings............ 363

MFB. *See* Median forebrain bundle

1-Methyl–4-phenyl–1,2,3,6-tetrahydropyridine. *See also* MPTP-treated mouse, primate

selective neurotoxin 366–367

safety issues.................. 354–355, 383, 388

mouse (*see* MPTP-treated mouse)

primate (*see* MPTP-treated primate)

zebrafish larvae

cell death................ 21, 22

neurotransmitter levels............ 22

Monoamine oxidase (MAO) genes

activity 22

Mouse. *See also* 6-Hydroxydopamine;
    MPTP-treated mouse
    balance beam/elevated bridge test
      (*see* Balance beam test)
    footprint test and gait analysis 76–79
    high-throughput phenotyping (*see* High-throughput
      mouse phenotyping)
    ES cells 245
    glutamate receptor function 233
    Huntington's disease 231
    intranigral grafted cells 235
    mammalian gene function 248
    PD-associated alleles 245
    operational sensitivity 66
    R6/2 230
    rotarod test (*see* Rotarod test)
    SHIRPA (*see* SHIRPA screen test)
MPTP-treated mouse
    degree of lesion 357–358
    lesion regimen 358–359
    motor behavior 65–86, 325–336, 353–370
    neuroplasticity 355–367
    neurotransmitter concentrations 357
    nigrostriatal dopaminergic neurons 357–358
    non-motor behavior 361–362
    parameters, lesioning regimen 357
    preparation and injections 355
    species, age and sex selection 355–356
    stained ventral mesencephalon section 358
    western immune-blotting 358
MPTP-treated primate
    African Green Monkey
      advantages and disadvantages 425
      controlling observer bias 410–411
      dyskinesia 424
      dosage differences 416–418
      eyeblink rate measurements 424
      intoxication 401–402
      Healthy Behavior scores 415
      low, mid and high 413–417
      Parkinsonism 418
      Parkscore 413–414
      Parkscore *vs.* Healthy Behaviors 413–418, 422
    behavioural effects 379–380, 385–386
    common marmoset
      advantages 384
      behavioural deficits 379–380, 385–386
      care regimes 385
      dyskinesia 387
      (*see also* L-dopa-induced dyskinesia)
      health effects 388–390
      laboratory comparisons 391–392
      MPTP administration protocols 383, 388, 392
      neuroprotection and resoration 394–395
      rating scales 378–383

cynomologous monkey 374–375, 380
MPTP-protocols 375, 383–384, 388, 392, 411
protocols and rating scales
    MPTP administration 411
    sample size 409–410
    variability, doses
      outcome and recovery 414
      motor and behavioral function 403–406, 412
      videotaped assessment 424–425
    rating scales of motor
      performance 378–383, 410–415
    rhesus monkey 374–375, 380
    squirrel monkey 375, 379
    vervet monkey 374–375, 380
MRI. *See* Magnetic resonance imaging
MSNs. *See* Medium spiny neurons
Multiple Systems Atrophy (MSA). *See* Volume II
    (Neuromethods 62)

N

Neurodegeneration
    C. elegans 43
    DA 39
    6-OHDA 44
    PD susceptibility genes 43–47
Neuroinflammation, PET
    dose-dependent displacement 185
    TSPO 184–185
Neurological evaluation, mice movement disorders
    balance beam/elevated bridge test 79–83
    footprint test and gait analysis 76–79
    rotarod 71–75
    SHIRPA screen 67–71
Neuroplasticity
    functional assessment
      biocytin injection, MSNs 365
      electrophysiology studies
        and biocytin labeling 362–365
      exercise paradigms 359–360
      fast-scan cyclic voltammetry 365–366
Neurotransmitter systems, zebrafish brain
    MAO genes 15–16
    Receptors 15
    tyrosine hydroxylase (TH) 15
Non-motor symptoms 112

O

Object recognition task 126–128
Ocularmotor systems. *See* Volume II (Neuromethods 62)
6-OHDA. *See* 6-Hydroxydopamine
Open field test 113–115
Open reading frame (ORF), 247
Optogenetic approaches
    axons and synaptic terminals 214
    ChR2 213

circuit analysis.. 213
coherent (laser) light source 214
corticostriatal and thalamostriatal
    microcircuits 213–214
dendritic topography 214
2P microscopy 215
somatic patch electrode............................ 214
Organelle imaging, neurons
dye loading 210
high-quality measurements..................... 211
membrane permeable cationic dyes 210
mitochondrial redox state 212–213
mito-roGFP probe 212
Parkinson's disease 210
redox status 212
regions of interest (ROIs)........................ 211
SNc dopaminergic neurons...................... 211
TMRM fluorescence intensity 211
ventral tegmental area (VTA)................. 212

P

Parkinson's disease (PD)
animal models.. 232
behavioural tests
    apomorphine.............................. 319–320
    drug-induced rotation (See Rotation)
    L-DOPA 325
    ledged beam test 328–329
    limb-use asymmetry................... 327–328
    somatosensory asymmetries sensory
        stimuli 330
    stimulus contact and removal........... 330
    tactile stimulation test.................... 329
    vibrissae-elicited limb placing.... 330–331
C. elegans 31–54
description 243–244
DJ-1................... 23–24, 37, 38, 127, 326, 330
drosophila.. 55–64
dyskinesia (See also L-dopa induced
    dyskinesia (LID))
endogenous proteins 248
gain-of-function 246
gene inactivation....................... 245–246, 248
genetic mouse models
    CreER^{T2} transgenic mice 259–262
    cell types and tissues 259
    EUCOMM-Tools programme 259
    inducible Cre activity...................... 250
    lymphoma cells 262
    transgenic mouse lines 228, 245, 259–260
    transgenic α –synuclein 58–59
    protocol.. 262
    reporter mice............................ 260–261
    Rosa26 locus 260
    ZFNs.. 249

genetic mutations
    conditional gene targeting vectors 256–258
    inducible expression vector cloning.... 251
    β-galactosidase reporter cassette 253–254
    large-scale mutagenesis programmes 254–256
    missense mutation 244
    ORF-coded protein 250
    plasmids.. 249
    potential duplication 254
    primers................................... 249–250
    RMCE positive ES cell clones 253
    Rosa26 locus.............................. 250–253
    somatic cells 253–254
    α-syn proteotoxicity...................... 45–47
    tamoxifen inducible gene expression cells 249
    tamoxifen inducible gene knockout 253–259
    targeting vector generation 259
    therapeutic targets and chemicals 38–39
    transgene activation 246–247
human movement studies
    basal ganglia..................................... 320
    motor system 320
    parkinsonian patients....................... 321
    skilled movements 321
    UPDRS 409, 423
MPTP (See 1-methyl–4-phenyl–1,2,3,6-
    tetrahydropyridine (MPTP))
MPTP-treated mouse (See MPTP treated mouse)
MPTP-treated primate (See MPTP-treated primate)
neurotoxic vs. genetic models................... 231
neuron loss.. 228, 317
nigrostriatal dopaminergic neuron........... 326
non-neuronal cell types, brain.................. 245
6-OHDA-treated mouse (See 6-Hydroxydopamine)
6-OHDA-treated rat See (6-Hydroxydopamine)
PET (See also Positron emission tomography)
    AAV5-TH 184
    dopamine neurotransmission system 183
    ligand role 184
    met-amphetamine 184
    pre-synaptic neuron 183
    radioligands selective targets.............. 183
susceptibility genes
    LRRK2 and PINK1 36
    α-synuclein....................................... 55
Zebrafish (See Zebrafish)
Partial lesion models of movement disorders................. 6–8
Patch-clamp recordings
    cell attached configuration.................. 226
    description 225
    infrared (IR) videomicroscopy........ 225–226
PET. See Positron emission tomography
Phototaxis assay................................... 59
PINK1.......................... 15, 23–25, 36–38, 326
2PLSM. See Two-photon laser scanning microscopy

2PLU caging molecules...............................209
  EPSP and calcium transient 208
  galvanometer mirrors.............................209
  glutamatergic synaptic inputs 208
  glutamate uncaging..............................210
  large-scale bath application 208
  photodamage 209
Pockels cell modulator...............................209
Positron emission tomography (PET)
  AIF
    description 167
    advantages.................................167
  animal handling
    anaesthesia protocol................. 169–172
  annihilation events
    circular set-up 162
    detection 511 keV photons...............162
    line of response (LOR) 162–163
    scattered and random coincidence 163–164
  attenuation correction
    fraction, photons.........................164
    methods...............................164–165
  biological applications
    assessing brain diseases in vivo..........183–185
    receptor-ligand interaction 181–183
  contribution...................................152
  data acquisition and camera............168–169
  data processing
    first-level analyses.....................177–178
    injected activity and radioligand mass.............178
    volume of interests (VOIs) 177–178
  image registration
    anatomical data......................176–177
    co-registration, Brainvisa/Anatomist
      software............................. 169, 176
    MRI brain atlases 177
    radiotracer-specific PET templates 177
    registration software 176
  metabolites and radiopharmaceuticals detection
    description 167
    gamma counter 168
    HPLC 168
    UHPLC, TLC and SPE...................168
  pharmacokinetic modelling
    absolute vs. biological quantification.................178
    arterial plasma concentration.............179–180
    case-by-case analysis.....................181
    graphical methods....................180–181
    multi-injection approach...........179–180
    non-metabolized radiotracer..............179
    rate equations, compartments 178
    simplified and reference tissue models.............180
  physics
    emission 511, keV photons 161
    isotope, mean distance 155, 161–162

    positron annihilation........................161
radiochemistry
  radionuclides production 154–155
  radiosynthesis 155–156
  radiotracer properties................153–154
  reaction schemes....................156–161
radioisotope production.............................168
receptor-ligand interaction
  benzodiazepine receptor ligand.................181–182
  dose-dependent displacement...................181–182
  in vivo characterisation 181
  positive linear correlations, agonists...........182–183
scanners
  commercially available PET 165, 166
  generations.................................165
  SIEMENS Concorde 220 microPET 175–176
  technological developments................175
  7 tesla Varian MRI scanner 175–176
  radiotracer-specific PET templates 177
single-photon external source/CT 170
structure, entire process 169
time-activity curve (TAC)
  measured input function 167
  pharmacokinetic compartmental modeling........178
  quantification level, PET 178
  receptor-ligand interaction 181
"translational" imaging method 152
transmission acquisition and emission scans
  scanning protocol non-human primate
    MPTP 170, 173–174
  scanning protocol, rodents.........170–172
  radionuclide counting statistics.........................170

R

Radiochemistry
  carbon–11....................................156–158
  constraint, positron-emitting nuclides....................154
  fluorine–18 158–161
  labelling position, isotope 153, 154
  labelling precursor..........................156
  pharmacokinetics.........................153–154
  positron emitters.........................154, 155
  radiation protection 155–156
  radionuclides production components.....................154
  radiosynthesis, dilution factor.................155
  radiotracer properties, factors,
    radioligand ability.............................153
  reaction schemes............................158–161
  reaction time factor.........................155
Radio frequency (RF) signals 136
Radiotracer, PET
  benzodiazepine receptor ligand 182
  chemical reaction...........................168
  co-registration 177
  electrophilic radiofluorination 159

in vivo characterization, receptor-ligand
  interaction.. 181
properties
  factors, radioligand ability................... 153
  labelling position, isotope 153, 154
  pharmacokinetics 153–154
scanning protocol
  non-human primate MPTP model 170, 173–174
  rodents.. 170–172
  "translational" imaging method 152
Rating scale. See L-DOPA-induced dyskinesia
  or MPTP-lesioned primate
Rat vs. human homology.................3–9, 317–324
  hand movements............................. 101–103
  learning... 102–103
  skilled reaching............................. 101, 104
Reach-to-eat. See Skilled reaching
Resting-state functional MRI (rs-fMRI) 142–143
RF signals. See Radio frequency signals
RNA interference (RNAi)
  neuronal cells
    description 47
    dopaminergic target gene.................... 48
    neuronal dysfunction 48
    non-neuronal somatic cells 48–49
    6-OHDA exposure............................. 49
    SID-1 49–50
  protein misfolding 40
  protein misfolding 40
  scoring aggregation............................. 42–43
  α-syn misfolding 40–41
  α-syn proteotoxicity 42–43
Rotarod test
  apparatus.......................71–77, 122, 289
  distracting stimuli 75
  footing, loss of 75
  methods...................... 73, 122–123, 289–290, 294, 360
    low variance data sets.......................... 73
    training trials 73, 122–123, 289–290, 294, 360
  motor function.. 71
  weight differences.................................... 75
Rotometer
  animal harness 303–304
  cam and pivot assembly 304
  data recording................................. 304–305
  description 301
  species issues..................................... 305
  test chamber 301–303
rs-fMRI. See Resting-state functional MRI

S

SHIRPA screen test
  methods
    force touch 69
    negative geotaxis 70
positional passivity................................. 68
rectangular test arena.......................... 68
righting reflex 70
toe pinch 70
visual placing.. 69
non-parametric analyses 71
scores, scale.. 68
Single pellet reaching
  apparatus dimensions................................ 89
  box.. 88
  frontal view 100
  gestures... 101–102
  video recording 89
Skilled reaching
  allied gestures................................. 101–102
  compensation assessment................. 103–104
  description 87
  disorders, human 104
  forelimb movement primitives................. 102
  humans 102
  independent forelimb............................ 288
  methods
    apparatus habituation................... 88–90
    establishment, hand dominance.......... 90
    feeding and food familiarization 88
    gestures... 93–95
    grasping movements 100–101
    hand-shaping transitions 98–99
    movement elements 95–98
    performance, end-point measures....... 90–91
    reaching trial................................ 91–93
    video recording 89
  motor learning 102–103
  outcome parameters............................... 288
  reaching tasks, types 88
  recovery, function................................ 103
  single pellet reaching task 101
Skill learning 321
Social discrimination test...................... 124–125
Somatosensory evoked potentials (SSEPs)........... 143, 144
Spinal cord systems. See Volume II (Neuromethods 62)
Spiny projection neurons (SPNs)
  classes 196
  dendritic anatomy
    deconvolution 200
    direct volume rendering 200–202
    limitations...................................... 202
    z-stack 197–200
  dendritic field diameter 197
SPN dendritic anatomy
  deconvolution 200
  direct volume rendering.................. 200–202
  limitations.................................... 202–203
  2PLSM.. 197
  z-stack 197–200

SSEPs. *See* Somatosensory evoked potentials
Stroke
    early phase
        ADC threshold and reduction 138–139
        MR angiography 140
        T1-weighted MRI 140
        T2 relaxation time 139–140
    ledged beam test 328–329
    limb-use asymmetry 327–328
    rodent models
        behavioral tests 326
        permanent occlusion 326
    somatosensory asymmetries
        sensory stimuli 330
        stimulus contact and removal 330
    structural damage
        chronic phase 140–141
        early phase 138–140
    tactile stimulation test 329
    T2-weighted imaging, chronic phase 140–141
    vibrissae-elicited limb placing 330–331

T

TAC. *See* PET; Time-activity curve
Tamoxifen
    conditional gene targeting vectors 256–258
    inducible expression vector cloning 251
    inducible gene expression cells 249
    β-galactosidase reporter cassette 253–254
    large-scale mutagenesis programmes 254–256
    ORF-coded protein 250
    plasmids 249
    potential duplication 254
    primers 249–250
    RMCE positive ES cell clones 253
    Rosa26 locus 250–253
    somatic cells 253–254
    targeting vector generation 259
Transgenic mice
    behavioural screens 65–86, 109–134
    blastocyst injection 253
    CreER^{T2} 247, 259
    gene expression 246
    production 250
    Tm1a mice 255
Transient receptor potential melastatin 7
    (TRPM7) gene 20–21
Translation deficits 8
    Parkinson's disease 23
    PINK1 15
    zebrafish 25
Translocator protein (TSPO), 184

TRPM7 gene. *See* Transient receptor potential
    melastatin 7 gene
TSPO. *See* Translocator protein
Two-photon laser scanning microscopy (2PLSM)
    imaging 208
    mitochondrial function 210
    scanning software 205
    software packages 204
    young adult mice 212

U

Unified Parkinson's disease rating scale
    (UPDRS) 409, 423

V

Validity
    construct 4
    face 5, 6
    Parkinson's disease 310, 317, 326, 425
    Huntington's disease (HD) 5–6, 74
    rodents 5, 6, 8, 105, 310, 317, 326
    rodent-human extrapolations 4
Vertical pole test 120–122
Vertebrate model organism. *See* Zebrafish
Vibrissae-elicited limb placing
    automatic placing response 330
    6-OHDA lesions 331
    unilateral nigrostriatal lesions 331

W

Western blot alpha-synuclein assays
    description, structural 58–59
    transgenic α–synuclein 58–59

Z

Zebrafish
    CNS systems and motor behaviors
        basal ganglia and cerebellum, role 14
        brainstem and spinal cord 16
        calcium imaging 14
        genetic approaches 14
        locomotor 18–20
        movement 16–18
        mutants 20–21
        neurotransmitter 15–16
        Purkinje cells 14
        telencephalon, ray-finned fish 12–13
        upper motor neuron and thalamus
            functions 12
    models of Parkinson's disease
        DJ1 expression and LRRK2

Mutations 20–24
MPTP 21–22
off-target effects............................... 23, 25
parkin mutants................................ 23, 25
PINK1 translation and mutant....................... 23–25

description 11–12
PD 21–25
Tol2 transposon system 26
zinc finger nuclease method 26, 249
Zinc-finger nucleases (ZFNs) 26, 249